D0010209

SECOND EDITION

Principles of Biostatistics

Marcello Pagano

Harvard School of Public Health

Kimberlee Gauvreau

Harvard Medical School

Duxbury
Thomson Learning™

Australia • Canada • Mexico • Singapore • Spain • United Kingdom • United States

Sponsoring Editor: *Carolyn Crockett*
Marketing Team: *Tom Ziolkowski, Beth Kroenke,*
   Samantha Cabaluna
Assistant Editor: *Seema Atwal*
Editorial Assistant: *Ann Day*
Production Editor: *Tom Novack*
Manuscript Editor: *Judith Abrahms*
Permissions Editor: *Connie Dowcett*
Interior Design: *John Edeen*

Cover Design: *Kelly Shoemaker*
Art Editor: *Jennifer Mackres*
Interior Illustration: *Pat Rogondino, and S. T. Associates Inc.*
Print Buyer: *Kris Waller*
Typesetting: *Omegatype Typography, Inc.*
Cover Printing: *R. R. Donnelley & Sons Co., Crawfordsville*
Printing and Binding: *R. R. Donnelley & Sons Co.,*
   Crawfordsville

For more information about this or any other Duxbury products, contact:
DUXBURY
511 Forest Lodge Road
Pacific Grove, CA 93950 USA
www.duxbury.com
1-800-423-0563 (Thomson Learning Academic Resource Center)

Printed in United States of America

10 9 8 7 6 5 4 3 2 1

Library of Congress Cataloging-in-Publication Data

ISBN 0-534-22902-6

Preface

This book was written for students of the health sciences and serves as an introduction to the study of biostatistics, or the use of numerical techniques to extract information from data and facts. Because numbers are more precise than words, they are particularly well suited for communicating scientific results.

However, just as one can lie with words, one can also lie with numbers. Indeed, numbers and lies have been linked for quite some time; there is even a book entitled *How to Lie with Statistics*. This association may owe its origin, or its affirmation at the very least, to the British prime minister Benjamin Disraeli. Disraeli is credited by Mark Twain as having said "There are three kinds of lies: lies, damned lies, and statistics." One has only to observe any modern political campaign to be convinced of the abuse of statistics. But enough about lies; this book adopts the position of Professor Frederick Mosteller, who said "It is easy to lie with statistics, but it is easier to lie without them."

Background

Principles of Biostatistics is aimed at students in the biological and health sciences who wish to learn modern research methods. It is based on a required course offered at the Harvard School of Public Health. In addition to these graduate students, a large number of health professionals from the Harvard medical area attend as well. The course is as old as the School itself, which attests to its importance. It spans 16 weeks of lectures and laboratory sessions. Each week includes two 50-minute lectures and one 2-hour lab. The entire class is together for the lectures, but is divided into smaller groups headed by teaching assistants for the lab sessions. These labs reinforce the material covered in the lectures, review the homework assignments, and introduce the computer into the course. We have included the lab materials—except those dealing with the homework assignments and specific computer commands—in the sections labeled Further Applications. These sections present either additional examples or a different perspective on the material covered in a chapter. They are designed to provoke discussion, although they are sufficiently complete for an individual who is not using the book as a course text to benefit from reading them.

This book has evolved to include topics that we believe can be covered at some depth in one American semester. Clearly, some choices had to be made; we hope that we have chosen well. In our course, we have sufficient time to cover most of the topics in the first 20 chapters. However, there is enough material presented to allow the instructor some flexibility. For example, some instructors may choose to omit the sections covering grouped data (Section 3.3), Chebychev's inequality (Section 3.4), and the Poisson distribution (Section 7.3), or the chapter on analysis of variance (Chapter 12), if they consider these concepts to be less important than others.

Structure

Some say that statistics is the study of variability and uncertainty. We believe there is truth to this adage, and have used it as a guide in dividing the book into three parts. The first five chapters deal with collections of numbers and ways in which to summarize, explore, and explain them. The next two chapters focus on probability and serve as an introduction to the tools needed for the subsequent investigation of uncertainty. It is only in the eighth chapter and thereafter that we distinguish between populations and samples and begin to investigate the inherent variability introduced by sampling, thus progressing to inference. We think that this modular introduction to the quantification of uncertainty is justified by the success achieved by our students. Postponing the slightly more difficult concepts until a solid foundation has been established makes it easier for the reader to comprehend them.

Data Sets and Examples

Throughout the text we have used data drawn from published studies to exemplify biostatistical concepts. Not only is real data more meaningful, it is usually more interesting as well. Of course, we do not wish to use examples in which the subject matter is too esoteric or too complex. To this end, we have been guided by the backgrounds and interests of our students—primarily topics in public health and clinical research—to choose examples that best illustrate the concepts at hand.

There is some risk involved in using published data. We cannot guarantee that all of the examples are honest and that the data were properly collected; for this we must rely on the reputations of our sources. We do not belittle the importance of this consideration. The value of our inference depends critically on the worth of the data, and we strongly recommend that a good deal of effort be expended on evaluating its quality. We assume that this is understood by the reader.

More than once we have used examples in which the population of the United States is broken down along racial lines. In reporting these official statistics we follow the lead of the government agencies that release them. We do not wish to reify this racial categorization, since in fact the observed differences may well be due to socioeconomic factors rather than the implied racial ones. One option would be to ignore these statistics; however, this would hide inequities which exist in our health system—inequities that need to be eliminated. We focus attention on the problem in the hope of stimulating interest in promoting solutions.

We have minimized the use of mathematical notation because of its well-deserved reputation of being the ultimate jargon. If used excessively, it can intimidate even the most ardent scholar. We do not wish to eliminate it entirely, however; it has been developed over the ages to be helpful in communicating results. We hope that in this respect we have written a succinct and understandable text.

Over and above their precision, there is something more to numbers—maybe a little magic—that makes them fun to study. The fun is in the conceptualization more than the calculations, and we are fortunate that we have the computer to do the drudge work. This allows students to concentrate on the ideas. In other words, the computer allows the instructor to teach the poetry of statistics and not the plumbing.

Computing

To take advantage of the computer, one needs a good statistical package. We use Stata, which is available from the Stata Corporation in College Station, Texas. We find this statistical package to be one of the best on the market today; it is user-friendly, accurate, powerful, reasonably priced, and works on a number of different platforms, including Windows, Unix, and Macintosh. Furthermore, the output from this package is acceptable to the Federal Drug Administration in New Drug Approval submissions. Other packages are available, and this book can be supplemented by any one of them. In this second edition, we also present output from SAS and Minitab in the Further Applications section of each chapter. We strongly recommend that some statistical package be used.

Some of the review exercises in the text require the use of the computer. To help the reader, we have included the data sets used in these exercises both in Appendix B and on a CD at the back of the book. The CD contains each data set in two different formats: an ASCII file (the "raw" suffix) and a Stata file (the "dta" suffix). There are also many exercises that do not require the computer. As always, active learning yields better results than passive observation. To this end, we cannot stress enough the importance of the review exercises, and urge the reader to attempt as many as time permits.

New to the Second Edition

This second edition includes revised and expanded discussions on many topics throughout the book, and additional figures to help clarify concepts. Previously used data sets, especially official statistics reported by government agencies, have been updated whenever possible. Many new data sets and examples have been included; data sets described in the text are now contained on the CD enclosed with the book. Tables containing exact probabilities for the binomial and Poisson distributions (generated by Stata) have been added to Appendix A. As previously mentioned, we now incorporate computer output from SAS and Minitab as well as Stata in the Further Applications sections. We have also added numerous new exercises, including questions reviewing the basic concepts covered in each chapter.

Acknowledgements

A debt of gratitude is owed a number of people: Harvard University President Derek Bok for providing the support which got this book off the ground, Dr. Michael K. Martin for calculating Tables A.3 through A.8 in Appendix A, and John-Paul Pagano for assisting in the editing of the first edition. We thank the individuals who reviewed the manuscript: Rick Chappell, University of Wisconsin; Dr. Todd G. Nick, University of Mississippi Medical Center; Al Bartolucci, University of Alabama at Birmingham; Bruce E. Trumbo, California State University, Hayward; James Godbold, The Mount Sinai School of Medicine of New York University; and Maureen Lahiff, University of California, Berkeley. Our thanks to the teaching assistants who have helped us teach the course and who have made many valuable suggestions. Probably the most deserving of thanks are the students who have taken the course over the years and who have tolerated us as we learned how to teach it. We are still learning.

Marcello Pagano
Kimberlee Gauvreau

Boston, Massachusetts

Contents

1

Introduction

In 1903, H. G. Wells hypothesized that statistical thinking would one day be as necessary for good citizenship as the ability to read and write. Statistics do play an important role in many decision-making processes. Before a new drug can be marketed, for instance, the United States Food and Drug Administration requires that it be subjected to a clinical trial, an experimental study involving human subjects. The data from this study must be compiled and analyzed to determine whether the drug is not only effective, but safe. In addition, the U.S. government's decisions regarding Social Security and public health programs rely in part on predictions about the longevity of the nation's population; consequently, it must be able to predict the number of years that each individual will live. Many other issues need to be addressed as well. Where should a government invest its resources if it wishes to reduce infant mortality? Does the use of a seat belt or an air bag decrease the chance of death in a motor vehicle accident? Should a mastectomy always be recommended to a patient with breast cancer? What factors increase the risk that an individual will develop coronary heart disease? To answer these questions and others, we rely on the methods of biostatistics.

The study of *statistics* explores the collection, organization, analysis, and interpretation of numerical data. The concepts of statistics may be applied to a number of fields that include business, psychology, and agriculture. When the focus is on the biological and health sciences, we use the term *biostatistics*.

Historically, statistics have been used to tell a story with numbers. Numbers often communicate ideas more succinctly than do words. The message carried by the following data is quite clear, for instance. In 1979, 48 persons in Japan, 34 in Switzerland, 52 in Canada, 58 in Israel, 21 in Sweden, 42 in Germany, 8 in England, and 10,728 in the United States were killed by handguns [1]. The power of these numbers is obvious; the point would be made even if we were to correct for differences in population size.

As a second example, consider the following quotation, taken from an editorial in *The Boston Globe* [2]:

> Lack of contraception is linked to an exceptionally high abortion rate in the Soviet Union—120 abortions for every 100 births, compared with 20 per 100 births in

> Great Britain, where access to contraception is guaranteed. Inadequate support for family planning in the United States has resulted in 40 abortions for every 100 births—a lower rate than the Soviet Union, but twice as high as most industrialized nations.

In this case, a great deal of information is contained in only three numbers: 120, 20, and 40. The statistics provide some insight into the consequences of differing attitudes toward family planning.

In both these examples, the numbers provide a concise summary of certain aspects of the situation being studied. Surely the numerical explanation of the handgun data is more illuminating than if we had been told that some people got killed in Japan, fewer in Switzerland, more in Canada, still more in Israel, but far fewer in Sweden, and so forth. Both examples deal with very complex situations, yet the numbers convey the essential information. Of course, no matter how powerful, no statistic will convince everyone that a given conclusion is true. The handgun data are often brushed away with the aphorism "Guns don't kill people, people do." This should not be surprising; after all, there are still members in the Flat Earth Society. The aim of a biostatistical study is to provide the numbers that contain information about a certain situation and to present them in such a way that valid interpretations are possible.

1.1 Overview of the Text

If we wish to study the effects of a new diet, we might begin by measuring the changes in body mass over time for all individuals who have been placed on the diet. Similarly, if we wanted to investigate the success of a certain therapy for treating prostate cancer, we would record the lengths of time that men treated with this therapy survive beyond diagnosis with the disease. These collections of numbers, however, can display a great deal of variability and are generally not very informative until we start combining them in some way. Descriptive statistics are methods for organizing and summarizing a set of data that help us to describe the attributes of a group or population. In Chapter 2, we examine tabular and graphical descriptive techniques. The graphical capabilities of computers have made this type of summarization more feasible than in the past, and a whole new mode of presentation is available for even the most modest analyses.

Chapter 3 goes beyond the graphical techniques presented in Chapter 2 and introduces numerical summary measures. By definition, a summary captures only a particular aspect of the data being studied; consequently, it is important to have an idea of how well the summary represents the set of measurements as a whole. For example, we might wish to know how long AIDS patients survive after diagnosis with one of the opportunistic infections that characterize the disease. If we calculate an average survival time, is this average then representative of all patients? Furthermore, how useful would the measure be for planning future health service needs? Chapter 3 investigates descriptive techniques that help us to answer questions such as these.

Data that take on only two distinct values require special attention. In the health sciences, one of the most common examples of this type of data is the categorization of being either alive or dead. If we denote the former state by 0 and the latter by 1, we are able to classify a group of individuals using these numbers and then to average the results. In this way, we can summarize the mortality associated with the group. Chapter 4 deals exclusively with measurements that assume only two values. The notion of dividing a group into smaller subgroups or classes based on a characteristic such as age or gender is introduced as well. We might wish to study the mortality of females separately from that of males, for example. Finally, this chapter investigates techniques that allow us to make valid comparisons among groups that may differ substantially in composition.

Chapter 5 introduces the life table, one of the most important techniques available for study in the health sciences. Life tables are used by public health professionals to characterize the well-being of a population, and by insurance companies to predict how long a particular individual will live. In this chapter, the study of mortality begun in Chapter 4 is extended to incorporate the actual time to death for each individual; this results in a more refined analysis. Knowing these times to death also provides a basis for calculating the survival curve for a population. This measure of longevity is used frequently in clinical trials designed to study the effects of various drugs and surgical treatments on survival time.

In summary, the first five chapters of the text demonstrate that the extraction of important information from a collection of numbers is not precluded by the variability among them. Despite this variability, the data often exhibit a certain regularity as well. For example, if we look at the annual mortality rates of teenagers in the United States for each of the last ten years, we do not see much variation in the numbers. Is this just a coincidence, or is it indicative of a natural underlying stability in the mortality rate? To answer questions such as this, we need to study the principles of probability.

Probability theory resides within what is known as an axiomatic system: we start with some basic truths and then build up a logical system around them. In its purest form, the system has no practical value. Its practicality comes from knowing how to use the theory to yield useful approximations. An analogy can be drawn with geometry, a subject that most students are exposed to relatively early in their schooling. Although it is impossible for an ideal straight line to exist other than in our imaginations, that has not stopped us from constructing some wonderful buildings based on geometric calculations. The same is true of probability theory: although it is not practical in its pure form, its basic principles—which we investigate in Chapter 6—can be applied to provide a means of quantifying uncertainty.

One important application of probability theory arises in diagnostic testing. Uncertainty is present because, despite their manufacturers' claims, no available tests are perfect. Consequently, there are a number of important questions that must be answered. For instance, can we conclude that every blood sample that tests positive for HIV actually harbors the virus? Furthermore, all the units in the Red Cross blood supply have tested negative for HIV; does this mean that there are no contaminated samples? If there are contaminated samples, how many might there be? To address questions such as these, we must rely on the average or long-term behavior of the diagnostic tests; probability theory allows us to quantify this behavior.

Chapter 7 extends the notion of probability and introduces some common probability distributions. These mathematical models are useful as a basis for the methods studied in the remainder of the text.

The early chapters of this book focus on the variability that exists in a collection of numbers. Subsequent chapters move on to another form of variability—the variability that arises when we draw a sample of observations from a much larger population. Suppose that we would like to know whether a new drug is effective in treating high blood pressure. Since the population of all people in the world who have high blood pressure is very large, it is extremely implausible that we would have either the time or the resources necessary to examine every person. In other situations, the population may include future patients; we might want to know how individuals who will ultimately develop a certain disease as well as those who currently have it will react to a new treatment. To answer these types of questions, it is common to select a sample from the population of interest and, on the basis of this sample, infer what would happen to the group as a whole.

If we choose two different samples, it is unlikely that we will end up with precisely the same sets of numbers. Similarly, if we study a group of children with congenital heart disease in Boston, we will get different results than if we study a group of children in Rome. Despite this difference, we would like to be able to use one or both of the samples to draw some conclusion about the entire population of children with congenital heart disease. The remainder of the text is concerned with the topic of statistical inference.

Chapter 8 investigates the properties of the sample mean or average when repeated samples are drawn from a population, thus introducing an important concept known as the central limit theorem. This theorem provides a foundation for quantifying the uncertainty associated with the inferences being made.

For a study to be of any practical value, we must be able to extrapolate its findings to a larger group or population. To this end, confidence intervals and hypothesis testing are introduced in Chapters 9 and 10. These techniques are essentially methods for drawing a conclusion about the population we have sampled, while at the same time having some knowledge of the likelihood that the conclusion is incorrect. These ideas are first applied to the mean of a single population. For instance, we might wish to estimate the mean concentration of a certain pollutant in a reservoir supplying water to the surrounding area, and then determine whether the true mean level is higher than the maximum concentration allowed by the Environmental Protection Agency. In Chapter 11, the theory is extended to the comparison of two population means; it is further generalized to the comparison of three or more means in Chapter 12. Chapter 13 continues the development of hypothesis testing concepts, but introduces techniques that allow the relaxation of some of the assumptions necessary to carry out the tests. Chapters 14, 15, and 16 develop inferential methods that can be applied to enumerated data or counts—such as the numbers of cases of sudden infant death syndrome among children put to sleep in various positions—rather than continuous measurements.

Inference can also be used to explore the relationships among a number of different attributes. If a full-term baby whose gestational age is 39 weeks is born weighing 4 kilograms, or 8.8 pounds, no one will be surprised. If the baby's gestational age is only 22 weeks, however, then his or her weight will be cause for alarm. Why? We know that birth weight tends to increase with gestational age, and, although it is extremely rare to find a baby weighing 4 kilograms at 22 weeks, it is not uncommon at 39 weeks. The study of the extent to which two factors are related is known as correlation analysis; this is the topic of Chapter 17. If we wish to predict the outcome of one factor based on the value of another, regression is the appropriate technique. Simple linear regression is investigated in Chapter 18, and is extended to the multiple regression setting—where two or more factors are used to predict a single outcome—in Chapter 19. If the outcome of interest can take on only two possible values, such as alive or dead, an alternative technique must be applied; logistic regression is explored in Chapter 20.

In Chapter 21, the inferential methods appropriate for life tables are introduced. These techniques enable us to draw conclusions about the mortality of a population based on a sample of individuals drawn from the group.

Finally, Chapter 22 examines an issue that is fundamental in inference—the concept of the representativeness of a sample. In any study, we need to be confident that the sample we choose provides an accurate picture of the population from which it is drawn. Several different methods for selecting representative samples are described. The notion of bias and various problems that can arise when choosing a sample are discussed as well. Common sense plays an important role in sampling, as it does throughout the entire book.

1.2   Review Exercises

1. Design a study aimed at investigating an issue that you believe might influence the health of the world. Briefly describe the data that you will require, how you will obtain them, how you intend to analyze the data, and the method you will use to present your results. Keep this study design and reread it after you have completed the text.

2. Consider the following quotation regarding rapid population growth [3]:

   512 million people were malnourished in 1986–1987, up from 460 million in 1979–1981.

   (a) Suppose that you agree with the point being made. Justify the use of these numbers.

   (b) Are you sure that the numbers are correct? Do you think it is possible that 513 million people were malnourished in 1986–1987, rather than 512 million?

3. In addition to stating that "the Chinese have eaten pasta since 1100 B.C.," the label on a box of pasta shells claims that "Americans eat 11 pounds of pasta per year," whereas "Italians eat 60 pounds per year." Do you believe that these statistics are accurate? Would you use these numbers as the basis for a nutritional study?

Bibliography

[1] McGervey, J. D., *Probabilities in Everyday Life*, Chicago: Nelson-Hall, 1986.

[2] "The Pill's Eastern Europe Debut," *The Boston Globe*, January 19, 1990, 10.

[3] United Nations Population Fund, "Family Planning: Saving Children, Improving Lives," New York: Jones & Janello.

2

Data Presentation

Every study or experiment yields a set of data. Its size can range from a few measurements to many thousands of observations. A complete set of data, however, will not necessarily provide an investigator with information that can easily be interpreted. For example, Table 2.1 lists by row the first 2560 cases of acquired immunodeficiency syndrome (AIDS) reported to the Centers for Disease Control and Prevention [1]. Each individual was classified as either suffering from Kaposi's sarcoma, designated by a 1, or not suffering from the disease, represented by a 0. (Kaposi's sarcoma is a tumor that affects the skin, mucous membranes, and lymph nodes.) Although Table 2.1 displays the entire set of outcomes, it is extremely difficult to characterize the data. We cannot even identify the relative proportions of 0s and 1s. Between the raw data and the reported results of the study lies some intelligent and imaginative manipulation of the numbers, carried out using the methods of descriptive statistics.

Descriptive statistics are a means of organizing and summarizing observations. They provide us with an overview of the general features of a set of data. Descriptive statistics can assume a number of different forms; among these are tables, graphs, and numerical summary measures. In this chapter, we discuss the various methods of displaying a set of data. Before we decide which technique is the most appropriate in a given situation, however, we must first determine what kind of data we have.

2.1 Types of Numerical Data

2.1.1 Nominal Data

In the study of biostatistics, we encounter many different types of numerical data. The different types have varying degrees of structure in the relationships among possible values. One of the simplest types of data is *nominal data*, in which the values fall into unordered categories or classes. As in Table 2.1, numbers are often used to represent the categories. In a certain study, for instance, males might be assigned the value 1 and females the value 0.

TABLE 2.1
Outcomes indicating whether an individual had Kaposi's sarcoma for the first 2560 AIDS patients reported to the Centers for Disease Control and Prevention in Atlanta, Georgia

00000000	00010100	00000010	00001000	00000001	00000000	10000000	00000000
00101000	00000000	00000000	00011000	00100001	01001100	00000000	00000010
00000001	00000000	00000010	01100000	00000000	00000100	00000000	00000000
00100010	00100000	00000101	00000000	00000000	00000001	00001001	00000000
00000000	00010000	00010000	00010000	00000000	00000000	00000000	00000000
00000000	00000000	00000000	00001000	00000000	00010000	10000000	00000000
00100000	00000000	00001000	00000010	00000000	00000100	00000000	00010000
00000000	00000000	00000100	00001000	00001000	00000101	00000000	01000000
00010000	00000000	00010000	01000000	00000000	00000000	00000101	00100000
00000000	00000000	00000100	00000000	01000100	00000000	00000001	10100000
00000100	00000000	00010000	00000000	00001000	00000000	00000010	00100000
00000000	00000000	00000000	10001000	00001000	00000000	01000000	00000000
00000000	00001100	00000000	00000000	10000011	00000001	11000000	00001000
00000000	00000000	00000000	00000000	01000000	00000001	00010001	00000000
10000000	00000000	01000000	00000000	00000000	01010100	00000000	00010100
00000000	00000000	00000000	00001010	00000101	00000000	00000000	00010000
00000000	00000000	00000000	00000001	00000100	00000000	00000000	00001000
11000000	00000100	00000000	00000000	00000000	00000000	00000000	00001000
11000000	00010010	00000000	00001000	00000000	00111000	00000001	01001100
00000000	01100000	00100010	10000000	00000000	00000010	00000001	00000000
01000010	01000100	00000000	00010000	00000000	01000000	00000001	00000000
01000000	00000001	00000000	10000000	01000000	00000000	00000000	00000100
00000000	00000000	01000010	00000000	00000000	00000000	00000000	00000000
00000000	00000010	00001010	00001001	10000000	00000000	00000010	00000000
00000000	01000000	00000000	00001000	00000000	01000000	00010000	00000000
00001000	01000010	01001111	00100000	00000000	00100000	00000000	10000001
00000001	00000000	01000000	00000000	00000000	00000000	00000000	01000000
00000000	00000000	00100000	01000000	00100000	00000000	00000011	00000000
01000000	00000100	10000001	00000001	00001000	00000100	00001000	00001000
00100000	00000000	00000000	00000000	00000010	01000001	00010011	00000000
00000000	10000000	10000000	00000000	00000000	00001000	01000000	00000000
00001000	00000000	01000010	00011000	00000001	00001001	00000000	00000001
01000010	01001000	01000000	00000010	00000000	10000000	00000100	00000000
00000010	00000000	00000000	00000010	00000000	00100100	00000000	10110100
00001100	00000100	00001010	00000000	00000000	00000000	00000000	00000000
00000010	00000000	00000000	00000000	00100000	10100000	00001000	00000000
01000000	00000000	00000000	00100000	00000000	01000001	00010010	00010001
00000000	00100000	00110000	00000000	00010000	00000000	00000100	00000000
00010100	00000000	00001001	00000001	00000000	00000000	00000000	00000000
00000010	00000100	01010100	10000001	00001000	00000000	00010010	00010000

Although the attributes are labeled with numbers rather than words, both the order and the magnitudes of the numbers are unimportant. We could just as easily let 1 represent females and 0 designate males. Numbers are used mainly for the sake of convenience; numerical values allow us to use computers to perform complex analyses of the data.

Nominal data that take on one of two distinct values—such as male and female—are said to be *dichotomous* or *binary*, depending on whether the Greek or the Latin root for *two* is preferred. However, not all nominal data need be dichotomous. Often there are three or more possible categories into which the observations can fall. For example, persons may be grouped according to their blood type, such that 1 represents type O, 2 is type A, 3 is type B, and 4 is type AB. Again, the sequence of these values is not important. The numbers simply serve as labels for the different blood types, just as the letters do. We must keep this in mind when we perform arithmetic operations on the data. An average blood type of 1.8 for a given population is meaningless. One arithmetic operation that can be interpreted, however, is the proportion of individuals that fall into each group. An analysis of the data in Table 2.1 shows that 9.6% of the AIDS patients suffered from Kaposi's sarcoma and 90.4% did not.

2.1.2   Ordinal Data

When the order among categories becomes important, the observations are referred to as *ordinal data*. For example, injuries may be classified according to their level of severity, so that 1 represents a fatal injury, 2 is severe, 3 is moderate, and 4 is minor. Here a natural order exists among the groupings; a smaller number represents a more serious injury. However, we are still not concerned with the magnitude of these numbers. We could have let 4 represent a fatal injury and 1 a minor one. Furthermore, the difference between a fatal injury and a severe injury is not necessarily the same as the difference between a moderate injury and a minor one, even though both pairs of outcomes are one unit apart. As a result, many arithmetic operations still do not make sense when applied to ordinal data.

Table 2.2 provides a second example of ordinal data; the scale displayed is used by oncologists to classify the performance status of patients enrolled in clinical trials [2]. A *clinical trial* is an experimental study involving human subjects. Its purpose is usually to facilitate the comparison of alternative treatments for some disease, such as cancer. Subjects are randomly allocated to the different treatment groups and then followed to a specified endpoint.

TABLE 2.2

Eastern Cooperative Oncology Group's classification of patient performance status

Status	Definition
0	Patient fully active, able to carry on all predisease performance without restriction
1	Patient restricted in physically strenuous activity but ambulatory and able to carry out work of a light or sedentary nature
2	Patient ambulatory and capable of all self-care but unable to carry out any work activities; up and about more than 50% of waking hours
3	Patient capable of only limited self-care; confined to bed or chair more than 50% of waking hours
4	Patient completely disabled; not capable of any self-care; totally confined to bed or chair

2.1.3   Ranked Data

In some situations, we have a group of observations that are first arranged from highest to lowest according to magnitude and then assigned numbers that correspond to each observation's place in the sequence. This type of data is known as *ranked data*. As an example, consider all possible causes of death in the United States. We could make a list of all of these causes, along with the number of lives that each one claimed in 1992. If the causes are ordered from the one that resulted in the greatest number of deaths to the one that caused the fewest and then assigned consecutive integers, the data are said to have been ranked. Table 2.3 lists the ten leading causes of death in the United States in 1992 [3]. Note that cerebrovascular diseases would be ranked third whether they caused 480,000 deaths or 98,000. In assigning the ranks, we disregard the magnitudes of the observations and consider only their relative positions. Even with this imprecision, it is amazing how much information the ranks contain. In fact, it is sometimes better to work with ranks than with the original data; this point is explored further in Chapter 13.

2.1.4   Discrete Data

For *discrete data*, both ordering and magnitude are important. In this case, the numbers represent actual measurable quantities rather than mere labels. In addition, discrete data are restricted to taking on only specified values—often integers or counts—that differ by fixed amounts; no intermediate values are possible. Examples of discrete data include the number of motor vehicle accidents in Massachusetts in a specified month, the number of times a woman has given birth, the number of new cases of tuberculosis reported in the United States during a one-year period, and the number of beds available in a particular hospital.

Note that for discrete data a natural order exists among the possible values. If we are interested in the number of times a woman has given birth, for instance, a larger number indicates that a woman has had more children. Furthermore, the difference between one and two births is the same as the difference between four and five births. Finally, the number of births is restricted to the nonnegative integers; a woman cannot give birth 3.4 times. Because it is meaningful to measure the distance between possible data

TABLE 2.3

Ten leading causes of death in the United States, 1992

Rank	Cause of Death	Total Deaths
1	Diseases of the heart	717,706
2	Malignant neoplasms	520,578
3	Cerebrovascular diseases	143,769
4	Chronic obstructive pulmonary diseases	91,938
5	Accidents and adverse effects	86,777
6	Pneumonia and influenza	75,719
7	Diabetes mellitus	50,067
8	Human immunodeficiency virus infection	33,566
9	Suicide	30,484
10	Homicide and legal intervention	25,488

values for discrete observations, arithmetic rules can be applied. However, the outcome of an arithmetic operation performed on two discrete values is not necessarily discrete itself. Suppose, for instance, that one woman has given birth three times, whereas another has given birth twice. The average number of births for these two women is 2.5, which is not itself an integer.

2.1.5   Continuous Data

Data that represent measurable quantities but are not restricted to taking on certain specified values (such as integers) are known as *continuous data*. In this case, the difference between any two possible data values can be arbitrarily small. Examples of continuous data include time, the serum cholesterol level of a patient, the concentration of a pollutant, and temperature. In all instances, fractional values are possible. Since we are able to measure the distance between two observations in a meaningful way, arithmetic operations can be applied. The only limiting factor for a continuous observation is the degree of accuracy with which it can be measured; consequently, we often see time rounded off to the nearest second and weight to the nearest pound or gram. The more accurate our measuring instruments, however, the greater the amount of detail that can be achieved in our recorded data.

At times we might require a lesser degree of detail than that afforded by continuous data; hence we occasionally transform continuous observations into discrete, ordinal, or even dichotomous ones. In a study of the effects of maternal smoking on newborns, for example, we might first record the birth weights of a large number of infants and then categorize the infants into three groups: those who weigh less than 1500 grams, those who weigh between 1500 and 2500 grams, and those who weigh more than 2500 grams. Although we have the actual measures of birth weight, we are not concerned with whether a particular child weighs 1560 grams or 1580 grams; we are only interested in the number of infants who fall into each category. From prior experience, we may not expect substantial differences among children within the very low birth weight, low birth weight, and normal birth weight groupings. Furthermore, ordinal data are often easier to handle than continuous data and thus simplify the analysis. There is a consequent loss of detail in the information about the infants, however. In general, the degree of precision required in a given set of data depends on the questions that are being studied.

Section 2.1 described a gradation of numerical data that ranges from nominal to continuous. As we progressed, the nature of the relationship between possible data values became increasingly complex. Distinctions must be made among the various types of data because different techniques are used to analyze them. As previously mentioned, it does not make sense to speak of an average blood type of 1.8; it does make sense, however, to refer to an average temperature of 24.55°C.

2.2   Tables

Now that we are able to differentiate among the various types of data, we must learn how to identify the statistical techniques that are most appropriate for describing each kind. Although a certain amount of information is lost when data are summarized, a great deal can also be gained. A *table* is perhaps the simplest means of summarizing a set of observations and can be used for all types of numerical data.

2.2.1 Frequency Distributions

One type of table that is commonly used to evaluate data is known as a *frequency distribution*. For nominal and ordinal data, a frequency distribution consists of a set of classes or categories along with the numerical counts that correspond to each one. As a simple illustration of this format, Table 2.4 displays the numbers of individuals (numerical counts) who did and did not suffer from Kaposi's sarcoma (classes or categories) for the first 2560 cases of AIDS reported to the Centers for Disease Control. A more complex example is given in Table 2.5, which specifies the numbers of cigarettes smoked per adult in the United States in various years [4].

To display discrete or continuous data in the form of a frequency distribution, we must break down the range of values of the observations into a series of distinct, nonoverlapping intervals. If there are too many intervals, the summary is not much of an improvement over the raw data. If there are too few, a great deal of information is lost. Although it is not necessary to do so, intervals are often constructed so that they all have equal widths; this facilitates comparisons among the classes. Once the upper and lower limits for each interval have been selected, the number of observations whose values fall within each pair of limits is counted, and the results are arranged as a table. As part of a National Health Examination Survey, for example, the serum cholesterol levels of 1067 25- to 34-year-old males were recorded to the nearest milligram per 100 milliliters [5]. The observations were then subdivided into intervals of equal width; the frequencies corresponding to each interval are presented in Table 2.6.

Table 2.6 gives us an overall picture of what the data look like; it shows how the values of serum cholesterol level are distributed across the intervals. Note that the ob-

TABLE 2.4
Cases of Kaposi's sarcoma for the first 2560 AIDS patients reported to the Centers for Disease Control in Atlanta, Georgia

Kaposi's Sarcoma	Number of Individuals
Yes	246
No	2314

TABLE 2.5
Cigarette consumption per person aged 18 or older, United States, 1900–1990

Year	Number of Cigarettes
1900	54
1910	151
1920	665
1930	1485
1940	1976
1950	3522
1960	4171
1970	3985
1980	3851
1990	2828

TABLE 2.6
Absolute frequencies of serum cholesterol levels for 1067 U.S. males, aged 25 to 34 years, 1976–1980

Cholesterol Level (mg/100 ml)	Number of Men
80–119	13
120–159	150
160–199	442
200–239	299
240–279	115
280–319	34
320–359	9
360–399	5
Total	1067

servations range from 80 to 399 mg/100 ml, with relatively few measurements at the ends of the range and a large proportion of the values falling between 120 and 279 mg/100 ml. The interval 160–199 mg/100 ml contains the greatest number of observations. Table 2.6 provides us with a much better understanding of the data than would a list of 1067 cholesterol level readings. Although we have lost some information—given the table, we can no longer recreate the raw data values—we have also extracted important information that helps us to understand the distribution of serum cholesterol levels for this group of males.

The fact that one kind of information is gained while another is lost holds true even for the simple dichotomous data in Tables 2.1 and 2.4. We might feel that we do not lose anything by summarizing these data and counting the numbers of 0s and 1s, but in fact we do. For example, if there is some type of trend in the observations over time—perhaps the proportion of AIDS patients with Kaposi's sarcoma is either increasing or decreasing as the epidemic matures—this information is lost in the summary.

Tables are most informative when they are not overly complex. As a general rule, tables and the columns within them should always be clearly labeled. If units of measurement are involved, such as mg/100 ml for the serum cholesterol levels in Table 2.6, they should be specified.

2.2.2 Relative Frequency

It is sometimes useful to know the proportion of values that fall into a given interval in a frequency distribution rather than the absolute number. The *relative frequency* for an interval is the proportion of the total number of observations that appears in that interval. The relative frequency is computed by dividing the number of values within an interval by the total number of values in the table. The proportion can be left as it is, or it can be multiplied by 100% to obtain the percentage of values in the interval. In Table 2.6, for example, the relative frequency in the 80–119 mg/100 ml class is $(13/1067) \times 100\% = 1.2\%$; similarly, the relative frequency in the 120–159 mg/100 ml class is $(150/1067) \times 100\% = 14.1\%$. The relative frequencies for all intervals in a table sum to 100%.

Relative frequencies are useful for comparing sets of data that contain unequal numbers of observations. Table 2.7 displays the absolute and relative frequencies of serum cholesterol level readings for the 1067 25- to 34-year-olds depicted in Table 2.6, as well as for a group of 1227 55- to 64-year-olds. Because there are more men in the older age group, it is inappropriate to compare the columns of absolute frequencies for the two sets of males. Comparing the relative frequencies is meaningful, however. We can see that in general, the older men have higher serum cholesterol levels than the younger men; the younger men have a greater proportion of observations in each of the intervals below 200 mg/100 ml, whereas the older men have a greater proportion in each class above this value.

The *cumulative relative frequency* for an interval is the percentage of the total number of observations that have a value less than or equal to the upper limit of the interval. The cumulative relative frequency is calculated by summing the relative frequencies for the specified interval and all previous ones. Thus, for the group of 25- to 34-year-olds in Table 2.7, the cumulative relative frequency of the second interval is

$1.2 + 14.1 = 15.3\%$; similarly, the cumulative relative frequency of the third interval is $1.2 + 14.1 + 41.4 = 56.7\%$. Like relative frequencies, cumulative relative frequencies are useful for comparing sets of data that contain unequal numbers of observations. Table 2.8 lists the cumulative relative frequencies for the serum cholesterol levels of the two groups of males in Table 2.7.

TABLE 2.7
Absolute and relative frequencies of serum cholesterol levels for 2294 U.S. males, 1976–1980

Cholesterol Level (mg/100 ml)	Ages 25–34		Ages 55–64	
	Number of Men	Relative Frequency (%)	Number of Men	Relative Frequency (%)
80–119	13	1.2	5	0.4
120–159	150	14.1	48	3.9
160–199	442	41.4	265	21.6
200–239	299	28.0	458	37.3
240–279	115	10.8	281	22.9
280–319	34	3.2	128	10.4
320–359	9	0.8	35	2.9
360–399	5	0.5	7	0.6
Total	1067	100.0	1227	100.0

TABLE 2.8
Relative and cumulative relative frequencies of serum cholesterol levels for 2294 U.S. males, 1976–1980

Cholesterol Level (mg/100 ml)	Ages 25–34		Ages 55–64	
	Relative Frequency (%)	Cumulative Relative Frequency (%)	Relative Frequency (%)	Cumulative Relative Frequency (%)
80–119	1.2	1.2	0.4	0.4
120–159	14.1	15.3	3.9	4.3
160–199	41.4	56.7	21.6	25.9
200–239	28.0	84.7	37.3	63.2
240–279	10.8	95.5	22.9	86.1
280–319	3.2	98.7	10.4	96.5
320–359	0.8	99.5	2.9	99.4
360–399	0.5	100.0	0.6	100.0

According to Table 2.7, older men tend to have higher serum cholesterol levels than younger men do. This is the sort of generalization we hear quite often; for instance, it might also be said that men are taller than women or that women live longer than men. The generalization about serum cholesterol does not mean that every 55- to 64-year-old male has a higher cholesterol level than every 25- to 34-year-old male, nor does it mean that the serum cholesterol level of every man increases with age. What the statement does imply is that for a given cholesterol level, the proportion of younger men with a reading less than or equal to this value is greater than the proportion of older men with a reading less than or equal to the value. This pattern is more obvious in Table 2.8 than it is in Table 2.7. For example, 56.7% of the 25- to 34-year-olds have a serum cholesterol level less than or equal to 199 mg/100 ml, whereas only 25.9% of the 55- to 64-year-olds fall into this category. Because the relative proportions for the two groups follow this trend in every interval in the table, the two distributions are said to be *stochastically ordered*. For any specified level, a larger proportion of the older men have serum cholesterol readings above this value than do the younger men; therefore, the distribution of levels for the older men is stochastically larger than the distribution for the younger men. This definition will start to make more sense when we encounter random variables and probability distributions in Chapter 7. At that point, the implications of this ordering will become more apparent.

2.3 Graphs

A second way to summarize and display data is through the use of *graphs*, or pictorial representations of numerical data. Graphs should be designed so that they convey the general patterns in a set of observations at a single glance. Although they are easier to read than tables, graphs often supply a lesser degree of detail. Once again, however, the loss of detail may be accompanied by a gain in understanding of the data. The most informative graphs are relatively simple and self-explanatory. Like tables, they should be clearly labeled, and units of measurement should be indicated.

2.3.1 Bar Charts

Bar charts are a popular type of graph used to display a frequency distribution for nominal or ordinal data. In a *bar chart*, the various categories into which the observations fall are presented along a horizontal axis. A vertical bar is drawn above each category such that the height of the bar represents either the frequency or the relative frequency of observations within that class. The bars should be of equal width and separated from one another so as not to imply continuity. As an example, Figure 2.1 is a bar chart that displays the data relating to cigarette consumption in the United States presented in Table 2.4. Note that when it is represented in the form of a graph, the trend in cigarette consumption over the years is even more apparent than it is in the table.

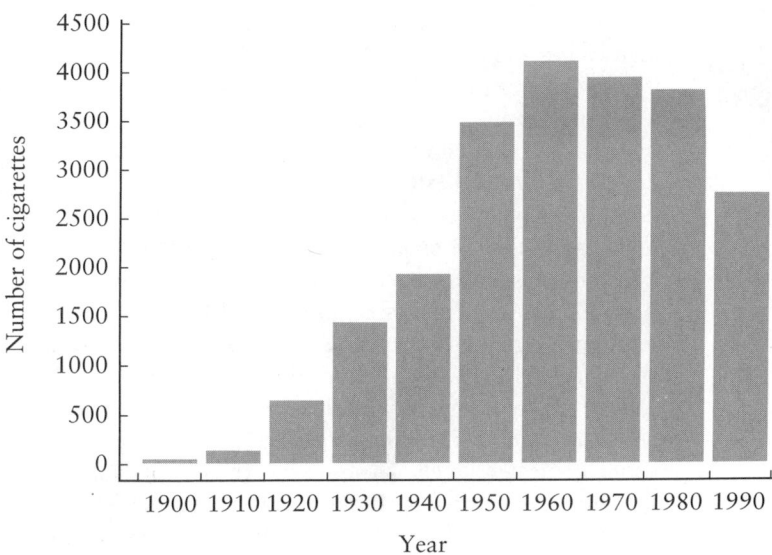

FIGURE 2.1
Bar chart: Cigarette consumption per person 18 years of age or older,
United States, 1900–1990

2.3.2   Histograms

Perhaps the most commonly used type of graph is the histogram. Whereas a bar chart is a pictorial representation of a frequency distribution for either nominal or ordinal data, a *histogram* depicts a frequency distribution for discrete or continuous data. The horizontal axis displays the true limits of the various intervals. The true limits of an interval are the points that separate it from the intervals on either side. For example, the boundary between the first two classes of serum cholesterol level in Table 2.5 is 119.5 mg/100 ml; it is the true upper limit of the interval 80–119 and the true lower limit of 120–159. The vertical axis of a histogram depicts either the frequency or the relative frequency of observations within each interval.

The first step in constructing a histogram is to lay out the scales of the axes. The vertical scale should begin at zero; if it does not, visual comparisons among the intervals may be distorted. Once the axes have been drawn, a vertical bar centered at the midpoint is placed over each interval. The height of the bar marks off the frequency associated with that interval. As an example, Figure 2.2 displays a histogram constructed from the serum cholesterol level data in Table 2.6.

In reality, the frequency associated with each interval in a histogram is represented not by the height of the bar above it but by the bar's area. Thus, in Figure 2.2, 1.2% of the total area corresponds to the 13 observations that lie between 79.5 and 119.5 mg/100 ml, and 14.1% of the area corresponds to the 150 observations between 119.5 and 159.5 mg/100 ml. The area of the entire histogram sums to 100%, or 1. Note that the proportion of the total area corresponding to an interval is equal to the relative frequency of that interval. As a result, a histogram displaying relative frequencies—such as Figure 2.3—will have the same shape as a histogram displaying absolute frequen-

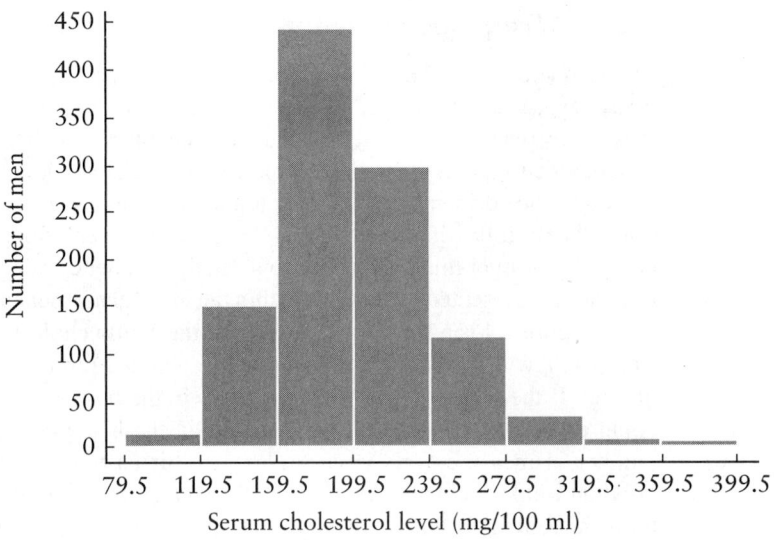

FIGURE 2.2

Histogram: Absolute frequencies of serum cholesterol levels for 1067 U.S. males, aged 25 to 34 years, 1976–1980

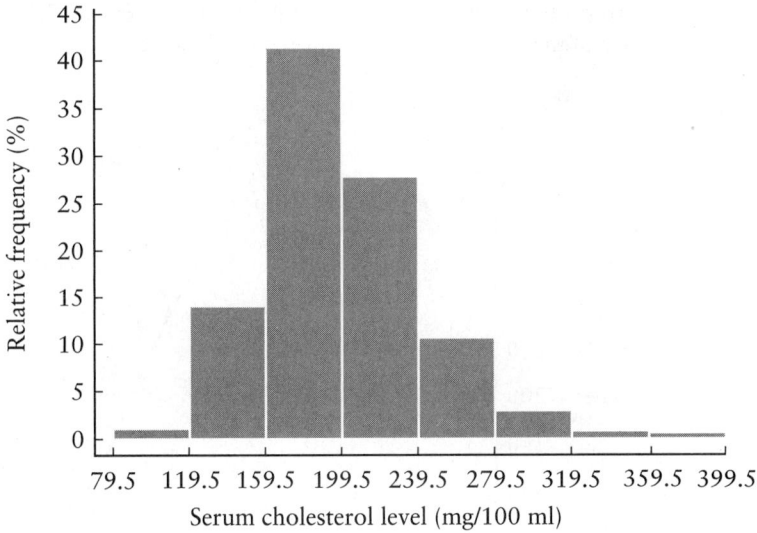

FIGURE 2.3

Histogram: Relative frequencies of serum cholesterol levels for 1067 U.S. males, aged 25 to 34 years, 1976–1980

cies. Because it is the area of each bar that represents the relative proportion of observations in an interval, care must be taken when constructing a histogram with unequal interval widths; the height must vary along with the width so that the area of each bar remains in proper proportion.

2.3.3 Frequency Polygons

The frequency polygon, another commonly used graph, is similar to the histogram in many respects. A *frequency polygon* uses the same two axes as a histogram. It is constructed by placing a point at the center of each interval such that the height of the point is equal to the frequency or relative frequency associated with that interval. Points are also placed on the horizontal axis at the midpoints of the intervals immediately preceding and immediately following the intervals that contain observations. The points are then connected by straight lines. As in a histogram, the frequency of observations for a particular interval is represented by the area within the interval and beneath the line segment.

Figure 2.4 is a frequency polygon of the serum cholesterol level data in Table 2.6. Compare it with the histogram in Figure 2.2, which is reproduced very lightly in the background. If the total number of observations in the data set were to increase steadily, we could decrease the widths of the intervals in the histogram and still have an adequate number of measurements in each class; in this case, the histogram and the frequency polygon would become indistinguishable. As they are, both types of graphs convey essentially the same information about the distribution of serum cholesterol levels for this population of men. We can see that the measurements are centered around 180 mg/100 ml, and drop off a little more quickly to the left of this value than they do to the right. Most of the observations lie between 120 and 280 mg/100 ml, and all are between 80 and 400 mg/100 ml.

Because they can easily be superimposed, frequency polygons are superior to histograms for comparing two or more sets of data. Figure 2.5 displays the frequency polygons of the serum cholesterol level data presented in Table 2.7. Since the older men tend

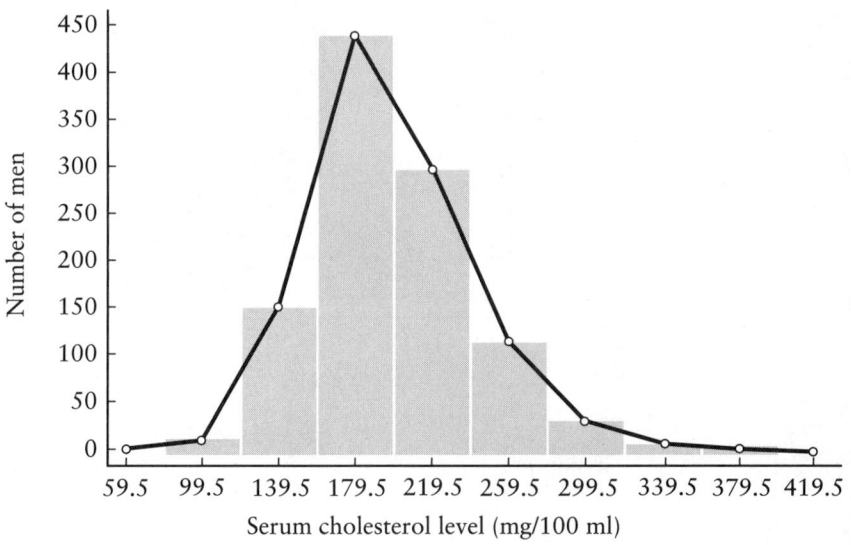

FIGURE 2.4
Frequency polygon: Absolute frequencies of serum cholesterol levels for
1067 U.S. males, aged 25 to 34 years, 1976–1980

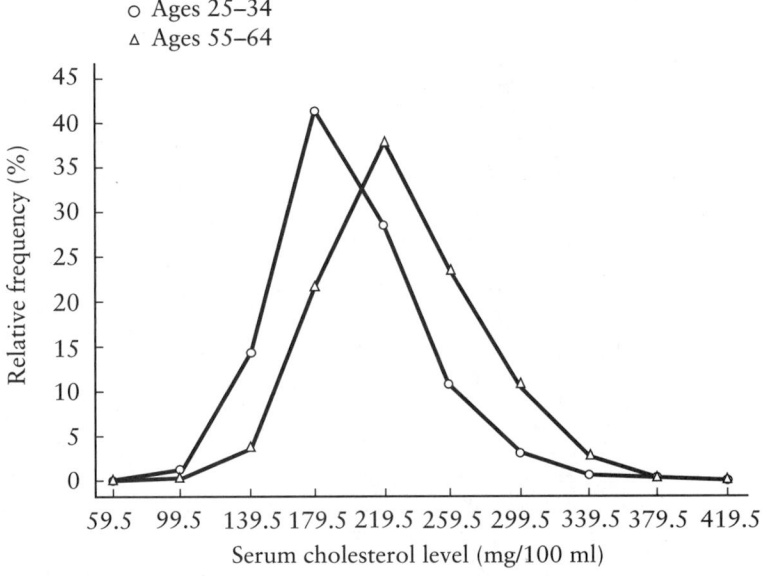

FIGURE 2.5
Frequency polygon: Relative frequencies of serum cholesterol levels for
2294 U.S. males, 1976–1980

to have higher serum cholesterol levels, their polygon lies to the right of the polygon for the younger men.

Although its horizontal axis is the same as that for a standard frequency polygon, the vertical axis of a *cumulative frequency polygon* displays cumulative relative frequencies. A point is placed at the true upper limit of each interval; the height of the point represents the cumulative relative frequency associated with that interval. The points are then connected by straight lines. Like frequency polygons, cumulative frequency polygons can be used to compare sets of data. This is illustrated in Figure 2.6. By noting that the cumulative frequency polygon for 55- to 64-year-old males lies to the right of the polygon for 25- to 34-year-old males for each value of serum cholesterol level, we can see that the distribution for older men is stochastically larger than the distribution for younger men.

Cumulative frequency polygons can also be used to obtain the *percentiles* of a set of data. Roughly, the 95th percentile is the value that is greater than or equal to 95% of the observations and less than or equal to the remaining 5%. Similarly, the 75th percentile is the value that is greater than or equal to 75% of the observations and less than or equal to the other 25%. This definition is only approximate, because taking 75% of an integer does not typically result in another integer; consequently, there is often some rounding or interpolation involved. In Figure 2.6, the 50th percentile of the serum cholesterol levels for the group of 25- to 34-year-olds—the value that is greater than or equal to half of the observations and less than or equal to the other half—is approximately 193 mg/100 ml; the 50th percentile for the 55- to 64-year-olds is about 226 mg/100 ml.

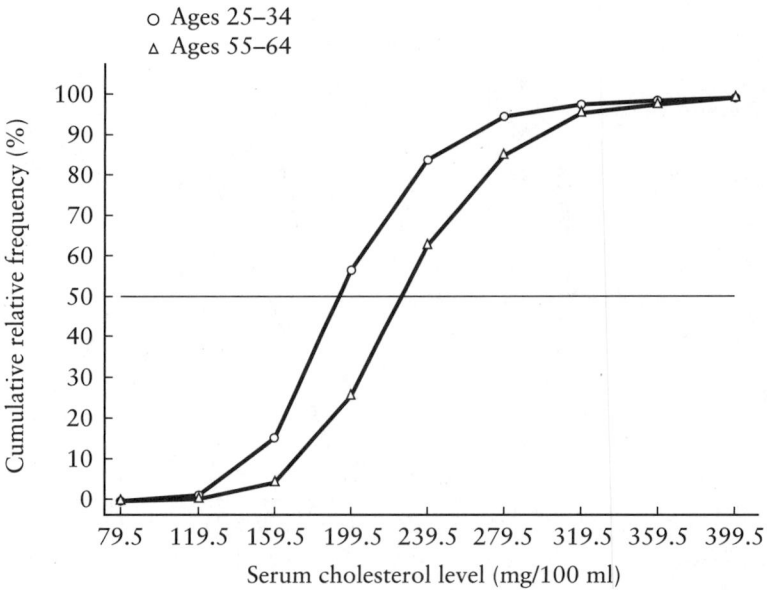

FIGURE 2.6
Cumulative frequency polygon: Cumulative relative frequencies of serum
cholesterol levels for 2294 U.S. males, 1976–1980

Percentiles are useful for describing the shape of a distribution. For example, if the 40th and 60th percentiles of a set of data lie an equal distance away from the midpoint, and the same is true of the 30th and 70th percentiles, the 20th and 80th, and all other pairs of percentiles that sum to 100, the data are *symmetric*; that is, the distribution of values has the same shape on each side of the 50th percentile. Alternatively, if there are a number of outlying observations on one side of the midpoint only, the data are said to be *skewed*. If these observations are smaller than the rest of the values, the data are skewed to the left; if they are larger than the other measurements, the data are skewed to the right. The various shapes that a distribution of data can assume are discussed further in Chapter 3.

2.3.4 One-Way Scatter Plots

Another type of graph that can be used to summarize a set of discrete or continuous observations is the one-way scatter plot. A *one-way scatter plot* uses a single horizontal axis to display the relative position of each data point in the group. As an example, Figure 2.7 depicts the crude death rates for all 50 states and the District of Columbia in 1992, from a low of 391.8 per 100,000 population in Alaska to a high of 1214.9 per 100,000 population in Washington, D.C. [3]. An advantage of the one-way scatter plot is that, since each observation is represented individually, no information is lost; a disadvantage is that the graph may be difficult to read if many data points lie close together.

391.8   1214.9

Rate per 100,000 population

FIGURE 2.7
One-way scatter plot: Crude death rates for the United States, 1992

2.3.5 Box Plots

Box plots are similar to one-way scatter plots in that they require a single axis; instead of plotting every observation, however, they display only a summary of the data [6]. Figure 2.8 is a box plot of the crude death rate data displayed in Figure 2.7. The central box—which is depicted vertically in Figure 2.8 but can also be horizontal—extends from the 25th percentile, 772.0 per 100,000, to the 75th percentile, 933.3 per 100,000. The 25th and 75th percentiles of a data set are called the *quartiles* of the data. The line running between the quartiles at 872.0 deaths per 100,000 population marks the 50th percentile of the data set; half the observations are less than or equal to 872.0 per 100,000, whereas the other half are greater than or equal to this value. If the 50th percentile lies approximately halfway between the two quartiles, this implies that the observations in the center of the data set are roughly symmetric.

The lines projecting out from the box on either side extend to the adjacent values of the plot. The *adjacent values* are the most extreme observations in the data set that are not more than 1.5 times the height of the box beyond either quartile. In Figure 2.8, 1.5 times the height of the box is $1.5 \times (933.3 - 772.0) = 242.0$ per 100,000 population. Therefore, the adjacent values are the smallest and largest observations in the data set that are not more extreme than $772.0 - 242.0 = 530.0$ and $933.3 + 242.0 = 1175.3$

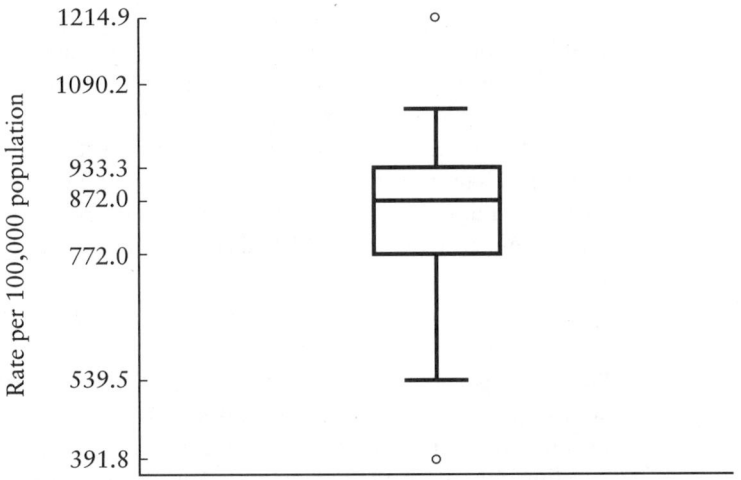

FIGURE 2.8
Box plot: Crude death rates for the United States, 1992

FIGURE 2.9
One-way scatter plot and box plot: Crude death rates for the United States, 1992

per 100,000 population respectively, or 539.5 per 100,000 and 1090.2 per 100,000. In fairly symmetric data sets, the adjacent values should contain approximately 99% of the measurements. All points outside this range are represented by circles; these observations are considered to be *outliers*, or data points that are not typical of the rest of the values.

It should be noted that the preceding explanation is merely one way to define a box plot; other definitions exist and exhibit varying degrees of complexity [7]. In addition, although a box plot conveys a fair amount of information about the distribution of a set of numbers, an even greater amount of information can be displayed by combining the one-way scatter plot and the box plot, as in Figure 2.9.

2.3.6 Two-Way Scatter Plots

Unlike the other graphs that we have discussed, a *two-way scatter plot* is used to depict the relationship between two different continuous measurements. Each point on the graph represents a pair of values; the scale for one quantity is marked on the horizontal axis, or *x*-axis, and the scale for the other on the vertical axis, or *y*-axis. For example, Figure 2.10 plots two simple measures of lung function—forced vital capacity (FVC) and forced expiratory volume in one second (FEV_1)—for 19 asthmatic subjects who participated in a study investigating the physical effects of sulphur dioxide [8]. Forced vital capacity is the volume of air that can be expelled from the lungs in six seconds, and forced expiratory volume in one second is the volume that can be expelled after one second of constant effort. Note that the individual represented by the point that is farthest to the left had an FEV_1 measurement of 2.0 liters and an FVC measurement of 2.8 liters. (Only 18 points are marked on the graph instead of 19 because two individuals had identical values of FVC and FEV_1; consequently, one point lies directly on top of another.) As might be expected, the graph indicates that there is a strong relationship between these two quantities; FVC increases in magnitude as FEV_1 increases.

2.3.7 Line Graphs

A *line graph* is similar to a two-way scatter plot in that it can be used to illustrate the relationship between continuous quantities. Once again, each point on the graph represents a pair of values. In this case, however, each value on the *x*-axis has a single corresponding measurement on the *y*-axis. Adjacent points are connected by straight lines. Most commonly, the scale along the horizontal axis represents time. Consequently, we are able to trace the chronological change in the quantity on the vertical axis over a specified period. As an example, Figure 2.11 displays the trends in the reported rates of

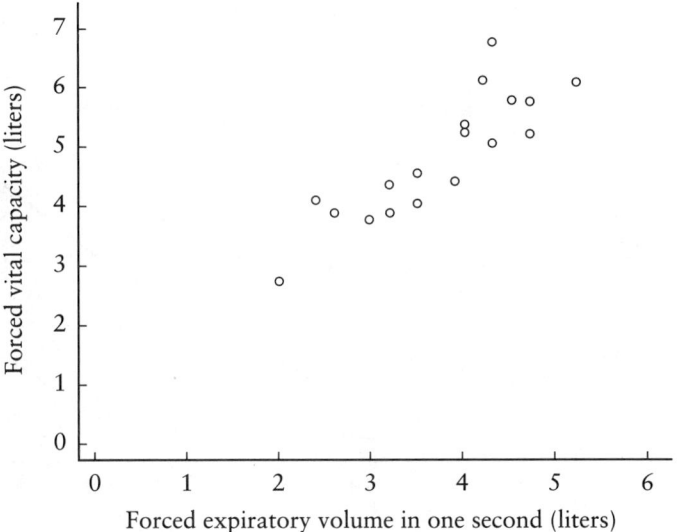

FIGURE 2.10

Two-way scatter plot: Forced vital capacity versus forced expiratory volume in one second for 19 asthmatic subjects

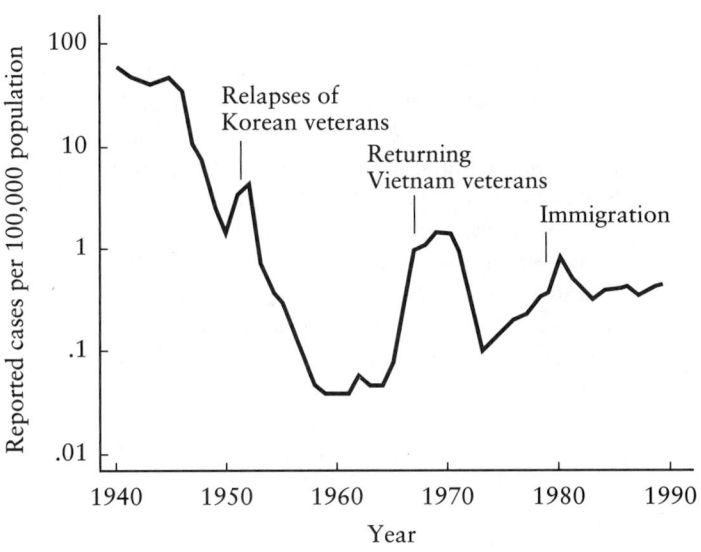

FIGURE 2.11

Line graph: Reported rates of malaria by year, United States, 1940–1989

malaria—including changes that are the result of identifiable sources—that occurred in the United States between 1940 and 1989 [9]. Note the log scale on the vertical axis; this scale allows us to depict a large range of observations while still showing the variation among the smaller values.

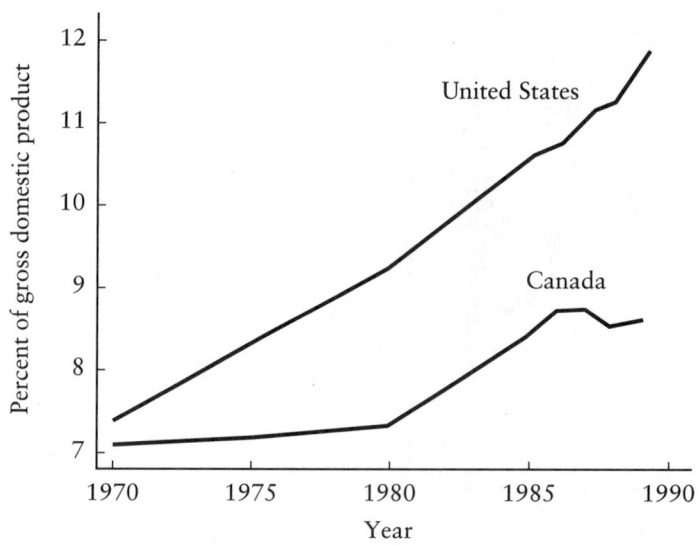

FIGURE 2.12
Line graph: Health care expenditures as a percentage of gross domestic product for the United States and Canada, 1970–1989

To compare two or more groups with respect to a given quantity, it is possible to plot more than one measurement along the y-axis. Suppose that we are concerned with the rising costs of health care. To investigate this problem, we might wish to compare the variations in cost that have occurred under two different health care systems in recent years. Figure 2.12 depicts the trends in health care expenditures in both the United States and Canada between 1970 and 1989 [10].

In this section, we have not attempted to examine all possible types of graphs. Instead, we have included only a selection of the more common ones. It should be noted that many other imaginative displays exist [11]. As a general rule, however, too much information should not be squeezed into a single graph. A relatively simple illustration is often the most effective.

2.4   Further Applications

Suppose that we wish to reduce the number of childhood deaths caused by injury. We first need to understand the nature of the problem. Displayed below is a set of data indicating the causes of death for 100 children between the ages of 5 and 9 who died as a result of injury [12]. The data are nominal: 1 represents a motor vehicle accident, 2 a drowning, 3 is a fire in the home, 4 a homicide, and 5 designates other causes, including suffocation, falls, and poisoning. Given these data, what are we able to conclude about childhood injury deaths?

```
1  5  3  1  2  4  1  3  1  5
2  1  1  5  3  1  2  1  4  1
4  1  3  1  5  1  2  1  1  2
5  1  1  5  1  5  3  1  2  1
2  3  1  1  2  1  5  1  5  1
1  2  5  1  1  1  3  4  1  1
1  1  2  1  1  2  1  1  2  3
3  3  1  5  2  3  5  1  3  4
1  1  2  4  5  4  1  5  1  5
5  1  1  5  1  1  5  1  1  5
```

Although the entire set of outcomes is available, it is extremely difficult to make any type of statement about these data. If we wished to summarize the observations, however, we could begin by constructing a frequency distribution. For nominal and ordinal data, a frequency distribution is a table made up of a list of categories or classes along with the numerical counts that correspond to each one. To construct a frequency distribution for the set of data shown above, we would begin by listing the various causes of death; we would then count the number of children who died as a result of each of these causes. The observations are displayed in frequency distribution format in Table 2.9. Using this table, we are able to see that 48 out of 100 of the injury deaths were the result of motor vehicle accidents, 14 were caused by drowning, 12 by house fires, 7 by homicide, and 19 by other causes.

Like nominal and ordinal data, discrete and continuous data can also be displayed in the form of a frequency distribution. To do this, we must subdivide the range of values of the outcomes into a series of distinct, nonoverlapping intervals. The numbers of observations that fall within each pair of limits are then counted and arranged in a table. Suppose that we are interested in studying the consequences of low birth weight among newborns in the United States. To put the magnitude of the problem into context, we first examine the distribution of birth weights for all infants born in 1986 [13]. We separate these observations into intervals of equal width; the corresponding frequencies are

TABLE 2.9
Injury deaths of 100 children between the ages of 5 and 9 years, United States, 1980–1985

Cause	Number of Deaths
Motor vehicle	48
Drowning	14
House fire	12
Homicide	7
Other	19
Total	100

displayed in Table 2.10. This table provides us with more information about the distribution of birth weights than would a list of 3,751,275 measurements. We can see that most of the observations lie between 2000 and 4499 grams; relatively few measurements fall outside this range. The intervals 3000–3499 and 3500–3999 grams contain the largest numbers of values.

After looking at the actual counts, we might also be interested in finding the relative frequency associated with each interval in the table. The relative frequency is the percentage of the total number of observations that lie within an interval. The relative frequencies for the birth weights displayed in Table 2.10—which we compute by dividing the number of values in the interval by the total number of measurements in the table and multiplying by 100—are shown in Table 2.11. The table indicates that $36.7 + 29.5 = 66.2\%$ of the birth weights are between 3000 and 3999 grams, and $4.3 + 15.9 + 36.7 + 29.5 + 9.2 = 95.6\%$ are between 2000 and 4499 grams. Only 2.5% of the children born in 1986 weighed less than 2000 grams.

In addition to tables, we can also use graphs to summarize and display a set of data. For example, we could illustrate the nominal data in Table 2.9 using the bar chart in Figure 2.13. The categories into which the observations fall are placed along the horizontal axis; the vertical bars represent the frequency of observations in each class. The graph emphasizes that a large proportion of childhood injury deaths are the result of motor vehicle accidents.

A *stacked bar chart* can be used to convey a greater amount of information in a single picture. In this type of graph, bars that represent the frequency of observations in two or more different subgroups are placed on top of one another. As an example, Figure 2.14 displays the mortality rates per 1000 births (the number of deaths for every 1000 births) in France for four categories of infants—those who were stillborn, those

TABLE 2.10
Absolute frequencies of birth weights for 3,751,275 infants born in the United States, 1986

Birth Weight (grams)	Number of Infants
0–499	4843
500–999	17,487
1000–1499	23,139
1500–1999	49,112
2000–2499	160,919
2500–2999	597,738
3000–3499	1,376,008
3500–3999	1,106,634
4000–4499	344,390
4500–4999	62,769
5000–5500	8236
Total	3,751,275

TABLE 2.11
Relative frequencies of birth weights for 3,751,275 infants born in the United States, 1986

Birth Weight (grams)	Relative Frequency (%)
0–499	0.1
500–999	0.5
1000–1499	0.6
1500–1999	1.3
2000–2499	4.3
2500–2999	15.9
3000–3499	36.7
3500–3999	29.5
4000–4499	9.2
4500–4999	1.7
5000–5500	0.2
Total	100.0

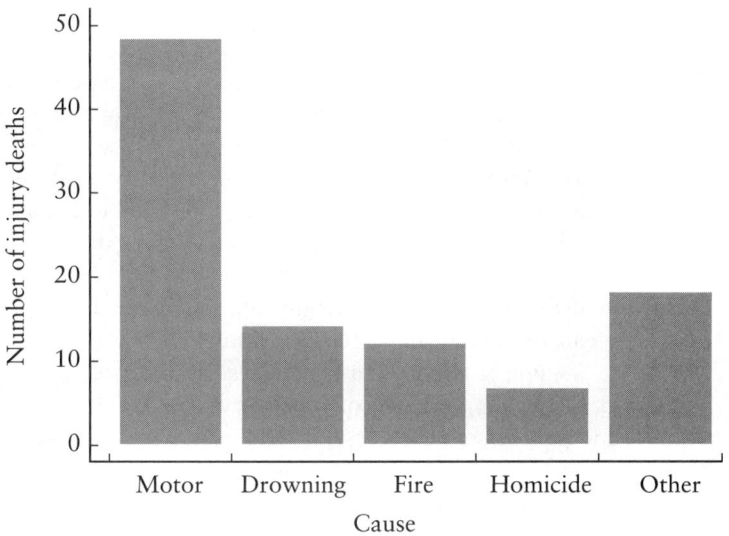

FIGURE 2.13
Injury deaths of 100 children between the ages of 5 and 9 years, United States, 1980–1985

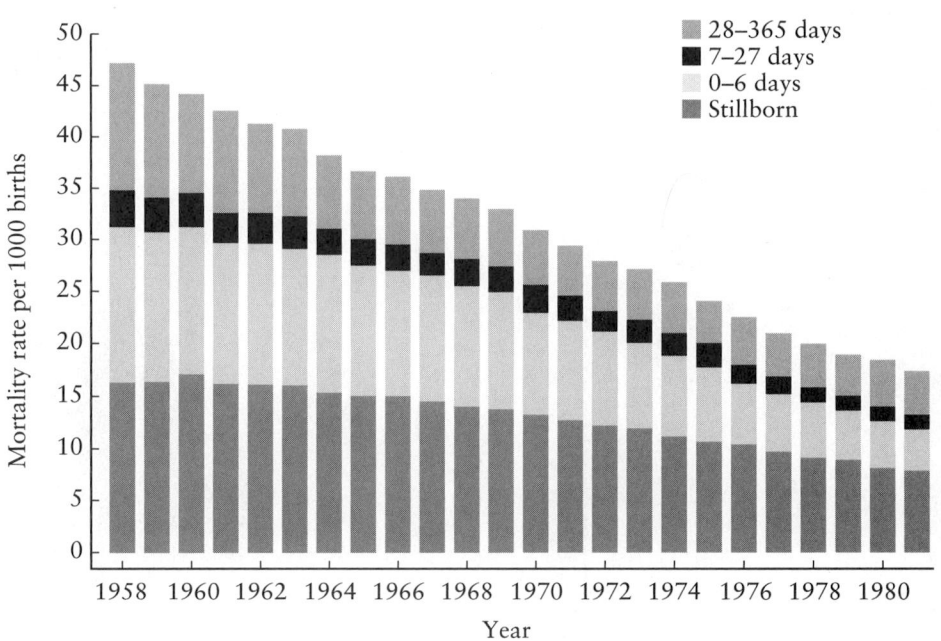

FIGURE 2.14
Infant and perinatal mortality in France, 1978–1981

who died less than one week after they were born, those who died between 7 and 27 days after they were born, and those who survived for more than 27 days but less than one year [14]. Since each of these rates is decreasing over time, the graph is able to make a powerful statement about the overall infant mortality.

Of the various graphical displays that can be used for discrete or continuous data, the histogram is perhaps the most common. Like a bar chart, a histogram is a pictorial representation of a frequency distribution. The horizontal axis displays the true limits of the intervals into which the observations fall; the vertical axis depicts the frequency or relative frequency of observations within each interval. As an example, Figure 2.15 is a histogram of the birth weight data summarized in Table 2.10. Looking at the graph, we can see that the data are skewed to the left.

A box plot is another type of graph often used for discrete or continuous data. The plot displays a summary of the observations using a single vertical or horizontal axis. Suppose that we are interested in comparing health care expenditures in 1989 for the 24 nations that make up the Organization for Economic Cooperation and Development. These expenditures are summarized as a percentage of gross domestic product in Figure 2.16, from a low of 5.1% in Greece to a high of 11.8% in the United States [10]. The three horizontal lines that make up the central box indicate that the 25th, 50th, and 75th percentiles of the data are 6.7%, 7.4%, and 8.3% respectively. The height of the box is the distance between the 25th and 75th percentiles, also known as the quartiles of the data. The lines extending from either side of the central box mark the most extreme observations that are not more than 1.5 times the height of the box beyond either quartile, or the adjacent values. In Figure 2.16, the adjacent values are 5.1% and 8.8%. The United States is an outlier, with a health care expenditure that is not typical of the rest of the data.

A line graph is one type of display that can be used to illustrate the relationship between two continuous measurements. Each point on the line represents a pair of val-

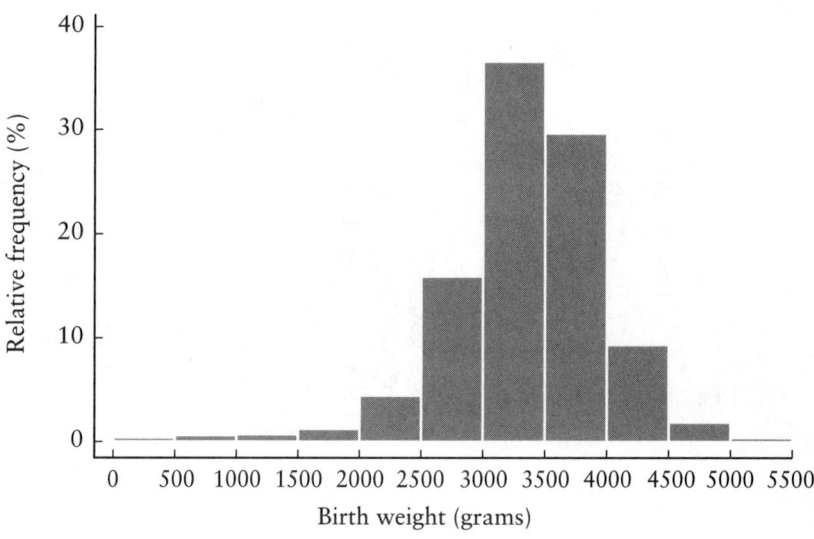

FIGURE 2.15
Relative frequencies of birth weights for 3,751,275 infants born in the
United States, 1986

ues; the line itself allows us to trace the change in the quantity on the *y*-axis that corresponds to a change along the *x*-axis. Figure 2.17, like Figure 2.1, depicts data relating to cigarette consumption in the United States. Note that the line graph shows more detail than the corresponding bar chart.

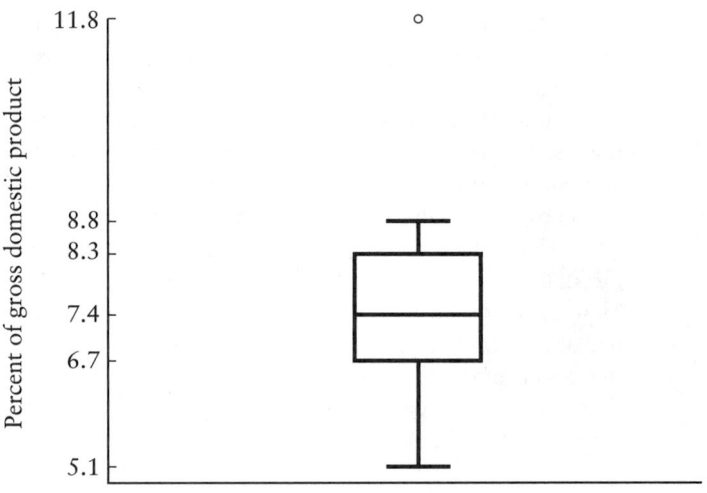

FIGURE 2.16
Health care expenditure as a percentage of gross domestic product for
24 nations, 1989

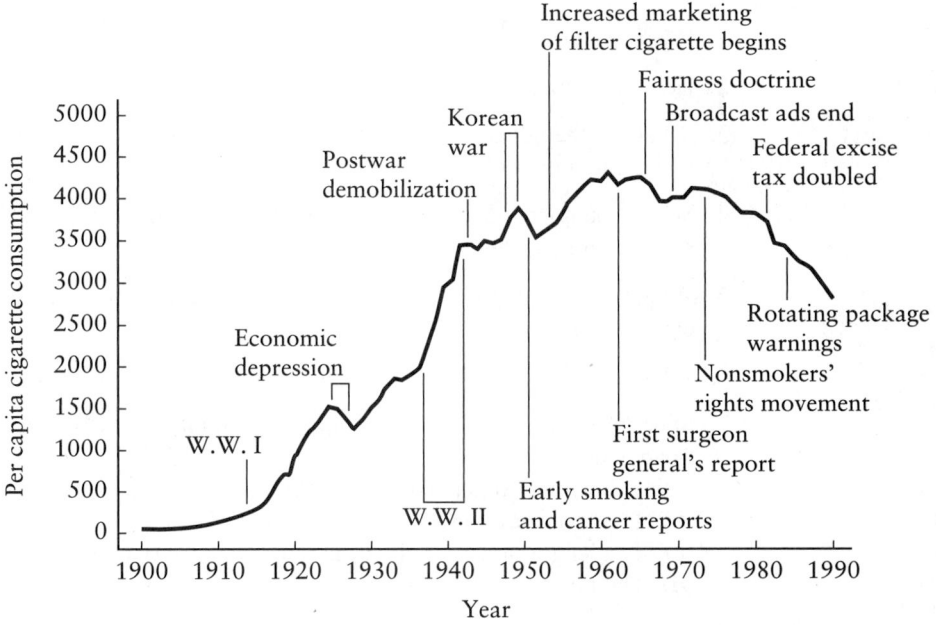

FIGURE 2.17
Cigarette consumption per person 18 years of age or older, United States,
1900–1990

Over the years, the use of computers in statistics has increased dramatically. As a result, many formerly time-consuming calculations can now be performed much more efficiently using a statistical package. A *statistical package* is a series of programs that have been designed to analyze numerical data. A variety of packages are available; in general, they differ with respect to the commands they use and the formats of the output they produce.

One statistical package that is both powerful and relatively easy to use is called *Stata*. Stata is an interactive program that helps us manage, display, and analyze data. Observations or measurements are saved in columns; each column is assigned a variable name. We then use these names to execute specific analytical procedures or commands. When it is appropriate, we reproduce output from Stata to illustrate what the computer is able to do. Since some readers may prefer to use other statistical packages, we incorporate output from Minitab and SAS as well.

Computers are particularly useful for constructing graphs. In fact, Figures 2.13 through 2.17 were all generated using Stata. To create Figure 2.17, we saved the years between 1900 and 1990 under the variable name `year` and saved the corresponding values of per capita cigarette consumption under the name `cigarett` (the final *e* is missing because a variable name can have at most eight letters). Stata plotted the points representing each pair of values. The points were connected and labels added using the appropriate commands.

2.5   Review Exercises

1. What are descriptive statistics?

2. How do ordinal data differ from nominal data?

3. What are the advantages and disadvantages of transforming continuous measurements into either discrete or ordinal ones?

4. When you construct a table, when might it be beneficial to use relative rather than absolute frequencies?

5. What types of graphs can be used to display nominal or ordinal observations? Discrete or continuous observations?

6. What are the percentiles of a set of data?

7. State whether each of the following observations is an example of discrete or continuous data.
   (a) The number of suicides in the United States in a specified year
   (b) The concentration of lead in a sample of water
   (c) The length of time that a cancer patient survives after diagnosis
   (d) The number of previous miscarriages an expectant mother has had

8. Listed on the following page are the per capita health care expenditures in 1989 for 23 of the 24 nations that make up the Organization for Economic Cooperation and Development [10]. (The per capita expenditure for Turkey was not available.)

Nation	Per Capita Expenditure (U.S.$)
Australia	1032
Austria	1093
Belgium	980
Britain	836
Canada	1683
Denmark	912
Finland	1067
France	1274
Germany	1232
Greece	371
Iceland	1353
Ireland	658
Italy	1050
Japan	1035
Luxembourg	1193
Netherlands	1135
New Zealand	820
Norway	1234
Portugal	464
Spain	644
Sweden	1361
Switzerland	1376
United States	2354

(a) Rank these countries according to per capita health care expenditure.
(b) Construct a histogram for the values of per capita expenditure.
(c) Describe the shape of the histogram.

9. The table below categorizes 10,614,000 office visits to cardiovascular disease specialists in the United States by the duration of each visit [15]. A duration of 0 minutes implies that the patient did not have face-to-face contact with the specialist.

Duration (minutes)	Number of Visits (thousands)
0	390
1–5	227
6–10	1023
11–15	3390
16–30	4431
31–60	968
61+	185
Total	10,614

The statement is made that office visits to cardiovascular disease specialists are most often between 16 and 30 minutes long. Do you agree with this statement? Why or why not?

10. The frequency distribution below displays the numbers of cases of pediatric AIDS reported in the United States between 1983 and 1989 [9].

Year	Number of Cases
1983	122
1984	250
1985	455
1986	848
1987	1412
1988	2811
1989	3098

Construct a bar chart showing the number of cases by year. What does the graph tell you about pediatric AIDS in this time period?

11. Listed below are the numbers of people who have been executed in the United States in each year since the 1976 Supreme Court decision allowing the death penalty to be carried out [16].

Year	Number of Executions	Year	Number of Executions
1976	0	1986	18
1977	1	1987	25
1978	0	1988	11
1979	2	1989	16
1980	0	1990	23
1981	1	1991	14
1982	2	1992	31
1983	5	1993	38
1984	21	1994	28
1985	18		

Use these data to create a bar chart of executions by year. How has the number of executions varied since 1976?

12. A study was conducted to examine gender and racial differences among individuals 65 years of age and older who suffered a hip fracture between 1984 and 1987 [17]. The data summarizing all hospital discharges covered by the Medicare program appear below.

Age Group	White Men	Black Men	White Women	Black Women
65–74	36,473	2295	103,105	3425
75–84	62,513	2902	233,047	6819
85–94	40,975	1659	189,459	5968
95+	4088	208	18,247	934

(a) Using these data, construct a stacked bar chart showing the number of hospital discharges following hip fracture by age group. (Each bar should consist of four separate sections representing white men, black men, white women, and black women.)

(b) How does the overall number of hip fractures vary with age?

(c) Based on the graph, what can you conclude about the relationship between gender and hip fracture?

13. In an investigation of the risk factors for cardiovascular disease, levels of serum cotinine—a metabolic product of nicotine—were recorded for a group of smokers and a group of nonsmokers [18]. The relevant frequency distributions are displayed below.

Cotinine Level (ng/ml)	Smokers	Nonsmokers
0–13	78	3300
14–49	133	72
50–99	142	23
100–149	206	15
150–199	197	7
200–249	220	8
250–299	151	9
300+	412	11
Total	1539	3445

(a) Is it fair to compare the distributions of cotinine levels for smokers and nonsmokers based on the absolute frequencies in each interval? Why or why not?

(b) Compute the relative frequencies of serum cotinine level readings for each of the two groups.

(c) Construct a pair of frequency polygons.

(d) Describe the shape of each polygon. What can you say about the distribution of recorded cotinine levels in each group?

(e) For all individuals in this study, smoking status is self-reported. Do you think any of the subjects might be misclassified? Why or why not?

14. The relative frequencies of blood lead concentrations for two groups of workers in Canada—one examined in 1979 and the other in 1987—are displayed below [19].

Blood Lead (μg/dl)	1979 (%)	1987 (%)
<20	11.5	37.8
20–29	12.1	14.7
30–39	13.9	13.1
40–49	15.4	15.3
50–59	16.5	10.5
60–69	12.8	6.8
70–79	8.4	1.4
≥80	9.4	0.4

(a) In which of the two years do the workers tend to have lower blood lead levels?

(b) Compute the cumulative relative frequencies for each group of workers. Use these data to construct a pair of cumulative frequency polygons.

(c) For which group of workers is the distribution of blood lead levels stochastically larger?

15. The reported numbers of live births in the United States for each month in the period January 1991 to December 1992 are listed below [20].

Month 1991	Number (thousands)	Month 1992	Number (thousands)
January	325	January	334
February	312	February	304
March	346	March	360
April	340	April	330
May	355	May	361
June	342	June	333
July	358	July	352
August	346	August	350
September	365	September	357
October	355	October	345
November	324	November	332
December	342	December	325

(a) Construct a line graph displaying the reported number of live births over time.

(b) Based on this two-year period, do you think the number of live births follows a seasonal pattern in the U.S.?

16. A frequency distribution for the serum zinc levels of 462 males between the ages of 15 and 17 is displayed below [21]. The observations are stored on the accompanying disk in the data set serzinc (Appendix B, Table B.1). The 462 serum zinc measurements, which were recorded in micrograms per deciliter, are saved under the variable name zinc.

United States Males, Ages 15–17	
Serum Zinc Level (μg/dl)	**Number of Males**
50–59	6
60–69	35
70–79	110
80–89	116
90–99	91
100–109	63
110–119	30
120–129	5
130–139	2
140–149	2
150–159	2

(a) Compute the relative frequency associated with each interval in the table. What can you conclude about this distribution of serum zinc levels?

(b) Produce a histogram of the data. The observations should be broken down into the 11 intervals of equal width specified in the frequency distribution above, from 50–59 to 150–159 μg/dl.

(c) Describe the shape of the histogram.

17. The percentages of low birth weight infants in various countries around the world are contained in the data set `unicef` [22] (Appendix B, Table B.2). The measurements themselves are saved under the variable name `lowbwt`.

(a) Construct a box plot for the percentages of low birth weight infants.

(b) Do the data appear to be skewed? If so, are they skewed to the right or to the left?

(c) Do the data contain any outlying observations?

18. The numbers of nursing home residents at least 65 years old per 1000 population 65 years of age and over for each state in the United States are contained in the data set `nurshome` [23] (Appendix B, Table B.3). The state names are saved under the variable name `state` and the numbers of nursing home residents per 1000 population under the variable name `resident`.

(a) Which state has the smallest number of nursing home residents per 1000 population 65 years of age and over? Which state has the largest number? What factors might influence the substantial amount of variability among different states?

(b) Construct a box plot for the number of nursing home residents per 1000 population.

(c) Are the observations symmetric or skewed? Are there any states that could be considered to be outliers?

(d) Display the number of nursing home residents per 1000 population using a histogram. Do you find this graph to be more or less informative than the box plot?

19. The declared concentrations of tar and nicotine for 35 brands of Canadian cigarettes are stored in a data set called `cigarett` [24] (Appendix B, Table B.4). The concentrations of tar per cigarette in milligrams are saved under the variable name `tar` and the corresponding concentrations of nicotine under the name `nicotine`.

(a) Produce a one-way scatter plot of the declared concentrations of tar per cigarette. Be sure to identify instances in which two or more measurements have the same value and therefore overlap.

(b) Describe the distribution of values.

(c) Construct a two-way scatter plot of the concentration of tar versus the concentration of nicotine. Label the axes appropriately.

(d) Does there appear to be a relationship between these two quantities?

20. The birth rates for unmarried women in the United States from 1940 to 1992 are saved in the data set `brate` [25] (Appendix B, Table B.5). The years are saved under the variable name `year`, and the numbers of live births per 1000 unmarried women 15 to 44 years of age are saved under the name `birthrt`.

(a) Create a line graph displaying the birth rates for unmarried women over time.

(b) Many people believe that the large number of children born to unmarried mothers is a relatively recent problem in our society. After seeing the line graph, do you agree?

Bibliography

[1] Centers for Disease Control and Prevention, *HIV/AIDS Surveillance Report*, Volume 5, Number 4, 1994.

[2] Oken, M. M., Creech, R. H., Tormey, D. C., Horton, J., Davis, T. E., McFadden, E. T., and Carbone, P. P., "Toxicity and Response Criteria of the Eastern Cooperative Oncology Group," *American Journal of Clinical Oncology*, Volume 5, December 1982, 649–655.

[3] National Center for Health Statistics, Kochanek, K. D., and Hudson, B. L., "Advance Report of Final Mortality Statistics, 1992," *Monthly Vital Statistics Report*, Volume 43, March 22, 1995.

[4] Garfinkel, L., and Silverberg, E., "Lung Cancer and Smoking Trends in the United States over the Past 25 Years," *Ca—A Cancer Journal for Clinicians*, Volume 41, May/June 1991, 137–145.

[5] National Center for Health Statistics, Fulwood, R., Kalsbeek, W., Rifkind, B., Russell-Briefel, R., Muesing, R., LaRosa, J., and Lippel, K., "Total Serum Cholesterol Levels of Adults 20–74 Years of Age: United States, 1976–1980," *Vital and Health Statistics*, Series 11, Number 236, May 1986.

[6] Spear, M. E., *Charting Statistics*, New York: McGraw-Hill, 1952.

[7] Tukey, J. W., *Exploratory Data Analysis*, Reading, Mass.: Addison-Wesley, 1977.

[8] Bethel, R. A., Sheppard, D., Geffroy, B., Tam, E., Nadel, J. A., and Boushey, H. A., "Effect of 0.25 ppm Sulphur Dioxide on Airway Resistance in Freely Breathing, Heavily Exercising, Asthmatic Subjects," *American Review of Respiratory Disease*, Volume 31, April 1985, 659–661.

[9] Centers for Disease Control, "Summary of Notifiable Diseases, United States, 1989," *Morbidity and Mortality Weekly Report*, Volume 38, October 5, 1990.

[10] Schieber, G. J., and Poullier, J. P., "International Health Spending: Issues and Trends," *Health Affairs*, Volume 10, Spring 1991, 106–116.

[11] Tufte, E. R., *The Visual Display of Quantitative Information*, Cheshire, Conn.: Graphics Press, 1983.

[12] Waller, A. E., Baker, S. P., and Szocka, A., "Childhood Injury Deaths: National Analysis and Geographic Variations," *American Journal of Public Health*, Volume 79, March 1989, 310–315.

[13] National Center for Health Statistics, "Advance Reports, 1986," *Supplements to the Monthly Vital Statistics Report*, Series 24, March 1990.

[14] Rumeau-Rouquette, C., "The French Perinatal Program: 'Born in France'," *Child Health and Development*, Volume 3, *Prevention of Perinatal Mortality and Morbidity*, New York: Karger, 1984.

[15] National Center for Health Statistics, Nelson, C., "Office Visits to Cardiovascular Disease Specialists, 1985," *Vital and Health Statistics*, Advance Data Report Number 171, June 23, 1989.

[16] Kuntz, T., "Killings, Legal and Otherwise, Around the U.S.," *The New York Times*, December 4, 1994, 3.

[17] Jacobsen, S. J., Goldberg, J., Miles, T. P., Brody, J. A., Stiers, W., and Rimm, A. A., "Race and Sex Differences in Mortality Following Fracture of the Hip," *American Journal of Public Health*, Volume 82, August 1992, 1147–1150.

[18] Wagenknecht, L. E., Burke, G. L., Perkins, L. L., Haley, N. J., and Friedman, G. D., "Misclassification of Smoking Status in the CARDIA Study: A Comparison of Self-Report with Serum Cotinine Levels," *American Journal of Public Health*, Volume 82, January 1992, 33–36.

[19] Yassi, A., Cheang, M., Tenenbein, M., Bawden, G., Spiegel, J., and Redekop, T., "An Analysis of Occupational Blood Lead Trends in Manitoba, 1979 Through 1987," *American Journal of Public Health*, Volume 81, June 1991, 736–740.

[20] National Center for Health Statistics, "Births, Marriages, Divorces, and Deaths for 1992," *Monthly Vital Statistics Report*, Volume 41, May 19, 1993.

[21] National Center for Health Statistics, Fulwood, R., Johnson, C. L., Bryner, J. D., Gunter, E. W., and McGrath, C. R., "Hematological and Nutritional Biochemistry Reference Data for Persons 6 Months–74 Years of Age: United States, 1976–1980," *Vital and Health Statistics*, Series 11, Number 232, December 1982.

[22] United Nations Children's Fund, *The State of the World's Children 1994*, New York: Oxford University Press.

[23] National Center for Health Statistics, *Health, United States, 1994 Chartbook*, May 1995.

[24] Kaiserman, M. J., and Rickert, W. S., "Carcinogens in Tobacco Smoke: Benzo[a]pyrene from Canadian Cigarettes and Cigarette Tobacco," *American Journal of Public Health*, Volume 82, July 1992, 1023–1026.

[25] National Center for Health Statistics, Ventura, S. J., "Births to Unmarried Mothers: United States, 1980–92," *Vital and Health Statistics*, Series 21, Number 53, June 1995.

3

Numerical Summary Measures

In the preceding chapter, we studied tables and graphs as methods of organizing, visually summarizing, and displaying a set of data. Although these techniques are extremely useful, they do not allow us to make concise, quantitative statements that characterize the distribution of values as a whole. In order to do this, we rely instead on *numerical summary measures*. Together, the various types of descriptive statistics can provide a great deal of information about a set of observations.

3.1 Measures of Central Tendency

The most commonly investigated characteristic of a set of data is its center, or the point about which the observations tend to cluster. Suppose that we are interested in examining the response to inhaled ozone and sulphur dioxide among adolescents suffering from asthma. Listed in Table 3.1 are the initial measurements of forced expiratory volume in 1 second for 13 subjects involved in such a study [1]. Recall that FEV_1 is the volume of air that can be expelled from the lungs after one second of constant effort. Before investigating the effect of pollutants on lung function, we might wish to determine the typical value of FEV_1 prior to exposure for the individuals in this group.

3.1.1 Mean

The most frequently used measure of central tendency is the arithmetic mean, or average. The *mean* is calculated by summing all the observations in a set of data and dividing by the total number of measurements. In Table 3.1, for example, we have 13 observations. If x is used to represent FEV_1, $x_1 = 2.30$ denotes the first in the series of observations; $x_2 = 2.15$, the second; and so on up through $x_{13} = 3.38$. In general, x_i

TABLE 3.1

Forced expiratory volumes in 1 second for 13 adolescents suffering from asthma

Subject	FEV$_1$ (liters)
1	2.30
2	2.15
3	3.50
4	2.60
5	2.75
6	2.82
7	4.05
8	2.25
9	2.68
10	3.00
11	4.02
12	2.85
13	3.38

refers to a single FEV$_1$ measurement, where i can take on any value from 1 to n, the total number of observations in the group. The mean of the observations in the data set—represented by $\bar{x}$, or x-bar—is

$$\bar{x} = \frac{1}{n} \sum_{i=1}^{n} x_i.$$

Note that we have used some mathematical shorthand. The uppercase Greek letter sigma, $\sum$, is the symbol for summation. The expression $\sum_{i=1}^{n} x_i$ means that we are to add up the values of all of the observations in the group, from x_1 to x_n. When $\sum$ appears in the text, the limits of summation are placed beside it; when it is displayed separately in an equation, the limits are above and below it. Both representations denote exactly the same thing. In some cases where it is clear that we are intended to sum all the observations in a data set, the limits may be dropped altogether. For the FEV$_1$ data,

$$\bar{x} = \frac{1}{13} \sum_{i=1}^{13} x_i$$

$$= \left(\frac{1}{13}\right)(2.30 + 2.15 + 3.50 + 2.60 + 2.75 + 2.82 + 4.05$$
$$+ 2.25 + 2.68 + 3.00 + 4.02 + 2.85 + 3.38)$$

$$= \frac{38.35}{13}$$

$$= 2.95 \text{ liters.}$$

The mean can be used as a summary measure for both discrete and continuous measurements. In general, however, it is not appropriate for either nominal or ordinal data. Recall that for these types of data, the numbers are merely labels, so that even if we choose to represent the blood types O, A, B, and AB by the numbers 1, 2, 3, and 4, an average blood type of 1.8 is meaningless. One exception to this rule applies when we have dichotomous data and the two possible outcomes are represented by the values 0 and 1. In this situation, the mean of the observations is equal to the proportion of 1s in the data set. For example, suppose that we wish to know the proportion of asthmatic adolescents in the previously mentioned study who are males. Listed in Table 3.2 are the relevant dichotomous data; the value 1 represents a male, and 0 designates a female. If we compute the mean of these observations, we find that

$$
\begin{aligned}
\bar{x} &= \frac{1}{n} \sum_{i=1}^{n} x_i \\
&= \left(\frac{1}{13}\right)(0 + 1 + 1 + 0 + 0 + 1 + 1 + 1 + 0 + 1 + 1 + 1 + 0) \\
&= \frac{8}{13} \\
&= 0.615.
\end{aligned}
$$

Therefore, 61.5% of the study subjects are males.

The method by which the mean is calculated takes into consideration the magnitude of every observation in a set of data. What happens when one observation has a value that is very different from the others? Suppose, for instance, that we had recorded the data in Table 3.1 on a computer disk and that this disk was accidentally x-rayed at

TABLE 3.2

Indicators of gender for 13 adolescents suffering from asthma

Subject	Gender
1	0
2	1
3	1
4	0
5	0
6	1
7	1
8	1
9	0
10	1
11	1
12	1
13	0

the airport; as a result, the FEV_1 measurement of subject 11 is now recorded as 40.2 rather than 4.02. The mean FEV_1 of all 13 subjects would then be calculated as

$$\bar{x} = \left(\frac{1}{13}\right)(2.30 + 2.15 + 3.50 + 2.60 + 2.75 + 2.82 + 4.05$$
$$+ 2.25 + 2.68 + 3.00 + 40.2 + 2.85 + 3.38)$$
$$= \frac{74.53}{13}$$
$$= 5.73 \text{ liters,}$$

which is nearly twice as large as it was before. Clearly, the mean is extremely sensitive to unusual values. In this particular example, we would have rightfully questioned an FEV_1 measurement of 40.2 liters and would have either corrected the error or separated this observation from the others. In general, however, the error might not be as obvious, or the unusual observation might not be an error at all. Since it is our intent to characterize an entire group of individuals, we might prefer to use a summary measure that is not as sensitive to every observation.

3.1.2   *Median*

One measure of central tendency that is not as sensitive to the value of each measurement is the median. The median can be used as a summary measure for ordinal observations as well as for discrete and continuous data. The *median* is defined as the 50th percentile of a set of measurements; if a list of observations is ranked from smallest to largest, half the values are greater than or equal to the median, whereas the other half are less than or equal to it. Therefore, if a set of data contains a total of n observations where n is odd, the median is the middle value, or the $[(n + 1)/2]$th largest measurement; if n is even, the median is usually taken to be the average of the two middlemost values, the $(n/2)$th and $[(n/2) + 1]$th observations. If we were to rank the 13 FEV_1 measurements listed in Table 3.1, for example, the following sequence would result:

2.15, 2.25, 2.30, 2.60, 2.68, 2.75, 2.82, 2.85, 3.00, 3.38, 3.50, 4.02, 4.05.

Since there are an odd number of observations in the list, the median is the $(13 + 1)/2 = 7$th observation, or 2.82. Seven of the measurements are less than or equal to 2.82 liters, and seven are greater than or equal to 2.82.

The calculation of the median takes into consideration only the ordering and relative magnitude of the observations in a set of data. In the situation where the FEV_1 of subject 11 was recorded as 40.2 rather than 4.02, the ranking of the measurements would change only slightly:

2.15, 2.25, 2.30, 2.60, 2.68, 2.75, 2.82, 2.85, 3.00, 3.38, 3.50, 4.05, 40.2.

As a result, the median FEV_1 would remain 2.82 liters. Unlike the mean, the median is said to be *robust;* that is, it is much less sensitive to unusual data points.

3.1.3   Mode

A third measure of central tendency is the mode; it can be used as a summary measure for all types of data. The *mode* of a set of values is the observation that occurs most frequently. The FEV_1 data in Table 3.1 do not have a unique mode, since each of the values occurs only once. The mode for the dichotomous data in Table 3.2 is 1. This value appears eight times, whereas 0 appears only five times.

The best measure of central tendency for a given set of data often depends on the way in which the values are distributed. If they are symmetric and *unimodal*—meaning that if we were to draw a histogram or a frequency polygon, there would be only one peak, as in the smoothed distribution pictured in Figure 3.1(a)—then the mean, the median, and the mode should all be roughly the same. If the distribution of values is symmetric but *bimodal*, so that the corresponding frequency polygon would have two peaks, as in Figure 3.1(b), then the mean and the median should again be approximately the same. Note, however, that this common value could lie between the two peaks, and hence be a measurement that is extremely unlikely to occur. A bimodal distribution often indicates that the population from which the values are taken actually consists of two distinct subgroups that differ in the characteristic being measured; in this situation, it might be better to report two modes rather than the mean or the median, or to treat the two subgroups separately. The data in Figure 3.1(c) are skewed to the right, and those in Figure 3.1(d) are skewed to the left. When the data are not symmetric, the median is often the best measure of central tendency. Because the mean is sensitive to extreme observations, it is pulled in the direction of the outlying data values and as a result might end up either excessively inflated or excessively deflated. Note that when the data are skewed to the right, the mean lies to the right of the median, and when they are skewed to the left, the mean lies to the left of the median.

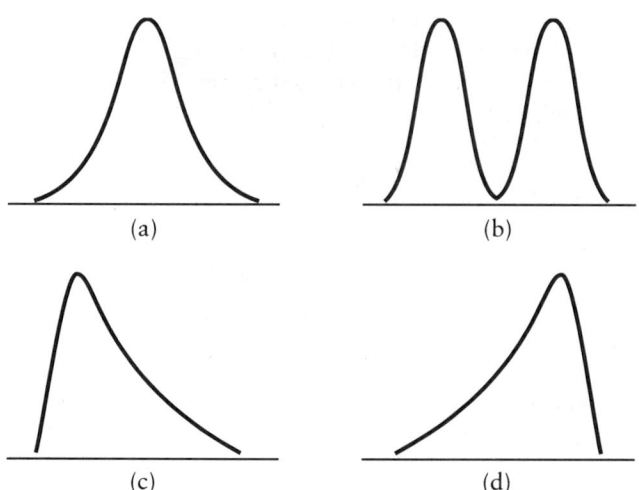

(a)   (b)

(c)   (d)

FIGURE 3.1
Possible distributions of data values

Regardless of the measure of central tendency used in a particular situation, it can be misleading to assume that this value is representative of all observations in the group. One example that illustrates this point was included in the November 17, 1991, episode of the popular news program "60 Minutes." The program contained a segment on diet and mortality that contrasted the French and American experiences. Although the French diet is extremely high in fat and cholesterol, France has a much lower rate of heart disease than the United States. This paradoxical difference was attributed to the French habit of drinking wine—red wine in particular—with meals. Studies have suggested that moderate alcohol consumption can lessen the risk of heart disease. While the per capita intake of wine in France is one of the highest in the world, the per capita intake in the United States is one of the lowest; the program implied that the French drink a moderate amount of wine each day, perhaps two or three glasses. The reality may be quite different, however. According to a wine industry survey conducted in 1990, more than half of all French adults never drink wine at all [2]. Of those who do, only 28% of males and 11% of females drink it daily. Obviously the distribution is far more variable than the "typical value" would suggest. Remember that when we summarize a set of data, information is always lost. Thus, although it is helpful to know where the center of a data set lies, this information is usually not sufficient to characterize an entire distribution of measurements.

As another example, in each of the two very different distributions of data values shown in Figure 3.2, the mean, median, and mode are equal. To know how good our measure of central tendency actually is, we need to have some idea about the variation among the measurements. Do all the observations tend to be quite similar and therefore lie close to the center, or are they spread out across a broad range of values?

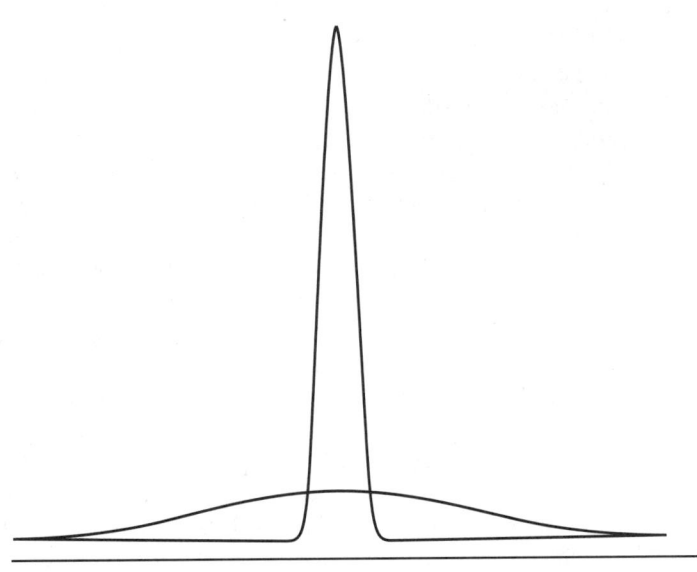

FIGURE 3.2
Two distributions with identical means, medians, and modes

3.2   Measures of Dispersion

3.2.1   Range

One number that can be used to describe the variability in a set of data is known as the range. The *range* of a group of measurements is defined as the difference between the largest observation and the smallest. Although the range is easy to compute, its usefulness is limited; it considers only the extreme values of a data set rather than the majority of the observations. Therefore, like the mean, it is highly sensitive to exceptionally large or exceptionally small values. The range for the FEV_1 data in Table 3.1 is $4.05 - 2.15 = 1.90$ liters. If the FEV_1 of subject 11 were recorded as 40.2 instead of 4.02, the range would be $40.2 - 2.15 = 38.05$ liters, a value 20 times as large. The ranges of values for the average annual sulphur dioxide concentration in the air of a number of cities around the world are displayed in Figure 3.3 [3].

3.2.2   Interquartile Range

A second measure of variability—one that is not as easily influenced by extreme values—is called the interquartile range. The *interquartile range* is calculated by subtracting the 25th percentile of the data from the 75th percentile; consequently, it encompasses the middle 50% of the observations. (Recall that the 25th and 75th percentiles of a data set are called the quartiles.) For the FEV_1 data in Table 3.1, the 75th percentile is 3.38. Note that three observations are greater than this value and nine are smaller. Similarly, the 25th percentile is 2.60. Therefore, the interquartile range is $3.38 - 2.60 = 0.78$ liters.

If a computer is not available, there are rules for finding the kth percentile of a set of data by hand, just as there are rules for finding the median. In that case, the rule used depends on whether the number of observations n is even or odd. We again begin by ranking the measurements from smallest to largest. If $nk/100$ is an integer, the kth percentile of the data is the average of the $(nk/100)$th and $(nk/100 + 1)$th largest observations. If $nk/100$ is not an integer, the kth percentile is the $(j + 1)$th largest measurement, where j is the largest integer that is less than $nk/100$. To find the 25th percentile of the 13 FEV_1 measurements, for example, we first note that $13(25)/100 = 3.25$ is not an integer. Therefore, the 25th percentile is the $3 + 1 = 4$th largest measurement (since 3 is the largest integer less than 3.25), or 2.60 liters. Similarly, $13(75)/100 = 9.75$ is not an integer, and the 75th percentile is the $9 + 1 = 10$th largest measurement, or 3.38 liters. The interquartile ranges of the numbers of episodes of various types of sexual behavior practiced by homosexual males before and after learning about AIDS—as well as their means, medians, and ranges—are presented in Figure 3.4 [4]. Note that the means are larger than the medians in all cases, indicating that the data are skewed and that there are a number of unusually large values that are causing the means to be inflated. The difference between the means and the medians is less evident after the men have learned about AIDS; education about the virus appears to have had a constraining effect on sexual behavior, especially in extreme cases.

Shown is the range of annual values at individual sites and the composite 5-year average for the city.

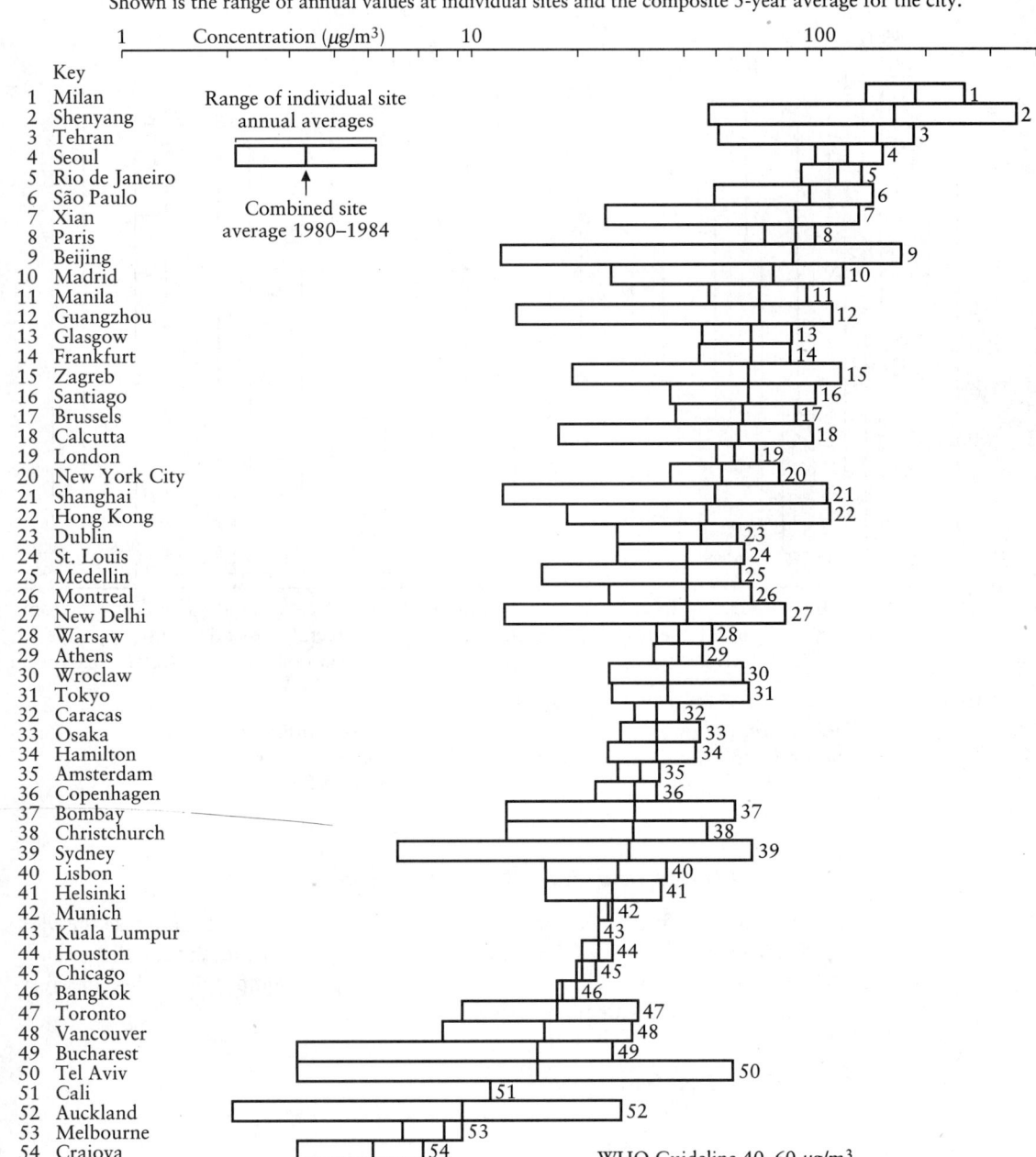

FIGURE 3.3

Summary of annual suphur dioxide averages, 1980–1984

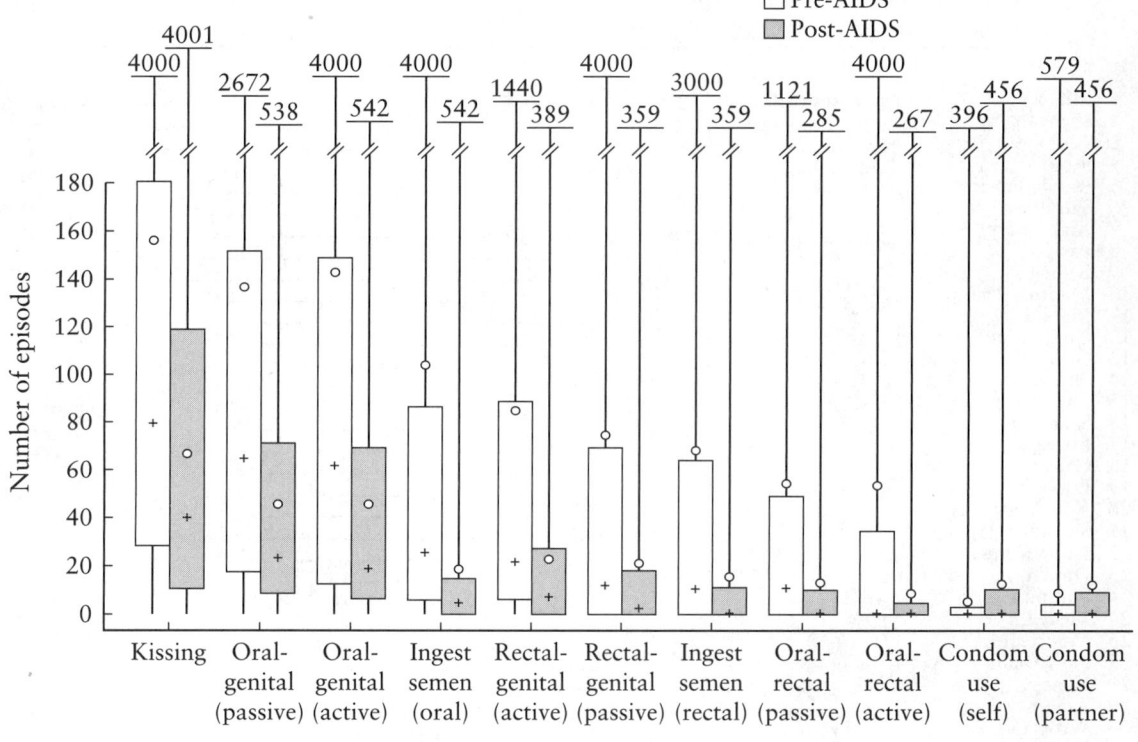

FIGURE 3.4

The median (+) and mean (○) annual frequency, interquartile range (boxes), and total range of engaging in specific acts during the year before hearing about AIDS and after hearing about AIDS

3.2.3   *Variance and Standard Deviation*

Another commonly used measure of dispersion for a set of data is known as the variance. The *variance* quantifies the amount of variability, or spread, around the mean of the measurements. To accomplish this, we might simply attempt to calculate the average distance of the individual observations from $\bar{x}$, or

$$\frac{1}{n} \sum_{i=1}^{n} (x_i - \bar{x}).$$

However, it can be shown mathematically that $\sum_{i=1}^{n} (x_i - \bar{x})$ is always equal to zero. By definition, the sum of the deviations from the mean of all observations less than $\bar{x}$ is equal to the sum of the deviations of all observations greater than $\bar{x}$; consequently, these two sums cancel each other out. To eliminate this problem, we might instead average the absolute values of the deviations from the mean, all of which are positive. While there is nothing wrong with this approach conceptually, the resulting summary measure does lack some important statistical properties and is rarely seen in the literature. A more widely used procedure is to square the deviations from the mean—recall that, like an absolute value, a squared quantity is always positive—and then find the average of the squared distances. This summary measure is the variance of the observations.

More explicitly, the variance is calculated by subtracting the mean of a set of values from each of the observations, squaring these deviations, adding them up, and dividing by 1 less than the number of observations in the data set. Representing the variance by s^2,

$$s^2 = \frac{1}{(n-1)} \sum_{i=1}^{n} (x_i - \bar{x})^2.$$

The reason for dividing by $n - 1$ rather than n will be discussed further in Chapter 9; we can, nonetheless, think of the variance as an average of squared deviations. For the 13 FEV$_1$ measurements presented in Table 3.1, the mean is 2.95 liters, and the squared deviations from the mean are given below.

Subject	x_i	$x_i - \bar{x}$	$(x_i - \bar{x})^2$
1	2.30	−0.65	0.4225
2	2.15	−0.80	0.6400
3	3.50	0.55	0.3025
4	2.60	−0.35	0.1225
5	2.75	−0.20	0.0400
6	2.82	−0.13	0.0169
7	4.05	1.10	1.2100
8	2.25	−0.70	0.4900
9	2.68	−0.27	0.0729
10	3.00	0.05	0.0025
11	4.02	1.07	1.1449
12	2.85	−0.10	0.0100
13	3.38	0.43	0.1849
Total	38.35	0.00	4.6596

Therefore, the variance is

$$s^2 = \frac{1}{(13-1)} \sum_{i=1}^{13} (x_i - 2.95)^2$$

$$= \frac{4.6596}{12}$$

$$= 0.39 \text{ liters}^2.$$

The *standard deviation* of a set of data is the positive square root of the variance. Thus, for the 13 FEV$_1$ measurements above, the standard deviation is equal to

$$s = \sqrt{s^2}$$

$$= \sqrt{0.39 \text{ liters}^2}$$

$$= 0.62 \text{ liters}.$$

In practice, the standard deviation is used more frequently than the variance. This is primarily because the standard deviation has the same units of measurement as the mean, rather than squared units. In a comparison of two groups of data, the group with the smaller standard deviation has the more homogeneous observations; the group with the larger standard deviation exhibits a greater amount of variability. The actual magnitude of the standard deviation depends on the values in the data set—what is large for one group of data may be small for another. In addition, because the standard deviation has units of measurement, it is meaningless to compare standard deviations for two unrelated quantities. Together, the mean and the standard deviation of a set of data can be used to summarize the characteristics of the entire distribution of values. We shall see how this works in Section 3.4.

3.2.4   Coefficient of Variation

While it may not make sense to compare standard deviations, it is possible to compare the variability among two or more sets of data representing different quantities with different units of measurement using a numerical summary measure known as the coefficient of variation. The *coefficient of variation* relates the standard deviation of a set of values to its mean; it is the ratio of s to $\bar{x}$ multiplied by 100 and is, therefore, a measure of relative variability. Because the standard deviation and the mean share the same units of measurement, the units cancel out and leave the coefficient of variation a dimensionless number. The coefficient of variation for the FEV_1 data is

$$CV = \frac{s}{\bar{x}} \times 100\%$$
$$= \frac{0.62}{2.95} \times 100\%$$
$$= 21.0\%.$$

It is difficult to assess whether this value, on its own, is large or small; the coefficient of variation is most useful for comparing two or more sets of data. Since it is independent of measurement units, it can be used to evaluate the relative variation between any two sets of observations. Although the coefficient of variation is still used as a summary measure in some circles, its statistical properties are not very good. As a result, its use is diminishing and should be discouraged.

3.3   Grouped Data

If we wish to count the change that we have in our pockets, there are two ways we can do it. The first is to consecutively sum the values of the coins as we retrieve them. The second is to begin by grouping the coins together by denomination, then multiplying the value of each coin by the number of coins of that denomination, and finally summing these values. For example, if we have 3 pennies, 4 nickels, 2 dimes, and 1 quarter, we have a total of

$$3(1) + 4(5) + 2(10) + 1(25) = 3 + 20 + 20 + 25$$
$$= 68 \text{ cents}.$$

The same procedure can be used to sum any given set of observations. For instance, consider the data in Table 3.3 [5]. Among patients with sickle cell disease—a hereditary form of anemia—regular blood transfusions are often used to prevent recurrent strokes after an initial cerebrovascular event. Long-term transfusion therapy, however, has its own associated risks and is not always advisable. Table 3.3 lists the durations of therapy for ten patients enrolled in a study investigating the effects of stopping the blood transfusions. We might be interested in determining the mean of these values.

3.3.1 Grouped Mean

The standard technique for finding the mean of the observations is to simply add the values and divide by $n = 10$. In this case, we find that

$$\bar{x} = \frac{1}{n} \sum_{i=1}^{n} x_i$$
$$= \left(\frac{1}{10}\right)(12 + 11 + 12 + 6 + 11 + 11 + 8 + 5 + 5 + 5)$$
$$= \frac{86}{10}$$
$$= 8.6 \text{ years}.$$

TABLE 3.3
Duration of transfusion therapy for ten patients with sickle cell disease

Subject	Duration (years)
1	12
2	11
3	12
4	6
5	11
6	11
7	8
8	5
9	5
10	5

Alternatively, we could have found the sum of the measurements by first grouping the observations with equal values; note that there are three 5s, one 6, one 8, three 11s, and two 12s. Therefore,

$$\sum_{i=1}^{10} x_i = 3(5) + 1(6) + 1(8) + 3(11) + 2(12)$$
$$= 15 + 6 + 8 + 33 + 24$$
$$= 86,$$

and

$$\bar{x} = \frac{86}{10}$$
$$= 8.6 \text{ years.}$$

We obtain the same mean regardless of the method we use.

The technique of grouping measurements that have equal values before calculating the mean has one distinct advantage over the standard method: this procedure can be applied to data that have been summarized in the form of a frequency distribution. Data that are organized in this way are often referred to as *grouped data*. Even if the original observations are no longer available—or perhaps never were, if the data were collected in grouped format to begin with—we might still be interested in calculating numerical summary measures for the data. An obstacle arises in that we do not know the values of the individual observations; however, we are able to determine the number of measurements that fall into each specified interval. This information can be used to calculate a *grouped mean*.

To compute the mean of a set of data arranged as a frequency distribution, we begin by assuming that all the values that fall into a particular interval are equal to the midpoint of that interval. Recall the serum cholesterol level data for the group of 25- to 34-year-old men that was presented in Table 2.6 [6]; these data are reproduced in Table 3.4. The first interval contains values that range from 80 to 119 mg/100 ml, with a midpoint of 99.5. We assume, therefore, that all 13 measurements within this interval take the value 99.5 mg/100 ml. Similarly, we assume that the 150 observations in the second interval, 120 to 159 mg/100 ml, all take the value 139.5 mg/100 ml. Because we are making these assumptions, our calculations are only approximate. Also, the results would change if we were to group the data in a different way.

To find the mean of the grouped data, we first sum the measurements by multiplying the midpoint of each interval by the corresponding frequency and adding these products; we then divide by the total number of values. Therefore,

$$\bar{x} = \frac{\sum_{i=1}^{k} m_i f_i}{\sum_{i=1}^{k} f_i},$$

where k is the number of intervals in the table, m_i is the midpoint of the ith interval, and f_i is the frequency associated with the ith interval. Note that the sum of the fre-

TABLE 3.4
Absolute frequencies of serum cholesterol levels for U.S. males, aged
25 to 34 years, 1976–1980

Cholesterol Level (mg/100 ml)	Number of Men
80–119	13
120–159	150
160–199	442
200–239	299
240–279	115
280–319	34
320–359	9
360–399	5
Total	1067

quencies, $\sum_{i=1}^{k} f_i$, is equal to the total number of observations, n. For the data in Table 3.4,

$$\bar{x} = \frac{\sum_{i=1}^{8} m_i f_i}{\sum_{i=1}^{8} f_i}$$

$$= \left(\frac{1}{1067}\right)[99.5(13) + 139.5(150) + 179.5(442) + 219.5(299)$$
$$+ 259.5(115) + 299.5(34) + 339.5(9) + 379.5(5)]$$

$$= \frac{212,166.5}{1067}$$

$$= 198.8 \text{ mg}/100 \text{ ml}.$$

The grouped mean is actually a weighted average of the interval midpoints; each mid-point is weighted by the frequency of observations within the interval.

3.3.2 Grouped Variance

After we have calculated the mean of a set of grouped data, we might also wish to find its variance or standard deviation. Again, we assume that all the observations falling within a particular interval are equal to the midpoint of that interval, m_i. The grouped variance of the data is

$$s^2 = \frac{\sum_{i=1}^{k} (m_i - \bar{x})^2 f_i}{\left[\sum_{i=1}^{k} f_i\right] - 1},$$

where all terms are defined as for the mean. Therefore, the grouped variance for the data in Table 3.4 is,

$$
\begin{aligned}
s^2 &= \frac{\sum_{i=1}^{8} (m_i - 198.8)^2 f_i}{\left[\sum_{i=1}^{8} f_i\right] - 1} \\
&= \left(\frac{1}{1067 - 1}\right)[(-99.3)^2(13) + (-59.3)^2(150) + (-19.3)^2(442) \\
&\qquad + (20.7)^2(299) + (60.7)^2(115) + (100.7)^2(34) \\
&\qquad + (140.7)^2(9) + (180.7)^2(5)] \\
&= \frac{2{,}058{,}342.8}{1066} \\
&= 1930.9 \ (\text{mg}/100 \ \text{ml})^2.
\end{aligned}
$$

Recall that the standard deviation is the square root of the variance; consequently, the grouped standard deviation of the serum cholesterol level data is

$$
\begin{aligned}
s &= \sqrt{1930.9 \ (\text{mg}/100 \ \text{ml})^2} \\
&= 43.9 \ \text{mg}/100 \ \text{ml}.
\end{aligned}
$$

3.4 Chebychev's Inequality

Once the mean and the standard deviation of a set of data have been calculated, these two numbers can be used to summarize the characteristics of the distribution of values as a whole. The mean tells us where the observations are centered; the standard deviation gives us an idea of the amount of dispersion around that center. Together, they can be used to construct an interval that captures a specified proportion of the observations in the data set.

It is often said that the mean plus or minus two standard deviations encompasses most of the data. If we know something about the shape of the distribution of values, this statement can be made more precise. When the data are symmetric and unimodal, for instance, we can say that approximately 67% of the observations lie in the interval $\bar{x} \pm 1s$, about 95% in the interval $\bar{x} \pm 2s$, and almost all the observations in the interval $\bar{x} \pm 3s$. This statement is known as the *empirical rule*. We will see the empirical rule again in Chapter 7, when we talk about theoretical distributions of data values.

Unfortunately, the empirical rule is an approximation that applies only when the data are symmetric and unimodal. If they are not, *Chebychev's inequality* can be used to summarize the distribution of values instead. Chebychev's inequality is less specific than the empirical rule, but it is true for any set of observations, no matter what its shape. It allows us to say that for any number k that is greater than or equal to 1, at least $[1 - (1/k)^2]$ of the measurements in the set of data lie within k standard deviations of their mean [7]. Given that $k = 2$, for example, at least

$$1 - \left(\frac{1}{2}\right)^2 = 1 - \left(\frac{1}{4}\right)$$
$$= \frac{3}{4}$$

of the values lie within two standard deviations of the mean. Equivalently, we could say that the interval $\bar{x} \pm 2s$ encompasses at least 75% of the observations in the group. This statement is true no matter what the values of $\bar{x}$ and s are. Similarly, if $k = 3$, at least

$$1 - \left(\frac{1}{3}\right)^2 = 1 - \left(\frac{1}{9}\right)$$
$$= \frac{8}{9}$$

of the observations lie within three standard deviations of the mean; therefore, $\bar{x} \pm 3s$ contains at least 88.9% of the measurements.

Chebychev's inequality is a conservative statement, more conservative than the empirical rule. It applies to the mean and standard deviation of any distribution of values, no matter what its shape. Returning to the FEV_1 data in Table 3.1, we can say that the interval

$$2.95 \pm (2 \times 0.62)$$

or

$$(1.71 , 4.19)$$

encompasses at least 75% of the observations, whereas the interval

$$2.95 \pm (3 \times 0.62)$$

or

$$(1.09 , 4.81)$$

contains at least 88.9%. In fact, both intervals contain all 13 of the measurements. Similarly, for the serum cholesterol level data in Table 3.4, we can state that the interval

$$198.8 \pm (2 \times 43.9)$$

or

$$(111.0 , 286.6)$$

contains at least 75% of the values, whereas the interval

$$198.8 \pm (3 \times 43.9)$$

or

$$(67.1 , 330.5)$$

contains at least 88.9%. Thus, although conservative, Chebychev's inequality allows us to use the mean and the standard deviation of **any** set of data—just two numbers—to describe the entire group.

3.5   *Further Applications*

In a study investigating the causes of death among persons with severe asthma, data were recorded for ten patients who arrived at the hospital in a state of respiratory arrest; breathing had stopped, and the subjects were unconscious. Table 3.5 lists the heart rates of the ten patients upon admission to the hospital [8]. How can we characterize this set of observations?

To begin, we might be interested in finding a typical heart rate for the ten individuals. The most commonly used measure of central tendency is the mean. To find the mean of these data, we simply sum all the observations and divide by $n = 10$. Therefore, for the data in Table 3.5,

$$\bar{x} = \frac{1}{n} \sum_{i=1}^{n} x_i$$

$$= \left(\frac{1}{10}\right)(167 + 150 + 125 + 120 + 150 + 150 + 40$$
$$+ 136 + 120 + 150)$$

$$= \frac{1308}{10}$$

$$= 130.8 \text{ beats per minute.}$$

TABLE 3.5

Heart rates for ten asthmatic patients in a state of respiratory arrest

Patient	Heart Rate (beats per minute)
1	167
2	150
3	125
4	120
5	150
6	150
7	40
8	136
9	120
10	150

The mean heart rate upon admission to the hospital is 130.8 beats per minute.

In this data set, the heart rate of patient 7 is considerably lower than those of the other subjects. What would happen if this observation were removed from the group? In this case,

$$\bar{x} = \left(\frac{1}{9}\right)(167 + 150 + 125 + 120 + 150 + 150 + 136 + 120 + 150)$$

$$= \frac{1268}{9}$$

$$= 140.9 \text{ beats per minute.}$$

The mean has increased by approximately ten beats per minute; this change demonstrates how much influence a single unusual observation can have on the mean.

A second measure of central tendency is the median, or the 50th percentile of the set of data. Ranking the measurements from smallest to largest, we have

40, 120, 120, 125, 136, 150, 150, 150, 150, 167.

Since there are an even number of observations, the median is taken to be the average of the two middlemost values. In this case, these values are the $10/2 = 5$th and the $(10/2) + 1 = 6$th largest observations. Consequently, the median of the data is $(136 + 150)/2 = 143$ beats per minute, a number quite a bit larger than the mean. Five observations are smaller than the median and five are larger.

The calculation of the median takes into account the ordering and the relative magnitudes of the observations. If we were again to remove patient 7, the ranking of heart rates would be

120, 120, 125, 136, 150, 150, 150, 150, 167.

There are nine observations in the list; the median is the $[(9 + 1)/2] = 5$th largest measurement, or 150 beats per minute. Although the median increases somewhat when patient 7 is removed, it does not change as much as the mean did.

The mode of a set of data is the observation that occurs most frequently. For the measurements in Table 3.5, the mode is 150 beats per minute; this is the only value that occurs four times.

Once we have found the center of a set of data, we often want to estimate the amount of variability among the observations as well; this allows us to quantify the degree to which the summary is representative of the group as a whole. One measure of dispersion that can be used is the range. The range of the data is the difference between the largest and smallest measurements. For the heart rates in Table 3.5, the range is $167 - 40 = 127$ beats per minute. Since the range considers only the most extreme observations in a data set, it is highly sensitive to outliers. If we were to remove patient 7 from the group, the range of the data would be only $167 - 120 = 47$ beats per minute.

The interquartile range of a set of data is defined as the 75th percentile minus the 25th percentile. If we were to construct a box plot using the data in Table 3.5—as we did in Figure 3.5—the interquartile range would be the height of the central box. (Note that for this particular set of measurements, the lower adjacent value is equal to the 25th percentile.) To find the 25th percentile of the data, we note that $nk/100 = 10(25)/100 = 2.5$ is not an integer. Therefore, the 25th percentile is the $2 + 1 = 3$rd

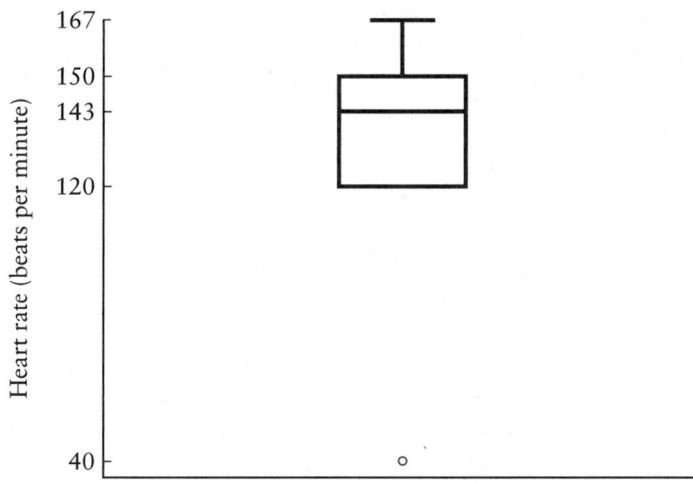

FIGURE 3.5

Heart rates for ten asthmatic patients in a state of respiratory arrest

largest measurement, or 120 beats per minute. Similarly, $10(75)/100 = 7.5$ is not an integer and the 75th percentile is the $7 + 1 = 8$th largest measurement, or 150 beats per minute. Subtracting these two values, the interquartile range for the heart rate data is $150 - 120 = 30$ beats per minute; this is the range of the middle 50% of the observations. The interquartile range is often used with the median to describe a distribution of values.

The most commonly used measures of dispersion for a set of data values are the variance and the standard deviation. The variance quantifies the amount of variability around the mean of the data; it is calculated by subtracting the mean from each of the measurements, squaring these deviations, summing them, and dividing by the total number of observations minus 1. The variance of the heart rates in Table 3.5 is

$$s^2 = \frac{1}{(10-1)} \sum_{i=1}^{10} (x_i - 130.8)^2$$

$$= \left(\frac{1}{9}\right)[(36.2)^2 + (19.2)^2 + (-5.8)^2 + (-10.8)^2 + (19.2)^2$$
$$+ (19.2)^2 + (-90.8)^2 + (5.2)^2 + (-10.8)^2 + (19.2)^2]$$

$$= \frac{11{,}323.6}{9}$$

$$= 1258.2 \text{ (beats per minute)}^2.$$

The standard deviation is the positive square root of the variance. It is used more frequently in practice because it has the same units of measurement as the mean. For the ten measures of heart rate, the standard deviation is

$$s = \sqrt{1258.2 \text{ (beats per minute)}^2}$$
$$= 35.5 \text{ beats per minute.}$$

The standard deviation is typically used with the mean to describe a set of values.

Now that we have some familiarity with numerical summary measures, consider the frequency distribution of birth weights in Table 3.6. These data were first presented in Table 2.9 [9]. Can we further summarize these data and make a concise statement about their distribution? Although we do not know the actual values of the 3,751,275 measurements of birth weight, we do know the numbers of observations that fall into each interval. We can, therefore, apply techniques for grouped data to obtain numerical summary measures for these observations.

To find the grouped mean, we begin by assuming that all the observations within a particular interval are equal to the midpoint of that interval. For example, we assume that the 4843 measurements in the first interval all have the value 249.5 grams, and that the 17,487 measurements in the second interval all have the value 749.5 grams. We then multiply each midpoint by the corresponding frequency in the interval, sum these products, and divide by the total number of observations. For the data in Table 3.6,

$$
\bar{x} = \frac{\sum_{i=1}^{11} m_i f_i}{\sum_{i=1}^{11} f_i}
$$

$$
= \left(\frac{1}{3,751,275}\right)[(249.5)(4843) + (749.5)(17,487) + (1249.5)(23,139)
$$
$$
+ (1749.5)(49,112) + (2249.5)(160,919)
$$
$$
+ (2749.5)(597,738) + (3249.5)(1,376,008)
$$
$$
+ (3749.5)(1,106,634) + (4249.5)(344,390)
$$
$$
+ (4749.5)(62,769) + (5249.5)(8236)]
$$

$$
= \frac{12,560,121,114.5}{3,751,275}
$$

$$
= 3348.2 \text{ grams.}
$$

TABLE 3.6

Absolute frequencies of birth weights for infants born in the United States, 1986

Birth Weight (grams)	Number of Infants
0–499	4843
500–999	17,487
1000–1499	23,139
1500–1999	49,112
2000–2499	160,919
2500–2999	597,738
3000–3499	1,376,008
3500–3999	1,106,634
4000–4499	344,390
4500–4999	62,769
5000–5499	8236
Total	3,751,275

The grouped mean is a weighted average of the interval midpoints.

In addition to calculating a measure of central tendency, we also can calculate a measure of dispersion for the frequency distribution. The grouped variance of the data in Table 3.6 is

$$
s^2 = \frac{\sum_{i=1}^{11} (m_i - 3348.2)^2 f_i}{\left[\sum_{i=1}^{11} f_i\right] - 1}
$$

$$
= \frac{1}{(3,751,275 - 1)} [(-3098.7)^2(4843) + (-2598.7)^2(17,487)
$$
$$
+ (-2098.7)^2(23,139) + (-1598.7)^2(49,112)
$$
$$
+ (-1098.7)^2(160,919) + (-598.7)^2(597,738)
$$
$$
+ (-98.7)^2(1,376,008) + (401.3)^2(1,106,634)
$$
$$
+ (901.3)^2(344,390) + (1401.3)^2(62,769)
$$
$$
+ (1901.3)^2(8236)]
$$

$$
= \frac{1,423,951,273,348.3}{3,751,274}
$$

$$
= 379,591.4 \text{ grams}^2.
$$

The standard deviation, which is the square root of the variance, is

$$
s = \sqrt{379,591.4 \text{ grams}^2}
$$
$$
= 616.1 \text{ grams}.
$$

Instead of working out all these numerical summary measures by hand, we could have used a computer to do the calculations for us. Table 3.7 shows the relevant output from Stata for the heart rate data in Table 3.5. Selected percentiles of the data are displayed on the left-hand side of the table. Using these values, we can determine the median and the interquartile range. The middle column contains the four smallest and four largest measurements; these allow us to calculate the range. The information on the right-hand side of the table includes the number of observations, the mean of the data, the standard deviation, and the variance.

Table 3.8 shows the corresponding output from Minitab. Note that Minitab provides the number of observations, the mean, the median, and the standard deviation of the measurements. The minimum and maximum values can be used to calculate the range, and the values labeled Q1 and Q3—the 25th and 75th percentiles, or quartiles—to compute the interquartile range. The section of the output called TRMEAN contains the 5% *trimmed mean* of the data. To compute the 5% trimmed mean, the observations are ranked. The smallest 5% and the largest 5% of the measurements are discarded; the remaining 90% are averaged. For the heart rate data, there are ten observations, and 5% of 10 is 0.5. Rounding this up to 1, the single smallest and single largest observations are removed. The mean is then calculated for the eight remaining measurements. Because potential outliers are eliminated when the data are trimmed, this type of mean is not influenced by exceptionally large or exceptionally small values to the same extent as the untrimmed mean.

TABLE 3.7
Stata output displaying numerical summary measures

		hrtrate			
	Percentiles	Smallest			
1%	40	40			
5%	40	120			
10%	80	120	Obs		10
25%	120	125	Sum of Wgt.		10
50%	143		Mean		130.8
		Largest	Std. Dev.		35.4708
75%	150	150			
90%	158.5	150	Variance		1258.178
95%	167	150	Skewness		-1.772591
99%	167	167	Kurtosis		5.479789

TABLE 3.8
Minitab output displaying numerical summary measures

	N	MEAN	MEDIAN	TRMEAN	STDEV	SEMEAN
HRTRATE	10	130.8	143.0	137.6	35.5	11.2
	MIN	MAX	Q1	Q3		
HRTRATE	40.0	167.0	120.0	150.0		

3.6   Review Exercises

1. Define and compare the mean, median, and mode as measures of central tendency.

2. Under what conditions is use of the mean preferred? The median? The mode?

3. Define and compare three commonly used measures of dispersion—the range, the interquartile range, and the standard deviation.

4. Is it possible to calculate numerical summary measures for observations that have been arranged in the form of a frequency distribution so that the original measurements are no longer available? Explain briefly. Why might personal information—such as annual income—be collected in this manner?

5. How is Chebychev's inequality useful for describing a set of observations? When can the empirical rule be used instead?

6. A study was conducted investigating the long-term prognosis of children who have suffered an acute episode of bacterial meningitis, an inflammation of the membranes enclosing the brain and spinal cord. Listed below are the times to the onset of seizure for 13 children who took part in the study [10]. In months, the measurements are:

   0.10   0.25   0.50   4   12   12   24   24   31   36   42   55   96

(a) Find the following numerical summary measures of the data.

    i. mean   iv. range
    ii. median   v. interquartile range
    iii. mode   vi. standard deviation

(b) Show that $\sum_{i=1}^{13} (x_i - \bar{x})$ is equal to 0.

7. In Massachusetts, eight individuals experienced an unexplained episode of vitamin D intoxication that required hospitalization; it was thought that these unusual occurrences might be the result of excessive supplementation of dairy milk [11]. Blood levels of calcium and albumin—a type of protein—for each subject at the time of hospital admission are provided below.

Calcium (mmol/l)	Albumin (g/l)
2.92	43
3.84	42
2.37	42
2.99	40
2.67	42
3.17	38
3.74	34
3.44	42

(a) Find the mean, median, standard deviation, and range of the recorded calcium levels.

(b) Compute the mean, median, standard deviation, and range of the given albumin levels.

(c) For healthy individuals, the normal range of calcium values is 2.12 to 2.74 mmol/l, while the range of albumin levels is 32 to 55 g/l. Do you believe that patients suffering from vitamin D intoxication have normal blood levels of calcium and albumin?

8. A study was conducted comparing female adolescents who suffer from bulimia to healthy females with similar body compositions and levels of physical activity. Listed below are measures of daily caloric intake, recorded in kilocalories per kilogram, for samples of adolescents from each group [12].

Daily Caloric Intake (kcal/kg)				
Bulimic			**Healthy**	
15.9	18.9	25.1	20.7	30.6
16.0	19.6	25.2	22.4	33.2
16.5	21.5	25.6	23.1	33.7
17.0	21.6	28.0	23.8	36.6
17.6	22.9	28.7	24.5	37.1
18.1	23.6	29.2	25.3	37.4
18.4	24.1	30.9	25.7	40.8
18.9	24.5		30.6	

(a) Find the median daily caloric intake for both the bulimic adolescents and the healthy ones.

(b) Compute the interquartile range for each group.

(c) Is a typical value of daily caloric intake larger for the individuals suffering from bulimia or for the healthy adolescents? Which group has a greater amount of variability in the measurements?

9. Figures 3.6 and 3.7 display the infant mortality rates for 111 nations on three continents: Africa, Asia, and Europe [13]. The infant mortality rate for a country is the number of deaths among children under one year of age in a given year divided by the total number of live births in that year. Figure 3.6 provides histograms that illustrate the distribution of infant mortality rates for each of the continents. Figure 3.7 displays the same data using one-way scatter plots and box plots.

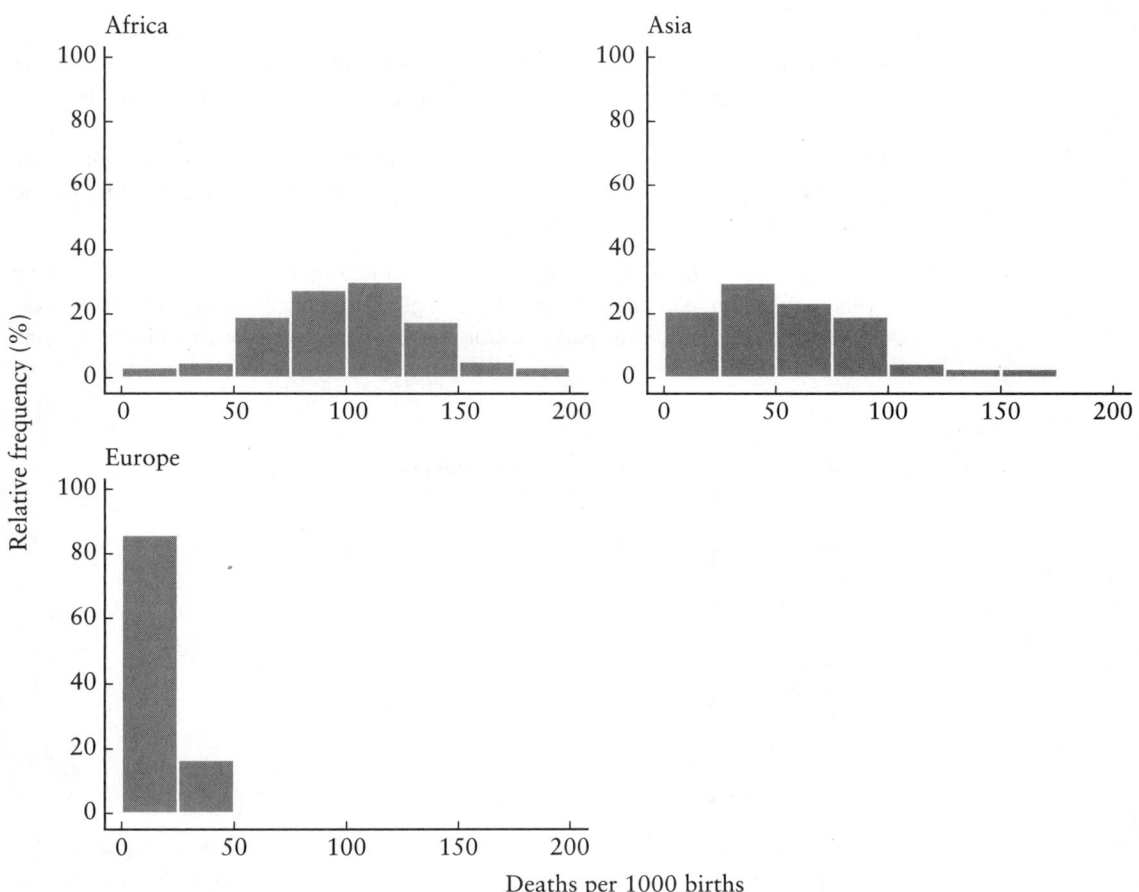

FIGURE 3.6
Histograms of infant mortality rates for Africa, Asia, and Europe, 1992

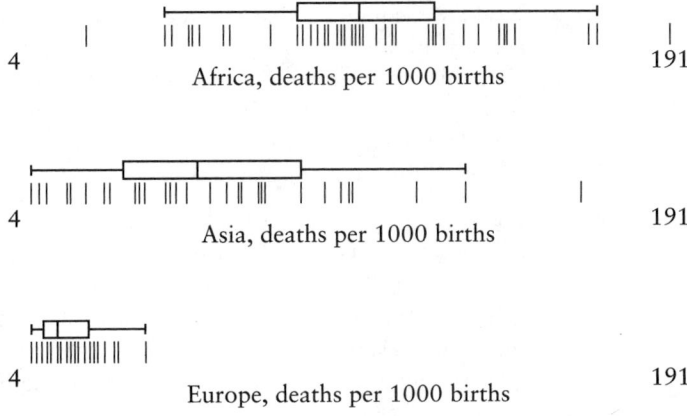

FIGURE 3.7
One-way scatter plots and box plots of infant mortality rates for Africa, Asia, and Europe, 1992

(a) Without doing any calculations, which continent would you expect to have the smallest mean? The largest median? The smallest standard deviation? Explain your reasoning.

(b) For Africa, would you expect the mean and the median infant mortality rates to be approximately equal? Would you expect the mean and the median to be equal for Asia? Why or why not?

10. Displayed below are a pair of frequency distributions containing serum cotinine levels for a group of cigarette smokers and a group of nonsmokers [14]. These measurements were recorded as part of a study investigating various risk factors for cardiovascular disease.

Cotinine Level (ng/ml)	Smokers	Nonsmokers
0–13	78	3300
14–49	133	72
50–99	142	23
100–149	206	15
150–199	197	7
200–249	220	8
250–299	151	9
300+	412	11
Total	1539	3445

(a) Compute the grouped mean and grouped standard deviation of the serum cotinine level measurements for both smokers and nonsmokers. For the last interval—300+ ng/ml—consider the midpoint of the interval to be 340 ng/ml.

(b) In which interval does the median serum cotinine level fall for smokers? For nonsmokers?

(c) Compare the distributions of serum cotinine levels for smokers and nonsmokers.

11. Serum zinc levels for 462 males between the ages of 15 and 17 are stored on your disk in a data set called `serzinc` (Appendix B, Table B.1); the serum zinc measurements in micrograms per deciliter are saved under the variable name `zinc` [15].

(a) Find the mean, median, standard deviation, range, and interquartile range of the data.

(b) Use Chebychev's inequality to describe the distribution of values.

(c) What percentage of the values would you expect to lie within 2 standard deviations of the mean? Within 3 standard deviations of the mean? What percentage of the 462 measurements actually fall within these ranges?

(d) Does the empirical rule do a better job of summarizing these serum zinc levels than Chebychev's inequality? Explain.

12. The percentages of low birth weight infants—defined as those weighing less than 2500 grams—for a number of nations around the world are saved under the variable name `lowbwt` in the data set `unicef` [13] (Appendix B, Table B.2).

(a) Compute the mean and the median of these observations.

(b) Compute the 5% trimmed mean.

(c) For this data set, which of these numbers would you prefer as a measure of central tendency? Explain.

13. The declared concentrations of nicotine in milligrams for 35 brands of Canadian cigarettes are saved under the variable name `nicotine` in the data set `cigarett` [16] (Appendix B, Table B.4).

(a) Find the mean and median concentrations of nicotine.

(b) Produce a histogram of the nicotine measurements. Describe the shape of the distribution of values.

(c) Which number do you think provides the best measure of central tendency for these concentrations, the mean or the median? Why?

14. Displayed below is a frequency distribution containing a summary of the resting systolic blood pressures for a sample of 35 patients with ischemic heart disease, or suppression of blood flow to the heart [17].

Blood Pressure (mm Hg)	Number of Patients
115–124	4
125–134	5
135–144	5
145–154	7
155–164	5
165–174	4
175–184	5
Total	35

(a) Compute the grouped mean and the grouped standard deviation of the data.

(b) The 35 measurements of systolic blood pressure are stored on your disk in a file called `ischemic` (Appendix B, Table B.6); the values are saved under the variable name `sbp`. Compute the ungrouped mean and the ungrouped standard deviation of these data.

(c) Are the grouped and the ungrouped numerical summary measures the same? Why or why not?

15. The data set `lowbwt` contains information recorded for a sample of 100 low birth weight infants—those weighing less than 1500 grams—born in two teaching hospitals in Boston, Massachusetts [18] (Appendix B, Table B.7). Measurements of systolic blood pressure are saved under the variable name `sbp`. The dichotomous random variable `sex` designates the gender of each child, with 1 representing a male and 0 a female.

(a) Construct a pair of box plots for the systolic blood pressure measurements—one for boys and one for girls. Compare the two distributions of values.

(b) Compute the mean and standard deviation of the systolic blood pressure measurements for males and for females. Which group has the larger mean? The larger standard deviation?

(c) Calculate the coefficient of variation corresponding to each gender. Is there any evidence that the amount of variability in systolic blood pressure differs for males and females?

Bibliography

[1] Koenig, J. Q., Covert, D. S., Hanley, Q. S., Van Belle, G., and Pierson, W. E., "Prior Exposure to Ozone Potentiates Subsequent Response to Sulfur Dioxide in Adolescent Asthmatic Subjects," *American Review of Respiratory Disease*, Volume 141, February 1990, 377–380.

[2] Prial, F. J., "Wine Talk," *The New York Times*, December 25, 1991, 29.

[3] United Nations Environment Programme, World Health Organization, "Urban Air Quality on Average," *Global Pollution and Health: Results of Health-Related Environmental Monitoring*, 1987.

[4] Martin, J. L., "The Impact of AIDS on Gay Male Sexual Behavior Patterns in New York City," *American Journal of Public Health*, Volume 77, May 1987, 578–581.

[5] Wang, W. C., Kovnar, E. H., Tonkin, I. L., Mulhern, R. K., Langston, J. W., Day, S. W., Schell, M. J., and Wilimas, J. A., "High Risk of Recurrent Stroke after Discontinuance of Five to Twelve Years of Transfusion Therapy in Patients with Sickle Cell Disease," *Journal of Pediatrics*, Volume 118, March 1991, 377–382.

[6] National Center for Health Statistics, Fulwood, R., Kalsbeek, W., Rifkind, B., Russell-Briefel, R., Muesing, R., LaRosa, J., and Lippel, K., "Total Serum Cholesterol Levels of Adults 20–74 Years of Age: United States, 1976–1980," *Vital and Health Statistics*, Series 11, Number 236, May 1986.

[7] Parzen, E., *Modern Probability Theory and Its Applications*, New York: Wiley, 1960.

[8] Molfino, N. A., Nannini, L. J., Martelli, A. N., and Slutsky, A. S., "Respiratory Arrest in Near-Fatal Asthma," *The New England Journal of Medicine*, Volume 324, January 31, 1991, 285–288.

[9] National Center for Health Statistics, "Advance Reports, 1986," *Supplements to the Monthly Vital Statistics Report*, Series 24, March 1990.

[10] Pomeroy, S. L., Holmes, S. J., Dodge, P. R., and Feigin, R. D., "Seizures and Other Neurologic Sequelae of Bacterial Meningitis in Children," *The New England Journal of Medicine*, Volume 323, December 13, 1990, 1651–1656.

[11] Jacobus, C. H., Holick, M. F., Shao, Q., Chen, T. C., Holm, I. A., Kolodny, J. M., Fuleihan, G. E. H., and Seely, E. W., "Hypervitaminosis D Associated with Drinking Milk," *The New England Journal of Medicine*, Volume 326, April 30, 1992, 1173–1177.

[12] Gwirtsman, H. E., Kaye, W. H., Obarzanek, E., George, D. T., Jimerson, D. C., and Ebert, M. H., "Decreased Caloric Intake in Normal-Weight Patients with Bulimia: Comparison with Female Volunteers," *American Journal of Clinical Nutrition*, Volume 49, January 1989, 86–92.

[13] United Nations Children's Fund, *The State of the World's Children 1994*, New York: Oxford University Press.

[14] Wagenknecht, L. E., Burke, G. L., Perkins, L. L., Haley, N. J., and Friedman, G. D., "Misclassification of Smoking Status in the CARDIA Study: A Comparison of Self-Report with Serum Cotinine Levels," *American Journal of Public Health*, Volume 82, January 1992, 33–36.

[15] National Center for Health Statistics, Fulwood, R., Johnson, C. L., Bryner, J. D., Gunter, E. W., and McGrath, C. R., "Hematological and Nutritional Biochemistry Reference Data for Persons 6 Months–74 Years of Age: United States, 1976–1980," *Vital and Health Statistics*, Series 11, Number 232, December 1982.

[16] Kaiserman, M. J., and Rickert, W. S., "Carcinogens in Tobacco Smoke: Benzo[a]pyrene from Canadian Cigarettes and Cigarette Tobacco," *American Journal of Public Health*, Volume 82, July 1992, 1023–1026.

[17] Miller, P. F., Sheps, D. S., Bragdon, E. E., Herbst, M. C., Dalton, J. L., Hinderliter, A. L., Koch, G. G., Maixner, W., and Ekelund, L. G., "Aging and Pain Perception in Ischemic Heart Disease," *American Heart Journal*, Volume 120, July 1990, 22–30.

[18] Leviton, A., Fenton, T., Kuban, K. C. K., and Pagano, M., "Labor and Delivery Characteristics and the Risk of Germinal Matrix Hemorrhage in Low Birth Weight Infants," *Journal of Child Neurology*, Volume 6, October 1991, 35–40.

4

Rates and Standardization

Demographic data and vital statistics are numbers that are used to characterize or describe a population. *Demographic data* include such information as the size of the population and its composition by gender, race, and age. *Vital statistics* describe the life of the population: they deal with births, deaths, marriages, divorces, and occurrences of disease. Researchers and public health professionals use both types of data to describe the health status of a population, to spot trends and make projections, and to plan for necessary services such as housing and medical care.

Vital statistics are also used to make comparisons among groups. For example, in order to evaluate health trends, we might want to compare the number of deaths in the United States in 1991 to the number of deaths in 1992. If we consider only the raw numbers of deaths in each year—2,169,518 in 1991 and 2,175,613 in 1992—it would be difficult to interpret the observed increase [1]. It is possible that there were fewer deaths in 1991 simply because the population base was smaller in that year; the smaller the population, the fewer deaths we would expect to see. On the other hand, it is also possible that we experienced the beginning of an epidemic in 1992 and that many people died as a result. How can we determine what is really happening?

4.1 Rates

To make comparisons among groups more meaningful, rates may be used instead of raw numbers. A *rate* is defined as the number of cases of a particular outcome of interest that occur over a given period of time divided by the size of the population in that time period. For example, we might be interested in the number of ear infections diagnosed for a specified group of elementary school students during a two-month period. Although they are often used interchangeably, the terms "rate" and "proportion" are not synonymous. A *proportion* is a ratio in which all individuals included in the numerator must also be included in the denominator, such as the fraction of women over the age of 60 who have ever suffered a heart attack. It has no units of measurement. A rate does incorporate units of measurement and is intrinsically dependent upon a measure of time.

Instead of comparing the total numbers of deaths in 1991 and 1992, we could compare the death rates for those years. A death rate, or *mortality rate*, is the number of deaths that occur during some time period, such as a calendar year, divided by the total population at risk during that time period. This type of rate is often expressed in terms of deaths per 1000 population or deaths per 100,000 population; the multiplier for a rate—whether 1000, 10,000, or 100,000—is usually chosen to minimize the number of decimal places in the reported result. If we were to compute the mortality rates for the United States, we would find the death rate in 1991 to be 860.3 per 100,000 population and the death rate in 1992 to be 852.9 per 100,000 population [1]. Although a greater number of people died in 1992, the death rate actually decreased.

One of the more commonly reported death rates for a given population is its *infant mortality rate*. This quantity is defined as the number of deaths during a calendar year among infants under one year of age divided by the total number of live births during that year. The infant mortality rate is one of the most important measures of the health status of a nation. Even though the numbers of births and deaths vary considerably from country to country, it is informative to compare mortality rates. The infant mortality rates for selected nations appear in Table 4.1 [2].

The mortality rates that we have considered up to this point have all been crude rates. A *crude rate* is a single number computed as a summary measure for an entire population; it disregards differences caused by age, gender, race, and other characteristics. As another example, Figure 4.1 displays trends in the crude marriage and divorce rates in the United States for the years 1950 through 1994 [3]; in each case, the crude rate is expressed as the number of events per 1000 population.

Factors such as age, gender, and race often do have a significant effect on the rates that describe vital statistics. Consider the mortality rates in Table 4.2 [1]: in addition to the crude death rate for 1992, the table displays death rates calculated for various subgroups

TABLE 4.1

Infant mortality rates for selected countries, 1992

Nation	Mortality Rate per 1000 Live Births	Nation	Mortality Rate per 1000 Live Births
Argentina	22	Italy	8
Australia	7	Japan	4
Brazil	54	Mexico	28
Canada	7	Philippines	46
China	35	Poland	14
Egypt	43	Russian Federation	28
Ethiopia	123	Saudi Arabia	35
Finland	6	Spain	8
France	7	Sweden	6
Greece	8	United Kingdom	7
India	83	United States	9
Israel	9	Venezuela	20

TABLE 4.2

Total deaths and death rates by age, race, and sex, United States, 1992

Age	All Races			White		
	Both Sexes	Male	Female	Both Sexes	Male	Female
	Number					
All ages	2,175,613	1,122,336	1,053,277	1,873,781	956,957	916,824
Under 1 year	34,628	19,545	15,083	22,164	12,625	9539
1–4 years	6764	3809	2955	4685	2690	1995
5–9 years	3739	2231	1508	2690	1605	1085
10–14 years	4454	2849	1605	3299	2093	1206
15–19 years	14,411	10,747	3664	10,308	7440	2888
20–24 years	20,137	15,460	4677	14,033	10,696	3337
25–29 years	24,314	18,032	6282	17,051	12,825	4226
30–34 years	34,167	24,863	9304	24,450	18,210	6240
35–39 years	42,089	29,641	12,448	30,127	21,690	8437
40–44 years	49,201	33,354	15,847	35,886	24,726	11,160
45–49 years	56,533	36,622	19,911	43,451	28,343	15,108
50–54 years	68,497	42,649	25,848	53,689	33,681	20,008
55–59 years	94,582	58,083	36,499	75,750	47,042	28,708
60–64 years	146,409	88,797	57,612	122,213	74,994	47,219
65–69 years	211,071	124,228	86,843	180,788	107,427	73,361
70–74 years	266,845	149,937	116,908	234,117	132,273	101,844
75–79 years	301,736	158,257	143,479	270,238	142,422	127,816
80–84 years	308,116	141,640	166,476	279,507	128,484	151,023
85 years and over ..	487,446	161,236	326,210	448,984	147,419	301,565
Not stated	474	356	118	351	272	79
	Death rate					
All ages	852.9	901.6	806.5	880.0	917.2	844.3
Under 1 year	865.7	956.6	770.8	701.8	780.9	618.7
1–4 years	43.6	48.0	39.0	38.1	42.6	33.3
5–9 years	20.4	23.7	16.8	18.3	21.3	15.2
10–14 years	24.6	30.7	18.2	22.8	28.2	17.2
15–19 years	84.3	122.4	44.0	75.6	106.0	43.3
20–24 years	105.7	159.4	50.1	91.0	135.4	44.3
25–29 years	120.5	178.0	62.5	103.2	153.3	51.9
30–34 years	153.5	224.0	83.3	132.4	195.8	68.1
35–39 years	199.5	282.8	117.2	171.2	245.5	96.3
40–44 years	261.6	359.1	166.5	226.3	312.2	140.6
45–49 years	368.0	485.7	254.6	328.6	432.5	226.5
50–54 years	568.2	728.1	417.1	518.6	663.4	379.3
55–59 years	902.1	1156.5	668.2	835.1	1071.5	613.4
60–64 years	1402.2	1815.2	1038.2	1334.9	1729.7	979.7
65–69 years	2114.8	2775.4	1577.7	2042.6	2688.5	1511.0
70–74 years	3146.8	4109.3	2419.9	3073.0	4012.4	2356.4
75–79 years	4705.9	6202.4	3716.8	4662.2	6148.8	3672.7
80–84 years	7429.1	9726.0	6186.1	7391.0	9700.5	6146.1
85 years and over ..	14,972.9	17,740.4	13,901.0	15,104.2	17,956.2	14,015.9

TABLE 4.2

(Continued)

Age	Black			American Indian			Asian or Pacific Islander		
	Both Sexes	Male	Female	Both Sexes	Male	Female	Both Sexes	Male	Female
					Number				
All ages	269,219	146,630	122,589	8953	5181	3772	23,660	13,568	10,092
Under 1 year	11,348	6298	5050	393	221	172	723	401	322
1–4 years	1799	965	834	127	67	60	153	87	66
5–9 years	894	529	365	54	33	21	101	64	37
10–14 years	982	633	349	61	48	13	112	75	37
15–19 years	3583	2923	660	206	155	51	314	229	85
20–24 years	5399	4246	1153	279	212	67	426	306	120
25–29 years	6559	4695	1864	293	228	65	411	284	127
30–34 years	8836	6083	2753	378	253	125	503	317	186
35–39 years	10,965	7308	3657	403	272	131	594	371	223
40–44 years	12,213	7949	4264	366	246	120	736	433	303
45–49 years	11,753	7493	4260	431	280	151	898	506	392
50–54 years	13,252	8021	5231	487	308	179	1069	639	430
55–59 years	16,727	9824	6903	668	392	276	1437	825	612
60–64 years	21,669	12,380	9289	719	408	311	1808	1015	793
65–69 years	27,011	14,946	12,065	818	454	364	2454	1401	1053
70–74 years	29,124	15,580	13,544	849	457	392	2755	1627	1128
75–79 years	27,875	13,782	14,093	799	422	377	2824	1631	1193
80–84 years	25,260	11,253	14,007	721	354	367	2628	1549	1079
85 years and over ..	33,856	11,646	22,210	900	370	530	3706	1801	1905
Not stated	114	76	38	1	1	—	8	7	1
					Death rate				
All ages	850.5	977.5	736.2	417.7	487.7	348.9	283.1	332.7	235.8
Under 1 year	1786.0	1957.9	1609.7	939.2	1057.5	821.2	439.8	477.7	400.2
1–4 years	73.2	77.6	68.7	72.0	74.7	69.3	26.9	29.9	23.8
5–9 years	32.1	37.5	26.6	25.1	30.1	19.8	15.4	19.1	11.5
10–14 years	35.3	44.9	25.4	28.3	44.0	*	16.9	22.2	11.3
15–19 years	135.5	218.4	50.5	110.8	163.7	55.9	49.7	70.6	27.6
20–24 years	200.7	321.0	84.3	149.7	218.0	75.2	57.4	80.8	33.1
25–29 years	241.3	361.7	131.3	160.2	245.2	72.4	53.8	75.4	32.8
30–34 years	316.0	464.4	185.2	203.2	275.3	132.8	61.4	79.9	44.1
35–39 years	427.0	609.6	267.1	240.8	334.0	152.4	77.6	101.5	55.8
40–44 years	570.7	803.2	370.7	257.3	355.9	164.1	110.4	139.6	85.0
45–49 years	762.4	1065.7	508.0	391.5	522.4	267.3	184.9	219.6	153.5
50–54 years	1054.9	1419.3	757.0	577.6	759.7	408.9	295.2	366.5	229.0
55–59 years	1579.0	2103.6	1165.4	997.2	1229.3	786.3	500.4	620.6	396.8
60–64 years	2204.1	2924.3	1659.5	1303.7	1574.4	1063.8	729.6	948.4	563.3
65–69 years	3075.9	4029.1	2378.8	1819.9	2219.3	1486.3	1189.4	1576.7	896.4
70–74 years	4278.6	5724.9	3315.3	2541.5	3145.9	2076.5	1872.3	2486.2	1380.5
75–79 years	5596.3	7502.0	4482.7	3434.9	4410.5	2753.2	3001.3	3882.7	2290.5
80–84 years	8400.8	10,969.8	7070.5	5133.1	6753.1	4168.6	5156.3	6461.7	3997.0
85 years and over ..	14,278.6	16,717.1	13,264.1	7726.0	9381.3	6878.7	10,841.3	12,628.8	9561.8

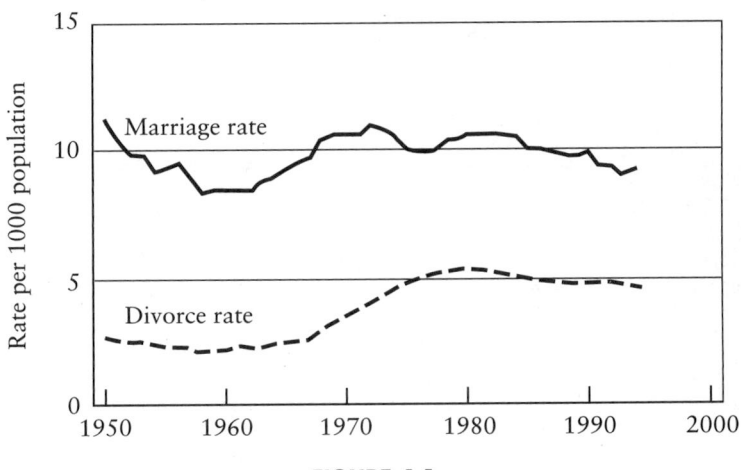

FIGURE 4.1

Trends in the crude marriage and divorce rates, United States, 1950–1994

of individuals in the United States. We can see that females tend to have lower death rates than males (why?), Asians and Pacific Islanders have lower death rates than members of other races, and, after age 5, death rate increases with age. Rates that are computed within relatively small, well-defined subgroups are called *specific rates*. Mortality rates calculated for individual age groups, for instance, are known as *age-specific death rates*.

4.2 Standardization of Rates

The National Health Interview Survey is an ongoing household survey of the civilian, noninstitutionalized population of the United States. It provided the following data enumerating hearing impairments due to injury reported by individuals 17 years of age and older in 1980–1981 [4].

Employment Status	Population	Impairments	Rate per 1000
Currently employed	98,917	552	5.58
Currently unemployed	7462	27	3.62
Not in the labor force	56,778	368	6.48
Total	163,157	947	5.80

The employment status categories are defined by the National Center for Health Statistics as follows: "Currently employed" individuals are persons who were working at any time during the two-week interview period; the "currently unemployed" comprise individuals looking for work and those who have been laid off; persons classified as "not in the labor force" include homemakers, volunteer workers, and individuals who have retired.

Judging from the crude data, the rate of hearing impairment due to injury seems to vary with employment status. The rate among individuals who are currently employed is 5.58 per 1000 population, whereas the rate among those not in the labor force is 6.48 per 1000 population. Of these two groups, it appears that individuals who are not in the labor force are at greater risk of hearing impairment due to injury than those who are currently employed. Is this a valid conclusion?

One problem that often arises when crude rates for distinct groups are compared is that the populations may differ substantially with respect to important characteristics such as age and sex. For example, if we have two populations from different geographical areas—one composed entirely of males and the other entirely of females—we can never be sure whether a difference in their mortality rates is due to location or to some effect of gender. In this situation, gender is referred to as a *confounder*. Because it is associated with both geographical area and death rate, it obscures the true relationship between these factors.

To determine whether it is fair to say that individuals who are not in the labor force have a higher risk of hearing impairment due to injury than those who are currently employed, it would be wise to check whether the two subpopulations have similar underlying structures. We therefore break each group down according to age.

Age	Currently Employed		Not in the Labor Force	
	Population	**Percent**	**Population**	**Percent**
17–44	67,987	68.7	20,760	36.6
45–64	27,592	27.9	15,108	26.6
65+	3338	3.4	20,910	36.8
Total	98,917	100.0	56,778	100.0

As we can see from the relative frequencies, the two groups differ in age composition: the individuals not included in the labor force are considerably older than those who are currently employed.

We next consider the age-specific impairment rates in the surveyed population as a whole.

Age	Population	Impairments	Rate per 1000
17–44	94,930	441	4.65
45–64	43,857	308	7.02
65+	24,370	198	8.12
Total	163,157	947	5.80

Note that the crude rate of hearing impairments due to injury in the entire surveyed population is actually a weighted average of the age-specific rates; in particular,

$$5.80 = \frac{(94{,}930)(4.65) + (43{,}857)(7.02) + (24{,}370)(8.12)}{163{,}157}.$$

The age-specific rates are weighted by the number of individuals in each group. Also observe that the rate of impairment increases with age. Age is a confounder in the relationship between hearing impairment and employment status; it is independently associated with each of these quantities. As a result, we cannot be sure whether the higher rate of impairment among individuals not in the labor force is the result of some inherent characteristic of the members of this group, or whether it is simply the effect of age.

To make a more accurate comparison between the two subpopulations, we should consider their age-specific impairment rates rather than their overall crude rates.

	Currently Employed			Not in the Labor Force		
Age	Population	Impairments	Rate per 1000	Population	Impairments	Rate per 1000
17–44	67,987	346	5.09	20,760	80	3.85
45–64	27,592	179	6.49	15,108	117	7.74
65+	3338	27	8.09	20,910	171	8.18
Total	98,917	552	5.58	56,778	368	6.48

It is difficult to draw a succinct conclusion based on these two sets of rates. Among individuals over the age of 45, the rate of hearing impairment due to injury is higher for persons not in the labor force than it is for the currently employed; among those between the ages of 17 and 44, the rate is much lower for persons not in the labor force. Comparing the crude impairment rates of the two subpopulations gave us an incomplete picture of the true situation.

Although the subgroup-specific rates provide a more accurate comparison among populations than the crude rates do, we could, if we had many subgroups, end up with an overwhelming number of rates to compare. It would be convenient to be able to summarize the entire situation with a single number calculated for each subpopulation, a number that adjusts for differences in composition. In practice, there are two ways to compute such a summary. The first is known as the direct method of standardization; the second is called the indirect method of standardization. Both strategies focus on the two components that factor into the calculation of a crude rate—the composition of the population and its subgroup-specific rates—and attempt to overcome the problem of confounding by holding one of these components constant across populations. Indices such as the consumer price index have a similar objective.

4.2.1 Direct Method of Standardization

The *direct method* of adjusting for differences among populations focuses on computing the overall rates that would result if, instead of having different distributions, all populations being compared were to have the same standard composition. The first step in applying this technique is to select the standard distribution. For the hearing impairment example, we use the total population questioned in the National Health Interview

Survey. We then calculate the numbers of impairments that would have occurred in each of the two employment status subgroups—the currently employed and those not in the labor force—assuming that each had this standard population distribution while retaining its own individual age-specific impairment rates.

	Total	Currently Employed		Not in the Labor Force	
Age	(1) Population	(2) Rate per 1000	(3) Expected Impairments	(4) Rate per 1000	(5) Expected Impairments
17–44	94,930	5.09	483.2	3.85	365.5
45–64	43,857	6.49	284.6	7.74	339.5
65+	24,370	8.09	197.2	8.18	199.3
Total	163,157		965.0		904.3

The expected numbers of impairments for the group of currently employed individuals are calculated by multiplying column (1) by column (2) and dividing by 1000; the expected numbers of impairments for those not in the labor force are obtained by multiplying column (1) by column (4) and dividing by 1000.

The age-adjusted impairment rate for each group is then calculated by dividing its total expected number of hearing impairments due to injury by the total standard population.

$$\text{Currently employed:} \qquad \frac{965.0}{163,157} = 5.91 \text{ per } 1000$$

$$\text{Not in the labor force:} \qquad \frac{904.3}{163,157} = 5.54 \text{ per } 1000$$

These age-adjusted rates are the impairment rates that would apply if both the currently employed and those not in the labor force had the same age distribution as the total surveyed population. After we control for the effect of age in this way, the adjusted impairment rate for individuals who are currently employed is higher than the adjusted rate for individuals not in the labor force. This is the opposite of what we observed when we looked at the crude rates, implying that these crude rates were indeed being influenced by the age structure of the underlying groups.

Note that the choice of a different standard age distribution—column (1) in the previous table—would have led to different adjusted impairment rates. This is not important, however, since an adjusted rate has no meaning by itself. It is merely a construct that is calculated based on a hypothetical standard distribution; unlike a crude or specific rate, it does not reflect the true impairment rate of any population. Adjusted rates have meaning only when we are comparing two or more groups, and it has been shown that trends among the groups are generally unaffected by the choice of a standard [5]. In general, its composition should not deviate radically from those of the groups being compared. If we were to choose a different but reasonable standard age distribution for the hearing impairment data—the U.S. population in 1980, for instance—the magnitude of the **difference** between the adjusted impairment rates of those currently employed and those not in the labor force should not change drastically, even if the rates themselves do; the currently employed should still have a slightly higher adjusted rate of impairment.

4.2.2   Indirect Method of Standardization

The *indirect method* of adjusting for differences in composition involves the use of a set of standard age-specific impairment rates along with the actual age composition of each subpopulation being compared. We again use the total surveyed population as the standard. This time, however, we calculate the number of impairments that would have occurred in the two population subgroups if each had taken on the age-specific impairment rates of the surveyed population as a whole while retaining its own individual age distribution.

	Total	Currently Employed		Not in the Labor Force	
Age	(1) Rate per 1000	(2) Population	(3) Expected Impairments	(4) Population	(5) Expected Impairments
17–44	4.65	67,987	316.1	20,760	96.5
45–64	7.02	27,592	193.7	15,108	106.1
65+	8.12	3338	27.1	20,910	169.8
Total	5.80	98,917	536.9	56,778	372.4

The expected numbers of impairments among those currently employed are calculated by multiplying column (1) by column (2) and dividing by 1000; the expected impairments for those not in the labor force are obtained by multiplying column (1) by column (4) and dividing by 1000.

We next divide the observed number of hearing impairments in each employment group by the total expected number of impairments. The quantity that results is known as the *standardized morbidity ratio*. If the data pertained to deaths rather than impairments, dividing the observed number of deaths by the expected number would give us the *standardized mortality ratio*.

$$\text{Currently employed:} \qquad \frac{552}{536.9} = 1.03$$

$$= 103\%$$

$$\text{Not in the labor force:} \qquad \frac{368}{372.4} = 0.99$$

$$= 99\%$$

These standardized morbidity ratios indicate that the group of currently employed individuals has a 3% higher impairment rate than the surveyed population as a whole, whereas the group not in the labor force has an impairment rate that is 1% lower than that of the total population. Recall that the total surveyed population also includes the group of individuals who are not currently employed.

The application of the indirect method often concludes with a comparison of the standardized ratios. We could, however, continue on and compute the actual age-adjusted impairment rates for each group. These rates are derived by multiplying the crude impairment rate for the total surveyed population by the appropriate standardized ratios.

$$\text{Currently employed:} \quad \frac{5.80}{1000} \times 1.03 = 5.97 \text{ per } 1000$$

$$\text{Not in the labor force:} \quad \frac{5.80}{1000} \times 0.99 = 5.74 \text{ per } 1000$$

With the effect of age removed, the group of currently employed individuals is again seen to have a slightly higher adjusted impairment rate than those not in the labor force. Although the rates themselves are different, this is the same conclusion that was reached when the direct method of standardization was applied.

4.2.3 Use of Standardized Rates

Standardized rates, and age-adjusted rates in particular, are encountered frequently in the study of vital statistics. One interesting example involves data taken from three different studies that investigate the relationship between mortality rate and smoking status [6]. Each study compares three groups of men: nonsmokers, cigarette smokers, and smokers of cigars and pipes. The crude death rates per 1000 person-years of experience are displayed below. These rates were calculated by dividing the total number of deaths in each group by the corresponding person-years of exposure and multiplying by 1000. A *person-year* is a unit of time that is defined as one person being followed for one year. If we were to follow ten different individuals for one year each, we would have a total of ten person-years; if we followed five people for two years each, we would also have ten person-years.

Smoking Group	Death Rate per 1000 Person-Years		
	Canada	Great Britain	United States
Nonsmokers	20.2	11.3	13.5
Cigarettes	20.5	14.1	13.5
Cigars/pipes	35.5	20.7	17.4

The studies conducted in Canada, Great Britain, and the United States all seem to carry the same message: individuals who smoke either cigars or pipes should give up smoking, or, at the very least, switch to cigarettes. Keep in mind, however, that these were all observational studies. In an *observational study*, the investigator has no control over the assignment of a treatment or exposure (smoking group, in the examples described above). Instead, the study subjects determine their own exposure status, and the investigator simply observes what happens to them. It is possible that the groups differed substantially in characteristics other than smoking status.

If we were to consider the age compositions of the various smoking groups, for example, we would find that they do differ considerably. To illustrate this, the mean ages for the three groups in each of the studies are displayed below.

	Mean Age (years)		
Smoking Group	**Canada**	**Great Britain**	**United States**
Nonsmokers	54.9	49.1	57.0
Cigarettes	50.5	49.8	53.2
Cigars/pipes	65.9	55.7	59.7

In general, smokers of cigars and pipes tend to be older than both nonsmokers and cigarette smokers.

Because of the differences in age, the mortality rates were standardized by separating the men into three different subclasses; the categories of age were selected so that roughly the same number of men fell into each class. If the group of nonsmokers is chosen to provide the standard age distribution, the adjusted death rates calculated using the direct method are shown below. (Note that these rates cannot be computed from the information provided.)

	Death Rate per 1000 Person-Years		
Smoking Group	**Canada**	**Great Britain**	**United States**
Nonsmokers	20.2	11.3	13.5
Cigarettes	28.3	12.8	17.7
Cigars/pipes	21.2	12.0	14.2

The adjusted rates for the nonsmokers are identical to the crude rates; this is to be expected, since this group was used as the standard distribution. Note that cigarette smoking appears much more dangerous than it did before. In addition, the adjusted death rates for cigar and pipe smokers are considerably lower than the crude rates. Thus, after adjusting for differences in age, we are led to a very different interpretation of the data.

As another example of the standardized rates that are encountered in the study of vital statistics, Figure 4.2 depicts the trends in age-adjusted mortality rates for 14 of the 15 leading causes of death in the United States [1]. The vertical bars represent changes in disease classification. The standardized rates were calculated by means of the direct method; they use the U.S. population in 1940 as the standard distribution. Although the

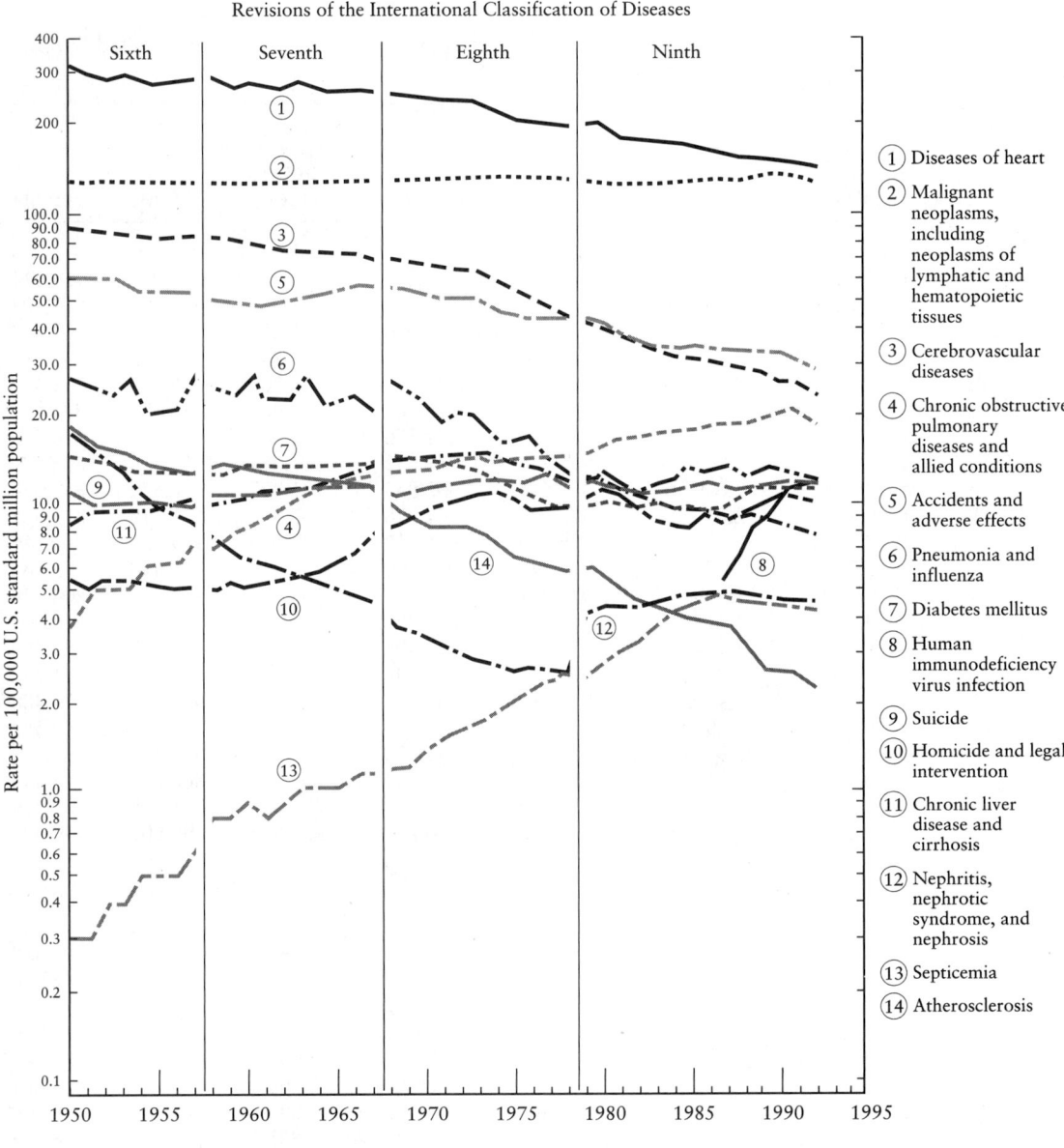

FIGURE 4.2
Age-adjusted death rates for 14 of the 15 leading causes of death, United
States, 1950–1992

adjusted rates have been decreasing for the most part, there are instances in which they
are on the rise. One that stands out is the dramatic jump in the death rate due to human
immunodeficiency virus infection. To illustrate this trend more clearly, Figures 4.3(a)
and (b) show the rise in the death rate due to HIV and AIDS relative to other leading

TABLE 4.3
Total deaths, death rates, and age-adjusted death rates by race and sex, United States, 1940, 1950, 1960, 1970, and 1975–1992

	All Races			White		
Year	Both Sexes	Male	Female	Both Sexes	Male	Female
	Number					
1992	2,175,613	1,122,336	1,053,277	1,873,781	956,957	916,824
1991	2,169,518	1,121,665	1,047,853	1,868,904	956,497	912,407
1990	2,148,463	1,113,417	1,035,046	1,853,254	950,812	902,442
1989	2,150,466	1,114,190	1,036,276	1,853,841	950,852	902,989
1988	2,167,999	1,125,540	1,042,459	1,876,906	965,419	911,487
1987	2,123,323	1,107,958	1,015,365	1,843,067	953,382	889,685
1986	2,105,361	1,104,005	1,001,356	1,831,083	952,554	878,529
1985	2,086,440	1,097,758	988,682	1,819,054	950,455	868,599
1984	2,039,369	1,076,514	962,855	1,781,897	934,529	847,368
1983	2,019,201	1,071,923	947,278	1,765,582	931,779	833,803
1982	1,974,797	1,056,440	918,357	1,729,085	919,239	809,846
1981	1,977,981	1,063,772	914,209	1,731,233	925,490	805,743
1980	1,989,841	1,075,078	914,763	1,738,607	933,878	804,729
1979	1,913,841	1,044,959	868,882	1,676,145	910,137	766,008
1978	1,927,788	1,055,290	872,498	1,689,722	920,123	769,599
1977	1,899,597	1,046,243	853,354	1,664,100	912,670	751,430
1976	1,909,440	1,051,983	857,457	1,674,989	918,589	756,400
1975	1,892,879	1,050,819	842,060	1,660,366	917,804	742,562
1970	1,921,031	1,078,478	842,553	1,682,096	942,437	739,659
1960	1,711,982	975,648	736,334	1,505,335	860,857	644,478
1950	1,452,454	827,749	624,705	1,276,085	731,366	544,719
1940	1,417,269	791,003	626,266	1,231,223	690,901	540,322
	Death rate					
1992	852.9	901.6	806.5	880.0	917.2	844.3
1991	860.3	912.1	811.0	886.2	926.2	847.7
1990	863.8	918.4	812.0	888.0	930.9	846.9
1989	871.3	926.3	818.9	893.2	936.5	851.8
1988	886.7	945.1	831.2	910.5	957.9	865.3
1987	876.4	939.3	816.7	900.1	952.7	849.6
1986	876.7	944.7	812.3	900.1	958.6	844.3
1985	876.9	948.6	809.1	900.4	963.6	840.1
1984	864.8	938.8	794.7	887.8	954.1	824.6
1983	863.7	943.2	788.4	885.4	957.7	816.4
1982	852.4	938.4	771.2	873.1	951.8	798.2
1981	862.0	954.0	775.0	880.4	965.2	799.8
1980	878.3	976.9	785.3	892.5	983.3	806.1
1979	852.2	957.5	752.7	865.2	963.3	771.8
1978	868.0	977.5	764.5	880.2	982.7	782.7
1977	864.4	978.9	756.0	874.6	983.0	771.3
1976	877.6	993.8	767.6	887.7	997.3	783.1
1975	878.5	1002.0	761.4	886.9	1004.1	775.1
1970	945.3	1090.3	807.8	946.3	1086.7	812.6
1960	954.7	1104.5	809.2	947.8	1098.5	800.9
1950	963.8	1106.1	823.5	945.7	1089.5	803.3
1940	1076.4	1197.4	954.6	1041.5	1162.2	919.4

TABLE 4.3

(Continued)

Year	Black			American Indian			Asian or Pacific Islander		
	Both Sexes	Male	Female	Both Sexes	Male	Female	Both Sexes	Male	Female
				Number					
1992	269,219	146,630	122,589	8953	5181	3772	23,660	13,568	10,092
1991	269,525	147,331	122,194	8621	4948	3673	22,173	12,727	9446
1990	265,498	145,359	120,139	8316	4877	3439	21,127	12,211	8916
1989	267,642	146,393	121,249	8614	5066	3548	20,042	11,688	8354
1988	264,019	144,228	119,791	7917	4617	3300	18,963	11,155	7808
1987	254,814	139,551	115,263	7602	4432	3170	17,689	10,496	7193
1986	250,326	137,214	113,112	7301	4365	2936	16,514	9795	6719
1985	244,207	133,610	110,597	7154	4181	2973	15,887	9441	6446
1984	235,884	129,147	106,737	6949	4117	2832	14,483	8627	5856
1983	233,124	127,911	105,213	6839	4064	2775	13,554	8126	5428
1982	226,513	125,610	100,903	6679	3974	2705	12,430	7564	4866
1981	228,560	127,296	101,264	6608	4016	2592	11,475	6908	4567
1980	233,135	130,138	102,997	6923	4193	2730	11,071	6809	4262
1979	220,818	124,433	96,385	6728	4171	2557	...	...	...
1978	221,340	124,663	96,677	6959	4343	2616	...	...	...
1977	220,076	123,894	96,182	6454	4019	2435	...	...	...
1976	219,442	123,977	95,465	6300	3883	2417	...	...	...
1975	217,932	123,770	94,162	6166	3838	2328	...	...	...
1970	225,647	127,540	98,107	5675	3391	2284	...	...	...
1960	196,010	107,701	88,309	4528	2658	1870	...	...	...
1950	169,606	92,004	77,602	4440	2497	1943	...	...	...
1940	178,743	95,517	83,226	4791	2527	2264	...	...	...
				Death rate					
1992	850.5	977.5	736.2	417.7	487.7	348.9	283.1	332.7	235.8
1991	864.9	998.7	744.5	407.2	471.2	343.9	277.3	325.6	231.1
1990	871.0	1008.0	747.9	402.8	476.4	330.4	283.3	334.3	234.3
1989	887.9	1026.7	763.2	430.5	510.7	351.3	280.9	334.5	229.4
1988	888.3	1026.1	764.6	411.7	485.0	339.9	282.0	339.0	227.4
1987	868.9	1006.2	745.7	410.7	483.8	339.0	278.9	338.3	222.0
1986	864.9	1002.6	741.5	409.5	494.9	325.9	276.2	335.1	219.9
1985	854.8	989.3	734.2	416.4	492.5	342.5	283.4	344.6	224.9
1984	836.1	968.5	717.4	419.6	502.7	338.4	275.9	336.5	218.1
1983	836.6	971.2	715.9	428.5	515.1	343.9	276.1	339.1	216.1
1982	823.4	966.2	695.5	434.5	522.9	348.1	271.3	338.3	207.4
1981	842.4	992.6	707.7	445.6	547.9	345.6	272.3	336.2	211.5
1980	875.4	1034.1	733.3	487.4	597.1	380.1	296.9	375.3	222.5
1979	839.3	999.6	695.3	...	...	...	...	...	...
1978	855.1	1016.8	709.5	...	...	...	...	...	...
1977	864.0	1026.0	718.0	...	...	...	...	...	...
1976	875.0	1041.6	724.5	...	...	...	...	...	...
1975	882.5	1055.4	726.1	...	...	...	...	...	...
1970	999.3	1186.6	829.2	...	...	...	...	...	...
1960	1038.6	1181.7	905.0	...	...	...	...	...	...
1950	...	...	...	...	...	...	...	...	...
1940	...	...	...	...	...	...	...	...	...

(Continued)

TABLE 4.3

(Continued)

Year	All Races			White		
	Both Sexes	Male	Female	Both Sexes	Male	Female
	Age-adjusted death rate					
1992	504.5	656.0	380.3	477.5	620.9	359.9
1991	513.7	669.9	386.6	486.8	634.4	366.3
1990	520.2	680.2	390.6	492.8	644.3	369.9
1989	528.0	689.3	397.3	499.6	652.2	376.0
1988	539.9	706.1	406.1	512.8	671.3	385.3
1987	539.2	706.8	404.6	513.7	674.2	384.8
1986	544.8	716.2	407.6	520.1	684.9	388.1
1985	548.9	723.0	410.3	524.9	693.3	391.0
1984	548.1	721.6	410.5	525.2	693.6	391.7
1983	552.5	729.4	412.5	529.4	701.6	393.3
1982	554.7	734.2	411.9	532.3	706.8	393.6
1981	568.6	753.8	420.8	544.8	724.8	401.5
1980	585.8	777.2	432.6	559.4	745.3	411.1
1979	577.0	768.6	423.1	551.9	738.4	402.5
1978	595.0	791.4	437.4	569.5	761.1	416.4
1977	602.1	801.3	441.8	575.7	770.6	419.6
1976	618.5	820.9	455.0	591.3	789.3	432.5
1975	630.4	837.2	462.5	602.2	804.3	439.0
1970	714.3	931.6	532.5	679.6	893.4	501.7
1960	760.9	949.3	590.6	727.0	917.7	555.0
1950	841.5	1001.6	688.4	800.4	963.1	645.0
1940	1076.1	1213.0	938.9	1017.2	1155.1	879.0

causes of death among men and women 25 to 44 years of age [7]. Table 4.3 displays the overall age-adjusted mortality rates for the United States by race and gender [1]. Again, these rates were calculated using the direct method of standardization. In general, the adjusted rates have been decreasing over time; note, however, that in all years females have lower death rates than males, and Asians and Pacific Islanders have lower death rates than persons in other racial groups.

In practical applications, we must know when to use an adjusted rate rather than crude or subgroup-specific rates. If there are no confounding factors such as age or gender, or if no comparisons between groups are required, crude rates are generally sufficient. An advantage of crude rates is that they give us information about the actual experience of a population. When one or more confounders are present, subgroup-specific rates are always suitable for comparisons; they provide the most detailed information about the groups. Adjusted rates—single numbers that summarize the situation in each population being considered—should be used only if one or more confounders are present and a comparison is still desired. It is important to keep in mind that the adjusted values themselves are meaningless.

TABLE 4.3

(Continued)

Year	Black			American Indian			Asian or Pacific Islander		
	Both Sexes	Male	Female	Both Sexes	Male	Female	Both Sexes	Male	Female
	Age-adjusted death rate								
1992	767.5	1026.9	568.4	453.1	579.6	343.1	285.8	364.1	220.5
1991	780.7	1048.8	575.1	441.8	562.6	335.9	283.2	360.2	218.3
1990	789.2	1061.3	581.6	445.1	573.1	335.1	297.6	377.8	228.9
1989	805.9	1082.8	594.3	475.7	622.8	353.4	295.8	378.9	225.2
1988	809.7	1083.0	601.0	456.3	585.7	343.2	300.2	385.4	226.5
1987	796.4	1063.6	592.4	456.7	580.8	351.3	297.0	386.2	221.3
1986	796.8	1061.9	594.1	451.4	591.6	328.4	296.7	385.3	220.3
1985	793.6	1053.4	594.8	468.2	602.6	353.3	305.7	396.9	228.5
1984	783.3	1035.9	590.1	476.9	614.2	347.3	299.4	386.0	223.0
1983	787.4	1037.5	595.3	485.9	634.0	360.1	298.9	388.6	218.0
1982	782.1	1035.4	585.9	494.3	634.6	371.6	293.6	389.2	212.8
1981	807.0	1068.8	602.7	514.0	676.7	368.5	293.2	382.3	213.9
1980	842.5	1112.8	631.1	564.1	732.5	414.1	315.6	416.6	224.6
1979	812.1	1073.3	605.0	...	...	...	...	...	...
1978	831.8	1093.9	622.7	...	...	...	...	...	...
1977	849.3	1112.1	639.6	...	...	...	...	...	...
1976	870.5	1138.3	654.5	...	...	...	...	...	...
1975	890.8	1163.0	670.6	...	...	...	...	...	...
1970	1044.0	1318.6	814.4	...	...	...	...	...	...
1960	1073.3	1246.1	916.9	...	...	...	...	...	...
1950	...	...	...	...	...	...	...	...	...
1940	...	...	...	...	...	...	...	...	...

When the direct method of standardization is applied, the subgroup-specific rates should have the same general trends in all groups being compared, as well as in the standard population. More explicitly, if we were to graph the subgroup-specific rates for two different groups, the trends would ideally be parallel, as in Figure 4.4(a). A somewhat less ideal situation occurs when the trends are not parallel but are comparable, as in Figure 4.4(b). When the subgroup-specific rates follow very different patterns in the two populations, as in Figure 4.4(c), direct standardization should not be attempted. In a situation such as this, a single number cannot capture the complex behavior of the rates; by manipulating the standard, we could make them show just about anything we chose. Therefore, instead of summarizing the data, we should simply report the subgroup-specific rates themselves.

The age-specific hearing impairment rates for the subpopulations of currently employed individuals and those not in the labor force are displayed in Figure 4.5. The graph most closely resembles Figure 4.4(b); the trends in the two groups are similar, although not parallel. In this case, we conclude that it was not inappropriate to use the direct method to standardize for age.

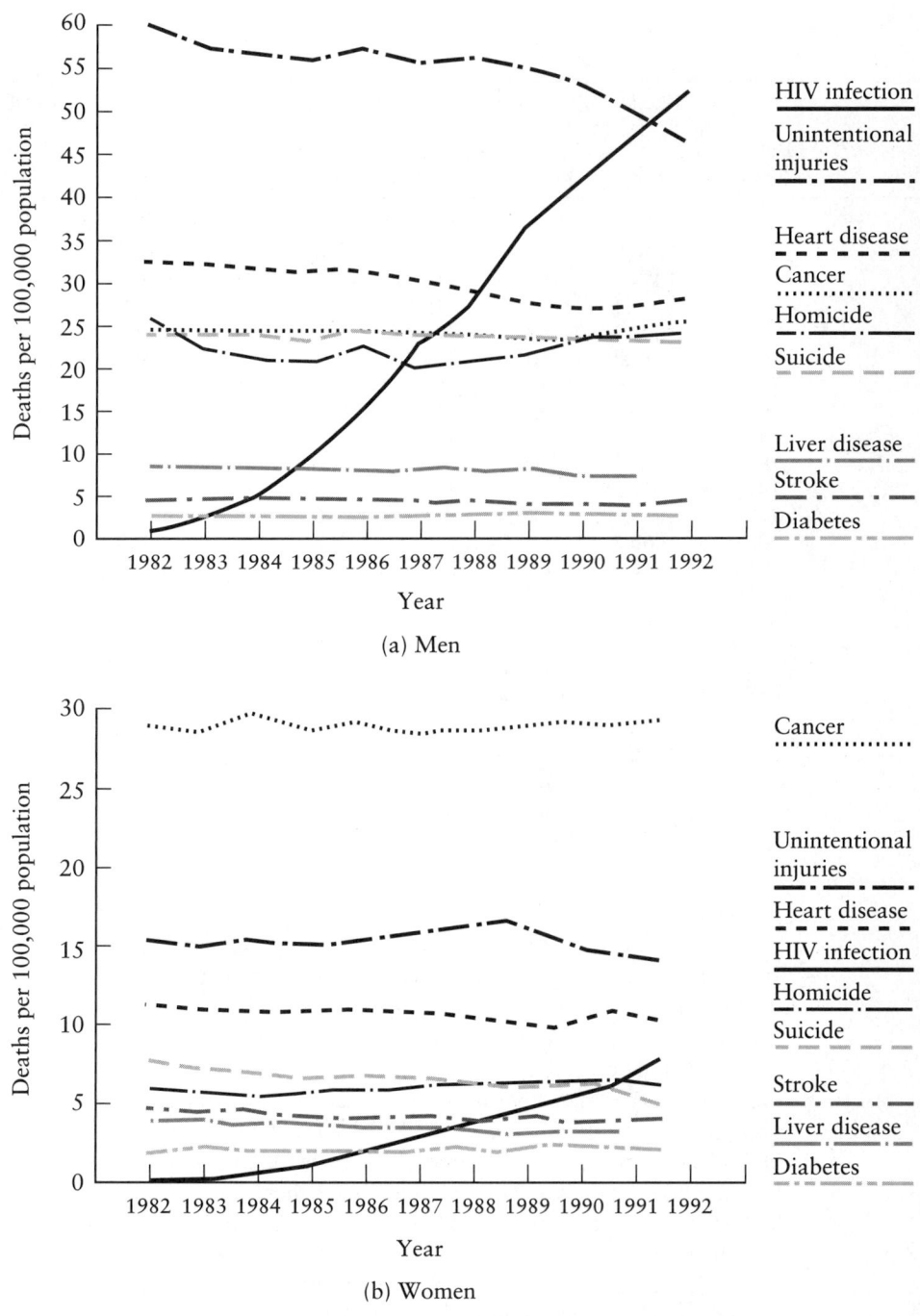

(a) Men

(b) Women

FIGURE 4.3
Death rates from leading causes of death for men and women 25 to 44
years of age, United States, 1982–1992

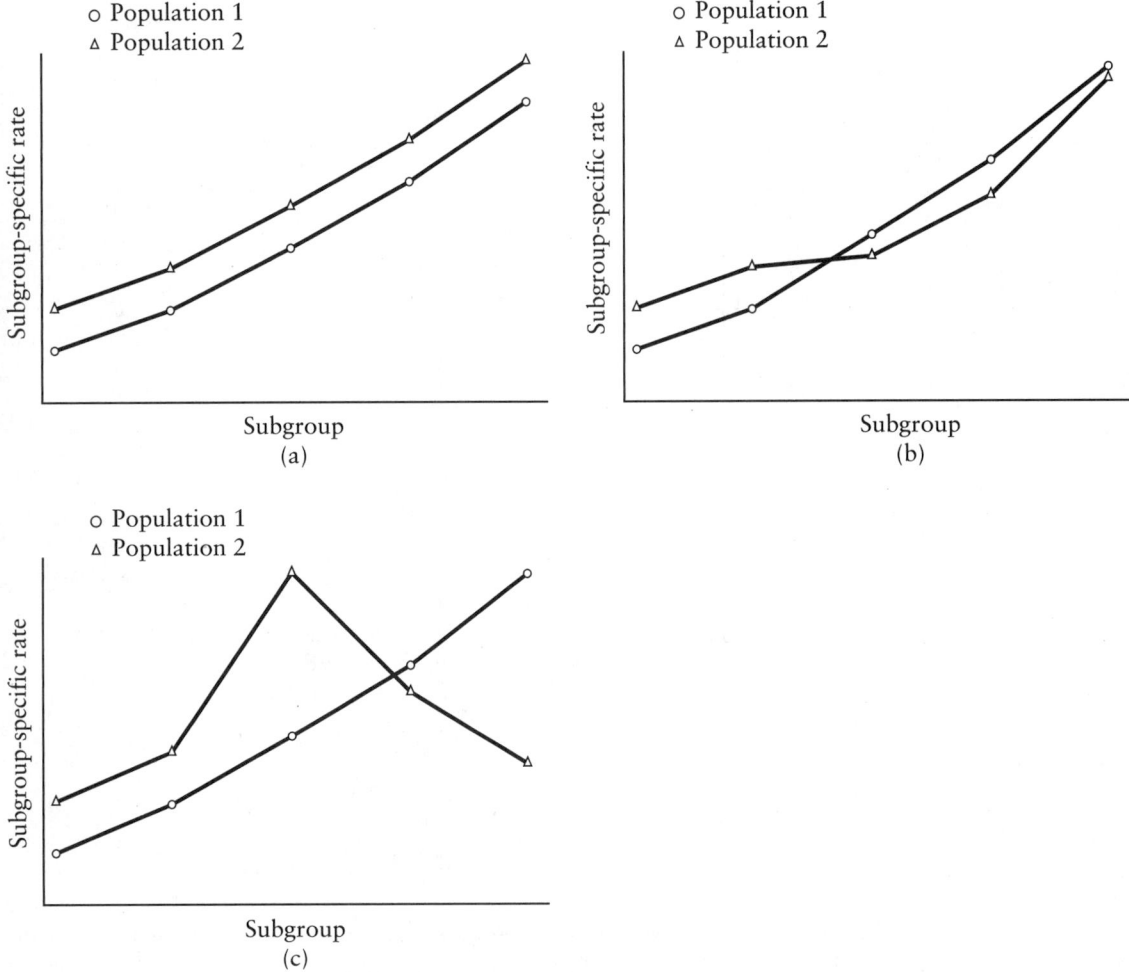

FIGURE 4.4
Trends in subgroup-specific rates for two populations

In practice, the direct method of standardization is used much more frequently than the indirect method. In fact, most adjusted rates reported by the National Center for Health Statistics in the United States apply the direct method and use the U.S. population in 1940 as the standard distribution [5]. However, the direct method does require that subgroup-specific rates be available for all populations being compared. In some situations, this is not feasible and the indirect method must be used instead. Also, if the subgroup-specific rates are available but have been calculated based on very small numbers—and are therefore subject to a great deal of fluctuation—the indirect method is preferred. In any event, the application of either method should lead to the same conclusions being drawn. If the conclusions differ, the situation needs to be investigated further.

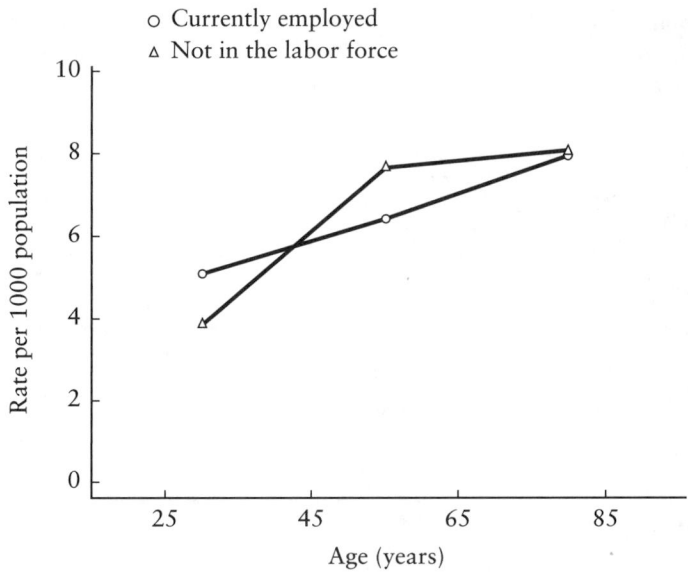

FIGURE 4.5
Age-specific hearing impairment rates for the currently employed and those not in the labor force, 1980–1981

4.3 Further Applications

Suppose that we are interested in comparing the infant mortality rates in two states, Colorado and Louisiana. How may this be accomplished? To begin, consider the following data, which specify the numbers of live births and deaths of infants under 1 year of age that occurred in each state in 1987 [8,9].

State	Live Births	Infant Deaths	Rate per 1000
Colorado	53,808	527	9.8
Louisiana	73,967	872	11.8

The crude data show that the infant mortality rate in Colorado is 9.8 per 1000 live births, whereas the rate in Louisiana is 11.8 per 1000 live births. It appears that infants born in Louisiana are more likely to die before they reach 1 year of age than those born in Colorado.

Before reporting this result, we might want to investigate our suspicion that race is a confounder in the relationship between state and infant mortality rate. Therefore, we first explore the underlying distributions of race in the two populations.

	Colorado		Louisiana	
Race	**Live Births**	**Percent**	**Live Births**	**Percent**
Black	3166	5.9	29,670	40.1
White	48,805	90.7	42,749	57.8
Other	1837	3.4	1548	2.1
Total	53,808	100.0	73,967	100.0

The relative frequencies indicate that the states do differ in racial composition. In Colorado, most of the live births are white children; in Louisiana, a much larger percentage of the infants are black.

We next examine the race-specific infant mortality rates in the entire United States population in 1987.

Race	**Live Births**	**Infant Deaths**	**Rate per 1000**
Black	641,567	11,461	17.9
White	2,992,488	25,810	8.6
Other	175,339	1137	6.5
Total	3,809,394	38,408	10.1

The crude infant mortality rate for the United States is a weighted average of the race-specific rates; note that

$$10.1 = \frac{(641{,}567)(17.9) + (2{,}992{,}488)(8.6) + (175{,}339)(6.5)}{3{,}809{,}394}.$$

From the table, we can see that the infant mortality rate is considerably higher among black children than it is among white children. Since race is associated with both state and infant mortality rate, it is a confounder in the relationship between these two quantities. Perhaps the higher crude infant mortality rate in Louisiana is a consequence of the greater proportion of black children who are born there.

A more accurate comparison between the two states can be made by looking at the race-specific infant mortality rates rather than the crude rates.

	Colorado			Louisiana		
Race	**Live Births**	**Infant Deaths**	**Rate per 1000**	**Live Births**	**Infant Deaths**	**Rate per 1000**
Black	3166	52	16.4	29,670	525	17.7
White	48,805	469	9.6	42,749	344	8.0
Other	1837	6	3.3	1548	3	1.9
Total	53,808	527	9.8	73,967	872	11.8

Among black children, the infant mortality rate is higher in Louisiana than it is in Colorado; among white children and children of other racial groups, however, the infant mortality rate is higher in Colorado. Although the race-specific rates provide the most detailed information about these two populations, it would be convenient to be able to summarize the entire situation with a pair of numbers—one for each state—that adjust for differences in racial composition.

4.3.1   Direct Method of Standardization

To apply the direct method of standardization, we first select a standard race distribution; in this example, we use all live births in the United States in 1987. We then calculate the number of infant deaths that would have occurred in each of the two states if both possessed this standard racial composition while retaining their own individual race-specific mortality rates.

	U.S.	Colorado		Louisiana	
Race	**Live Births**	**Rate per 1000**	**Expected Deaths**	**Rate per 1000**	**Expected Deaths**
Black	641,567	16.4	10,521.7	17.7	11,355.7
White	2,992,488	9.6	28,727.9	8.0	23,939.9
Other	175,339	3.3	578.6	1.9	333.1
Total	3,809,394		39,828.2		35,628.7

The expected numbers of deaths are calculated by multiplying the total number of live births in the U.S. for a particular racial group by the race-specific rates for each state and dividing by 1000.

The race-adjusted infant mortality rate for each state is then calculated by dividing its total expected number of infant deaths by the number of live births in the standard population.

$$\text{Colorado:} \quad \frac{39,828.2}{3,809,394} = 10.5 \text{ per } 1000$$

$$\text{Louisiana:} \quad \frac{35,628.7}{3,809,394} = 9.4 \text{ per } 1000$$

These race-adjusted rates are the infant mortality rates that would apply if the live births in both Colorado and Louisiana had the same distribution of race as the United States as a whole. Although the crude infant mortality rate for Louisiana is higher than the crude rate for Colorado, the adjusted infant mortality rate in Colorado is larger after we control for the effect of race.

In this example, was it appropriate to calculate the race-adjusted infant mortality rates for these two populations using the direct method of standardization? To attempt to answer this question, consider the plot of race-specific mortality rates for Colorado

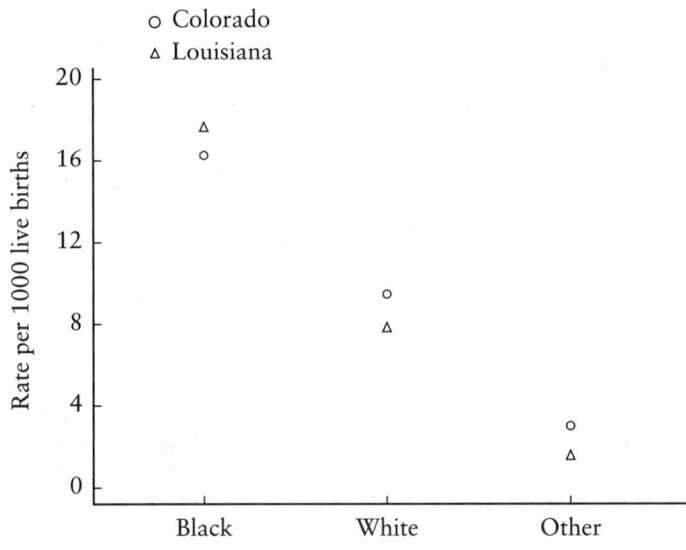

FIGURE 4.6

Race-specific infant mortality rates for Colorado and Louisiana, 1987

and Louisiana presented in Figure 4.6. Since race is not a continuous measurement, we cannot connect the points with straight lines. Observe, however, that the two sets of rates follow the same trend: for both states, black children have the highest infant mortality rate, followed by white children, and then by children from other racial groups. Therefore, we conclude that it is acceptable to use the direct method to standardize for race.

Rather than work out the race-adjusted infant mortality rates on our own, we could have used Stata to do the calculations for us. Table 4.4 displays the appropriate output. On the left-hand side of the upper portion of the table, the columns labeled "Pop." and "Cases" contain the numbers of live births and infant deaths for each racial group in the state of Colorado. To the right of these columns are the relative frequencies associated with the numbers of live births and the race-specific infant mortality rates. The column titled "Std. Pop. Dst[P]" contains the relative frequencies corresponding to all live births in the United States in 1987. Directly beneath this are both the crude and the race-adjusted infant mortality rates. Note that the adjusted rate for Colorado, 0.01047, is just 10.5 per 1000 live births carried out one decimal place farther. The central portion of the table contains the same data for Louisiana. The bottom summarizes the important information for each state.

4.3.2 Indirect Method of Standardization

To apply the indirect method of standardization, we select a standard set of race-specific infant mortality rates and apply these to the actual numbers of live births in each state. As a standard, we again choose the live births in the United States in 1987. We then calculate the numbers of infant deaths that would have occurred in each state if

TABLE 4.4
Stata output illustrating the direct method of standardization

```
state = Colorado
                         —Unadjusted—            Std.
                          Pop.    Stratum        Pop.
Stratum      Pop.   Cases Dist.   Rate[s]    Dst[P]     s*P

Black        3166     52  0.059   0.0164     0.168    0.0028
Other        1837      6  0.034   0.0033     0.046    0.0002
White       48805    469  0.907   0.0096     0.786    0.0075

Totals:     53808    527          Adjusted Cases:      563.1
                                     Crude Rate:    0.00979
                                  Adjusted Rate:    0.01047
                     95% Conf. Interval:   [0.01045,  0.01048]

state = Louisiana
                         —Unadjusted—            Std.
                          Pop.    Stratum        Pop.
Stratum      Pop.   Cases Dist.   Rate[s]    Dst[P]     s*P

Black       29670    525  0.401   0.0177     0.168    0.0030
Other        1548      3  0.021   0.0019     0.046    0.0001
White       42749    344  0.578   0.0080     0.786    0.0063

Totals:     73967    872          Adjusted Cases:      694.6
                                     Crude Rate:    0.01179
                                  Adjusted Rate:    0.00939
                     95% Conf. Interval:   [0.00939,  0.00940]

Summary of Study Populations:
     state      N       Crude    Adj. Rate   Confidence Interval

  Colorado    53808   0.009794   0.010465   [0.010451,  0.010479]
  Louisiana   73967   0.011789   0.009391   [0.009386,  0.009395]
```

they were to take on the race-specific infant mortality rates of the United States as a whole while retaining their own individual distributions of race.

	U.S.	Colorado		Louisiana	
Race	**Rate per 1000**	**Live Births**	**Expected Deaths**	**Live Births**	**Expected Deaths**
Black	17.9	3166	56.7	29,670	531.1
White	8.6	48,805	419.7	42,749	367.6
Other	6.5	1837	11.9	1548	10.1
Total	10.1	53,808	488.3	73,967	908.8

The expected numbers of deaths are calculated by multiplying the race-specific rates in the U.S. by the numbers of live births in each state and dividing by 1000.

The standardized mortality ratio for each state is obtained by dividing the observed number of infant deaths by the total expected number of deaths.

$$\text{Colorado:} \quad \frac{527}{488.3} = 1.08$$
$$= 108\%$$

$$\text{Louisiana:} \quad \frac{872}{908.8} = 0.96$$
$$= 96\%$$

These standardized mortality ratios indicate that Colorado has an 8% higher infant mortality rate than the United States as a whole, whereas Louisiana has a 4% lower infant mortality rate.

Finally, we calculate the race-adjusted infant mortality rate for each state by multiplying its standardized mortality ratio by the crude infant mortality rate in the standard population, 10.1 per 1000 live births.

$$\text{Colorado:} \quad \frac{10.1}{1000} \times 1.08 = 10.9 \text{ per } 1000$$

$$\text{Louisiana:} \quad \frac{10.1}{1000} \times 0.96 = 9.7 \text{ per } 1000$$

After controlling for the effect of race, Colorado again has a higher adjusted infant mortality rate than Louisiana. Although the rates themselves differ, this is the same conclusion that was reached when the direct method of standardization was applied.

4.4   Review Exercises

1. What are demographic data and vital statistics? How can they be used to describe the health status of a population?

2. What is the difference between a rate and a proportion?

3. What is a confounder?

4. How does the direct method of standardization differ from the indirect method? When would you use one rather than the other?

5. How does the choice of a standard population distribution affect the results of the standardization process?

6. Under what circumstances should crude, specific, and adjusted rates each be used?

7. The total numbers of deaths in the United States in various years are presented below [1].

Year	Number of Deaths
1990	2,148,463
1980	1,989,841
1970	1,921,031
1960	1,711,982
1950	1,452,454
1940	1,417,269

The statement is made that since the number of deaths has been increasing over the years, the population as a whole must be growing less healthy. Do you agree with this statement? Why or why not?

8. The following data were reported for the state of Massachusetts in 1992 [10].

	Number
Population	6,060,943
Live births	87,202
Deaths	
Total	53,804
Under 1 year	569

Compute the following rates:
(a) Crude birth rate
(b) Crude death rate
(c) Infant mortality rate

9. In the table below, the numbers of live births and infant deaths in the United States in 1983 are categorized by birth weight [11].

Birth Weight (grams)	Live Births	Infant Deaths
2500^+	3,385,912	15,349
1500–2499	204,534	6136
750–1499	31,246	7283
500–749	7594	5815
<500	4444	3937
Unknown	5383	1163
Total	3,639,113	39,683

(a) Compute the infant mortality rate for each category of birth weight.
(b) What can you conclude about the relationship between infant mortality and birth weight?
(c) Do you think that a large proportion of the infants for whom birth weight is unknown are likely to weigh less than 1500 grams? Why or why not?

10. In a news report investigating factors that affect the human life span, the leading causes of death at various points in a person's lifetime were examined. For each of the four age groups 15 to 24 years, 25 to 44 years, 45 to 64 years, and 65 years and above, the numbers of deaths per 100,000 population in the United States are presented below for the top five causes of death [12].

	Deaths per 100,000 Population by Age Group (years)			
Cause of Death	**15–24**	**25–44**	**45–64**	**65⁺**
Accidents	45.8	35.4	32.4	—
AIDS	—	20.3	—	—
Cancer	5.1	26.2	290.0	1085.1
Chronic respiratory ailments	—	—	28.0	225.8
Heart disease	2.6	19.0	241.5	1949.2
Homicide	16.9	—	—	—
Pneumonia and influenza	—	—	—	217.5
Stroke	—	—	32.5	408.8
Suicide	13.3	14.8	—	—

(a) For the 15- to 24-year-olds, construct a bar chart displaying the age-specific mortality rates for each of the five leading causes of death in that group. (Instead of the number or percentage of deaths, the height of each bar should reflect the mortality rate corresponding to a particular cause.) Construct a similar bar chart for each of the other age groups.

(b) Describe the ways in which the factors that affect mortality change throughout a person's lifetime. Which causes of death are influential across all age groups? Which have a significant impact on only one or two of the given groups?

11. In 1954, a study was undertaken to test the effectiveness of the poliomyelitis vaccine developed by Dr. Jonas Salk. Poliomyelitis is a communicable illness caused by the poliovirus; it ranges in severity from a mild infection to fatal paralytic disease. The clinical trial consisted of two separate parts, one of which involved randomly assigning first-, second-, and third-graders in the United States and Canada to one of two treatment groups. Members of the first group received the Salk vaccine, while those in the second group were given a placebo, an inert substance that was physically indistinguishable from the true treatment. A total of 749,236 children were eligible for this portion of the study; 401,974 eventually received either the vaccine or the placebo [13]. Listed below are the numbers of reported cases of disease for both the vaccine and the placebo groups. The reported cases are divided into true instances of polio and incorrect diagnoses. The true cases are further broken down depending on whether the disease was paralytic or nonparalytic.

			Polio			Not
Group	**Number of Children**	**Reported Cases**	**Total**	**Paralytic**	**Nonparalytic**	**Polio**
Vaccine	200,745	82	57	33	24	25
Placebo	201,229	162	142	115	27	20

(a) For both the vaccine and the placebo groups, calculate the rates per 100,000 children for all reported cases, for all true instances of polio and incorrect diagnoses, and for paralytic and nonparalytic disease.

(b) Based on the results of this part of the study, does it appear that the Salk poliomyelitis vaccine helped to prevent disease?

12. Between the years 1984 and 1987, the crude death rate for females in the United States increased steadily. At the same time, the age-adjusted death rate decreased. The relevant data are shown below [1].

Year	Crude Death Rate per 100,000 Population	Age-Adjusted Death Rate per 100,000 Population
1987	816.7	404.6
1986	812.3	407.6
1985	809.1	410.3
1984	794.7	410.5

Explain how the age-adjusted rate could fall while the crude rate was rising.

13. Figure 4.7 displays the crude and age-adjusted mortality rates for the United States from 1940 to 1993 [1]. The age-adjusted death rates were calculated using the direct method of standardization; the U.S. population in 1940 was used as the standard. Although both rates are decreasing over time, the adjusted rate is falling off more rapidly than the crude rate. How do you explain this?

14. Listed in the table below are the total numbers of births that occurred in the United States on each day of the week in 1991 [14]. Also included in the table are the "indices of occurrence" for standard vaginal births and for cæsarian-section births. The index of occurrence for a given day—Monday, for instance—is the average number of births that occurred on a Monday divided by the average number of births for all days of the week combined. The computations were performed separately for vaginal and cæsarian-section deliveries. The index of occurrence can be thought of as the actual average number of births corresponding to a given day divided by the expected average assuming that all days are identical. It is interpreted in the same manner as a standardized morbidity ratio.

Day of the Week	Number of Births	Index of Occurrence	
		Vaginal	Cæsarean Section
Sunday	466,706	85.9	58.7
Monday	601,244	101.2	108.0
Tuesday	651,952	106.8	117.5
Wednesday	626,733	105.2	113.2
Thursday	628,656	105.3	114.0
Friday	635,814	104.9	120.8
Saturday	499,802	90.6	67.5

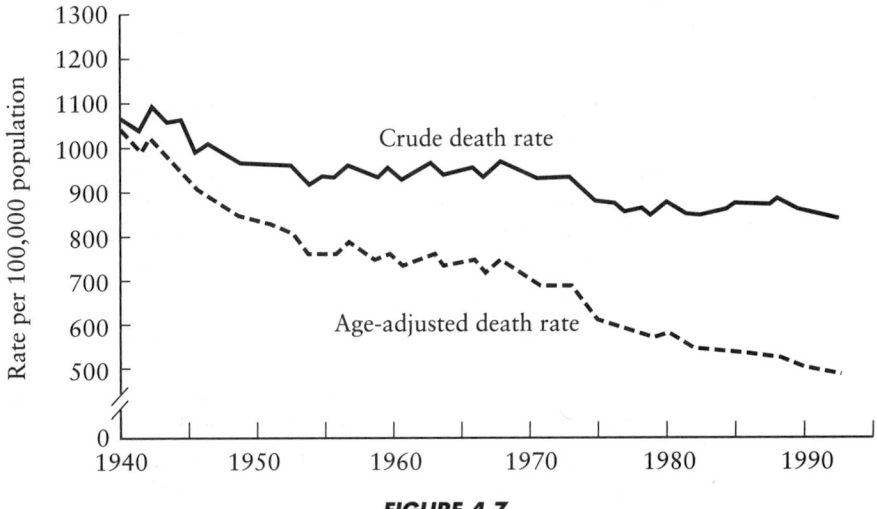

FIGURE 4.7
Crude and age-adjusted mortality rates for the United States, 1940–1993

(a) Treating the total numbers of births for each day of the week as a set of seven observations, calculate the mean total number of births for a typical day. Do you feel that the mean is an adequate measure of central tendency for these data? Why or why not?

(b) Construct a line graph displaying the index of occurrence across days of the week, by method of delivery. (There should be separate lines for vaginal and cæsarian-section births.) Is a diurnal pattern evident for these data? What do you think might account for this trend?

(c) Is the pattern different for vaginal and cæsarian deliveries?

(d) What does it mean if an index of occurrence (or standardized morbidity ratio) has a value greater than 100%? What if it is less than 100%?

15. Consider the following data comparing cancer mortality in the United States in 1940 and in 1986 [15,16].

Age	1940 Population (thousands)	1940 Deaths	1986 Population (thousands)	1986 Deaths
0–4	10,541	494	18,152	666
5–14	22,431	667	33,860	1165
15–24	23,922	1287	39,021	2115
25–34	21,339	3696	42,779	5604
35–44	18,333	11,198	33,070	14,991
45–54	15,512	26,180	22,815	37,800
55–64	10,572	39,071	22,232	98,805
65–74	6377	44,328	17,332	146,803
75+	2643	31,279	11,836	161,381
Total	131,670	158,200	241,097	469,330

(a) Compute the crude cancer mortality rates for 1940 and 1986 and compare these rates.

(b) For both 1940 and 1986, calculate the proportion of the total population in each age group. Describe the way the two populations differ with respect to age composition.

(c) Compute the age-specific cancer death rates for each population. Is there a relationship between age and death rate?

(d) Does it appear to be necessary to control for the effect of age when comparing cancer death rates in the two populations? Why or why not?

(e) Using the U.S. population in 1940 as the standard distribution, apply the direct method of standardization. What are the age-adjusted cancer mortality rates for 1940 and 1986?

(f) How does the age-adjusted death rate differ from the crude death rate in each of these populations? Explain these results.

(g) Using suitably scaled axes, plot the age-specific cancer mortality rates versus age for both 1940 and 1986. Comment on the extent to which it was appropriate to standardize for age using the direct method.

(h) Using the age-specific cancer mortality rates for 1940 as the standard, apply the indirect method to compute the standard mortality ratios for 1940 and 1986.

(i) How does the 1986 population compare with the 1940 population in terms of cancer mortality rate?

(j) Calculate the age-adjusted cancer mortality rates for 1940 and 1986. Do the results obtained using the indirect method correspond to those obtained when the direct method was applied?

16. In 1940, a statistician at the Bureau of the Census was interested in comparing the death rate in Maine to the death rate in South Carolina. The table below contains some relevant data from 1930 [17].

	Maine		South Carolina	
Age	**Population**	**Deaths**	**Population**	**Deaths**
0–4	75,037	1543	205,076	4905
5–9	79,727	148	240,750	446
10–14	74,061	104	222,808	410
15–19	68,683	153	211,345	901
20–24	60,575	224	166,354	1073
25–34	105,723	413	219,327	1910
35–44	101,192	552	191,349	2377
45–54	90,346	980	143,509	2862
55–64	72,478	1476	80,491	2667
65–74	46,614	2433	40,441	2486
75+	22,396	3056	16,723	2364
Total	796,832	11,082	1,738,173	22,401

The age-specific population sizes and numbers of deaths for each state are saved under the variable names popn and deaths in the data set dthrate (Appendix B,

Table B.8). Identifiers of state and age group are saved under the variable names `state` and `age` respectively.

(a) What are the crude mortality rates in Maine and in South Carolina in 1930?

(b) Do these two states differ with respect to age composition? Explain.

(c) Compute the age-specific mortality rates for each state.

(d) Do you believe that it is necessary to control for the effect of age when comparing mortality rates for these two states? Why or why not?

(e) Displayed below are the relative frequencies by age group for the entire United States population in 1940. These percentages are saved under the variable name `popn` in the data set `us1940` (Appendix B, Table B.8b).

Age	Percent
0–4	8.01
5–9	8.11
10–14	8.92
15–19	9.37
20–24	8.80
25–34	16.21
35–44	13.92
45–54	11.78
55–64	8.03
65–74	4.84
75+	2.01
Total	100.00

Using the U.S. population as the standard, apply the direct method of standardization. What are the age-adjusted death rates for Maine and South Carolina?

(f) Plot the age-specific mortality rates versus age for each of the two states. Based on this graph, was it appropriate to standardize for age using the direct method?

Bibliography

[1] National Center for Health Statistics, Kochanek, K. D., and Hudson, B. L., "Advance Report of Final Mortality Statistics, 1992," *Monthly Vital Statistics Report*, Volume 43, Number 6, March 22, 1995.

[2] United Nations Children's Fund, *The State of the World's Children 1994*, New York: Oxford University Press.

[3] National Center for Health Statistics, Singh, G. K., Mathews, T. J., Clarke, S. C., Yannicos, T., and Smith, B. L., "Annual Summary of Births, Marriages, Divorces, and Deaths, United States, 1994," *Monthly Vital Statistics Report*, Volume 43, Number 13, October 23, 1995.

[4] National Center for Health Statistics, Collins, J. G., "Types of Injuries and Impairments Due to Injuries, United States," *Vital and Health Statistics*, Series 10, Number 159, November 1986.

[5] Centers for Disease Control, Curtin, L. R., and Klein, R. J., "Direct Standardization (Age-Adjusted Death Rates)," *Healthy People 2000 Statistical Notes*, Number 6—Revised, March 1995.

[6] Cochran, W. G., "The Effectiveness of Adjustment by Subclassification in Removing Bias in Observational Studies," *Biometrics*, Volume 24, June 1968, 295–313.

[7] Centers for Disease Control, "Update: Mortality Attributable to HIV Infection Among Persons Aged 25–44 Years—United States, 1991–1992," *Morbidity and Mortality Weekly Report*, Volume 42, Number 45, November 19, 1993.

[8] National Center for Health Statistics, *Vital Statistics of the United States*, *1987*, Volume I—Natality, 1989.

[9] National Center for Health Statistics, *Vital Statistics of the United States*, *1987*, Volume II—Mortality, Part A, 1990.

[10] Massachusetts Department of Public Health, *1992 Annual Report*, *Vital Statistics of Massachusetts*, Public Document 1, December 1994.

[11] Overpeck, M. D., Hoffman, H. J., and Prager, K., "The Lowest Birth-Weight Infants and the U.S. Mortality Rate: NCHS 1983 Linked Birth/Infant Death Data," *American Journal of Public Health*, Volume 82, March 1992, 441–444.

[12] Foreman, J., "Making Age Obsolete: Scientists See Falling Barriers to Human Longevity," *The Boston Globe*, September 27, 1992, 1, 28–29.

[13] Meier, P., "Polio Trial: An Early Efficient Clinical Trial," *Statistics in Medicine*, Volume 9, Number 1/2, January–February 1990, 13–16.

[14] National Center for Health Statistics, "Advance Report of Maternal and Infant Health Data From Birth Certificates, 1991," *Monthly Vital Statistics Report*, Volume 42, Number 11(S), May 11, 1994.

[15] National Center for Health Statistics, *Vital Statistics of the United States*, *1986*, Volume II—Mortality, Part B, 1988.

[16] United States Department of Commerce, *Vital Statistics of the United States*, *1940*, Part II—Natality and Mortality Data Tabulated by Place of Residence, 1943.

[17] National Center for Health Statistics, *Vital Statistics Rates in the United States, 1900–1940*, Chapters I–IV, reprinted 1972.

5

Life Tables

It is almost always impossible to predict how long a particular individual will live, let alone to predict the span of life for each person in a population of millions. In spite of this, it is the job of policy planners to assess and describe the health and longevity of an entire nation. For many years, life tables have been used as a means of summarizing the health status of a group of individuals. Like the techniques that we have studied in previous chapters, they are descriptive in nature. *Life tables* identify the death rates experienced by a population over a given period of time. They have many practical applications: life tables are used to analyze the mortality of a particular population, to make international comparisons, to compute insurance premiums and annuities, and to predict survival. They have even been used, rather inanely, in a debate regarding the existence of a biological limit to human life [1].

By following a fictional cohort of individuals—usually a group of 100,000 persons—from birth until the last individual in the cohort has died, a life table describes the mortality experience of the group over a specified period of time. As an example, Table 5.1 displays the 1992 life table for the United States [2]. The table assumes that we observe 100,000 persons from birth until death and that, as they pass through each year of life, the individuals die at the rates experienced by the U.S. population in 1992. The number 100,000 is chosen to simplify calculations; it is arbitrary, and we would end up with the same results no matter what number we chose.

5.1   Computation of the Life Table

5.1.1   Column 1

Before describing some of its applications, we begin by examining the way in which the life table is computed. The first column of the table displays the age interval. The age interval represents the period of life between age x and age $x + n$, where n is the

TABLE 5.1
Abridged life table for the total population, United States, 1992

Age Interval	Proportion Dying	Of 100,000 Born Alive		Stationary Population		Average Remaining Lifetime
Period of Life between Two Exact Ages Stated in Years (1)	Proportion of Persons Alive at Beginning of Age Interval Dying during Interval (2)	Number Living at Beginning of Age Interval (3)	Number Dying during Age Interval (4)	In the Age Interval (5)	In This and All Subsequent Age Intervals (6)	Average Number of Years of Life Remaining at Beginning of Age Interval (7)
x to $x+n$	$_nq_x$	l_x	$_nd_x$	$_nL_x$	T_x	$\overset{\circ}{e}_x$
0–1	0.00851	100,000	851	99,275	7,577,757	75.8
1–5	0.00172	99,149	171	396,195	7,478,482	75.4
5–10	0.00102	98,978	101	494,615	7,082,287	71.6
10–15	0.00121	98,877	120	494,152	6,587,672	66.6
15–20	0.00418	98,757	413	492,848	6,093,520	61.7
20–25	0.00528	98,344	519	490,448	5,600,672	56.9
25–30	0.00601	97,825	588	487,654	5,110,224	52.2
30–35	0.00765	97,237	744	484,369	4,622,570	47.5
35–40	0.01001	96,493	966	480,187	4,138,201	42.9
40–45	0.01305	95,527	1247	474,740	3,658,014	38.3
45–50	0.01822	94,280	1718	467,420	3,183,274	33.8
50–55	0.02799	92,562	2591	456,739	2,715,854	29.3
55–60	0.04421	89,971	3978	440,481	2,259,115	25.1
60–65	0.06800	85,993	5848	416,137	1,818,634	21.1
65–70	0.10084	80,145	8082	381,393	1,402,497	17.5
70–75	0.14673	72,063	10,574	334,799	1,021,104	14.2
75–80	0.21189	61,489	13,029	275,667	686,305	11.2
80–85	0.31480	48,460	15,255	204,369	410,638	8.5
85 and over ..	1.00000	33,205	33,205	206,269	206,269	6.2

length of the interval. Therefore, 0–1 designates the one-year period of life from birth until an individual's first birthday. Interval 1–5 represents the time from the first birthday until the fifth birthday, a period of four years. All other age intervals span five years, except for the last. This is an open-ended interval representing the entire period of life beyond the 85th birthday.

For the sake of convenience, abridged life tables—such as Table 5.1—are used most often in practice. Abridged tables display data in terms of five-year age intervals. A complete life table would have an entry for each year; the complete life table calculated for the U.S. population in 1979–1981 is shown in Table 5.8, at the end of this chapter [3]. In the United States, complete life tables are constructed every ten years using decennial census data. In practice, abridged tables are often used solely for pedagogical purposes. Note from Table 5.8, however, that within each of the five-year intervals between ages 5 and 85, mortality is relatively constant. Therefore, the abridged table does not discard a great deal of information. This is not true for the very young; mortality in the first year of life is quite different from that in years 2 through 4. At the other end of the scale, the consolidation of everyone over 85 years of age into a single group is influenced in part by tradition. In the past, life expectancy was considerably less than 85 years, and only a small proportion of individuals lived beyond that age.

5.1.2 Column 2

The second column of the life table, represented by $_nq_x$, lists the proportion of individuals alive at the beginning of the age interval x to $x + n$ who die at some time during the interval. This quantity is also called the *hazard function*; it can be calculated from the age-specific death rates for the United States population in 1992, such as those provided in Table 4.2. For example,

$$_1q_0 = \text{proportion of individuals alive at birth who die before}$$
$$\text{their first birthday}$$
$$= 865.7 \text{ per } 100,000 \qquad \text{(from Table 4.2)}$$
$$= 0.008657$$
$$\approx 0.00851.$$

The numbers in the two tables do not match exactly because they were calculated in slightly different ways. Table 4.2 lists the estimated overall death rate for a given age interval, whereas Table 5.1 displays a weighted average calculated by breaking the interval down into smaller pieces. (In fact, the hazard is more properly defined as the death rate in an infinitesimally small interval.) Note that

$$_\infty q_{85} = \text{proportion of individuals alive on their}$$
$$\text{85th birthday who die during the time period}$$
$$\text{after their 85th birthday}$$
$$= 1.00000,$$

since death is inevitable.

The other proportions in column 2 are somewhat more difficult to calculate. If we were to plot the age-specific death rates for the U.S. population for each individual year—the hazard function—from birth until age 90, the result would be Figure 5.1. From the graph, a single rate can be chosen to summarize each age interval in column 1. For an abridged table, this value is used to approximate the proportion of individuals dying within the interval. For example, consider the age group 1–5. From Table 4.2, the

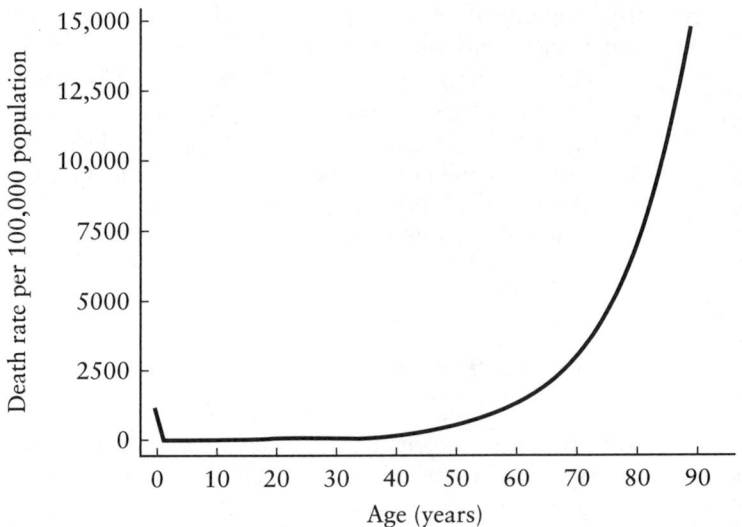

FIGURE 5.1
Age-specific death rates from birth until age 90 for the U.S. population,
1979–1981

approximate age-specific death rate for this interval is 43.6 per 100,000 population, or 0.000436. Since this is the estimated mortality rate for **each** year within the four-year period, the total proportion of individuals dying between their first and fifth birthdays is roughly

$$_4q_1 = 4 \times 0.000436$$
$$= 0.001744$$
$$\approx 0.00172.$$

Again, the numbers in the two tables do not match exactly. Similarly,

$$_5q_5 = 5 \times 0.000204$$
$$= 0.001020,$$

$$_5q_{10} = 5 \times 0.000246$$
$$= 0.001230$$
$$\approx 0.00121,$$

and so on. The estimates calculated using data from Table 4.2 are accurate if the age-specific death rates are relatively constant throughout an entire interval. In theory, we should actually compound the mortality each year rather than simply multiply by the number of years; in practice, however, the approximation is usually adequate.

5.1.3   Columns 3 and 4

The third column of the life table, l_x, displays the number of individuals out of the original 100,000 who are still alive on their xth birthday. Note that column headings with a single subscript refer to conditions at the beginning of the age interval; those with two subscripts specify conditions within the interval. Hence the fourth column, $_nd_x$, lists the number out of the l_x alive at the beginning of the interval x to $x + n$ who die at some time during that interval. The computations of the third and fourth columns are interdependent; they also rely on column 2. To illustrate this point, l_0, the number of individuals born into the cohort, is equal to 100,000. The number who die before their first birthday is the total number born alive multiplied by the proportion of individuals who die during the age interval 0–1, or

$$_1d_0 = 100,000 \times {_1q_0}$$
$$= 100,000 \times 0.00851$$
$$= 851.$$

Therefore, the number of individuals out of the original 100,000 who live to see their first birthday is

$$l_1 = 100,000 - 851$$
$$= 99,149.$$

Similarly,

$$_4d_1 = 99,149 \times {_4q_1}$$
$$= 99,149 \times 0.00172$$
$$= 171,$$
$$l_5 = 99,149 - 171$$
$$= 98,978,$$
$$_5d_5 = 98,978 \times {_5q_5}$$
$$= 98,978 \times 0.00102$$
$$= 101,$$

and

$$l_{10} = 98,978 - 101$$
$$= 98,877.$$

In general, the number of individuals alive at the beginning of a particular interval is equal to the number who were alive at the beginning of the previous interval minus the number who died during that interval, or

$$l_{x+n} = l_x - {}_nd_x.$$

If we plot l_x against age x, as in Figure 5.2, we can observe the manner in which the number of survivors decreases over time. The number of individuals dying during an age interval is derived by multiplying the number alive at the beginning of the interval by the proportion of persons dying during that interval, or

$$_nd_x = l_x \times {}_nq_x.$$

5.1.4   Column 5

Column 5 of the life table, represented by $_nL_x$, is known as the *stationary population* within an interval. Demographers find this concept useful; it may be interpreted in the following way. Suppose that a cohort of 100,000 individuals is born each year. Furthermore, assume that in each cohort the proportion of individuals dying within the age interval x to $x + n$ is given by $_nq_x$ in column 2. As a result, the 1992 age-specific death rates apply to every cohort. If no migration occurred and a sufficient number of years were allowed to pass, we would reach a stationary population: the number of persons

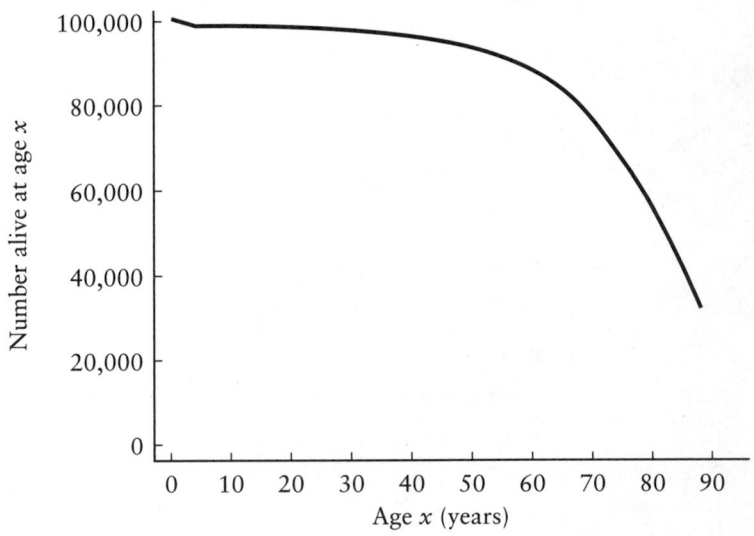

FIGURE 5.2
Number of individuals alive at age x from birth until age 85 for the U.S.
population, 1992

alive in any given age group would never change. As soon as one individual left an interval—either by dying or by growing older and entering the subsequent interval—his or her place would be taken by someone from the preceding age group. Consequently, a census taken at any arbitrary point in time would enumerate the same population with the same distribution among age groups.

The stationary population $_nL_x$ has a second interpretation as well. It may be thought of as the total time in years lived during the age interval x to $x + n$ by the l_x individuals alive at the beginning of the interval. Consider, for example, the age interval 1–5. There are $l_1 = 99{,}149$ individuals alive at the beginning of this interval. Of these 99,149, $l_5 = 98{,}978$ are still alive on their fifth birthday. Since each of these 98,978 individuals survives throughout the entire four-year period, they contribute $98{,}978 \times 4 = 395{,}912$ person-years to the total time lived during the interval. There are also $_4d_1 = 171$ persons who die at some point during the interval. Each of these individuals contributes some amount of person-time before he or she dies, something greater than zero but less than four years. If we sum the person-years contributed by each of the individuals who die during the interval—information that is not provided in the abridged life table—and add this to the 395,912 person-years lived by the individuals who survive the entire four years, we have a total of $_4L_1 = 396{,}195$ years lived during the age interval 1–5.

Note that the procedure used to derive this total is identical to the technique for summing a set of observations that was presented in Chapter 3. We begin by grouping individuals according to the number of years that they have contributed to the interval. For each group, we then multiply the number of individuals by the associated number of years. Finally, we sum these quantities to obtain the total time lived during the interval. Once again, all calculations are based on the assumption that the 1992 age-specific death rates hold throughout the entire lifetime of the cohort.

5.1.5 Column 6

The sixth column of the life table, T_x, displays the total stationary population in the age interval x to $x + n$ and all subsequent intervals. In other words, it is the total number of person-years lived beyond their xth birthday by the l_x individuals alive on that birthday. It is obtained by summing column 5 from the bottom up; for example,

$$T_{65} = {_5L_{65}} + {_5L_{70}} + {_5L_{75}} + {_5L_{80}} + {_\infty L_{85}}$$
$$= 381{,}393 + 334{,}799 + 275{,}667 + 204{,}369 + 206{,}269$$
$$= 1{,}402{,}497.$$

5.1.6 Column 7

Finally, column 7 of the life table, $\mathring{e}_x$, is the average number of years of life remaining for an individual who is alive at age x. It is calculated by dividing the total number of person-years lived beyond the xth birthday, T_x, by the number of individuals who survive to age x or beyond; dividing by l_x eliminates the dependency on the size of the original cohort. As an example,

$$\overset{\circ}{e}_0 = \text{average number of years of life remaining at birth}$$

$$= \frac{T_0}{l_0}$$

$$= \frac{7,577,757}{100,000}$$

$$= 75.8$$

and

$$\overset{\circ}{e}_1 = \text{average number of years of life remaining after the first birthday}$$

$$= \frac{T_1}{l_1}$$

$$= \frac{7,478,482}{99,149}$$

$$= 75.4.$$

If $\overset{\circ}{e}_1$ is larger than $\overset{\circ}{e}_0$—which was the case in the United States as recently as 1976—this reflects a high infant mortality rate. It implies that if an infant is strong enough to survive the first year of life, the average remaining lifetime on his or her first birthday is greater than it was at birth.

5.2   Applications of the Life Table

Life tables have a number of practical applications. As one example, life tables can be used to predict the chance that a person will live to a particular age x. Suppose we wish to know the probability that an individual will survive from birth until age 65, or, equivalently, the proportion of persons who do so. Of the 100,000 individuals who were born into the 1992 cohort, 80,145 are still alive on their 65th birthday. Therefore, the proportion of persons surviving until age 65 is

$$\frac{l_{65}}{l_0} = \frac{80,145}{100,000}$$

$$= 0.80145,$$

or about 80.1%. The probability that a 50-year-old will reach his or her 65th birthday is the number of individuals alive on that birthday divided by the number alive on their 50th birthday, or

$$\frac{l_{65}}{l_{50}} = \frac{80,145}{92,562}$$

$$= 0.86585,$$

approximately 86.6%. Note that the chance of surviving until age 65 increases if a person has already made it through the first 50 years. This increase in probability—from 80.1% to 86.6%—is important to an individual calculating insurance rates. The concept of probability will be discussed further in Chapter 6; for now, we consider "probability" to be synonymous with "proportion."

To illustrate another application of the life table, average remaining lifetime, $\overset{\circ}{e}_0$ in particular, is often used to describe the health status of a population [4]. The average life expectancies at birth in the United States are broken down by race and gender in Table 5.2 [2]. Because of its value as a population summary measure, average remaining lifetime can also be used to make international comparisons. The average life expectancies at birth for selected nations are provided in Table 5.3 [5]. In all the countries included, females have a greater life expectancy than males; there are only a few countries in the world for which the opposite is true.

TABLE 5.2

Life expectancy at birth by race and sex, United States, 1940, 1950, 1960, 1970–1992

| | All Races | | | White | | | All Other | | | | | |
| | | | | | | | Total | | | Black | | |
Year	Both Sexes	Male	Female	Both Sexes	Male	Female	Both Sexes	Male	Female	Both Sexes	Male	Female
1992	75.8	72.3	79.1	76.5	73.2	79.8	71.8	67.7	75.7	69.6	65.0	73.9
1991	75.5	72.0	78.9	76.3	72.9	79.6	71.5	67.3	75.5	69.3	64.6	73.8
1990	75.4	71.8	78.8	76.1	72.7	79.4	71.2	67.0	75.2	69.1	64.5	73.6
1989	75.1	71.7	78.5	75.9	72.5	79.2	70.9	66.7	74.9	68.8	64.3	73.3
1988	74.9	71.4	78.3	75.6	72.2	78.9	70.8	66.7	74.8	68.9	64.4	73.2
1987	74.9	71.4	78.3	75.6	72.1	78.9	71.0	66.9	75.0	69.1	64.7	73.4
1986	74.7	71.2	78.2	75.4	71.9	78.8	70.9	66.8	74.9	69.1	64.8	73.4
1985	74.7	71.1	78.2	75.3	71.8	78.7	71.0	67.0	74.8	69.3	65.0	73.4
1984	74.7	71.1	78.2	75.3	71.8	78.7	71.1	67.2	74.9	69.5	65.3	73.6
1983	74.6	71.0	78.1	75.2	71.6	78.7	70.9	67.0	74.7	69.4	65.2	73.5
1982	74.5	70.8	78.1	75.1	71.5	78.7	70.9	66.8	74.9	69.4	65.1	73.6
1981	74.1	70.4	77.8	74.8	71.1	78.4	70.3	66.2	74.4	68.9	64.5	73.2
1980	73.7	70.0	77.4	74.4	70.7	78.1	69.5	65.3	73.6	68.1	63.8	72.5
1979	73.9	70.0	77.8	74.6	70.8	78.4	69.8	65.4	74.1	68.5	64.0	72.9
1978	73.5	69.6	77.3	74.1	70.4	78.0	69.3	65.0	73.5	68.1	63.7	72.4
1977	73.3	69.5	77.2	74.0	70.2	77.9	68.9	64.7	73.2	67.7	63.4	72.0
1976	72.9	69.1	76.8	73.6	69.9	77.5	68.4	64.2	72.7	67.2	62.9	71.6
1975	72.6	68.8	76.6	73.4	69.5	77.3	68.0	63.7	72.4	66.8	62.4	71.3
1974	72.0	68.2	75.9	72.8	69.0	76.7	67.1	62.9	71.3	66.0	61.7	70.3
1973	71.4	67.6	75.3	72.2	68.5	76.1	66.1	62.0	70.3	65.0	60.9	69.3
1972	71.2	67.4	75.1	72.0	68.3	75.9	65.7	61.5	70.1	64.7	60.4	69.1
1971	71.1	67.4	75.0	72.0	68.3	75.8	65.6	61.6	69.8	64.6	60.5	68.9
1970	70.8	67.1	74.7	71.7	68.0	75.6	65.3	61.3	69.4	64.1	60.0	68.3
1960	69.7	66.6	73.1	70.6	67.4	74.1	63.6	61.1	66.3	...	...	...
1950	68.2	65.6	71.1	69.1	66.5	72.2	60.8	59.1	62.9	...	...	...
1940	62.9	60.8	65.2	64.2	62.1	66.6	53.1	51.5	54.9	...	...	...

TABLE 5.3

Average life expectancies at birth for selected countries, 1992

Nation	$\overset{\circ}{e}_0$	Females as a Percentage of Males
Argentina	71	110
Australia	77	109
Brazil	66	109
Canada	77	105
China	71	109
Egypt	61	104
Ethiopia	47	107
Finland	76	111
France	77	111
Greece	77	107
India	60	101
Israel	76	105
Italy	77	109
Japan	79	108
Mexico	70	110
Philippines	65	106
Poland	72	113
Russian Federation	69	—
Saudi Arabia	69	104
Spain	77	108
Sweden	78	108
United Kingdom	76	107
United States	76	109
Venezuela	70	109

In addition to its use as a population summary measure, average remaining lifetime is also used as a basis for the computation of life insurance premiums. Recall, however, that the formula $\overset{\circ}{e}_x = T_x/l_x$ is derived based on the stationary population of columns 5 and 6. All calculations depend on the assumption that the age-specific death rates for 1992 remain constant throughout the entire lifetime of a cohort born in that year. In reality, the cohort members would be subject to the $_1q_0$ for 1992 during their first year of life, the $_1q_1$ for 1993 during their second year of life, the $_1q_2$ for 1994 during their third year of life, and so on. Since the age-specific death rates for future years are unknown, in the life table they are approximated by the 1992 death rates. As shown in Figure 5.3, however, the death rates for almost all age groups have been decreasing in recent years [2]. If this trend continues, the average remaining lifetime calculated in column 7 is actually an underestimate of the true average remaining lifetime. Consequently, since individuals will live longer than expected and continue to pay premiums throughout their lifetimes, insurance companies that use $\overset{\circ}{e}_x$ to predict survival will end

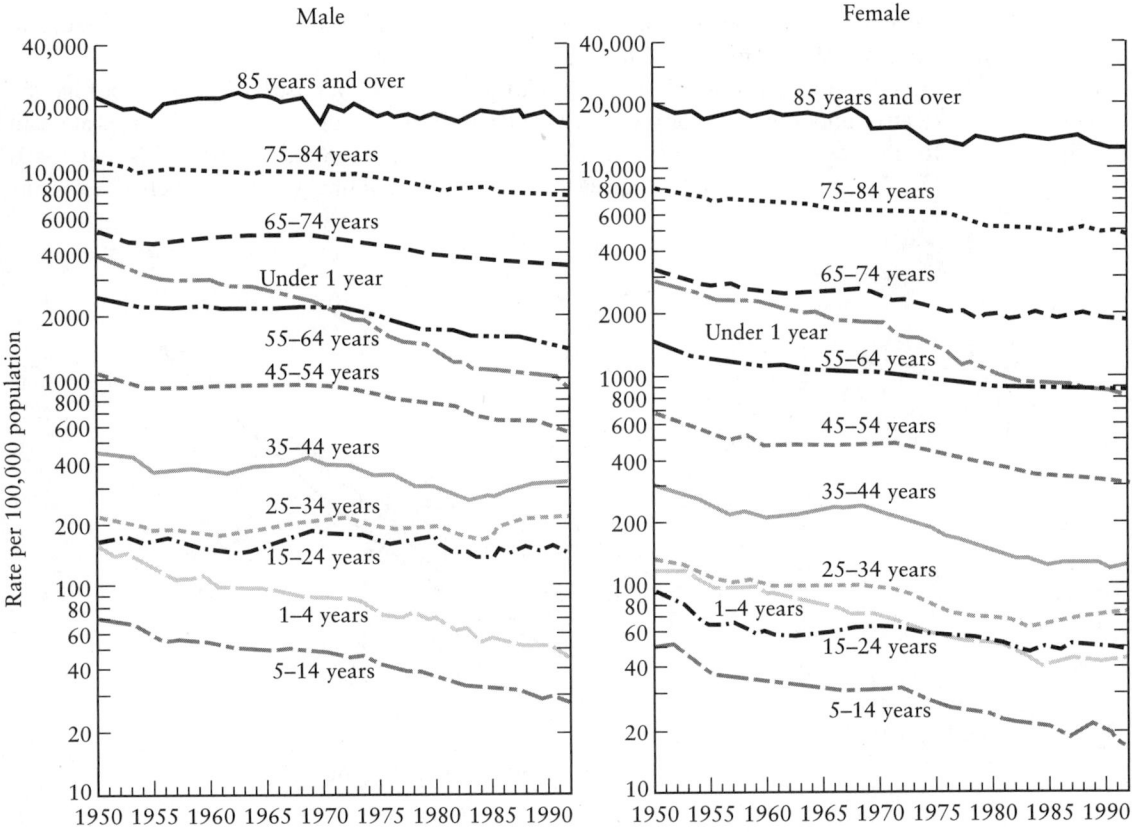

FIGURE 5.3
Death rates by age and sex, United States, 1950–1992

up increasing their profits. In contrast, government and public health organizations that use $\overset{\circ}{e}_x$ to plan for future services could run into serious problems.

5.3   Years of Potential Life Lost

The methodology of life tables can also be used to quantify premature mortality. The improvement in survival in England and Wales during the last century is illustrated in Figure 5.4 [6]. The age- and gender-specific mortality rates in both 1851 and 1951 are displayed. The experience in the United States has been similar; in particular, the most significant reductions in mortality have been made in the younger age groups. The older groups do not show much improvement. This is largely a reflection of advances brought about by the eradication of a number of infectious "childhood" diseases, as well as great improvements in nutrition, housing, and sanitation. In other words, the reductions in mortality result from advances in the prevention of disease. Note that the difference in survival between males and females is a relatively recent phenomenon.

To do their jobs effectively, health policy workers must be able to determine the relative importance of various causes of death. Table 5.4 displays the mean ages at death for the seven leading causes of mortality in the United States in the 1930s and 1940s [7]. We can see that for the causes that tend to affect younger persons, such as accidents

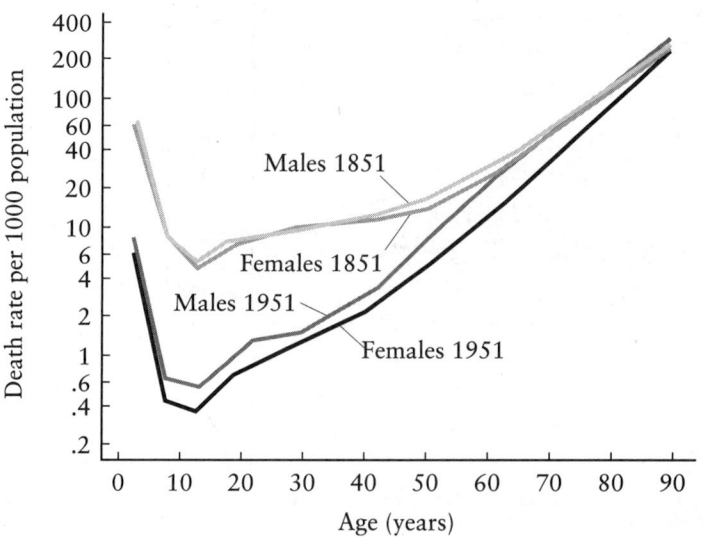

FIGURE 5.4

Age- and sex-specific death rates per 1000 population, England and Wales, 1851 and 1951

TABLE 5.4

Mean age at death in years for the seven leading causes of mortality, United States, 1930–1945

Cause of Death	1930	1935	1940	1945
Cerebrovascular*	67.5	67.7	68.5	69.1
Diseases of the heart	64.5	65.9	67.1	67.6
Nephritis	64.0	65.5	66.7	66.8
Cancer	61.4	62.1	62.5	63.0
Pneumonia/influenza	40.1	43.7	47.2	46.2
Accidents	41.3	43.8	45.5	45.8
Tuberculosis	37.4	39.9	41.7	43.7
All others	35.1	38.5	42.0	43.5
All causes combined	48.8	52.2	55.8	57.4

*Intracranial lesions of vascular origin

and tuberculosis, the average age at death shows greater improvement over the 15-year period than it does for causes that affect primarily the elderly, such as cancer. This difference is not incorporated into the calculation of overall mortality rates, however, since each death counts the same as any other. The majority of deaths occur among older persons; therefore, mortality data are dominated by diseases of the elderly.

Incidentally, it should be noted that studies of mortality usually rely on information provided by death certificates. Such data must be treated with caution. Death certificates are legal, not medical, documents. Furthermore, procedures for filling out death certificates vary over time and across different cultures, and the certificates are often incorrect when the deaths of older people are involved.

When used for a relatively stable group over a short period of time, the average age at death can be quite informative. However, like other crude summary statistics, it may be affected by a variety of confounding factors. As a general descriptive measure of survival, the average age at death was discredited long ago in favor of the life table method. A related concept, the *years of potential life lost*, or *YPLL*, was introduced in the early part of this century [8]. Rather than taking into account the years of life lived by an individual, the YPLL considers the amount of life that is lost as the result of a premature death. This has the effect of placing more importance on life that is lost to diseases of the young.

There are a number of different ways of defining YPLL. The one that is used most commonly—and that is reported by the Centers for Disease Control and Prevention—excludes from consideration the deaths of all individuals over the age of 65 [9]. Other persons are allotted 65 years at birth. The years of potential life lost by an individual are the number of years fewer than 65 that the person actually lives. If a 50-year-old man dies of heart disease, for instance, he has lost 15 years of potential life before the age of 65. The total YPLL for a population is obtained by summing the contributions from each individual.

Years of potential life lost can be contrasted with the more common crude mortality rate that we discussed in the previous chapter. Recall that the crude mortality rate assigns equal weight to each death and averages over all persons in a population. YPLL, in contrast, places more value on the deaths of younger individuals; in fact, deaths occurring beyond a certain age are excluded from consideration.

A second method of calculating YPLL is based on the claim that the potential life lost by an individual is equal to his or her average remaining lifetime at the age at which death occurs [9]. According to the U.S. life table for 1992, a 50-year-old man is expected to have an average of 29.3 years of life remaining. If this individual were to die of heart disease, he would be losing 29.3 years of potential life rather than 15. This method again counts the quantity of life lost, but it does not arbitrarily assign each person a maximum of 65 years. As a result, it continues to value individuals who live beyond this age.

To compare each of the three measures described—the crude mortality rate, the YPLL based on average remaining lifetime, and the YPLL prior to age 65—the ten leading causes of death in the United States in 1984 are ranked in Table 5.5 [9]. According to the crude death rates, heart disease was the leading cause of mortality in the United States in that year; however, if we concentrate only on deaths occurring before the age of 65, unintentional injury appears to have been the leading cause. Both methods indicate that malignant neoplasms, or cancer, were the second leading cause of death. Note that in general, the ranking generated by the YPLL based on life expectancy lies somewhere between the rankings produced by the other two methods.

As previously mentioned, the Centers for Disease Control and Prevention choose to report the years of potential life lost based on the allotment of 65 years, thus discounting lives lost beyond that age. Table 5.6 displays the years of potential life lost for persons who died in the United States in 1986 and 1987 [10]. The total YPLL dropped from 5016 years per 100,000 U.S. residents in 1986 to 4949 years per 100,000 residents

TABLE 5.5

Rankings of the ten leading causes of death by method of calculation, United States, 1984

Rank	Crude Mortality Rate	YPLL (Life Expectancy)	YPLL (Age 65)
1	Heart disease	Heart disease	Unintentional injuries
2	Malignant neoplasms	Malignant neoplasms	Malignant neoplasms
3	Cerebrovascular disease	Unintentional injuries	Heart disease
4	Unintentional injuries	Suicide/homicide	Suicide/homicide
5	COPD*	Cerebrovascular disease	Congenital anomalies
6	Pneumonia/influenza	COPD*	Prematurity
7	Suicide/homicide	Congenital anomalies	SIDS†
8	Diabetes	Pneumonia/influenza	Cerebrovascular disease
9	Cirrhosis	Cirrhosis	Cirrhosis
10	Congenital anomalies	Diabetes	Pneumonia/influenza

*Chronic obstructive pulmonary diseases and allied conditions
†Sudden Infant Death Syndrome

TABLE 5.6

Years of potential life lost before age 65 by cause of death, United States, 1986 and 1987

Cause of Death	YPLL 1986	YPLL 1987
All causes	12,093,486	12,045,778
Unintentional injuries	2,358,426	2,295,710
Malignant neoplasms	1,832,210	1,837,742
Heart disease	1,557,041	1,494,227
Suicide/homicide	1,360,508	1,289,223
Congenital anomalies	661,117	642,551
Prematurity	428,796	422,813
HIV infection	246,823	357,536
SIDS	340,431	286,733
Cerebrovascular disease	246,131	246,479
Chronic liver disease	231,558	228,145
Pneumonia/influenza	175,386	166,775
Pulmonary disease	128,590	123,260
Diabetes mellitus	121,117	119,155

in 1987. Overall, the years of life lost to most major causes of death declined. There were small increases in the number of years lost to malignant neoplasms and to cerebrovascular disease. The number of years of potential life lost as a result of HIV infection, however, increased by an amazing 45%.

With a little algebra, we are able to show that for a particular cause of mortality, the mean age at death and the years of potential life lost before age 65 are closely related. Suppose that in a given population there are N_{65} individuals who die before age 65 from the specified cause. Denote by m_{65} the average age at death of these persons. In this case,

$$\text{YPLL} = N_{65}(65 - m_{65}).$$

Note that YPLL increases if m_{65} decreases, meaning that individuals dying from the particular cause are doing so at an earlier age, or if N_{65} increases, implying that more persons are dying as a result of the cause. If we divide YPLL by N_{65} to obtain the average number of years lost by each individual, the resulting quantity is simply

$$\frac{\text{YPLL}}{N_{65}} = 65 - m_{65},$$

or 65 minus the mean age at death.

When we compare years of potential life lost across different groups or populations, the same problems exist as when we compare either crude mortality rates or the mean ages at death; in particular, both the sizes of the populations and their age structures may vary. To compensate for differing population sizes, we can calculate a YPLL rate per 1000 persons. Furthermore, a technique similar to the direct method of standardization has been proposed, in which a standard population is used to adjust YPLL for differing age compositions [11]. Although these approaches are sensible, they do not seem to have found favor and are not frequently applied.

5.4 *Further Applications*

Table 5.7 is an abridged life table for the U.S. population in 1990 [12]. Suppose that we wish to fill in the missing entries.

Recall that column 1 of the life table contains the age interval x to $x + n$, while column 2 shows the proportion of persons alive at the beginning of the interval who die some time during the interval. Column 3 of the table, l_x, lists the number of individuals out of the original cohort of 100,000 who are still alive on their xth birthday. Column 4, $_n d_x$, displays the number out of the l_x alive at the beginning of the age interval x to $x + n$ who die at some time during the interval. The calculations of columns 3 and 4 are interdependent.

The first missing entry in the table, l_{70}—the number of individuals out of the original 100,000 who are still alive at age 70—is equal to the number of persons who were

TABLE 5.7
Abridged life table for the total population, United States, 1990

Age Interval	Proportion Dying	Of 100,000 Born Alive		Stationary Population		Average Remaining Lifetime
Period of Life between Two Exact Ages Stated in Years (1)	Proportion of Persons Alive at Beginning of Age Interval Dying during Interval (2)	Number Living at Beginning of Age Interval (3)	Number Dying during Age Interval (4)	In the Age Interval (5)	In This and All Subsequent Age Intervals (6)	Average Number of Years of Life Remaining at Beginning of Age Interval (7)
x to $x+n$	$_nq_x$	l_x	$_nd_x$	$_nL_x$	T_x	$\overset{\circ}{e}_x$
0–1	0.0093	100,000	927	99,210	7,535,219	75.4
1–5	0.0018	99,073	183	395,863	7,436,009	75.1
5–10	0.0011	98,890	118	494,150	7,040,146	71.2
10–15	0.0013	98,780	127	493,654	6,545,996	66.3
15–20	0.0044	98,653	430	492,290	6,052,342	61.3
20–25	0.0055	98,223	539	489,794	5,580,052	56.6
25–30	0.0062	97,684	607	486,901	5,070,258	51.9
30–35	0.0077	97,077	743	483,571	4,583,357	47.2
35–40	0.0099	96,334	952	479,425	4,099,786	42.6
40–45	0.0126	95,382	1203	474,117	3,620,361	38.0
45–50	0.0187	94,179	1759	466,820	3,146,244	33.4
50–55	0.0290	92,420	2685	455,809	2,679,424	29.0
55–60	0.0457	89,735	4101	439,012	2,223,615	24.8
60–65	0.0706	85,634	6044	413,879	1,784,603	20.8
65–70	0.1029	79,590	8186	378,369	1,370,724	17.2
70–75	0.1519			330,846		
75–80	0.2211			270,129		
80–85	0.3239			197,857		
85 and over ..	1.0000			193,523		

alive at age 65 minus the number who died between their 65th and 70th birthdays; in particular,

$$l_{70} = l_{65} - {_5d_{65}}$$
$$= 79,590 - 8186$$
$$= 71,404.$$

The number of persons dying during the age interval 70–75, $_5d_{70}$, is equal to the number of individuals alive on their 70th birthday multiplied by the proportion of persons dying during the interval; therefore,

$$_5d_{70} = l_{70} \times {_5q_{70}}$$
$$= 71,404 \times 0.1519$$
$$= 10,846.$$

Similarly, we can fill in the remainder of columns 3 and 4 by computing

$$l_{75} = l_{70} - {}_5d_{70}$$
$$= 71{,}404 - 10{,}846$$
$$= 60{,}558,$$

$$_5d_{75} = l_{75} \times {}_5q_{75}$$
$$= 60{,}558 \times 0.2211$$
$$= 13{,}389,$$

$$l_{80} = l_{75} - {}_5d_{75}$$
$$= 60{,}558 - 13{,}389$$
$$= 47{,}169,$$

$$_5d_{80} = l_{80} \times {}_5q_{80}$$
$$= 47{,}169 \times 0.3239$$
$$= 15{,}278,$$

$$l_{85} = l_{80} - {}_5d_{80}$$
$$= 47{,}169 - 15{,}278$$
$$= 31{,}891,$$

and

$$_\infty d_{85} = l_{85} \times {}_\infty q_{85}$$
$$= 31{,}891 \times 1.0000$$
$$= 31{,}891.$$

Column 5 of the life table, $_nL_x$, contains the stationary population, or the total number of person-years lived during the age interval x to $x + n$ by the l_x individuals alive at the beginning of the interval. Column 6, T_x, specifies the total number of person-years lived beyond the xth birthday. It is calculated by summing column 5 from the bottom up. Therefore, we have

$$T_{85} = {}_\infty L_{85}$$
$$= 193{,}523,$$

$$T_{80} = {}_5L_{80} + {}_\infty L_{85}$$
$$= {}_5L_{80} + T_{85}$$
$$= 197{,}857 + 193{,}523$$
$$= 391{,}380,$$

$$T_{75} = {}_5L_{75} + T_{80}$$
$$= 270{,}129 + 391{,}380$$
$$= 661{,}509,$$

and

$$T_{70} = {}_5L_{70} + T_{75}$$
$$= 330{,}846 + 661{,}509$$
$$= 992{,}355.$$

Column 7 of the life table, $\overset{\circ}{e}_x$, shows the average number of years of life remaining to an individual who is alive at age x. It is calculated by taking the total number of person-years lived beyond the xth birthday and dividing by the number of persons still alive on that birthday; thus,

$$\overset{\circ}{e}_{70} = \frac{T_{70}}{l_{70}}$$
$$= \frac{992{,}355}{71{,}404}$$
$$= 13.9,$$

$$\overset{\circ}{e}_{75} = \frac{T_{75}}{l_{75}}$$
$$= \frac{661{,}509}{60{,}558}$$
$$= 10.9,$$

$$\overset{\circ}{e}_{80} = \frac{T_{80}}{l_{80}}$$
$$= \frac{391{,}380}{47{,}169}$$
$$= 8.3,$$

and

$$\overset{\circ}{e}_{85} = \frac{T_{85}}{l_{85}}$$
$$= \frac{193{,}523}{31{,}891}$$
$$= 6.1.$$

A person who reaches the age of 85 has an average of 6.1 years of life remaining.

Note that the average remaining lifetime at birth, $\overset{\circ}{e}_0$, is not the same as the average age at death. Consider the following data comparing England with France and Sweden [13].

Country		Mean Duration of Life	Mean Age at Death	One Death in
England	(1841)	41 years	29 years	46 living
France	(1817–1831)	40 years	34 years	42 living
Sweden	(1801–1805)	39 years	31 years	41 living

The table was explained as follows:

> The average age of the persons who died, or the "mean age at death," was 34 years in France, 31 years in Sweden, and 29 years in England; yet we know that the "expectation of life" is greater in England than in Sweden or France. A Society that granted life annuities to children in England would have to make 40 annual payments on an average, and only 38 in Sweden.

The mean duration of life is calculated using the life table method. The mean age at death, however, is simply the average age of all persons dying in the specified time period; its value depends on the age composition of the population involved.

Returning to the U.S. life table for 1990, suppose that we wish to know the chance that a newborn infant will live to see his or her 75th birthday. Of the $l_0 = 100,000$ persons born into the cohort, $l_{75} = 60,558$ are still alive at age 75. Therefore, the probability of living from birth until age 75 is

$$\frac{l_{75}}{l_0} = \frac{60,558}{100,000}$$
$$= 0.606$$
$$= 60.6\%.$$

Note that this is just the proportion of individuals who live to be 75. If the cohort member has already reached 30 years of age, the probability of living until age 75 becomes

$$\frac{l_{75}}{l_{30}} = \frac{60,558}{97,077}$$
$$= 0.624$$
$$= 62.4\%.$$

If he or she is 70 years old, then the probability of surviving another five years is

$$\frac{l_{75}}{l_{70}} = \frac{60,558}{71,404}$$
$$= 0.848$$
$$= 84.8\%.$$

The longer a person has already lived, the greater his or her chance of surviving until age 75.

5.5 Review Exercises

1. Describe two practical applications of the life table.

2. What is a hazard function?

3. Explain the concept of the stationary population within an interval of a life table.

4. Why have the biggest reductions in mortality over the past hundred years been made in the relatively younger age groups rather than in the older ones?

5. How does the years of potential life lost differ from the crude mortality rate as a descriptive measure of survival?

6. A study exploring the lives of Canadian males in various states of health presents an interesting application of the life table [14]. When constructing their table, the authors divided the person-years lived during a given age interval, $_nL_x$, into person-years lived inside institutions and person-years lived outside institutions. The life table can now be used to determine the average remaining years of life outside institutions.

(1) Age	(2) Number Alive	(3) Person-Years of Life Lived	(4) Rate of Institutionalization	(5) Person-Years in Institutions
0	100,000	1,475,419	0.0020387	3008
15	98,004	972,823	0.0029173	2838
25	96,408	1,896,306	0.0027395	5195
45	92,509	1,699,064	0.0059150	10,050
65	72,274	1,037,206	0.0526520	54,611

(6) Person-Years Outside Institutions	(7) Total Person-Years Outside Institutions from Age x on	(8) Average Remaining Years Outside Institutions
1,472,411		
969,985		
1,891,111		
1,689,014		
982,595		

Columns 2 and 3 list the values of l_x and $_nL_x$ from the 1978 life table for Canadian males. The rates of institutionalization displayed in column 4 were obtained from the Canadian Health Survey.

(a) Explain how columns 5 and 6 are derived.

(b) Complete columns 7 and 8 of the table. Note that column 8 represents the average remaining years of life outside institutions for all men involved in the study.

7. In young adult life, it has been noted that males have a higher rate of accidental and violent death than do females.

(a) Explain the effect this would have on a male life table relative to the corresponding female life table with respect to the age-specific death rates for the young adult years.

(b) How might this affect the average life expectancies for males versus females in these age groups?

8. Listed below are the average remaining lifetimes at various ages for males from two countries—Malawi, which has one of the lowest average life expectancies at birth, and Iceland, which has one of the highest [15].

	Average Remaining Lifetime (years)	
Age	Malawi	Iceland
0	40.9	73.7
1	47.9	73.3
5	60.8	69.5
15	55.0	59.8
25	47.3	50.6
35	40.3	41.2
45	34.0	31.9
55	27.7	23.4

(a) In Malawi, the average life expectancy at age 1 is seven years greater than it is at birth. What phenomenon is causing this difference?

(b) On the basis of comparing the average remaining lifetimes at birth, you might assume that the age-specific death rates in Malawi are higher than those in Iceland at all ages. This is not the case, however. Explain how you know this by comparing the average life expectancies at subsequent ages.

9. The table below shows the mean expectations of life in years at three different ages for males and females in Sweden over a period of two centuries [16].

Time Period	Gender	Mean Expectation at Age		
		0	60	80
1755–1776	Male	33.20	12.24	4.27
	Female	35.70	13.08	4.47
1856–1860	Male	40.48	13.12	3.12
	Female	44.15	14.04	4.91
1936–1940	Male	64.30	16.35	5.25
	Female	66.90	17.19	5.49
1971–1975	Male	72.07	17.65	6.08
	Female	77.65	21.29	7.28

This table has been used as evidence in the argument that the human lifespan is getting longer. Discuss the table, giving your interpretation of the varying time trends in expectation of life at different ages and for males versus females.

10. Listed below are selected values of l_x, the number of individuals alive at age x, taken from the complete life tables for white females in the U.S. population in 1909–1911 and 1969–1971.

Age	Number of Survivors out of 100,000 Live Births	
	1909–1911	1969–1971
0	100,000	100,000
15	83,093	97,902
45	69,341	94,649
75	26,569	63,290

(a) Compute the probabilities of surviving from birth to age 15, from age 15 to age 45, and from age 45 to age 75 in each cohort.

(b) If the probability of surviving from age x to age $x + n$ is p_1 in 1909–1911 and p_2 in 1969–1971, the relative percent improvement in survival over the 60-year period is $(p_2 - p_1)/p_1$. Which age group has the greatest relative percent improvement in survival?

11. Refer to Table 5.1, the 1992 life table for the United States [2].
   (a) What is the chance of surviving from birth until age 80?
   (b) If an individual reaches his or her 50th birthday, what is the probability of that person's surviving to age 80?
   (c) What is the chance of surviving from birth until age 10? Age 30? Age 50?
   (d) Given that a child has reached his or her first birthday, what is the probability that that individual will survive to age 10? Age 30? Age 50?
   (e) What is the chance that a 25-year-old will survive 10 years? A 45-year-old? A 65-year-old?
   (f) What is the probability that a 10-year-old will survive 20 years? 40 years? 60 years?

12. The number of deaths caused by cancer in 1984 and the corresponding years of potential life lost are displayed below for both men and women [17]. In this example, the years of potential life lost for a specified individual are defined as the remaining life expectancy at the age at which death occurs.

	Number of Deaths	YPLL
Men	242,763	3,284,558
Women	210,687	3,596,723

Although a greater number of men died as a result of cancer, women lost more years of potential life. Explain how this could occur.

13. To compute the U.S. life table for 1940, we assume that the 1940 age-specific death rates remain constant throughout the entire lifetime of a cohort born in that year. These death rates are listed in column 2 of the table below. In theory, however, any set of age-specific mortality rates could be used. The true mortality rates experi-

enced by the cohort born in 1940 through age 49 are provided in column 3 of the table. For ages 50 and above, the 1989 rates are used. Note the differences between the two columns.

(1) Age Interval x to $x + n$	(2) Proportion Dying (1940) $_nq_x$	(3) Proportion Dying (True) $_nq_x$
0–1	0.0549	0.0549
1–5	0.0115	0.0101
5–10	0.0055	0.0038
10–15	0.0050	0.0028
15–20	0.0085	0.0056
20–25	0.0119	0.0053
25–30	0.0139	0.0077
30–35	0.0169	0.0077
35–40	0.0218	0.0123
40–45	0.0301	0.0106
45–50	0.0427	0.0247
50–55	0.0624	0.0300
55–60	0.0896	0.0473
60–65	0.1270	0.0728
65–70	0.1812	0.1055
70–75	0.2704	0.1568
75–80	0.3946	0.2288
80–85	0.5941	0.3445
85+	1.0000	1.0000

(a) Without doing any calculations, predict which set of mortality rates will yield the longer life expectancy at birth.

(b) Complete the life tables corresponding to each set of rates. Assume that all deaths within an interval occur at its midpoint, except for the last interval, in which all deaths occur at age 91. Thus show that, even with the simplifications and approximations mentioned, the discrepancy between the two life expectancies at birth is more than five years.

(c) What might be the ramifications of this difference in life expectancy?

14. Table 5.2 shows the average life expectancies at birth from 1940 to 1992, broken down by gender and race [2]. The data for males and females of all races are contained in the data set lifeexp (Appendix B, Table B.9). The life expectancies for males are saved under the variable name male and those for females under the name female.

(a) Using these data, construct a line graph displaying the trends in $\overset{\circ}{e}_0$ over time for males versus females.

(b) How has average life expectancy at birth changed over the years? Has the gap between males and females been widening or narrowing?

15. To explore potential differences in life expectancy that may exist between racial groups, the average life expectancies at birth from 1970 to 1989 for four subsets of the U.S. population are contained in the data set liferace [18] (Appendix B, Table B.10). The life expectancies for white males are saved under the variable

name wmale, those for black males under the name bmale, those for white females under wfemale, and those for black females under bfemale.

(a) Construct a line graph displaying the variation in $\overset{\circ}{e}_0$ over time for each of the four different groups.

(b) Describe the differences in average life expectancy between males and females. Describe the differences between blacks and whites.

(c) From 1984 through 1989, the average life expectancy for the entire U.S. population increased by 0.5 year. What happened to the life expectancy for white males over this time period? for black males? for white females? for black females? What do you think may be causing this difference?

16. The average life expectancies at birth in the years 1960 and 1992 for a number of countries around the world are saved under the variable names life60 and life92 in the data set unicef [5] (Appendix B, Table B.2).

(a) Construct a box plot for the average life expectancies at birth in 1960. Describe the distribution of values.

(b) Construct a box plot for the life expectancies at birth in 1992. How did the distribution of values change?

(c) If the average life expectancy at birth in 1960 is represented by $\overset{\circ}{e}_{1960}$ and the average life expectancy in 1992 is denoted by $\overset{\circ}{e}_{1992}$, the relative percent improvement over the 32-year period is defined as $(\overset{\circ}{e}_{1992} - \overset{\circ}{e}_{1960})/\overset{\circ}{e}_{1960}$. Which country displayed the largest relative percent improvement in life expectancy? Which country showed the least improvement?

TABLE 5.8
Complete life table for the total population, United States, 1979–1981

Age Interval	Proportion Dying	Of 100,000 Born Alive		Stationary Population		Average Remaining Lifetime
Period of Life between Two Exact Ages (1)	Proportion of Persons Alive at Beginning of Age Interval Dying during Interval (2)	Number Living at Beginning of Age Interval (3)	Number Dying during Age Interval (4)	In the Age Interval (5)	In This and All Subsequent Age Intervals (6)	Average Number of Years of Life Remaining at Beginning of Age Interval (7)
x to $x+n$	$_nq_x$	l_x	$_nd_x$	$_nL_x$	T_x	$\overset{\circ}{e}_x$
Days						
0–1	.00463	100,000	· 463	273	7,387,758	73.88
1–7	.00246	99,537	245	1635	7,387,485	74.22
7–28	.00139	99,292	138	5708	7,385,850	74.38
28–365	.00418	99,154	414	91,357	7,380,142	74.43
Years						
0–1	.01260	100,000	1260	98,973	7,387,758	73.88
1–2	.00093	98,740	92	98,694	7,288,785	73.82
2–3	.00065	98,648	64	98,617	7,190,091	72.89
3–4	.00050	98,584	49	98,560	7,091,474	71.93
4–5	.00040	98,535	40	98,515	6,992,914	70.97

TABLE 5.8

(Continued)

Age Interval	Proportion Dying	Of 100,000 Born Alive		Stationary Population		Average Remaining Lifetime
Period of Life between Two Exact Ages (1)	Proportion of Persons Alive at Beginning of Age Interval Dying during Interval (2)	Number Living at Beginning of Age Interval (3)	Number Dying during Age Interval (4)	In the Age Interval (5)	In This and All Subsequent Age Intervals (6)	Average Number of Years of Life Remaining at Beginning of Age Interval (7)
x to $x+n$	$_nq_x$	l_x	$_nd_x$	$_nL_x$	T_x	$\overset{\circ}{e}_x$

Years (cont.)

5–6	.00037	98,495	36	98,477	6,894,399	70.00
6–7	.00033	98,459	33	98,442	6,795,922	69.02
7–8	.00030	98,426	30	98,412	6,697,480	68.05
8–9	.00027	98,396	26	98,383	6,599,068	67.07
9–10	.00023	98,370	23	98,358	6,500,685	66.08
10–11	.00020	98,347	19	98,338	6,402,327	65.10
11–12	.00019	98,328	19	98,319	6,303,989	64.11
12–13	.00025	98,309	24	98,297	6,205,670	63.12
13–14	.00037	98,285	37	98,266	6,107,373	62.14
14–15	.00053	98,248	52	98,222	6,009,107	61.16
15–16	.00069	98,196	67	98,163	5,910,885	60.19
16–17	.00083	98,129	82	98,087	5,812,722	59.24
17–18	.00095	98,047	94	98,000	5,714,635	58.28
18–19	.00105	97,953	102	97,902	5,616,635	57.34
19–20	.00112	97,851	110	97,796	5,518,733	56.40
20–21	.00120	97,741	118	97,682	5,420,937	55.46
21–22	.00127	97,623	124	97,561	5,323,255	54.53
22–23	.00132	97,499	129	97,435	5,225,694	53.60
23–24	.00134	97,370	130	97,306	5,128,259	52.67
24–25	.00133	97,240	130	97,175	5,030,953	51.74
25–26	.00132	97,110	128	97,046	4,933,778	50.81
26–27	.00131	96,982	126	96,919	4,836,732	49.87
27–28	.00130	96,856	126	96,793	4,739,813	48.94
28–29	.00130	96,730	126	96,667	4,643,020	48.00
29–30	.00131	96,604	127	96,541	4,546,353	47.06
30–31	.00133	96,477	127	96,414	4,449,812	46.12
31–32	.00134	96,350	130	96,284	4,353,398	45.18
32–33	.00137	96,220	132	96,155	4,257,114	44.24
33–34	.00142	96,088	137	96,019	4,160,959	43.30
34–35	.00150	95,951	143	95,880	4,064,940	42.36
35–36	.00159	95,808	153	95,731	3,969,060	41.43
36–37	.00170	95,655	163	95,574	3,873,329	40.49
37–38	.00183	95,492	175	95,404	3,777,755	39.56
38–39	.00197	95,317	188	95,224	3,682,351	38.63
39–40	.00213	95,129	203	95,027	3,587,127	37.71

(Continued)

TABLE 5.8
(Continued)

Age Interval	Proportion Dying	Of 100,000 Born Alive		Stationary Population		Average Remaining Lifetime
Period of Life between Two Exact Ages (1)	Proportion of Persons Alive at Beginning of Age Interval Dying during Interval (2)	Number Living at Beginning of Age Interval (3)	Number Dying during Age Interval (4)	In the Age Interval (5)	In This and All Subsequent Age Intervals (6)	Average Number of Years of Life Remaining at Beginning of Age Interval (7)
x to $x + n$	$_nq_x$	l_x	$_nd_x$	$_nL_x$	T_x	$\overset{\circ}{e}_x$
Years (cont.)						
40–41	.00232	94,926	220	94,817	3,492,100	36.79
41–42	.00254	94,706	241	94,585	3,397,283	35.87
42–43	.00279	94,465	264	94,334	3,302,698	34.96
43–44	.00306	94,201	288	94,057	3,208,364	34.06
44–45	.00335	93,913	314	93,756	3,114,307	33.16
45–46	.00366	93,599	343	93,427	3,020,551	32.27
46–47	.00401	93,256	374	93,069	2,927,124	31.39
47–48	.00442	92,882	410	92,677	2,834,055	30.51
48–49	.00488	92,472	451	92,246	2,741,378	29.65
49–50	.00538	92,021	495	91,773	2,649,132	26.79
50–51	.00589	91,526	540	91,256	2,557,359	27.94
51–52	.00642	90,986	584	90,695	2,466,103	27.10
52–53	.00699	90,402	631	90,086	2,375,408	26.28
53–54	.00761	89,771	684	89,430	2,285,322	25.46
54–55	.00830	89,087	739	88,717	2,195,892	24.65
55–56	.00902	88,348	797	87,950	2,107,175	23.85
56–57	.00978	87,551	856	87,122	2,019,225	23.06
57–58	.01059	86,695	919	86,236	1,932,103	22.29
58–59	.01151	85,776	987	85,283	1,845,867	21.52
59–60	.01254	84,789	1063	84,258	1,760,584	20.76
60–61	.01368	83,726	1145	83,153	1,676,326	20.02
61–62	.01493	82,581	1233	81,965	1,593,173	19.29
62–63	.01628	81,348	1324	80,686	1,511,208	18.58
63–64	.01767	80,024	1415	79,316	1,430,522	17.88
64–65	.01911	78,609	1502	77,859	1,351,206	17.19
65–66	.02059	77,107	1587	76,314	1,273,347	16.51
66–67	.02216	75,520	1674	74,683	1,197,033	15.85
67–68	.02389	73,846	1764	72,964	1,122,350	15.20
68–69	.02585	72,082	1864	71,150	1,049,386	14.56
69–70	.02806	70,218	1970	69,233	978,236	13.93
70–71	.03052	68,248	2083	67,206	909,003	13.32
71–72	.03315	66,165	2193	65,069	841,797	12.72
72–73	.03593	63,972	2299	62,823	776,728	12.14
73–74	.03882	61,673	2394	60,476	713,905	11.58
74–75	.04184	59,279	2480	58,039	653,429	11.02

TABLE 5.8

(Continued)

Age Interval	Proportion Dying	Of 100,000 Born Alive		Stationary Population		Average Remaining Lifetime
Period of Life between Two Exact Ages (1)	Proportion of Persons Alive at Beginning of Age Interval Dying during Interval (2)	Number Living at Beginning of Age Interval (3)	Number Dying during Age Interval (4)	In the Age Interval (5)	In This and All Subsequent Age Intervals (6)	Average Number of Years of Life Remaining at Beginning of Age Interval (7)
x to $x+n$	$_nq_x$	l_x	$_nd_x$	$_nL_x$	T_x	$\overset{\circ}{e}_x$
Years *(cont.)*						
75–76	.04507	56,799	2560	55,520	595,390	10.48
76–77	.04867	54,239	2640	52,919	539,870	9.95
77–78	.05274	51,599	2721	50,238	486,951	9.44
78–79	.05742	48,878	2807	47,475	436,713	8.93
79–80	.06277	46,071	2891	44,626	389,238	8.45
80–81	.06882	43,180	2972	41,694	344,612	7.98
81–82	.07552	40,208	3036	38,689	302,918	7.53
82–83	.08278	37,172	3077	35,634	264,229	7.11
83–84	.09041	34,095	3083	32,553	228,595	6.70
84–85	.09842	31,012	3052	29,486	196,042	6.32
85–86	.10725	27,960	2999	26,461	166,556	5.96
86–87	.11712	24,961	2923	23,500	140,095	5.61
87–88	.12717	22,038	2803	20,636	116,595	5.29
88–89	.13708	19,235	2637	17,917	95,959	4.99
89–90	.14728	16,598	2444	15,376	78,042	4.70
90–91	.15868	14,154	2246	13,031	62,666	4.43
91–92	.17169	11,908	2045	10,886	49,635	4.17
92–93	.18570	9863	1831	8948	38,749	3.93
93–94	.20023	8032	1608	7228	29,801	3.71
94–95	.21495	6424	1381	5733	22,573	3.51
95–96	.22976	5043	1159	4463	16,840	3.34
96–97	.24338	3884	945	3412	12,377	3.19
97–98	.25637	2939	754	2562	8965	3.05
98–99	.26868	2185	587	1892	6403	2.93
99–100	.28030	1598	448	1374	4511	2.82
100–101	.29120	1150	335	983	3137	2.73
101–102	.30139	815	245	692	2154	2.64
102–103	.31089	570	177	481	1462	2.57
103–104	.31970	393	126	330	981	2.50
104–105	.32786	267	88	223	651	2.44
105–106	.33539	179	60	150	428	2.38
106–107	.34233	119	41	99	278	2.33
107–108	.34870	78	27	64	179	2.29
108–109	.35453	51	18	42	115	2.24
109–110	.35988	33	12	27	73	2.20

Bibliography

[1] Fries, J. F., "Aging, Natural Death, and the Compression of Morbidity," *The New England Journal of Medicine*, Volume 303, July 17, 1980, 130–135.

[2] National Center for Health Statistics, Kochanek, K. D., and Hudson, B. L., "Advance Report of Final Mortality Statistics, 1992," *Monthly Vital Statistics Report*, Volume 43, Number 6, March 22, 1995.

[3] National Center for Health Statistics, *United States Decennial Life Tables for 1979–1980*, Volume I, Number 1, August 1985.

[4] Vandenbroucke, J. P., "Survival and Expectation of Life from the 1400s to the Present: A Study of the Knighthood Order of the Golden Fleece," *American Journal of Epidemiology*, Volume 122, December 1985, 1007–1015.

[5] United Nations Children's Fund, *The State of the World's Children 1994*, New York: Oxford University Press.

[6] Taylor, I., and Knowelden, J., *Principles of Epidemiology*, London: Churchill, 1957.

[7] Dickinson, F. G., and Welker, E. L., *What Is the Leading Cause of Death? Two New Measures*, Bulletin 64, Chicago: American Medical Association, 1948.

[8] Dempsey, M., "Decline in Tuberculosis: The Death Rate Fails to Tell the Entire Story," *American Review of Tuberculosis*, Volume 86, August 1947, 157–164.

[9] Centers for Disease Control, "Premature Mortality in the United States: Public Health Issues in the Use of Years of Potential Life Lost," *Morbidity and Mortality Weekly Report*, Volume 35, Number 2S, December 19, 1986.

[10] Centers for Disease Control, "YPLL Before Age 65: United States, 1987," *Morbidity and Mortality Weekly Report*, Volume 38, Number 2, January 20, 1989.

[11] Haenszel, W., "A Standardized Rate for Mortality Defined in Units of Lost Years of Life," *American Journal of Public Health*, Volume 40, January 1950, 17–26.

[12] National Center for Health Statistics, "Advance Report of Final Mortality Statistics, 1990," *Monthly Vital Statistics Report*, Volume 41, Number 7, January 7, 1993.

[13] Farr, W., *Vital Statistics: A Memorial Volume of Selections from the Reports and Writings of William Farr*, London: The Sanitary Institute of Great Britain, 1883.

[14] Wilkins, R., and Adams, O. B., "Health Expectancy in Canada, Late 1970s: Demographic, Regional, and Social Dimensions," *American Journal of Public Health*, Volume 73, September 1983, 1073–1080.

[15] United Nations, "Expectation of Life at Specified Ages for Each Sex," *Demographic Yearbook 1982*, New York, 1984.

[16] Medawar, P., *The Strange Case of the Spotted Mice, and Other Classic Essays on Science*, Oxford: Oxford University Press, 1996.

[17] Horm, J. W., and Sondik, E. J., "Person-Years of Life Lost Due to Cancer in the United States, 1970 and 1984," *American Journal of Public Health*, Volume 79, November 1989, 1490–1493.

[18] Kochanek, K. D., Maurer, J. D., and Rosenberg, H. M., "Why Did Black Life Expectancy Decline from 1984 through 1989 in the United States?," *American Journal of Public Health*, Volume 84, June 1994, 938–944.

6

Probability

In the previous chapters, we studied ways in which descriptive statistics can be used to organize and summarize a set of data. However, in addition to describing a group of observations, we might also be interested in investigating how the information contained in the sample can be used to infer the characteristics of the population from which it was drawn. Before we can do this, we must first lay the groundwork. The foundation for statistical inference is the theory of probability. In Chapter 5, we used the term *probability* as a synonym for *proportion*. Before we can give a more precise definition, we must first explain the concept of an event.

6.1 Operations on Events and Probability

An *event* is the basic element to which probability can be applied. It is the result of an observation or experiment, or the description of some potential outcome. For example, we might consider the event that a 30-year-old woman lives to see her 70th birthday, or the event that the same woman is diagnosed with cervical cancer before she reaches the age of 40. Another event might be that a particular nuclear power plant experiences a meltdown within the next ten years. An event either occurs or does not occur. In the study of probability, events are represented by uppercase letters such as A, B, and C.

A number of different operations can be performed on events. The *intersection* of two events A and B, denoted $A \cap B$, is defined as the event "both A and B." For example, let A represent the event that a 30-year-old woman lives to see her 70th birthday, and B the event that her 30-year-old husband is still alive at age 70. The intersection of A and B would be the event that both the 30-year-old woman and her husband are alive at age 70.

The *union* of A and B, denoted $A \cup B$, is the event "either A or B, or both A and B." In the example mentioned above, the union of A and B would be the event that either the 30-year-old woman or her 30-year-old husband lives to age 70, or that they both live to be 70 years of age.

The *complement* of an event A, denoted A^C or $\overline{A}$, is the event "not A." Consequently, A^C is the event that the 30-year-old woman dies before she reaches the age of 70.

These three operations—the intersection, the union, and the complement—can be used to describe even the most complicated situations in terms of simple events. To help make this idea more concrete, a picture called a *Venn diagram* is a useful device for depicting the relationships among events. In Figure 6.1, for example, the area within each box represents all the outcomes that could possibly occur. Inside the boxes, the circles labeled A represent the subset of outcomes for which a 30-year-old woman lives to be 70 years of age, and those labeled B denote the outcomes for which her 30-year-old husband survives to age 70. The intersection of A and B is represented by the area in which the two circles overlap; this area is shaded in Figure 6.1(a). The union of A and B is shaded in Figure 6.1(b) and is the area that is either A or B or both. The complement of A, as shown in Figure 6.1(c), is everything inside the box that lies outside of A.

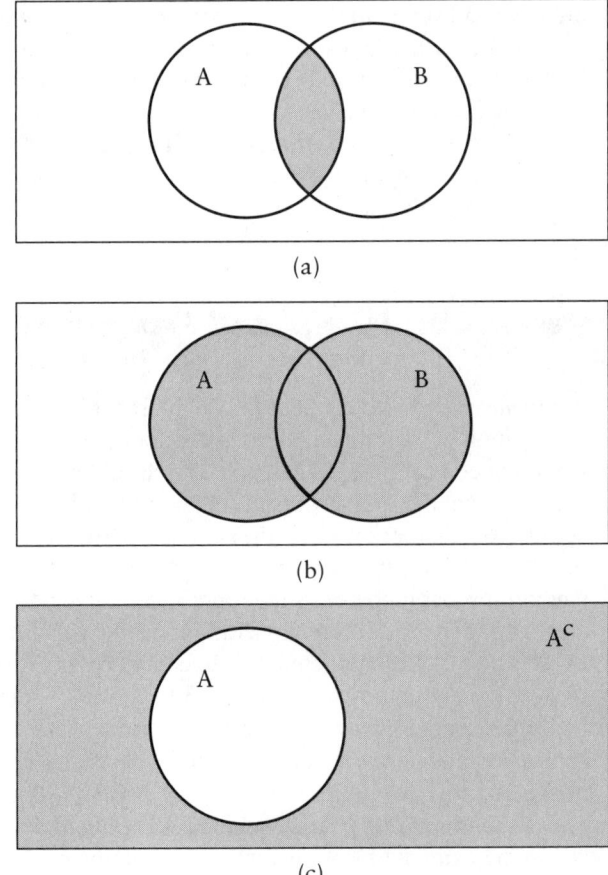

(a)

(b)

(c)

FIGURE 6.1
Venn diagrams representing the operations on events

We are now ready to discuss the concept of probability. As a mathematical system, probability theory is well defined. Since we wish to apply this theory, however, we need a practical working definition. Many definitions of probability have been proposed; the one presented here is called the *frequentist definition*. It states:

> If an experiment is repeated n times under essentially identical conditions, and if the event A occurs m times, then as n grows large, the ratio m/n approaches a fixed limit that is the probability of A;
>
> $$P(A) = \frac{m}{n}.$$

In other words, the *probability* of an event A is its relative frequency of occurrence—or the proportion of times the event occurs—in a large number of trials repeated under virtually identical conditions. The practical nature of this definition causes it to be somewhat vague, although it does work quite well.

As an application of the frequentist definition, we can determine the probability that a newborn infant will live to see his or her first birthday. Consult Table 5.1, the 1992 life table for the United States population [1]. Among the 100,000 individuals born into this cohort—we consider these infants to be the "experiments"—the event of surviving the first year of life occurs 99,149 times. Therefore,

$$P(\text{a child survives his or her first year}) = \frac{99,149}{100,000}$$
$$= 0.99149.$$

We assume that 100,000 repetitions is a large enough number to satisfy the frequentist definition of probability.

The numerical value of a probability lies between 0 and 1. If a particular event happens with certainty, it occurs in each of the n trials and has probability $n/n = 1$. Let A again represent the event that a 30-year-old woman lives to be 70. In this case,

$$P(A \cup A^C) = P(\text{either } A \text{ or } A^C \text{ or both})$$
$$= P(\text{a 30-year-old woman lives to age 70 or}$$
$$\text{she does not live to age 70})$$
$$= 1,$$

since it is certain that the woman will either live or die. In Figure 6.1(c), A and A^C together fill up the entire box. Furthermore, note that it is impossible for A and A^C to occur simultaneously. If an event can never happen, it has probability $0/n = 0$; hence

$$P(A \cap A^C) = P(A \text{ and } A^C)$$
$$= P(\text{a 30-year-old woman lives to age 70 and}$$
$$\text{she does not live to age 70})$$
$$= 0.$$

An event that can never occur is called the *null event* and is represented by the symbol ϕ. Therefore, $A \cap A^C = \phi$. Most events have probabilities somewhere between 0 and 1.

Using the frequentist definition of the probability of an event A, we can calculate the probability of the complementary event A^C in a straightforward manner. If an experiment is repeated n times under essentially identical conditions and the event A occurs m times, the event A^C, or not A, must occur $n - m$ times. Therefore, for large n,

$$P(A^C) = \frac{n - m}{n}$$

$$= 1 - \frac{m}{n}$$

$$= 1 - P(A).$$

The probability that a newborn does not survive the first year of life is 1 minus the probability that he or she does, or

$$1 - 0.99149 = 0.00851.$$

Two events A and B that cannot occur simultaneously are said to be *mutually exclusive* or *disjoint*. If A is the event that a newborn's birth weight is under 2000 grams and B the event that it is between 2000 and 2499 grams, for example, the events A and B are mutually exclusive. A child cannot be in both weight groups at the same time. By definition, $A \cap B = \phi$ and $P(A \cap B) = 0$. In Figure 6.2, the nonoverlapping circles represent mutually exclusive events.

When two events are mutually exclusive, the *additive rule of probability* states that the probability that either of the two events will occur is equal to the sum of the probabilities of the individual events; more explicitly,

$$P(A \cup B) = P(A) + P(B).$$

Suppose we know that the probability that a newborn's birth weight is under 2000 grams is 0.025 and the probability that it is between 2000 and 2499 grams is 0.043. The probability that either of these two events will occur, or, equivalently, the probability that the child weighs less than 2500 grams, is

$$P(A \cup B) = 0.025 + 0.043$$

$$= 0.068.$$

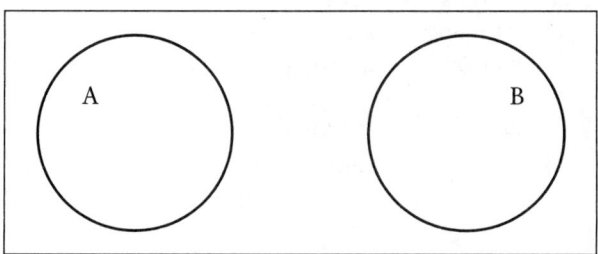

FIGURE 6.2
Venn diagram representing two mutually exclusive events

The additive rule can be extended to the case of three or more mutually exclusive events. If $A_1, A_2, \ldots$, and A_n are n events such that $A_1 \cap A_2 = \phi, A_1 \cap A_3 = \phi, A_2 \cap A_3 = \phi$, and so on for all possible pairs, then

$$P(A_1 \cup A_2 \cup \cdots \cup A_n) = P(A_1) + P(A_2) + \cdots + P(A_n).$$

If the events A and B are not mutually exclusive, as in Figure 6.1(b), the additive rule no longer applies. Let A be the event that a newborn's birth weight is under 2000 grams and B the event that it is under 2500 grams. Since the two events can occur simultaneously—consider a child whose birth weight is 1850 grams—there is some area in which they will overlap. If we were to simply sum the probabilities of the individual events, this area of overlap would be counted twice. When two events are not mutually exclusive, therefore, the probability that either of the events will occur is equal to the sum of the individual probabilities minus the probability of their intersection:

$$P(A \cup B) = P(A) + P(B) - P(A \cap B).$$

6.2   Conditional Probability

We are often interested in determining the probability that an event B will occur given that we already know the outcome of another event A. Does the prior occurrence of A cause the probability of B to change? For instance, instead of finding the probability that a person will live to the age of 65, we might wish to know the probability that the individual will survive for the next five years given that he or she has already reached the age of 60. In this case, we are dealing with a *conditional probability*. The notation $P(B \mid A)$ is used to represent the probability of the event B given that event A has already occurred.

The *multiplicative rule of probability* states that the probability that two events A and B will both occur is equal to the probability of A multiplied by the probability of B given that A has already occurred. This can be expressed as

$$P(A \cap B) = P(A) P(B \mid A).$$

Since it is arbitrary which event we call A and which we call B, we can also write

$$P(A \cap B) = P(B) P(A \mid B).$$

Dividing both sides of the first equation by $P(A)$, we find the formula for a conditional probability to be

$$P(B \mid A) = \frac{P(A \cap B)}{P(A)},$$

given that $P(A) \neq 0$. Similarly, we have

$$P(A \mid B) = \frac{P(A \cap B)}{P(B)},$$

given that $P(B) \neq 0$.

If A is the event that an individual is alive at 60 years of age, and B is the event that he or she survives to 65, then $A \cap B$ is the event that the person is alive at age 60 and also at 65. If someone is alive at 65, he or she must have been alive at 60 as well; therefore, $A \cap B$ is simply the event that the individual survives to see his or her 65th birthday. According to the 1992 life table for the U.S. population,

$$P(A) = P(\text{an individual lives to be 60 years of age})$$
$$= \frac{85,993}{100,000}$$
$$= 0.85993.$$

In other words, the event A occurs 85,993 times out of 100,000 trials. Similarly,

$$P(A \cap B) = P(\text{an individual lives to be 65 years of age})$$
$$= \frac{80,145}{100,000}$$
$$= 0.80145.$$

Therefore,

$$P(B \mid A) = P(\text{an individual lives to be 65 years of age} \mid \text{he or she is alive at 60})$$
$$= \frac{P(A \cap B)}{P(A)}$$
$$= \frac{0.80145}{0.85993}$$
$$= 0.9320.$$

An equivalent way to calculate this probability would be to start with the 85,993 persons alive at age 60 and note that the event of surviving to age 65 occurs 80,145 times in these 85,993 trials. Hence,

$$P(B \mid A) = \frac{80,145}{85,993}$$
$$= 0.9320.$$

If a person lives to be 60, his or her chance of surviving to age 65 is greater than it was at birth.

When we are concerned with two events such that the outcome of one event has no effect on the occurrence or nonoccurrence of the other, the events are said to be *independent*. If A and B are independent events,

$$P(A \mid B) = P(A)$$

and

$$P(B \mid A) = P(B).$$

In this special case, the multiplicative rule of probability may be written

$$P(A \cap B) = P(A) P(B).$$

It is important to note that the terms *independent* and *mutually exclusive* do not mean the same thing. If A and B are independent and event A occurs, the outcome of B is not affected. Event B might occur or it might not occur, and $P(B \mid A) = P(B)$. If A and B are mutually exclusive, however, and event A occurs, event B cannot occur. By definition, $P(B \mid A) = 0$.

6.3   Bayes' Theorem

Chapter 4 included a presentation of data collected in the National Health Interview Survey of 1980–1981 [2]. The data pertained to hearing impairments due to injury reported by individuals 17 years of age and older. The 163,157 persons included in the study were subdivided into three mutually exclusive categories: the currently employed, the currently unemployed, and those not in the labor force.

Employment Status	Population	Impairments
Currently employed	98,917	552
Currently unemployed	7462	27
Not in the labor force	56,778	368
Total	163,157	947

Let E_1 be the event that an individual included in the survey is currently employed, E_2 the event that he or she is currently unemployed, and E_3 the event that the individual is not in the labor force. If we assume that these numbers are large enough to satisfy the frequentist definition of probability, then, from the data provided, we find that

$$P(E_1) = \frac{98,917}{163,157}$$
$$= 0.6063,$$

$$P(E_2) = \frac{7462}{163,157}$$
$$= 0.0457,$$

and

$$P(E_3) = \frac{56,778}{163,157}$$
$$= 0.3480.$$

If S is the event that an individual in the study is currently employed or currently unemployed or not in the labor force,

$$S = E_1 \cup E_2 \cup E_3.$$

Since the three categories are mutually exclusive, the additive rule of probability may be applied:

$$
\begin{aligned}
P(S) &= P(E_1 \cup E_2 \cup E_3) \\
&= P(E_1) + P(E_2) + P(E_3) \\
&= 0.6063 + 0.0457 + 0.3480 \\
&= 1.0000.
\end{aligned}
$$

When the probabilities of mutually exclusive events sum to 1, the events are said to be *exhaustive*; in this case, there are no other possible outcomes. Therefore, every person included in the survey must fall into one of the three groups.

Now let H be the event that an individual has a hearing impairment due to injury. Overall,

$$
\begin{aligned}
P(H) &= \frac{947}{163,157} \\
&= 0.0058.
\end{aligned}
$$

Looking at each employment status subgroup separately,

$$
\begin{aligned}
P(H \mid E_1) &= P(\text{an individual has a hearing impairment} \mid \text{he or she is currently employed}) \\
&= \frac{552}{98,917} \\
&= 0.0056,
\end{aligned}
$$

$$
\begin{aligned}
P(H \mid E_2) &= P(\text{an individual has a hearing impairment} \mid \text{he or she is currently unemployed}) \\
&= \frac{27}{7462} \\
&= 0.0036,
\end{aligned}
$$

and

$$
\begin{aligned}
P(H \mid E_3) &= P(\text{an individual has a hearing impairment} \mid \text{he or she is not in the labor force}) \\
&= \frac{368}{56,778} \\
&= 0.0065.
\end{aligned}
$$

The probability of having a hearing impairment is smallest among the currently unemployed and largest among those not in the labor force.

Note that H, the event that an individual has a hearing impairment due to injury, may be expressed as the union of three mutually exclusive events: $E_1 \cap H$, the event that an individual is currently employed and has a hearing impairment; $E_2 \cap H$, the event that he or she is currently unemployed and has a hearing impairment; and $E_3 \cap H$, the event that the individual is not in the labor force and has a hearing impairment. Thus,

$$H = (E_1 \cap H) \cup (E_2 \cap H) \cup (E_3 \cap H).$$

Everyone with a hearing impairment can be placed in one and only one of these three categories. Since the categories are mutually exclusive, we can apply the additive rule; therefore,

$$P(H) = P[(E_1 \cap H) \cup (E_2 \cap H) \cup (E_3 \cap H)]$$
$$= P(E_1 \cap H) + P(E_2 \cap H) + P(E_3 \cap H).$$

This is sometimes called the *law of total probability*.

Now, applying the multiplicative rule to each term on the right-hand side of the equation separately and plugging in the previously calculated probabilities,

$$P(H) = P(E_1 \cap H) + P(E_2 \cap H) + P(E_3 \cap H)$$
$$= P(E_1) P(H \mid E_1) + P(E_2) P(H \mid E_2) + P(E_3) P(H \mid E_3)$$
$$= 0.0034 + 0.0002 + 0.0023$$
$$= 0.0059.$$

These calculations are summarized in the table below, where i, the subscript of the event E, takes values from 1 to 3.

Event E_i	$P(E_i)$	$P(H \mid E_i)$	$P(E_i)P(H \mid E_i)$
E_1	0.6063	0.0056	0.0034
E_2	0.0457	0.0036	0.0002
E_3	0.3480	0.0065	0.0023
$P(H)$			0.0059

If we ignore the rounding error in these computations, the value 0.0059 is the number we originally generated as the probability that an individual has a hearing impairment due to injury,

$$P(H) = \frac{947}{163,157}$$
$$= 0.0058.$$

The more complicated method of calculation, using the expression

$$P(H) = P(E_1)\,P(H \mid E_1) + P(E_2)\,P(H \mid E_2) + P(E_3)\,P(H \mid E_3),$$

may be useful when we are unable to calculate $P(H)$ directly.

Suppose that we now change our perspective and attempt to find $P(E_1 \mid H)$, the probability that an individual is currently employed given that he or she has a hearing impairment. The multiplicative rule of probability states that

$$P(E_1 \cap H) = P(H)\,P(E_1 \mid H);$$

hence,

$$P(E_1 \mid H) = \frac{P(E_1 \cap H)}{P(H)}.$$

Applying the multiplicative rule to the numerator of the right-hand side of the equation, we have

$$P(E_1 \mid H) = \frac{P(E_1)\,P(H \mid E_1)}{P(H)}.$$

Using the identity that was derived above,

$$P(H) = P(E_1)\,P(H \mid E_1) + P(E_2)\,P(H \mid E_2) + P(E_3)\,P(H \mid E_3),$$

results in

$$P(E_1 \mid H) = \frac{P(E_1)\,P(H \mid E_1)}{P(E_1)\,P(H \mid E_1) + P(E_2)\,P(H \mid E_2) + P(E_3)\,P(H \mid E_3)}.$$

This rather daunting expression is known as *Bayes' theorem*. Substituting in the numerical values of all probabilities,

$$P(E_1 \mid H) = \frac{(0.6063)(0.0056)}{(0.6063)(0.0056) + (0.0457)(0.0036) + (0.3480)(0.0065)}$$

$$= 0.583.$$

The probability that an individual is currently employed given that he or she has a hearing impairment due to injury is approximately 0.583. In this particular example, the result can be checked directly by looking at the original data. Among the 947 persons with hearing impairments, 552 are currently employed. Therefore,

$$P(E_1 \mid H) = \frac{552}{947}$$

$$= 0.583.$$

Bayes' theorem is not restricted to situations in which individuals fall into one of three distinct subgroups. If $A_1, A_2, \ldots,$ and A_n are n mutually exclusive and exhaustive events such that

$$P(A_1 \cup A_2 \cup \cdots \cup A_n) = P(A_1) + P(A_2) + \cdots + P(A_n)$$
$$= 1,$$

Bayes' theorem states that

$$P(A_i \mid B) = \frac{P(A_i)\,P(B \mid A_i)}{P(A_1)\,P(B \mid A_1) + \cdots + P(A_n)\,P(B \mid A_n)}$$

for each i, $1 \le i \le n$.

Bayes' theorem is valuable because it allows us to recalculate a probability based on some new information. In the National Health Interview Survey example, we know that

$$P(E_1) = P(\text{an individual is currently employed})$$
$$= 0.6063.$$

If we are then given an additional piece of information—the knowledge that a particular individual has a hearing impairment due to injury, for instance—does our assessment of the probability that he or she is currently employed change? We observed that it does. Using Bayes' theorem, we found that

$$P(E_1 \mid H) = P(\text{an individual is currently employed} \mid \text{he or she has a hearing}$$
$$\text{impairment})$$
$$= 0.5832.$$

Once we are told that someone has a hearing impairment, the probability that he or she is currently employed decreases slightly.

6.4 *Diagnostic Tests*

Bayes' theorem is often employed in issues of diagnostic testing or screening. *Screening* is the application of a test to individuals who have not yet exhibited any clinical symptoms in order to classify them with respect to their probability of having a particular disease. Those who test positive are considered to be more likely to have the disease and are usually subjected either to further diagnostic procedures or to treatment. Screening is most often employed by health care professionals in situations where the early detection of disease would contribute to a more favorable prognosis for the individual or for the population in general. Bayes' theorem allows us to use probability to evaluate the associated uncertainties.

6.4.1   Sensitivity and Specificity

Suppose that we are interested in two mutually exclusive and exhaustive states of health: D_1 is the event that an individual has a particular disease, and D_2 the event that he or she does not have the disease. We could use the more succinct notation defined earlier—namely, D and D^C—but we wish to emphasize that the situation may be generalized to include three or more events. Let T^+ represent a positive screening test result. We would like to find $P(D_1 \mid T^+)$, the probability that a person with a positive test result actually does have the disease.

Cervical cancer is a disease for which the chance of containment is high given that it is detected early. The Pap smear is a widely accepted screening procedure that can detect a cancer that is as yet asymptomatic; it has been credited with being primarily responsible for the decreasing death rate due to cervical cancer in recent years. An on-site proficiency test conducted in 1972, 1973, and 1978 assessed the competency of technicians who scan Pap smear slides for abnormalities [3]. Technicians in 306 cytology labs in 44 states were tested.

Overall, 16.25% of the tests performed on women with cancer resulted in false negative outcomes. A *false negative* occurs when the test of a woman who has cancer of the cervix incorrectly indicates that she does not. Therefore, in this study,

$$P(\text{test negative} \mid \text{cancer}) = 0.1625.$$

The other $100 - 16.25 = 83.75\%$ of the women who had cervical cancer did in fact test positive; as a result,

$$P(\text{test positive} \mid \text{cancer}) = 0.8375.$$

The probability of a positive test result given that the individual tested actually has the disease is called the *sensitivity* of a test. In this study, the sensitivity of the Pap smear was 0.8375.

Not all the women tested actually suffered from cervical cancer. In fact, 18.64% of the tests were *false positive* outcomes; this implies that

$$P(\text{test positive} \mid \text{no cancer}) = 0.1864.$$

The *specificity* of a test is the probability that its result is negative given that the individual tested does not have the disease. In this study, the specificity of the Pap smear was

$$P(\text{test negative} \mid \text{no cancer}) = 1 - 0.1864$$
$$= 0.8136.$$

6.4.2   Applications of Bayes' Theorem

Now that we have examined the accuracy of the Pap smear among women who have cervical cancer and women who do not, we can investigate the question that is of primary concern to the individuals being tested and to the health care professionals involved in the screening: What is the probability that a woman with a Pap smear that is positive for cancer actually does have the disease? Let D_1 represent the event that a woman has cervical cancer and D_2 the event that she does not. Also, let T^+ represent a positive Pap smear. We wish to compute $P(D_1 \mid T^+)$. Using Bayes' theorem, we can write

$$P(D_1 \mid T^+) = \frac{P(D_1 \cap T^+)}{P(T^+)}$$

$$= \frac{P(D_1)\,P(T^+ \mid D_1)}{P(D_1)\,P(T^+ \mid D_1) + P(D_2)\,P(T^+ \mid D_2)}.$$

We already know that $P(T^+ \mid D_1) = 0.8375$ and $P(T^+ \mid D_2) = 0.1864$. We must now find $P(D_1)$ and $P(D_2)$.

$P(D_1)$ is the probability that a woman suffers from cervical cancer. It can also be interpreted as the proportion of women who have cancer of the cervix at a given point in time, or the *prevalence* of the disease. One source reports that the rate of cases of cervical cancer among women studied in 1983–1984 was 8.3 per 100,000 [4]. Using these data,

$$P(D_1) = 0.000083.$$

$P(D_2)$ is the probability that a woman does not have cervical cancer. Since D_2 is the complement of D_1,

$$P(D_2) = 1 - P(D_1)$$
$$= 1 - 0.000083$$
$$= 0.999917.$$

Substituting these probabilities into Bayes' theorem,

$$P(D_1 \mid T^+) = \frac{0.000083 \times 0.8375}{(0.000083 \times 0.8375) + (0.999917 \times 0.1864)}$$
$$= 0.000373.$$

$P(D_1 \mid T^+)$, the probability of disease given a positive test result, is called the *predictive value* of a positive test. Here, it tells us that for every 1,000,000 positive Pap smears, only 373 represent true cases of cervical cancer.

Bayes' theorem can also be used to calculate the predictive value of a negative test. If T^- represents a negative test result, the negative predictive value, or the probability of no disease given a negative test result, is equal to

$$P(D_2 \mid T^-) = \frac{P(D_2)\,P(T^- \mid D_2)}{P(D_2)\,P(T^- \mid D_2) + P(D_1)\,P(T^- \mid D_1)}$$

$$= \frac{0.999917 \times 0.8136}{(0.999917 \times 0.8136) + (0.000083 \times 0.1625)}$$

$$= 0.999983.$$

Therefore, for every 1,000,000 women with negative Pap smears, 999,983 do not have cervical cancer. Figure 6.3 illustrates the results of the entire diagnostic testing process. Note that all numbers have been rounded to the nearest integer.

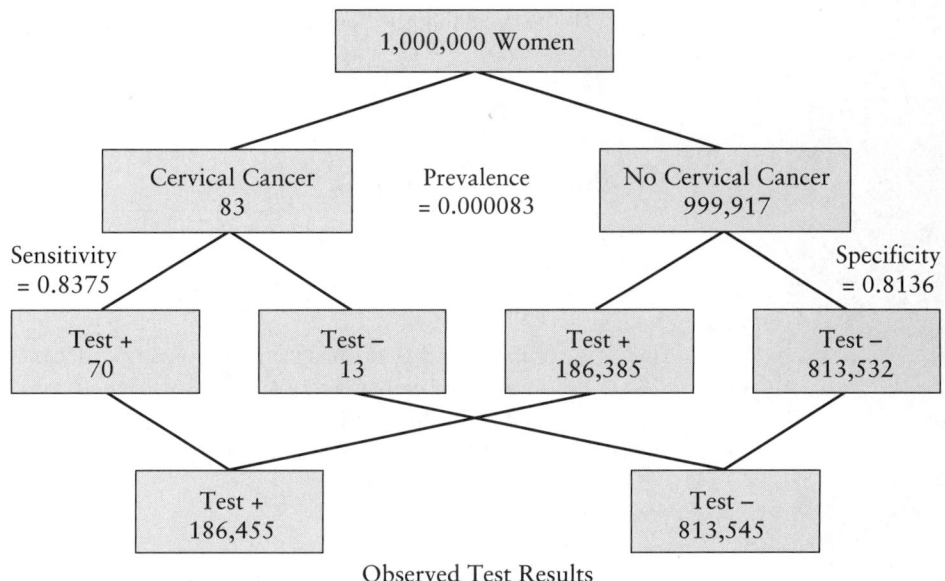

Observed Test Results

FIGURE 6.3
Performance of the Pap smear as a diagnostic test for cervical cancer

Although the Pap smear is widely accepted as a screening test for cervical cancer, its previously assumed high rate of accuracy is now being questioned. A number of studies estimate the proportion of false negative results to be in the range of 20% to 40%, or even as large as 89% [5,6]. The proportion of false positive results has been found to be as high as 86%. Some of the lab errors are due to poor cell sampling techniques or to inadequate preparation of the specimens; others result from the fatigue suffered by lab technicians who must examine a large number of slides each day.

As a second example of the application of Bayes' theorem in diagnostic testing, consider the following data. Among the 1820 subjects in a study, 30 suffered from tuberculosis and 1790 did not [7]. Chest x-rays were administered to all individuals; 73 had a positive x-ray—implying that there was significant evidence of inflammatory disease—whereas the results of the other 1747 were negative. The data for this study are presented in the table below. What is the probability that a randomly selected individual has tuberculosis given that his or her x-ray is positive?

X-ray	Tuberculosis		Total
	No	**Yes**	**Total**
Negative	1739	8	1747
Positive	51	22	73
Total	1790	30	1820

Let D_1 represent the event that an individual is suffering from tuberculosis and D_2 the event that he or she is not. These two events are mutually exclusive and exhaustive. In addition, let T^+ represent a positive x-ray. We wish to find $P(D_1 \mid T^+)$, the probability that an individual who tests positive for tuberculosis actually has the disease. This is the positive predictive value of the x-ray. Using Bayes' theorem, we can write

$$P(D_1 \mid T^+) = \frac{P(D_1)\, P(T^+ \mid D_1)}{P(D_1)\, P(T^+ \mid D_1) + P(D_2)\, P(T^+ \mid D_2)}.$$

Therefore, to solve for $P(D_1 \mid T^+)$, we must first know $P(D_1)$, $P(D_2)$, $P(T^+ \mid D_1)$, and $P(T^+ \mid D_2)$.

$P(D_1)$ is the probability that an individual in the general population has tuberculosis. Since the 1820 individuals in the study described above were not chosen from the population at random, the prevalence of disease cannot be obtained from the information in the table. In 1987, however, there were 9.3 cases of tuberculosis per 100,000 population [8]. With the spread of the human immunodeficiency virus (HIV), this number has increased markedly, but for this exercise we can estimate

$$P(D_1) = 0.000093.$$

$P(D_2)$ is the probability that an individual does not have tuberculosis. Since D_2 is the complement of D_1,

$$
\begin{aligned}
P(D_2) &= 1 - P(D_1) \\
&= 1 - 0.000093 \\
&= 0.999907.
\end{aligned}
$$

$P(T^+ \mid D_1)$ is the probability of a positive x-ray given that an individual has tuberculosis—the sensitivity of the test. In this study, the sensitivity of the x-ray is

$$
\begin{aligned}
P(T^+ \mid D_1) &= \frac{22}{30} \\
&= 0.7333.
\end{aligned}
$$

$P(T^+ \mid D_2)$, the probability of a positive x-ray given that a person does not have tuberculosis, is the complement of the specificity. Therefore,

$$
\begin{aligned}
P(T^+ \mid D_2) &= 1 - P(T^- \mid D_2) \\
&= 1 - \frac{1739}{1790} \\
&= 1 - 0.9715 \\
&= 0.0285.
\end{aligned}
$$

Using all this information, we can now calculate the probability that an individual suffers from tuberculosis given that he or she has a positive x-ray; this probability is

$$P(D_1 \mid T^+) = \frac{P(D_1)\,P(T^+ \mid D_1)}{P(D_1)\,P(T^+ \mid D_1) + P(D_2)\,P(T^+ \mid D_2)}$$

$$= \frac{(0.000093)(0.7333)}{(0.000093)(0.7333) + (0.999907)(0.0285)}.$$

$$= 0.00239.$$

For every 100,000 positive x-rays, only 239 signal true cases of tuberculosis.

Note that before an x-ray is taken, an individual who is randomly selected from the U.S. population has a

$$9.3/100,000 = 0.000093 = 0.0093\%$$

chance of having tuberculosis. This is called the *prior probability*. After an x-ray is taken and the result is positive, the same individual has a

$$239/100,000 = 0.00239 = 0.239\%$$

chance of being afflicted with tuberculosis. This is the *posterior probability*. The posterior probability takes into account a new piece of information—the positive test result. Although 99,761/100,000 persons with a positive x-ray do not actually have the disease, we have greatly increased the chance of properly diagnosing tuberculosis. Since $0.00239/0.000093 = 25.7$, the probability that an individual with a positive x-ray has tuberculosis is 25.7 times greater than the probability for a person randomly selected from the population.

6.4.3 ROC Curves

Diagnosis is an imperfect process. In theory, it is desirable to have a test that is both highly sensitive and highly specific. In reality, however, such a procedure is usually not possible. Many tests are actually based on a clinical measurement that can assume a range of values; in this case, there is an inherent trade-off between sensitivity and specificity.

Consider Table 6.1. This table displays data from a kidney transplant program in which renal allografts were performed [9]. The level of serum creatinine, a chemical compound found in the blood and measured in milligrams percent, was used as a diagnostic tool for detecting potential transplant rejection. An increased creatinine level is often associated with subsequent organ failure.

If we use a level greater than 2.9 mg % as an indicator of imminent rejection, the test has a sensitivity of 0.303 and a specificity of 0.909. To increase the sensitivity, we could lower the arbitrary cutoff point that distinguishes a positive test result from a negative one; if we use 1.2 mg %, for example, a much greater proportion of the results would be designated positive. In this case, we would rarely fail to identify a patient who will reject the organ. At the same time, we would increase the probability of a false pos-

TABLE 6.1

Sensitivity and specificity of serum creatinine level for predicting transplant rejection

Serum Creatinine (mg %)	Sensitivity	Specificity
1.2	0.939	0.123
1.3	0.939	0.203
1.4	0.909	0.281
1.5	0.818	0.380
1.6	0.758	0.461
1.7	0.727	0.535
1.8	0.636	0.649
1.9	0.636	0.711
2.0	0.545	0.766
2.1	0.485	0.773
2.2	0.485	0.803
2.3	0.394	0.811
2.4	0.394	0.843
2.5	0.364	0.870
2.6	0.333	0.891
2.7	0.333	0.894
2.8	0.333	0.896
2.9	0.303	0.909

itive result, thereby decreasing the specificity. By increasing the specificity we would hardly ever misclassify a person who is not going to reject the organ, and, in turn, we would decrease the sensitivity. In general, a sensitive test is most useful when the failure to detect a disease as early as possible has dangerous consequences; a specific test is important in situations where a false positive result is harmful.

The relationship between sensitivity and specificity may be illustrated using a graph known as a *receiver operator characteristic (ROC) curve*. An ROC curve is a line graph that plots the probability of a true positive result—or the sensitivity of the test— against the probability of a false positive result for a range of different cutoff points. These graphs were first used in the field of communications. As an example, Figure 6.4 displays an ROC curve for the data shown in Table 6.1. When an existing diagnostic test is being evaluated, this type of graph can be used to help assess the usefulness of the test and to determine the most appropriate cutoff point. The dashed line in Figure 6.4 corresponds to a test that gives positive and negative results by chance alone; such a test has no inherent value. The closer the line to the upper left-hand corner of the graph, the more accurate the test. Furthermore, the point that lies closest to this corner is usually chosen as the cutoff that maximizes both sensitivity and specificity simultaneously.

6.4.4 Calculation of Prevalence

In addition to being used in applications involving Bayes' theorem, diagnostic testing or screening can also be used to calculate the prevalence of disease in a specified population.

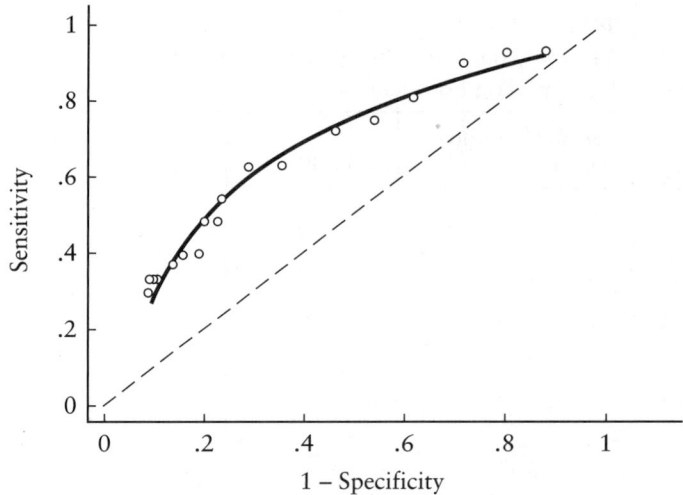

FIGURE 6.4

ROC curve for serum creatinine level as a predictor of transplant rejection

For example, the New York State Department of Health initiated a program to screen all infants born over a 28-month period for HIV. Since maternal antibodies cross the placenta, the presence of antibodies in an infant signals infection in the mother. Because the tests are performed anonymously, however, no verification of the results is possible. The reported outcomes of the statewide screening are presented in Table 6.2 [10].

Let n^+ represent the number of newborns who test positive and n the total number of infants screened. In each region of New York, HIV seroprevalence—or $P(H)$, where H is the event that a mother is infected with the virus—is calculated as n^+/n. In

TABLE 6.2

Percentage of HIV-positive newborns by region for the state of New York, December 1987–March 1990

Region	Number Positive	Total Tested	Percent Positive
New York State exclusive of NYC	601	346,522	0.17
NYC suburban	329	120,422	0.27
Mid-Hudson Valley	71	29,450	0.24
Upstate urban	119	88,088	0.14
Upstate rural	82	108,562	0.08
New York City	3650	294,062	1.24
Manhattan	799	50,364	1.59
Bronx	998	58,003	1.72
Brooklyn	1352	104,613	1.29
Queens	424	67,474	0.63
Staten Island	77	13,608	0.57

Manhattan, for example, 50,364 infants were tested and 799 of the results were positive. In this borough,

$$\frac{n^+}{n} = \frac{799}{50,364}$$
$$= 0.0159.$$

There is a problem here, however—the quantity n^+/n actually represents $P(T^+)$, the probability of a positive test result. If the screening test were perfect, $P(H)$ and $P(T^+)$ would be identical. The test is not infallible, however; both false positive and false negative results are possible. In fact, applying the law of total probability followed by the multiplicative rule, the true probability of a positive test is

$$P(T^+) = P(T^+ \cap H) + P(T^+ \cap H^C)$$
$$= P(T^+ \mid H)\,P(H) + P(T^+ \mid H^C)\,P(H^C)$$
$$= P(T^+ \mid H)\,P(H) + [1 - P(T^- \mid H^C)][1 - P(H)].$$

Note that a positive test result can occur in two different ways: either the mother is infected with HIV or she is not. In addition to the prevalence of infection, this equation incorporates both the sensitivity and specificity of the diagnostic test.

If n^+/n is the probability of a positive test result, how do we compute the prevalence of HIV? Using the expression for $P(T^+)$, we can solve for the true quantity of interest. After some algebraic manipulation, we find that

$$P(H) = \frac{P(T^+) - P(T^+ \mid H^C)}{P(T^+ \mid H) - P(T^+ \mid H^C)}$$
$$= \frac{(n^+/n) - P(T^+ \mid H^C)}{P(T^+ \mid H) - P(T^+ \mid H^C)}.$$

Since the prevalence of HIV infection is also a probability, its value must lie between 0 and 1. Let us examine the expression for $P(H)$. For any screening test of value,

$$P(T^+ \mid H) > P(T^+ \mid H^C);$$

the probability of a positive test result among individuals who are infected with HIV is higher than the probability among individuals who are not. As a result, the denominator of the ratio is positive. For $P(H)$ to be greater than 0, the numerator is required to be positive as well. Consequently, we must have

$$\frac{n^+}{n} > P(T^+ \mid H^C)$$
$$= 1 - P(T^- \mid H^C).$$

The proportion of positive test results in the entire population must be greater than the proportion of positive results among those who are not infected with HIV. Note that the specificity of the screening test plays a critical role in the calculation of prevalence; if the prevalence is very low, it may not be detected by a test with inadequate specificity.

Return to the data in Table 6.2. We do not know the sensitivity and specificity of the diagnostic procedure used, although we can be sure that the test was not perfect. Suppose, however, that the sensitivity of the screening test is 0.99 and that its specificity is 0.998; these values represent the higher end of the range of possible values. Also, recall that the probability of a positive test result in Manhattan is 0.0159. As a result, the prevalence of HIV infection in this borough would be calculated as

$$P(H) = \frac{0.0159 - (1 - 0.998)}{0.99 - (1 - 0.998)}$$
$$= 0.0141,$$

which is lower than the probability of a positive test result. For the upstate urban region of New York,

$$\frac{n^+}{n} = \frac{119}{88,088}$$
$$= 0.0014,$$

and

$$P(H) = \frac{0.0014 - (1 - 0.998)}{0.99 - (1 - 0.998)}$$
$$= -0.0006.$$

Even with a specificity as high as 0.998, the prevalence is found to be negative. Obviously, this is a nonsensical result; it most likely occurred because the testing procedure was not accurate enough to measure the very low prevalence of HIV in this region.

6.5   The Relative Risk and the Odds Ratio

The concept of relative risk is often useful when we want to compare the probabilities of disease in two different groups or situations. The *relative risk*, abbreviated RR, is the chance that a member of a group receiving some exposure will develop disease relative to the chance that a member of an unexposed group will develop the same disease. It is defined as the probability of disease in the exposed group divided by the probability of disease in the unexposed group, or

$$RR = \frac{P(\text{disease} \mid \text{exposed})}{P(\text{disease} \mid \text{unexposed})}.$$

Consider a study that examines the risk factors for breast cancer among women participating in the first National Health and Nutrition Examination Survey [11]. In a *cohort study* such as this, the exposure is measured at the onset of the investigation. Groups of individuals with and without the exposure—subjects without the exposure are often called *controls*—are followed to look for cases of disease. In this breast cancer study, a woman is considered to be "exposed" if she first gave birth at age 25 or older. In a sample of 4540 women who gave birth to their first child before the age of 25, 65 developed breast cancer. Of the 1628 women who first gave birth at age 25 or older, 31 were diagnosed with breast cancer. If we assume that the numbers are large enough to satisfy the frequentist definition of probability, the relative risk of developing breast cancer is

$$
\begin{aligned}
RR &= \frac{P(\text{disease} \mid \text{exposed})}{P(\text{disease} \mid \text{unexposed})} \\
&= \frac{31/1628}{65/4540} \\
&= 1.33.
\end{aligned}
$$

A relative risk of 1.33 implies that women who first give birth at a later age are 33% more likely to develop breast cancer than women who give birth at an earlier age. In Chapter 15, we will explain how to determine whether this is an important difference.

In general, a relative risk of 1.0 indicates that the probabilities of disease in the exposed and unexposed groups are identical; consequently, an association between the exposure and the disease does not exist. A relative risk greater than 1.0 implies that there is an increased risk of disease among those with the exposure, whereas a value less than 1.0 suggests that there is a decreased risk of developing disease among the exposed individuals.

Note that the value of the relative risk is independent of the magnitudes of the relevant probabilities; only the ratio of these probabilities is important. This is especially useful when we are concerned with unlikely events. In the United States, for example, the probability that a man over the age of 35 dies of lung cancer is 0.002679 for current smokers and 0.000154 for nonsmokers [12]. The relative risk of death for smokers versus nonsmokers, however, is

$$
\begin{aligned}
RR &= \frac{0.002679}{0.000154} \\
&= 17.4.
\end{aligned}
$$

Similarly, the probability that a woman over the age of 35 dies of lung cancer is 0.001304 for current smokers and 0.000121 for nonsmokers; the relative risk is

$$
\begin{aligned}
RR &= \frac{0.001304}{0.000121} \\
&= 10.8.
\end{aligned}
$$

Even though we are dealing with very low-probability events, the relative risk allows us to see that smoking has a large effect on the likelihood that a particular individual will die from lung cancer.

Another commonly used measure of the relative probabilities of disease is the *odds ratio*, or *relative odds*. If an event takes place with probability p, the *odds* in favor of the event are $p/(1-p)$ to 1. If $p = 1/2$, for instance, the odds are $(1/2)/(1/2) = 1$ to 1. In this case, the event is equally likely either to occur or not to occur. If $p = 2/3$, the odds of the event are $(2/3)/(1/3) = 2$ to 1; the probability that the event occurs is twice as large as the probability that it does not. Similarly, if for every 100,000 individuals there are 9.3 cases of tuberculosis, the odds of a randomly selected person's having the disease are

$$\frac{(9.3/100,000)}{(99,990.7/100,000)} = 0.00009301 \text{ to } 1.$$

Conversely, we know that if the odds in favor of an event are a to b, the probability that the event will occur is $a/(a + b)$. The odds ratio is defined as the odds of disease among exposed individuals divided by the odds of disease among the unexposed, or

$$OR = \frac{P(\text{disease} \mid \text{exposed})/[1 - P(\text{disease} \mid \text{exposed})]}{P(\text{disease} \mid \text{unexposed})/[1 - P(\text{disease} \mid \text{unexposed})]}.$$

It can also be defined as the odds of exposure among diseased individuals divided by the odds of exposure among those who are not diseased, or

$$OR = \frac{P(\text{exposure} \mid \text{diseased})/[1 - P(\text{exposure} \mid \text{diseased})]}{P(\text{exposure} \mid \text{nondiseased})/[1 - P(\text{exposure} \mid \text{nondiseased})]}.$$

Mathematically, these two definitions for the relative odds can be shown to be equivalent.

Consider the following data, taken from another study of the risk factors for breast cancer. This one is a case-control study that examines the effects of the use of oral contraceptives [13]. In a *case-control study*, investigators start by identifying groups of individuals with the disease (the cases) and without the disease (the controls). They then go back in time to determine whether the exposure in question was present or absent for each individual. Among the 989 women in the study who had breast cancer, 273 had previously used oral contraceptives and 716 had not. Of the 9901 women who did not have breast cancer, 2641 had used oral contraceptives and 7260 had not. In a case-control study, the proportions of subjects with and without the disease are chosen by the investigator; therefore, the probabilities of disease in the exposed and unexposed groups cannot be determined. However, we can calculate the probability of exposure for both cases and controls. Consequently, using the second definition for the odds ratio,

$$OR = \frac{P(\text{exposure} \mid \text{diseased})/[1 - P(\text{exposure} \mid \text{diseased})]}{P(\text{exposure} \mid \text{nondiseased})/[1 - P(\text{exposure} \mid \text{nondiseased})]}$$

$$= \frac{(273/989)/(1 - 273/989)}{(2641/9901)/(1 - 2641/9901)}$$

$$= \frac{(273/989)/(716/989)}{(2641/9901)/(7260/9901)}$$

$$= \frac{273/716}{2641/7260}$$

$$= 1.05.$$

These data imply that women who have used oral contraceptives have an odds of developing breast cancer that is only 1.05 times the odds of nonusers. Again, we will interpret this result in Chapter 15. As with the relative risk, however, an odds ratio of 1.0 indicates that exposure does not have an effect on the probability of disease.

The relative risk and the odds ratio are two different measures that attempt to explain the same phenomenon. Although the relative risk might seem more intuitive, the odds ratio has better statistical properties; these properties will be clarified later in the text. In any event, for rare diseases, the odds ratio is a close approximation of the relative risk. To see this, if

$$P(disease \mid exposed) \approx 0$$

and

$$P(disease \mid unexposed) \approx 0,$$

then

$$1 - P(disease \mid exposed) \approx 1$$

and

$$1 - P(disease \mid unexposed) \approx 1.$$

Therefore,

$$
\begin{aligned}
OR &= \frac{P(disease \mid exposed)/[1 - P(disease \mid exposed)]}{P(disease \mid unexposed)/[1 - P(disease \mid unexposed)]} \\[2mm]
&\approx \frac{P(disease \mid exposed)/1}{P(disease \mid unexposed)/1} \\[2mm]
&= \frac{P(disease \mid exposed)}{P(disease \mid unexposed)} \\[2mm]
&= RR.
\end{aligned}
$$

When we use odds ratios and relative risks, care must always be taken to put the information obtained into context. As we mentioned earlier, the numerical values of these measures do not reflect the magnitudes of the probabilities used to calculate them. To illustrate this point, a third study of breast cancer—this one investigating the effects of hormone use in postmenopausal women—concluded that women who have used hormone therapy for five to nine years have an odds of developing invasive breast cancer that is 1.46 times the odds for women who have never used hormones [14]. This seems to be quite a substantial increase in risk. However, the effect of this increase depends on the background probability of disease for the women who were not exposed to hormone therapy. It has been reported that a 60-year-old woman has a 3.59% chance of developing breast cancer in the next ten years [15]. In this case, we would have

$$OR = 1.46$$

$$= \frac{P(\text{cancer} \mid \text{hormone use})/[1 - P(\text{cancer} \mid \text{hormone use})]}{P(\text{cancer} \mid \text{no use})/[1 - P(\text{cancer} \mid \text{no use})]}$$

$$= \frac{P(\text{cancer} \mid \text{hormone use})/[1 - P(\text{cancer} \mid \text{hormone use})]}{0.0359/(1 - 0.0359)},$$

and thus that

$$P(\text{cancer} \mid \text{hormone use}) = 0.0516.$$

Even though the odds ratio of 1.46 is relatively large, the change in the actual probability of disease from 3.59% to 5.16% is not as alarming.

Figure 6.5 illustrates the relationship between the probability of a given outcome and the odds ratio. In the graph, p_u represents the background probability of disease in

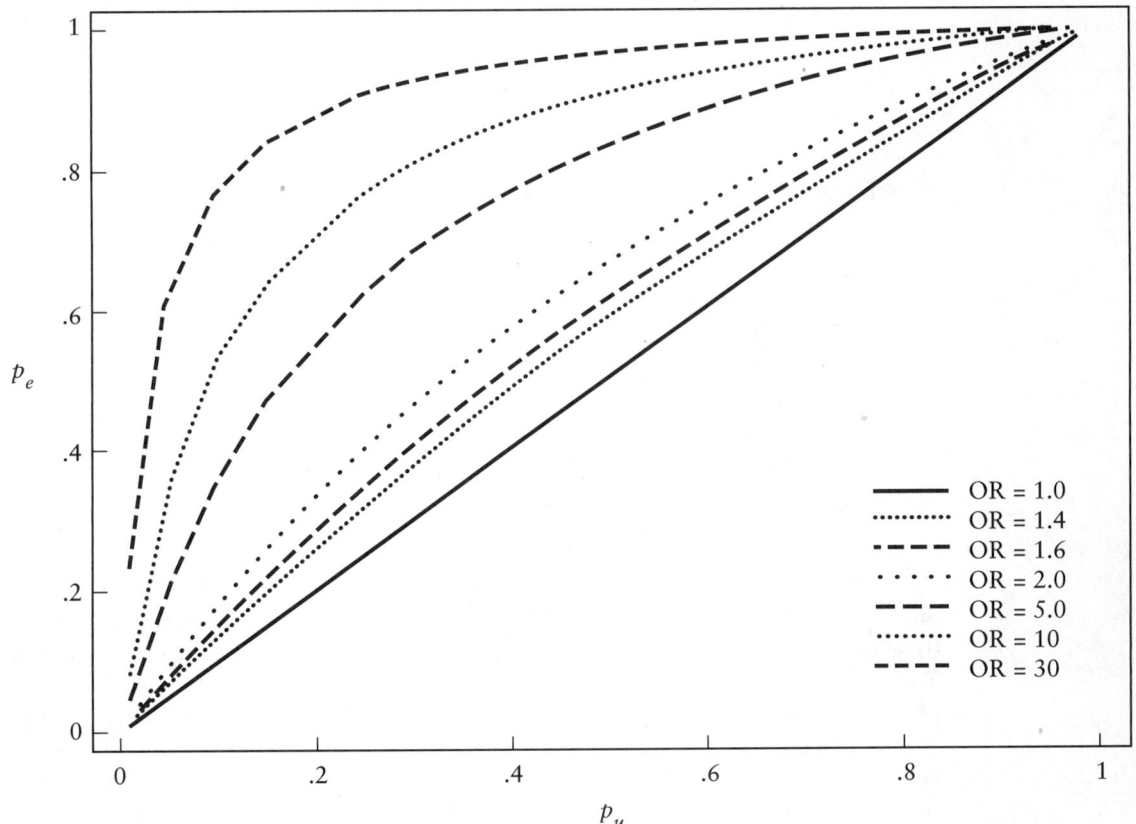

FIGURE 6.5

Relationship between the probabilities of an event in an exposed and an unexposed group and the odds ratio

TABLE 6.3
Relationships among the probabilities of an event in an exposed
and an unexposed group, the odds ratio, and the relative risk

Odds Ratio	p_u	p_e	Relative Risk
1.2	0.01	0.012	1.20
1.2	0.05	0.059	1.19
1.2	0.25	0.286	1.14
1.2	0.50	0.545	1.09
1.4	0.01	0.014	1.39
1.4	0.05	0.069	1.37
1.4	0.25	0.318	1.27
1.4	0.50	0.583	1.17
2.0	0.01	0.020	1.98
2.0	0.05	0.095	1.90
2.0	0.25	0.400	1.60
2.0	0.50	0.667	1.33

the unexposed group, and p_e is the higher probability of disease in the group that has undergone some exposure. If the odds ratio is equal to 1.0, p_u and p_e must be the same, no matter what the value of p_u. On the other hand, if the odds ratio is equal to 2.0, our interpretation depends on the value of p_u. For instance, if the probability of disease in the unexposed group is 0.05, the probability of disease among the exposed is 0.095, an increase of 90%. If the probability of disease among the unexposed is 0.50, however, the probability among the exposed is 0.667, an increase of only 33%. Table 6.3 shows the relationships among p_u, p_e, the odds ratio, and the relative risk for a variety of background probabilities.

6.6 Further Applications

Suppose that we are interested in determining the probability that a woman who becomes pregnant will give birth to a boy. In 1992, there were 4,065,014 registered births in the United States [16]. Of these infants, 2,081,287 were boys and 1,983,727 were girls. Therefore, if a randomly selected woman were to become pregnant, we could calculate the probability that her child will be a boy as

$$P(B) = P(\text{child will be a boy})$$
$$= \frac{2,081,287}{4,065,014}$$
$$= 0.512.$$

We can discuss the probability that the child will be a boy only before the woman actually becomes pregnant; after conception, the sex of the fetus has been determined and the concept of probability no longer applies.

The complement of the event that the child is a boy is the event that it is a girl. Consequently,

$$P(G) = P(\text{child will be a girl})$$
$$= 1 - P(B)$$
$$= 1 - 0.512$$
$$= 0.488.$$

Since the child is classified as either a boy or a girl, these two events are mutually exclusive. When events are mutually exclusive, the additive rule of probability states that the probability that either of the events will occur is equal to the sum of the probabilities of the individual events; thus,

$$P(B \cup G) = P(B) + P(G)$$
$$= 0.512 + 0.488$$
$$= 1.000.$$

The sum of the probabilities of these two events is 1. This indicates that the events are exhaustive. The child must be classified as either a boy or a girl; there are no other possible outcomes.

Now suppose that we randomly select two women from the population and that they both become pregnant. What is the probability that both children will be boys? We know that the two events are independent: the gender of the first woman's child has no effect on the gender of the second woman's child. Therefore, using the multiplicative rule of probability for independent events and representing the event that both children will be boys by $B_1 \cap B_2$,

$$P(B_1 \cap B_2) = P(B_1)\,P(B_2)$$
$$= (0.512)(0.512)$$
$$= 0.262.$$

There are three other possible events: $B_1 \cap G_2$, the first woman's child will be a boy and the second woman's child a girl; $G_1 \cap B_2$, the first woman will have a girl and the second a boy; and $G_1 \cap G_2$, both children will be girls. The probabilities of these events are

$$P(B_1 \cap G_2) = P(B_1)\,P(G_2)$$
$$= (0.512)(0.488)$$
$$= 0.250,$$
$$P(G_1 \cap B_2) = P(G_1)\,P(B_2)$$
$$= (0.488)(0.512)$$
$$= 0.250,$$

and

$$P(G_1 \cap G_2) = P(G_1) P(G_2)$$
$$= (0.488)(0.488)$$
$$= 0.238.$$

Note that these four probabilities sum to 1.

If we choose three women from the population and each one becomes pregnant, what is the probability that all three children will be girls? The concept of independence can be extended to three or more different events; in this case, the gender of one woman's child does not affect the gender of either of the other children. The multiplicative rule of probability for independent events states that the probability that all three of the children will be girls is

$$P(G_1 \cap G_2 \cap G_3) = P(G_1) P(G_2) P(G_3)$$
$$= (0.488)(0.488)(0.488)$$
$$= 0.116.$$

Returning to the example in which we select only two women, what is the probability that both children will be boys given that at least one child is a boy? The chance that a particular event will occur given that another event has already taken place is known as a conditional probability. Representing the event that at least one child is a boy by A and applying the formula for a conditional probability,

$$P(B_1 \cap B_2 \mid A) = P(\text{both children will be boys} \mid \text{at least one boy})$$
$$= \frac{P[(B_1 \cap B_2) \cap A]}{P(A)}$$
$$= \frac{P(B_1 \cap B_2)}{P(A)}.$$

The event that both children are boys and at least one is a boy is simply the event that both children are boys. We already know that $P(B_1 \cap B_2) = 0.262$. What is $P(A)$, the probability that there is at least one boy? Note that this event can occur in three different ways—either both children will be boys, the first will be a boy and the second a girl, or the first will be a girl and the second a boy. Since these three events are mutually exclusive, we apply the additive rule to find

$$P(A) = P[(B_1 \cap B_2) \cup (B_1 \cap G_2) \cup (G_1 \cap B_2)]$$
$$= P(B_1 \cap B_2) + P(B_1 \cap G_2) + P(G_1 \cap B_2)$$
$$= 0.262 + 0.250 + 0.250$$
$$= 0.762.$$

Therefore,

$$P(B_1 \cap B_2 \mid A) = \frac{P(B_1 \cap B_2)}{P(A)}$$

$$= \frac{0.262}{0.762}$$

$$= 0.344.$$

If we know that at least one child is a boy, the probability that both children will be boys increases from 0.262 to 0.344.

At first glance, this result may appear counterintuitive. We are told that one child is a boy; consequently, we might expect the probability that the other child is a boy to be simply $P(B) = 0.512$. Instead, we have calculated the probability to be 0.344. The important point to keep in mind is that we did not specify which of the two children was a boy. In this example, order is important. When we deal with probabilities, the seemingly obvious answer is not always correct; each problem must be considered carefully.

Conditional probabilities often come into play when we are working with life tables. Suppose we would like to know the probability that a person will live to be 80 years of age given that he or she is now 40. Let A represent the event that the individual is 40 years old and B the event that he or she lives to be 80. Using Table 5.1, the 1992 life table for the U.S. population,

$$P(A) = P(\text{a person lives to be 40 years of age})$$

$$= \frac{95{,}527}{100{,}000}$$

$$= 0.95527,$$

and

$$P(A \cap B) = P(\text{a person lives to be 40 and lives to be 80})$$

$$= P(\text{a person lives to be 80})$$

$$= \frac{48{,}460}{100{,}000}$$

$$= 0.48460.$$

Therefore,

$$P(B \mid A) = P(\text{a person lives to be 80} \mid \text{he or she is now 40})$$

$$= \frac{P(A \cap B)}{P(A)}$$

$$= \frac{0.48460}{0.95527}$$

$$= 0.5073.$$

If an individual lives to be 40, his or her chance of surviving to age 80 is greater than it was at birth.

If A_1 and A_2 are mutually exclusive and exhaustive events, so that

$$P(A_1 \cup A_2) = P(A_1) + P(A_2)$$
$$= 1,$$

Bayes' theorem states that

$$P(A_1 \mid B) = \frac{P(A_1)\,P(B \mid A_1)}{P(A_1)\,P(B \mid A_1) + P(A_2)\,P(B \mid A_2)}.$$

Bayes' theorem is important in diagnostic testing. It relates the predictive value of a test to its sensitivity and specificity, as well as to the prevalence of disease in the population being tested.

Consider the following data, taken from a study that investigates the accuracy of three brands of home pregnancy tests [17]. Let A_1 represent the event that a woman is pregnant, A_2 the event that she is not pregnant, and T^+ a positive home pregnancy test result. The average sensitivity for the three brands of test kits is 80%; therefore,

$$P(T^+ \mid A_1) = 0.80.$$

Consequently, the probability of a false negative result is

$$P(T^- \mid A_1) = 1 - 0.80$$
$$= 0.20.$$

The specificity of the home pregnancy tests is found to be 68%; therefore,

$$P(T^- \mid A_2) = 0.68$$

and the probability of a false positive is

$$P(T^+ \mid A_2) = 1 - 0.68$$
$$= 0.32.$$

What is the probability that a woman with a positive home test kit result is actually pregnant?

Suppose that in the population being tested, $P(A_1) = 0.60$; that is, 60% of the women who use the home pregnancy tests are actually pregnant. Since A_2 is the complement of A_1, the probability that a woman is not pregnant is

$$P(A_2) = 1 - P(A_1)$$
$$= 1 - 0.60$$
$$= 0.40.$$

Using Bayes' theorem, the predictive value of a positive test is

$$P(A_1 \mid T^+) = \frac{P(A_1)\,P(T^+ \mid A_1)}{P(A_1)\,P(T^+ \mid A_1) + P(A_2)\,P(T^+ \mid A_2)}$$

$$= \frac{(0.60)(0.80)}{(0.60)(0.80) + (0.40)(0.32)}$$

$$= 0.79.$$

Therefore, a positive home test kit result increases the probability that a woman in this population is pregnant from 0.60 to 0.79.

What is the probability that a woman is not pregnant given that her home test result is negative? Again applying Bayes' theorem, the predictive value of a negative test is

$$P(A_2 \mid T^-) = \frac{P(A_2)\,P(T^- \mid A_2)}{P(A_2)\,P(T^- \mid A_2) + P(A_1)\,P(T^- \mid A_1)}$$

$$= \frac{(0.40)(0.68)}{(0.40)(0.68) + (0.60)(0.20)}$$

$$= 0.69.$$

A negative home test kit result increases the probability that a woman is not pregnant from 0.40 to 0.69.

When we wish to compare the probabilities of a specified event in two different groups, the concept of relative risk is often useful. Consider Figure 6.6. This bar chart illustrates the risks of lung cancer among women who have smoked 21 or more cigarettes per day relative to women who have never smoked [12,18]. For the group of persons who have stopped smoking within the past two years, for example,

$$RR = \frac{P(\text{lung cancer} \mid \text{quit within past two years})}{P(\text{lung cancer} \mid \text{nonsmoker})}$$

$$= 32.4.$$

It is rather surprising that this relative risk is higher than the corresponding risk for current smokers; however, this group contains many sick people who were forced to quit because of their illness. Given time, the risk of lung cancer among even heavy smokers gradually decreases after an individual stops smoking.

The odds ratio is another measure that is often used to compare the probabilities of an event in two different groups. Unlike the relative risk, which compares the probabilities directly, however, the odds ratio (as its name would suggest) relates the odds of the event in the two populations. For women who have smoked 21 or more cigarettes per day but have quit within the past two years, the odds of developing lung cancer relative to the odds for women who have never smoked would be calculated as

$$OR = \frac{P(\text{lung cancer} \mid \text{quit})/[1 - P(\text{lung cancer} \mid \text{quit})]}{P(\text{lung cancer} \mid \text{nonsmoker})/[1 - P(\text{lung cancer} \mid \text{nonsmoker})]}.$$

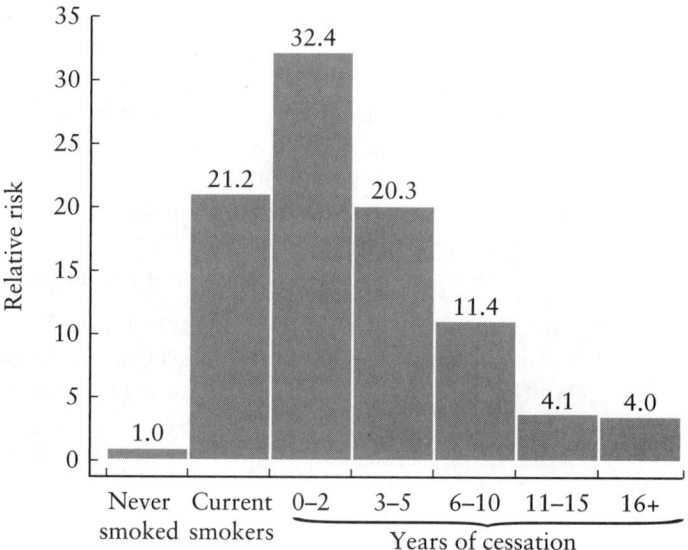

FIGURE 6.6
Relative risks of lung cancer in female ex-smokers, 21 or more cigarettes a day

For rare diseases such as lung cancer, the odds ratio is a close approximation to the relative risk.

6.7   Review Exercises

1. What is the frequentist definition of probability?

2. What are the three basic operations that can be performed on events?

3. Explain the difference between mutually exclusive and independent events.

4. What is the value of Bayes' theorem? How is it applied in diagnostic testing?

5. What would happen if you tried to increase the sensitivity of a diagnostic test?

6. How can the probabilities of disease in two different groups be compared?

7. Let A represent the event that a particular individual is exposed to high levels of carbon monoxide and B the event that he or she is exposed to high levels of nitrogen dioxide.
   (a) What is the event $A \cap B$?
   (b) What is the event $A \cup B$?
   (c) What is the complement of A?
   (d) Are the events A and B mutually exclusive?

8. For Mexican American infants born in Arizona in 1986 and 1987, the probability that a child's gestational age is less than 37 weeks is 0.142 and the probability that

his or her birth weight is less than 2500 grams is 0.051 [19]. Furthermore, the probability that these two events occur simultaneously is 0.031.

(a) Let A be the event that an infant's gestational age is less than 37 weeks and B the event that his or her birth weight is less than 2500 grams. Construct a Venn diagram to illustrate the relationship between events A and B.

(b) Are A and B independent?

(c) For a randomly selected Mexican American newborn, what is the probability that A or B or both occur?

(d) What is the probability that event A occurs given that event B occurs?

9. Consider the following natality statistics for the U.S. population in 1992 [16]. According to these data, the probabilities that a randomly selected woman who gave birth in 1992 was in each of the following age groups are as follows:

Age	Probability
<15	0.003
15–19	0.124
20–24	0.263
25–29	0.290
30–34	0.220
35–39	0.085
40–44	0.014
45–49	0.001
Total	1.000

(a) What is the probability that a woman who gave birth in 1992 was 24 years of age or younger?

(b) What is the probability that she was 40 or older?

(c) Given that the mother of a particular child was under 30 years of age, what is the probability that she was not yet 20?

(d) Given that the mother was 35 years of age or older, what is the probability that she was under 40?

10. The probabilities associated with the expected principal source of payment for hospital discharges in the United States in the year 1990 are listed below [20].

Principal Source of Payment	Probability
Private insurance	0.387
Medicare	0.345
Medicaid	0.116
Other govt. program	0.033
Self-payment	0.058
Other/No charge	0.028
Not stated	0.033
Total	1.000

(a) What is the probability that the principal source of payment for a given hospital discharge is the patient's private insurance?

(b) What is the probability that the principal source of payment is Medicare, Medicaid, or some other government program?

(c) Given that the principal source of payment is a government program, what is the probability that it is Medicare?

11. Looking at the United States population in 1993, the probability that an adult between the ages of 45 and 64 does not have health insurance coverage of any kind is 0.123 [21].

(a) Suppose that you randomly select a 47-year-old woman and an unrelated 59-year-old man from this population. What is the probability that both individuals are uninsured?

(b) What is the probability that both adults have health insurance coverage?

(c) If five unrelated adults between the ages of 45 and 64 are chosen from the population, what is the probability that all five are uninsured?

12. Consult Table 5.1, the 1992 abridged life table for the United States [1].

(a) What is the probability that a newborn infant will live to see his or her fifth birthday?

(b) What is the probability that an individual who is 60 years old will survive for the next ten years?

(c) Consider a man and woman who are married and who are both 60 years of age. What is the probability that both the woman and her husband will be alive on their 70th birthdays? Assume that the two events are independent.

(d) What is the probability that either the wife or the husband, but not both, will be alive at age 70?

13. One study has reported that the sensitivity of the mammogram as a screening test for detecting breast cancer is 0.85, while its specificity is 0.80 [22].

(a) What is the probability of a false negative test result?

(b) What is the probability of a false positive result?

(c) In a population in which the probability that a woman has breast cancer is 0.0025, what is the probability that she has cancer given that her mammogram is positive?

14. The National Institute for Occupational Safety and Health has developed a case definition of carpal tunnel syndrome—an affliction of the wrist—that incorporates three criteria: symptoms of nerve involvement, a history of occupational risk factors, and the presence of physical exam findings [23]. The sensitivity of this definition as a test for carpal tunnel syndrome is 0.67; its specificity is 0.58.

(a) In a population in which the prevalence of carpal tunnel syndrome is estimated to be 15%, what is the predictive value of a positive test result?

(b) How does this predictive value change if the prevalence is only 10%? If it is 5%?

(c) Construct a diagram—like the one in Figure 6.3—illustrating the results of the diagnostic testing process. Assume that you start with a population of 1,000,000 people and that the prevalence of carpal tunnel syndrome is 15%.

15. The following data are taken from a study investigating the use of a technique called radionuclide ventriculography as a diagnostic test for detecting coronary artery disease [24].

Test	Disease Present	Disease Absent	Total
Positive	302	80	382
Negative	179	372	551
Total	481	452	933

(a) What is the sensitivity of radionuclide ventriculography in this study? What is its specificity?

(b) For a population in which the prevalence of coronary artery disease is 0.10, calculate the probability that an individual has the disease given that he or she tests positive using radionuclide ventriculography.

(c) What is the predictive value of a negative test?

16. The table below displays data taken from a study comparing self-reported smoking status with measured serum cotinine level [25]. As part of the study, cotinine level was used as a diagnostic tool for predicting smoking status; the self-reported status was considered to be true. For a number of different cutoff points, the observed sensitivities and specificities are given below.

Cotinine Level (ng/ml)	Sensitivity	Specificity
5	0.971	0.898
7	0.964	0.931
9	0.960	0.946
11	0.954	0.951
13	0.950	0.954
14	0.949	0.956
15	0.945	0.960
17	0.939	0.963
19	0.932	0.965

(a) As the cutoff point is raised, how does the probability of a false positive result change? How does the probability of a false negative result change?

(b) Use these data to construct a receiver operator characteristic curve.

(c) Based on this graph, what value of serum cotinine level would you choose as an optimal cutoff point for predicting smoking status? Why?

17. Table 6.2 shows the percentages of HIV-positive newborns for various regions in the state of New York [10].

(a) In Brooklyn, what is the probability of a positive test result?

(b) Assume that the sensitivity of the screening test used is 0.99 and that its specificity is 0.998. What is the prevalence of HIV infection in this borough?

(c) What is the prevalence of HIV infection in the Bronx?

18. For several different methods of contraception, the probabilities that a currently married woman experiences an unplanned pregnancy during the first year of use are given below [26].

Method of Contraception	Probability of Pregnancy
None	0.431
Diaphragm	0.149
Condom	0.106
IUD	0.071
Pill	0.037

For each method listed, calculate the relative risk of pregnancy for women using the method versus women not using any type of protection. How does the risk change with respect to method of contraception?

19. A community-based study of respiratory illness during the first year of life was conducted in North Carolina. As part of this study, a group of children were classified according to family socioeconomic status. The numbers of children in each group who experienced persistent respiratory symptoms are shown below [27].

Socioeconomic Status	Number of Children	Number with Symptoms
Low	79	31
Middle	122	29
High	192	27

(a) Use these data to compute the probability of suffering from persistent respiratory symptoms in each socioeconomic group. Assume that the numbers are large enough to satisfy the frequentist definition of probability.
(b) Calculate the odds of experiencing persistent respiratory symptoms for both the middle and low socioeconomic groups relative to the high socioeconomic group.
(c) Does there appear to be an association between socioeconomic status and respiratory symptoms?

20. A study was conducted investigating the use of fasting capillary glycemia (FCG)—the level of glucose in the blood for individuals who have not eaten in a specified number of hours—as a screening test for diabetes [28]. FCG cutoff points ranging from 3.9 to 8.9 mmol/liter were examined; the sensitivities and specificities of the test corresponding to these different levels are contained in the data set `diabetes` (Appendix B, Table B.11). The levels of FCG are saved under the variable name `fcg`, the sensitivities under `sens`, and the specificities under `spec`.
(a) How does the sensitivity of the screening test change as the cutoff point is raised from 3.9 to 8.9 mmol/l? How does the specificity change?

(b) Use these data to construct a receiver operator characteristic curve.

(c) The investigators who conducted this study chose an FCG level of 5.6 mmol/liter as the optimal cutoff point for predicting diabetes. Do you agree with this choice? Why or why not?

Bibliography

[1] National Center for Health Statistics, Kochanek, K. D., and Hudson, B. L., "Advanced Report of Final Mortality Statistics, 1992," *Monthly Vital Statistics Report*, Volume 43, Number 6, March 22, 1995.

[2] National Center for Health Statistics, Collins, J. G., "Types of Injuries and Impairments Due to Injuries, United States," *Vital and Health Statistics*, Series 10, Number 159, November 1986.

[3] Yobs, A. R., Swanson, R. A., and Lamotte, L. C., "Laboratory Reliability of the Papanicolaou Smear," *Obstetrics and Gynecology*, Volume 65, February 1985, 235–244.

[4] Devesa, S. S., Silverman, D. T., Young, J. L., Pollack, E. S., Brown, C. C., Horm, J. W., Percy, C. L., Myers, M. H., McKay, F. W., and Fraumeni, J. F., "Cancer Incidence and Mortality Trends Among Whites in the United States, 1949–1984," *Journal of the National Cancer Institute*, Volume 79, October 1987, 701–770.

[5] Henig, R. M., "Is the Pap Test Valid?," *The New York Times Magazine*, May 28, 1989, 37–38.

[6] Fahey, M. T., Irwig, L., and Macaskill, P., "Meta-analysis of Pap Test Accuracy," *American Journal of Epidemiology*, Volume 141, April 1, 1995, 680–689.

[7] Yerushalmy, J., Harkness, J. T., Cope, J. H., and Kennedy, B. R., "The Role of Dual Reading in Mass Radiography," *American Review of Tuberculosis*, Volume 61, April 1950, 443–464.

[8] Centers for Disease Control, "A Strategic Plan for the Elimination of Tuberculosis in the United States," *Morbidity and Mortality Weekly Report*, Volume 38, Number 16, April 28, 1989.

[9] DeLong, E. R., Vernon, W. B., and Bollinger, R. R., "Sensitivity and Specificity of a Monitoring Test," *Biometrics*, Volume 41, December 1985, 947–958.

[10] Novick, L. F., Glebatis, D. M., Stricof, R. L., MacCubbin, P. A., Lessner, L., and Berns, D. S., "Newborn Seroprevalence Study: Methods and Results," *American Journal of Public Health*, Volume 81, May 1991, 15–21.

[11] Carter, C. L., Jones, D. Y., Schatzkin, A., and Brinton, L. A., "A Prospective Study of Reproductive, Familial, and Socioeconomic Risk Factors for Breast Cancer Using NHANES I Data," *Public Health Reports*, Volume 104, January–February 1989, 45–49.

[12] Garfinkel, L., and Silverberg, E., "Lung Cancer and Smoking Trends in the United States Over the Past 25 Years," *Ca—A Cancer Journal for Clinicians*, Volume 41, May/June 1991, 137–145.

[13] Hennekens, C. H., Speizer, F. E., Lipnick, R. J., Rosner, B., Bain, C., Belanger, C., Stampfer, M. J., Willett, W., and Peto, R., "A Case-Control Study of Oral Contraceptive Use and Breast Cancer," *Journal of the National Cancer Institute*, Volume 72, January 1984, 39–42.

[14] Colditz, G. A., Hankinson, S. E., Hunter, D. J., Willett, W. C., Manson, J. E., Stampfer, M. J., Hennekens, C., Rosner, B., and Speizer, F. E., "The Use of Estrogens and Progestins and the Risk of Breast Cancer in Postmenopausal Women," *The New England Journal of Medicine*, Volume 332, June 15, 1995, 1589–1593.

[15] Feuer, E. J., Wun, L. M., Boring, C. C., Flanders, W. D., Timmel, M. J., and Tong, T., "The Lifetime Risk of Developing Breast Cancer," *Journal of the National Cancer Institute*, Volume 85, June 2, 1993, 892–897.

[16] National Center for Health Statistics, Ventura, S. J., Martin, J. A., Taffel, S. M., Mathews, T. J., and Clarke, S. C., "Advanced Report of Final Natality Statistics, 1992," *Monthly Vital Statistics Report*, Volume 43, Number 5, October 25, 1994.

[17] Doshi, M. L., "Accuracy of Consumer Performed In-home Tests for Early Pregnancy Detection," *American Journal of Public Health*, Volume 76, May 1986, 512–514.

[18] Garfinkel, L., and Stellman, S. D., "Smoking and Lung Cancer in Women: Findings in a Prospective Study," *Cancer Research*, Volume 48, December 1, 1988, 6951–6955.

[19] Balcazar, H., "The Prevalence of Intrauterine Growth Retardation in Mexican Americans," *American Journal of Public Health*, Volume 84, March 1994, 462–465.

[20] National Center for Health Statistics, Graves, E. J., "Expected Principal Source of Payment for Hospital Discharges: United States, 1990," *Vital and Health Statistics*, Advance Data Report Number 220, November 12, 1992.

[21] National Center for Health Statistics, *Health, United States, 1994 Chartbook*, May 1995.

[22] Hulka, B. S., "Cancer Screening: Degrees of Proof and Practical Application," *Cancer*, Volume 62, Supplement to October 15, 1988, 1776–1780.

[23] Katz, J. N., Larson, M. G., Fossel, A. H., and Liang, M. H., "Validation of a Surveillance Case Definition of Carpal Tunnel Syndrome," *American Journal of Public Health*, Volume 81, February 1991, 189–193.

[24] Begg, C. B., and McNeil, B. J., "Assessment of Radiologic Tests: Control of Bias and Other Design Considerations," *Radiology*, Volume 167, May 1988, 565–569.

[25] Wagenknecht, L. E., Burke, G. L., Perkins, L. L., Haley, N. J., and Friedman, G. D., "Misclassification of Smoking Status in the CARDIA Study: A Comparison of Self-Report with Serum Cotinine Levels," *American Journal of Public Health*, Volume 82, January 1992, 33–36.

[26] Grady, W. R., Hayward, M. D., and Yagi, J., "Contraceptive Failure in the United States: Estimates from the 1982 National Survey of Family Growth," *Family Planning Perspectives*, Volume 18, September/October 1986, 200–209.

[27] Margolis, P. A., Greenberg, R. A., Keyes, L. L., LaVange, L. M., Chapman, R. S., Denny F. W., Bauman, K. E., and Boat, B. W., "Lower Respiratory Illness in Infants and Low Socioeconomic Status," *American Journal of Public Health*, Volume 82, August 1992, 1119–1126.

[28] Bortheiry, A. L., Malerbi, D. A., and Franco, L. J., "The ROC Curve in the Evaluation of Fasting Capillary Blood Glucose as a Screening Test for Diabetes and IGT," *Diabetes Care*, Volume 17, November 1994, 1269–1272.

7

Theoretical Probability Distributions

Any characteristic that can be measured or categorized is called a *variable*. If a variable can assume a number of different values such that any particular outcome is determined by chance, it is a *random variable*. We have already looked at a number of different random variables in previous chapters, although we did not use this term. In Chapter 2, for instance, the serum cholesterol level of a 25- to 34-year-old male in the United States is a random variable; in Chapter 3, forced expiratory volume in 1 second for an adolescent suffering from asthma is another. Random variables are typically represented by uppercase letters such as X, Y, and Z. A *discrete random variable* can assume only a finite or countable number of outcomes. One example is marital status: an individual can be single, married, divorced, or widowed. Another example would be the number of ear infections an infant develops during his or her first year of life. A *continuous random variable*, such as weight or height, can take on any value within a specified interval or continuum.

7.1 Probability Distributions

Every random variable has a corresponding probability distribution. A *probability distribution* applies the theory of probability to describe the behavior of the random variable. In the discrete case, it specifies all possible outcomes of the random variable along with the probability that each will occur. In the continuous case, it allows us to determine the probabilities associated with specified ranges of values.

For example, let X be a discrete random variable that represents the birth order of each child born to a woman residing in the United States [1]. If a child is a woman's first-born, $X = 1$; if it is her second child, $X = 2$. To construct a probability distribution for X, we list each of the values x that the random variable can assume, along with $P(X = x)$ for each one. This has been done in Table 7.1. The outcomes $X = 8$, $X = 9$, and so on for the countable integers have been grouped together and called "8^+." Note

TABLE 7.1

Probability distribution of a random variable X representing the birth order of children born in the United States

x	$\mathbf{P}(X = x)$
1	0.416
2	0.330
3	0.158
4	0.058
5	0.021
6	0.009
7	0.004
8$^+$	0.004
Total	1.000

that we use an uppercase X to denote the random variable and a lowercase x to represent the outcome of a particular child.

Table 7.1 resembles the frequency distributions introduced in Chapter 2. For a sample of observations, a frequency distribution displays each observed outcome and the number of times it appears in the set of data. The frequency distribution sometimes includes the relative frequency of each outcome as well. For a discrete random variable, a probability distribution lists each possible outcome and its corresponding probability. The probabilities represent the relative frequency of occurrence of each outcome x in a large number of trials repeated under essentially identical conditions; equivalently, they can be thought of as the relative frequencies associated with an infinitely large sample. They tell us which values are more likely to occur than others. Since all possible values of the random variable are taken into account, the outcomes are exhaustive; therefore, the sum of their probabilities must be 1.

In many cases, we can display a probability distribution by means of either a graph or a mathematical formula. Figure 7.1, for example, is a histogram of the probability distribution shown in Table 7.1. The area of each vertical bar represents $P(X = x)$, the probability associated with that particular outcome of the random variable; the total area of the histogram is equal to 1.

The probability distribution of X can be used to make statements about the possible outcomes of the random variable. Suppose we wish to know the probability that a randomly chosen newborn is its mother's fourth child. Using the information in Table 7.1, we observe that $P(X = 4) = 0.058$. What is the probability that the infant is its mother's first or second child? Applying the additive rule of probability for mutually exclusive events,

$$P(X = 1 \text{ or } X = 2) = P(X = 1) + P(X = 2)$$
$$= 0.416 + 0.330$$
$$= 0.746.$$

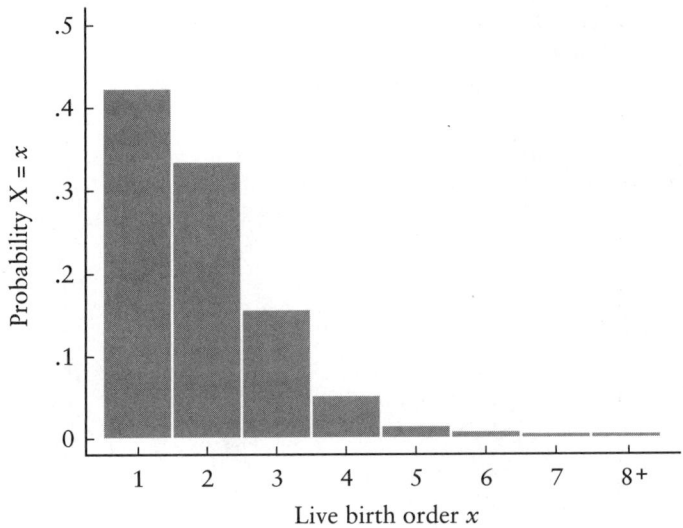

FIGURE 7.1
Probability distribution of a random variable representing the birth order of children born in the United States

If a random variable can take on a large number of values, a probability distribution may not be a useful way to summarize its behavior. As with a frequency distribution of grouped data, however, we can describe a probability distribution using a measure of central tendency and a measure of dispersion. The average value assumed by a random variable is known as the *population mean*; the dispersion of the values relative to this mean is the *population variance*. Furthermore, the square root of the population variance is the *population standard deviation*.

The probability distribution of the birth order of children born to women in the United States was generated based on the experience of the U.S. population in 1986. Probabilities that are calculated from a finite amount of data are called *empirical probabilities*. The probability distributions of many other variables of interest, however, can be determined based on theoretical considerations. Distributions of this kind are known as *theoretical probability distributions*.

7.2 The Binomial Distribution

Consider a dichotomous random variable Y. By definition, the variable Y must assume one of two possible values; these mutually exclusive outcomes could represent life and death, male and female, or sickness and health. For simplicity, they are often referred to as "failure" and "success." A random variable of this type is known as a *Bernoulli random variable*.

As an example, let Y be a random variable that represents smoking status; $Y = 1$ if an adult is currently a smoker and $Y = 0$ if he or she is not. The two outcomes of Y are mutually exclusive and exhaustive. In 1987, 29% of the adults in the United States smoked cigarettes, cigars, or pipes [2]; therefore, we can enumerate the probabilities associated with the respective outcomes of Y as

$$P(Y = 1) = p$$
$$= 0.29$$

and

$$P(Y = 0) = 1 - p$$
$$= 1.00 - 0.29$$
$$= 0.71.$$

These two equations completely describe the probability distribution of the dichotomous random variable Y; supposing that smoking habits have not changed since 1987 (an unreasonable assumption, perhaps), if we were to travel around the United States observing whether particular adults are smokers or nonsmokers, Y would assume the value 1 approximately 29% of the time and the value 0 the remaining 71% of the time. Recall that the proportion of times that a dichotomous random variable takes on the value 1 is equal to its population mean.

Suppose that we now randomly select two individuals from the population of adults in the United States. If we introduce a new random variable X that represents the number of persons in the pair who are current smokers, X can take on three possible values: 0, 1, or 2. Both of the individuals selected do not smoke, or one smokes and the other does not, or both smoke. The smoking statuses of the two persons chosen are independent; therefore, we can apply the multiplicative rule to find the probability that X will take on a particular value.

Outcome of Y		Probability of These Outcomes	Number of Smokers X
First Person	Second Person		
0	0	$(1 - p)(1 - p)$	0
1	0	$p(1 - p)$	1
0	1	$(1 - p)p$	1
1	1	pp	2

Substituting in the value of p, we find that

$$P(X = 0) = (1 - p)^2$$
$$= (0.71)^2$$
$$= 0.504,$$

$$P(X = 1) = p(1 - p) + (1 - p)p$$
$$= 2p(1 - p)$$
$$= 2(0.29)(0.71)$$
$$= 0.412,$$

and

$$P(X = 2) = p^2$$
$$= (0.29)^2$$
$$= 0.084.$$

Note that there are two possible situations in which one adult smokes and the other does not: either $Y = 1$ for the first individual and $Y = 0$ for the second, or $Y = 0$ for the first and $Y = 1$ for the second. The two outcomes are mutually exclusive, and we apply the additive rule of probability to find $P(X = 1)$. Also note that because all possible outcomes of X are considered, their probabilities must sum to 1; that is,

$$P(X = 0) + P(X = 1) + P(X = 2) = 0.504 + 0.412 + 0.084$$
$$= 1.000.$$

The probability distribution of the discrete random variable X described above is a special case of the *binomial distribution*. In general, if we have a sequence of n independent Bernoulli trials—or, equivalently, n independent outcomes of the Bernoulli random variable Y—each with a probability of "success" p, then the total number of successes X is a binomial random variable. The fixed numbers n and p are called the parameters of the distribution. *Parameters* are numerical quantities that summarize the characteristics of a probability distribution. In the preceding example, the parameters are $n = 2$, since two individuals are selected, and $p = 0.29$, because the probability that any randomly chosen adult is a current smoker is 0.29. The binomial distribution involves three assumptions:

1. There are a fixed number of trials n, each of which results in one of two mutually exclusive outcomes.
2. The outcomes of the n trials are independent.
3. The probability of success p is constant for each trial.

The binomial distribution can be used to describe a variety of situations, such as the number of siblings who will inherit a certain genetic trait from their parents or the number of patients who experience an adverse reaction to a new drug.

Suppose we were to continue with the preceding example by randomly selecting three adults from the population instead of two. In this case, X would be a binomial random variable with parameters $n = 3$ and $p = 0.29$.

	Outcome of Y		Probability of	Number of
First Person	**Second Person**	**Third Person**	**These Outcomes**	**Smokers X**
0	0	0	$(1-p)(1-p)(1-p)$	0
1	0	0	$p(1-p)(1-p)$	1
0	1	0	$(1-p)p(1-p)$	1
0	0	1	$(1-p)(1-p)p$	1
1	1	0	$pp(1-p)$	2
1	0	1	$p(1-p)p$	2
0	1	1	$(1-p)pp$	2
1	1	1	ppp	3

Substituting in the value of p,

$$P(X = 0) = (1-p)^3$$
$$= (0.71)^3$$
$$= 0.358,$$

$$P(X = 1) = p(1-p)^2 + p(1-p)^2 + p(1-p)^2$$
$$= 3(0.29)(0.71)^2$$
$$= 0.439,$$

$$P(X = 2) = p^2(1-p) + p^2(1-p) + p^2(1-p)$$
$$= 3(0.29)^2(0.71)$$
$$= 0.179,$$

and

$$P(X = 3) = p^3$$
$$= (0.29)^3$$
$$= 0.024.$$

These equations describe the probability distribution of X. The random variable X can take on four possible values, and

$$P(X = 0) + P(X = 1) + P(X = 2) + P(X = 3)$$
$$= 0.358 + 0.439 + 0.179 + 0.024$$
$$= 1.000.$$

Note that $P(X = 1)$ and $P(X = 2)$ both involve the summation of three terms; if we have a total of three individuals, there are exactly three ways in which one of them can be a smoker and three ways in which any two of them can be smokers.

If we proceed with our example and select a total of *n* adults from the population, the probability that exactly *x* of them smoke can be written as

$$P(X = x) = \frac{n!}{x!(n-x)!}p^x(1-p)^{n-x}$$

$$= \binom{n}{x}p^x(1-p)^{n-x}$$

$$= \binom{n}{x}(0.29)^x(0.71)^{n-x}$$

where $n = 1, 2, 3, \ldots$ and $x = 0, 1, \ldots n$. This is the general expression for the probability distribution of a binomial random variable *X* where *X* is the number of smokers in a sample of size *n*. Given a total of *n* adults, $n!$—or *n factorial*—allows us to calculate the number of ways in which the *n* individuals can be ordered; note that we have *n* choices for the first position, $n - 1$ choices for the second position, and so on. Thus,

$$n! = n(n-1)(n-2) \cdots (3)(2)(1).$$

By definition, $0!$ is equal to 1. The expression

$$\binom{n}{x} = \frac{n!}{x!(n-x)!}$$

is the *combination* of *n* objects chosen *x* at a time; it represents the number of ways in which *x* objects can be selected from a total of *n* objects without regard to order. For example, if we were to randomly choose three individuals from the adult population, there would be

$$\binom{3}{0} = \frac{3!}{0!(3-0)!} = \frac{6}{(1)(6)} = 1$$

way in which we could select zero smokers; in particular, all three adults would have to be nonsmokers. There would be

$$\binom{3}{1} = \frac{3!}{1!(3-1)!} = \frac{6}{(1)(2)} = 3$$

ways in which we could choose one smoker, since the smoker could be the first person, the second person, or the third person. Similarly, there would be

$$\binom{3}{2} = \frac{3!}{2!(3-2)!} = \frac{6}{(2)(1)} = 3$$

ways in which we could select two smokers, and only

$$\binom{3}{3} = \frac{3!}{3!(3-3)!} = \frac{6}{(6)(1)} = 1$$

way for all three adults to be smokers. Therefore, as we found earlier,

$$P(X = 0) = \binom{3}{0}p^0(1-p)^{3-0}$$
$$= 1(0.29)^0(0.71)^3$$
$$= 0.358,$$

$$P(X = 1) = \binom{3}{1}p^1(1-p)^{3-1}$$
$$= 3(0.29)(0.71)^2$$
$$= 0.439,$$

$$P(X = 2) = \binom{3}{2}p^2(1-p)^{3-2}$$
$$= 3(0.29)^2(0.71)$$
$$= 0.179,$$

and

$$P(X = 3) = \binom{3}{3}p^3(1-p)^{3-3}$$
$$= 1(0.29)^3(0.71)^0$$
$$= 0.024.$$

Rather than perform these calculations by hand—and assuming that we do not have one of the many computer programs written to calculate them for us—we can use Table A.1 in Appendix A to obtain the binomial probabilities for selected values of n and p. The number of trials n appears in the first column on the left-hand side of the table for $n \leq 20$. The number of successes x is in the second column and takes integer values from 0 to n. The probability p appears in the row across the top. For specified values of n, x, and p, the entry in the body of the table represents

$$P(X = x) = \binom{n}{x}p^x(1-p)^{n-x}.$$

Once again, suppose that we randomly choose three individuals from the adult population and wish to find the probability that exactly two of them are smokers. We first locate $n = 3$ on the left-hand side of the table, then select the row that corresponds to $x = 2$. Rounding the probability $p = 0.29$ to 0.3, we find the column corresponding to $p = 0.3$. This allows us to approximate the probability that exactly two of the three adults are smokers by 0.189. (This result differs from 0.179, the probability calculated above, because it was necessary to round the value of p.)

What if we choose three adults from the population and wish to find the probability that exactly two of them are nonsmokers? In this case, we want to determine the binomial probability corresponding to $n = 3$, $x = 2$, and $p = 0.71$. Even if we round p to 0.7, however, Table A.1 does not contain any value of p greater than 0.5. The way to solve this problem is to realize that if two of the three individuals are nonsmokers, the other one must be a smoker. Therefore, we simply use the table to find $P(X = 1 \mid n = 3, p = 0.3)$, which is mathematically equivalent to $P(X = 2 \mid n = 3, p = 0.7)$.

In addition to the probabilities of individual outcomes, we can also compute the numerical summary measures associated with a probability distribution. For example, the mean value of a binomial random variable X—or the average number of "successes" in repeated samples of size n—is obtained by multiplying the number of independent Bernoulli trials by the probability of success at each trial; hence, the mean value of X is equal to np. The variance of X is $np(1 - p)$. These expressions were derived using a method analogous to that for finding the mean and variance of grouped data [3]. Applying the formulas, if we were to select repeated samples of size $n = 10$ from the adult population, the mean number of smokers per sample would be

$$np = 10(0.29)$$
$$= 2.9,$$

and the standard deviation would be

$$\sqrt{np(1 - p)} = \sqrt{10(0.29)(0.71)}$$
$$= \sqrt{2.059}$$
$$= 1.4.$$

Upon inspection, the expression for the variance of a binomial random variable seems quite reasonable. The quantity $np(1 - p)$ is largest when p is equal to 0.5; it decreases as p approaches 0 or 1. When p is very large or very small, nearly all the outcomes take the same value—for example, almost everyone smokes or almost everyone does not smoke—and the variability among outcomes is small. In contrast, if half the population takes the value 0 and the other half takes the value 1, it will be more difficult to predict any particular outcome; in this case, the variability is relatively large.

Figure 7.2 is a graph of the probability distribution of X—the number of smokers—for which $n = 10$ and $p = 0.29$. Because all the possible outcomes of X are considered, the area represented by the vertical bars sums to 1. Figure 7.3 is the probability distribution of another binomial random variable for which $n = 10$ and $p = 0.71$. Note that the distribution is skewed to the right when $p < 0.5$ and skewed to the left when $p > 0.5$. If $p = 0.5$, as is the case in Figure 7.4, the probability distribution is symmetric.

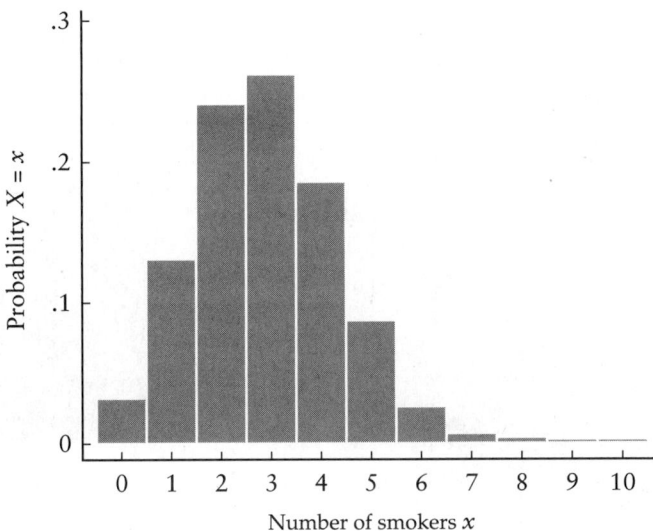

FIGURE 7.2
Probability distribution of a binomial random variable for which $n = 10$ and $p = 0.29$

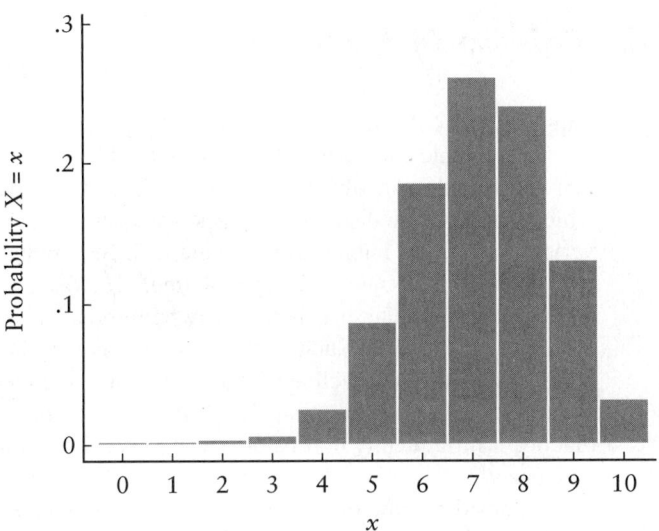

FIGURE 7.3
Probability distribution of a binomial random variable for which $n = 10$ and $p = 0.71$

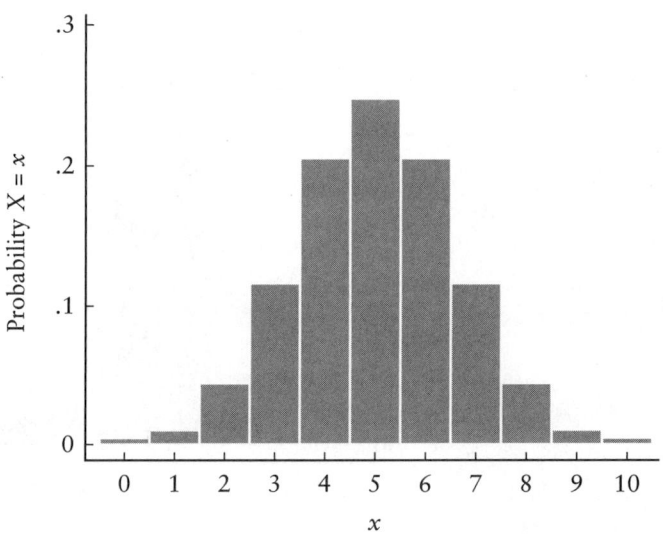

FIGURE 7.4
Probability distribution of a binomial random variable for which $n = 10$ and $p = 0.50$

7.3   The Poisson Distribution

Suppose that X is a random variable that represents the number of individuals involved in a motor vehicle accident each year. In the United States, the probability that a particular individual is involved is 0.00024 [4]. Technically, this is a binomial situation in which there are two distinct outcomes—accident or no accident. Note, however, that n is very large; we are interested in the entire U.S. population. When n becomes large, the combination of n objects taken x at a time, $n!/x!(n-x)!$, is very tedious to evaluate. As a result, the binomial distribution may be impractical to use as a basis for calculations. However, in situations such as this—when n is very large and p is very small—the binomial distribution is well approximated by another theoretical probability distribution, called the Poisson distribution. The *Poisson distribution* is used to model discrete events that occur infrequently in time or space; hence, it is sometimes called the *distribution of rare events*.

Consider a random variable X that represents the number of occurrences of some event of interest over a given interval. Since X is a count, it can theoretically assume any integer value between 0 and infinity. Let λ (the Greek letter lambda) be a constant that denotes the average number of occurrences of the event in an interval. If the probability that X assumes the value x is

$$P(X = x) = \frac{e^{-\lambda}\lambda^x}{x!},$$

X is said to have a Poisson distribution with parameter λ. The symbol e represents a constant that is approximated by 2.71828; in fact, e is the base of the natural logarithms. Like the binomial distribution, the Poisson distribution involves a set of underlying assumptions:

1. The probability that a single event occurs within an interval is proportional to the length of the interval.

2. Within a single interval, an infinite number of occurrences of the event are theoretically possible. We are not restricted to a fixed number of trials.

3. The events occur independently both within the same interval and between consecutive intervals.

The Poisson distribution can be used to model the number of ambulances needed in a city in a given night, the number of particles emitted from a specified amount of radioactive material, or the number of bacterial colonies growing in a Petri dish.

Recall that the mean of a binomial random variable is equal to np and that its variance is $np(1 - p)$. When p is very small, $1 - p$ is close to 1 and $np(1 - p)$ is approximately equal to np. In this case, the mean and the variance of the distribution are identical and can be represented by the single parameter λ. The property that the mean is equal to the variance is an identifying characteristic of the Poisson distribution.

Suppose that we are interested in determining the number of people in a population of 10,000 who will be involved in a motor vehicle accident each year. The mean number of persons involved would be

$$\lambda = np$$
$$= (10,000)(0.00024)$$
$$= 2.4;$$

this is also the variance. The probability that no one in this population will be involved in an accident in a given year is

$$P(X = 0) = \frac{e^{-2.4}(2.4)^0}{0!}$$
$$= 0.091.$$

The probability that exactly one person will be involved is

$$P(X = 1) = \frac{e^{-2.4}(2.4)^1}{1!}$$
$$= 0.218.$$

Similarly,

$$P(X = 2) = \frac{e^{-2.4}(2.4)^2}{2!}$$
$$= 0.261,$$

$$P(X = 3) = \frac{e^{-2.4}(2.4)^3}{3!}$$
$$= 0.209,$$

$$P(X = 4) = \frac{e^{-2.4}(2.4)^4}{4!}$$
$$= 0.125,$$

$$P(X = 5) = \frac{e^{-2.4}(2.4)^5}{5!}$$
$$= 0.060,$$

and

$$P(X = 6) = \frac{e^{-2.4}(2.4)^6}{6!}$$
$$= 0.024.$$

Since the outcomes of X are mutually exclusive and exhaustive,

$$P(X \geq 7) = 1 - P(X < 7)$$
$$= 1 - (0.091 + 0.218 + 0.261 + 0.209$$
$$+ 0.125 + 0.060 + 0.024)$$
$$= 0.012.$$

Instead of performing the calculations by hand—or using a computer program—we can use Table A.2 in Appendix A to obtain Poisson probabilities for selected values of λ. The number of successes x appears in the first column on the left-hand side of the table; λ is in the row across the top. For specified values of x and λ, the entry in the table represents

$$P(X = x) = \frac{e^{-\lambda}\lambda^x}{x!}.$$

In a population of 10,000 people, what is the probability that exactly three of them will be involved in a motor vehicle accident in a given year? We begin by locating $x = 3$ in the first column of Table A.2. Rounding 2.4 up to 2.5, we find the column corresponding to $\lambda = 2.5$. The table tells us that we can approximate the probability that exactly three individuals are involved in an accident by 0.214. (Again this result differs from 0.209, the probability calculated above, because it was necessary to round the value of the parameter λ in order to use the table.)

Figure 7.5 is a graph of the probability distribution of X, the number of individuals in the population involved in a motor vehicle accident each year. The area represented by the vertical bars sums to 1. As shown in Figure 7.6, the Poisson distribution is highly skewed for small values of λ; as λ increases, the distribution becomes more symmetric.

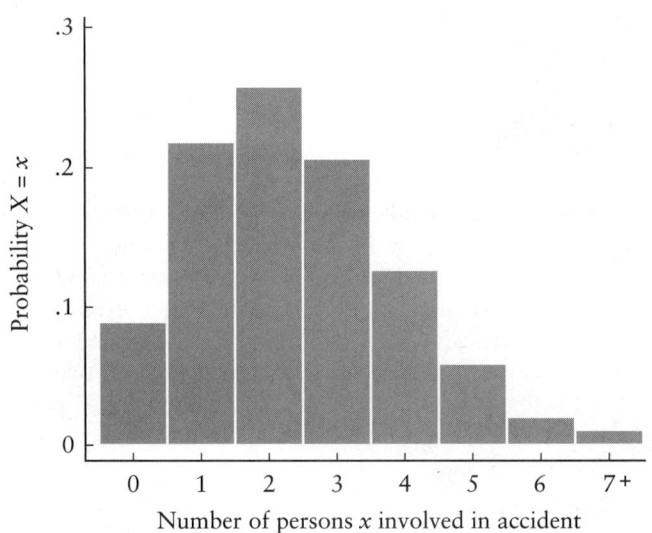

FIGURE 7.5
Probability distribution of a Poisson random variable for which $\lambda = 2.4$

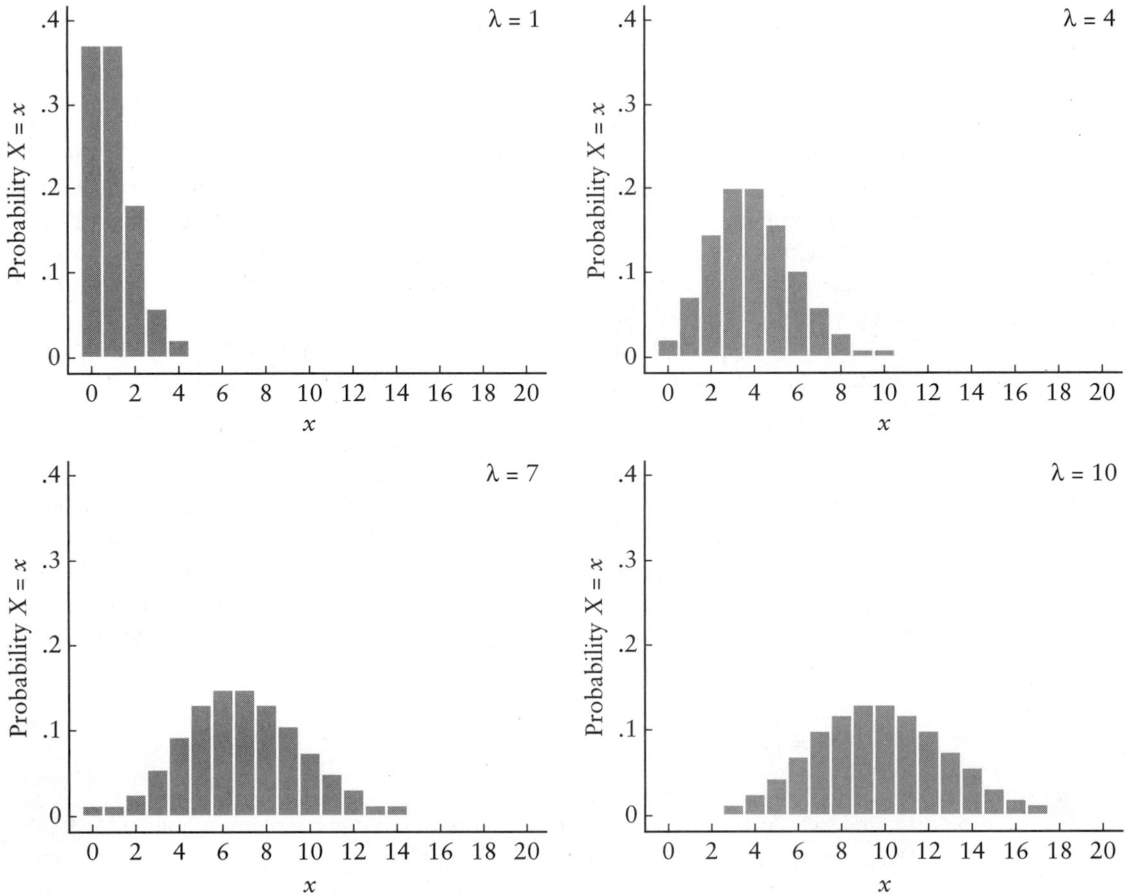

FIGURE 7.6
Probability distributions of Poisson random variables for various values of λ

7.4 The Normal Distribution

When it follows either a binomial or a Poisson distribution, a random variable X is restricted to taking on integer values only. Under different circumstances, however, the outcomes of a random variable may not be limited to integers or counts. Suppose that X represents height. Rarely is an individual exactly 67 inches tall or exactly 68 inches tall; theoretically, X can assume an infinite number of intermediate values, such as 67.04 inches or 67.8352 inches. In fact, between any two possible outcomes of X we can always find a third value. Although we could argue philosophically that we can measure only discrete outcomes due to the limitations of our measuring instruments—perhaps we can measure height only to the nearest tenth of an inch—treating such a variable as if it were continuous allows us to take advantage of powerful mathematical results.

As we have seen, the probability distribution of a discrete random variable is represented by an equation for $P(X = x)$, the probability that the random variable X will take on the specific value x. For a binomial random variable with parameters n and p, for instance,

$$P(X = x) = \binom{n}{x} p^x (1 - p)^{n-x}.$$

These probabilities can be plotted against x, as in Figure 7.4. Suppose that the number of possible outcomes of X were to become very large and the widths of the corresponding intervals very small. In Figure 7.7, for example, $n = 30$ and $p = 0.50$. In general, if the number of possible values of X approaches infinity while the widths of the intervals approach zero, the graph will increasingly resemble a smooth curve. A smooth curve is used to represent the probability distribution of a continuous random variable; the curve is called a *probability density*.

For any graph that illustrates a discrete probability distribution, the area represented by the vertical bars sums to 1. For a probability density, the total area beneath the curve must also be 1. Because a continuous random variable X can take on an uncountably infinite number of values, the probability associated with any particular one of them is equal to 0. However, the probability that X will assume some value in the interval enclosed by the outcomes x_1 and x_2 is equal to the area beneath the curve that lies between these two values.

The most common continuous distribution is the *normal distribution*, also known as the *Gaussian distribution* or the *bell-shaped curve*. Its shape is that of a binomial distribution for which p is constant but n approaches infinity, or a Poisson distribution for which λ approaches infinity. Its probability density is given by the equation

$$f(x) = \frac{1}{\sqrt{2\pi}\sigma} e^{-\frac{1}{2}\left(\frac{x-\mu}{\sigma}\right)^2},$$

where $-\infty < x < \infty$. The symbol π (pi) represents a constant approximated by 3.14159. The normal curve is unimodal and symmetric about its mean μ (mu); in this special case, the mean, median, and mode of the distribution are all identical. The standard de-

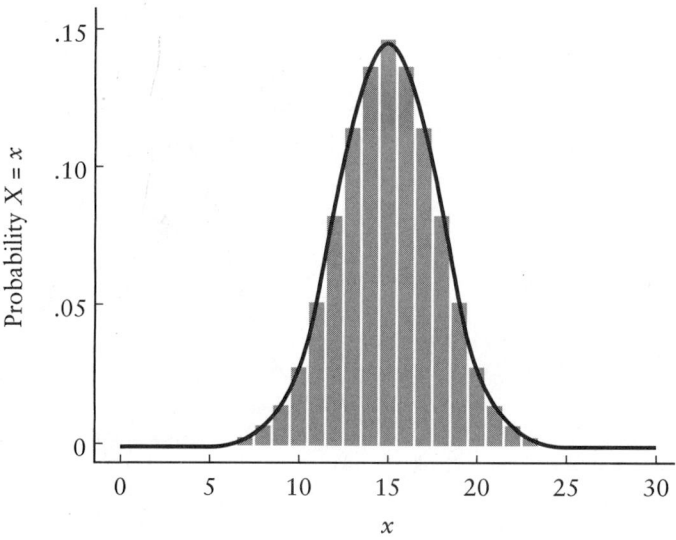

FIGURE 7.7
Probability distribution of a binomial random variable for which $n = 30$
and $p = 0.50$

viation, represented by σ (sigma), specifies the amount of dispersion around the mean. Together, the two parameters μ and σ completely define a normal curve.

The value of the normal distribution will become more apparent when we begin to work with the sampling distribution of the mean. For now, however, it is important to note that many random variables of interest—including blood pressure, serum cholesterol level, height, and weight—are approximately normally distributed. The normal curve can thus be used to estimate probabilities associated with these variables. For example, in a population in which serum cholesterol level is normally distributed with mean μ and standard deviation σ, we might wish to find the probability that a randomly chosen individual has a serum cholesterol level greater than 250 mg/100 ml. Perhaps this knowledge will help us to plan for future cardiac services. Since the total area beneath the normal curve is equal to 1, we can estimate the probability in question by determining the proportion of the area under the curve that lies to the right of the point $x = 250$, or $P(X > 250)$. This can be done using a computer program or a table of areas calculated for the normal curve.

Since a normal distribution could have an infinite number of possible values for its mean and standard deviation, it is impossible to tabulate the area associated with each and every normal curve. Instead, only a single curve is tabulated—the special case for which $\mu = 0$ and $\sigma = 1$. This curve is known as the *standard normal distribution*. Figure 7.8 illustrates the standard normal curve, and Table A.3 in Appendix A displays the areas in the upper tail of the distribution. Outcomes of the random variable Z are denoted by z; the whole number and tenths decimal place of z are listed in the column to the left of the table, and the hundredths decimal place is shown in the row across the top. For a particular value of z, the entry in the body of the table specifies the area

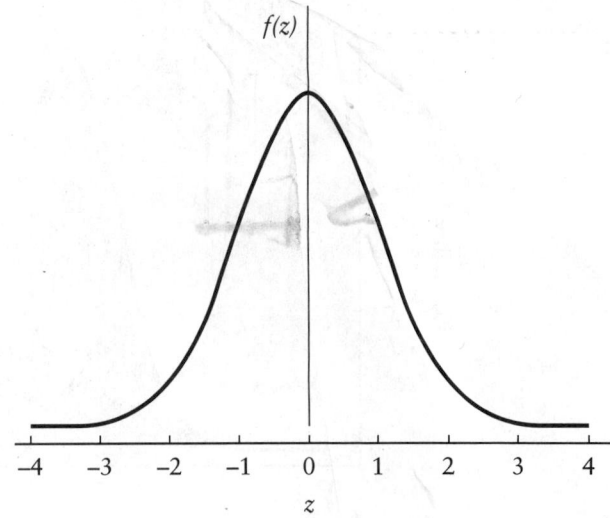

FIGURE 7.8

The standard normal curve for which $\mu = 0$ and $\sigma = 1$

beneath the curve to the right of z, or $P(Z > z)$. Some sample values of z and their corresponding areas are as follows:

z	Area in Right Tail
0.00	0.500
1.65	0.049
1.96	0.025
2.58	0.005
3.00	0.001

Thus, for instance, $P(Z > 2.58) = 0.005$. Since the standard normal distribution is symmetric about $z = 0$, the area under the curve to the right of z is equal to the area to the left of $-z$.

$-z$	Area in Left Tail
0.00	0.500
-1.65	0.049
-1.96	0.025
-2.58	0.005
-3.00	0.001

Suppose that we wish to know the area under the standard normal curve that lies between $z = -1.00$ and $z = 1.00$; since $\mu = 0$ and $\sigma = 1$, this is the area contained in the interval $\mu \pm 1\sigma$, illustrated in Figure 7.9. Equivalently, it is $P(-1 \leq Z \leq 1)$. Looking at Table A.3, we see that the area to the right of $z = 1.00$ is $P(Z > 1) = 0.159$. Therefore, the area to the left of $z = -1.00$ must be 0.159 as well. The events that $Z > 1$ and $Z < -1$ are mutually exclusive; consequently, applying the additive rule of probability, the sum of the area to the right of 1 and to the left of -1 is

$$P(Z > 1) + P(Z < -1) = 0.159 + 0.159$$
$$= 0.318.$$

Since the total area under the curve is equal to 1, the area between -1 and 1 must be

$$P(-1 \leq Z \leq 1) = 1 - [P(Z > 1) + P(Z < -1)]$$
$$= 1 - 0.318$$
$$= 0.682.$$

Therefore, for the standard normal distribution, approximately 68.2% of the area beneath the curve lies within ± 1 standard deviation from the mean.

We might also wish to calculate the area under the standard normal curve that is contained in the interval $\mu \pm 2\sigma$, or $P(-2 \leq Z \leq 2)$. This area is illustrated in Figure 7.10. Table A.3 indicates that the area to the right of $z = 2.00$ is 0.023; the area to the left of $z = -2.00$ is 0.023 as well. Therefore, the area between -2.00 and 2.00 must be

$$P(-2 \leq Z \leq 2) = 1 - [P(Z > 2) + P(Z < -2)]$$
$$= 1.000 - [0.023 + 0.023]$$
$$= 0.954.$$

Approximately 95.4% of the area under the standard normal curve lies within ± 2 standard deviations from the mean. The two preceding calculations form the basis of the empirical rule described in Section 3.4, which states that if a distribution of values is

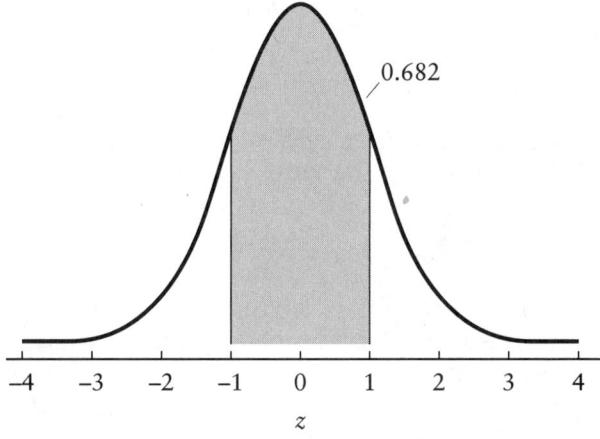

FIGURE 7.9
The standard normal curve, area between $z = -1.00$ and $z = 1.00$

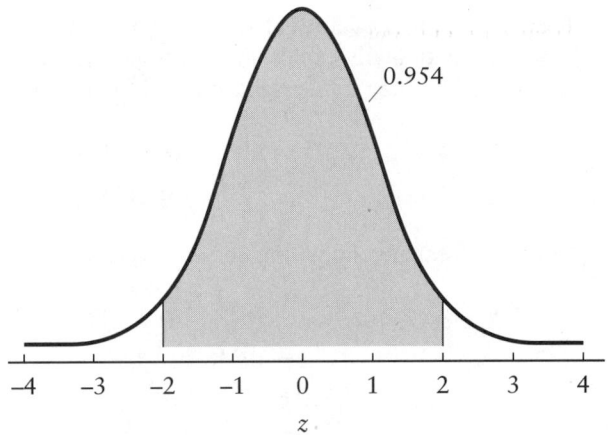

FIGURE 7.10

The standard normal curve, area between $z = -2.00$ and $z = 2.00$

symmetric and unimodal, approximately 67% of the observations lie within one standard deviation of the mean and about 95% lie within two standard deviations.

Table A.3 can also be used the other way around. For example, we might wish to find the value of z that cuts off the upper 10% of the standard normal distribution, or the value of z for which $P(Z > z) = 0.10$. Locating 0.100 in the body of the table, we observe that the corresponding value of z is 1.28. Therefore, 10% of the area under the standard normal curve lies to the right of $z = 1.28$; this area is shown in Figure 7.11. Similarly, another 10% of the area lies to the left of $z = -1.28$.

Now suppose that X is a normal random variable with mean 2 and standard deviation 0.5. Subtracting 2 from X would give us a normal random variable that has mean 0; as shown in Figure 7.12, the whole distribution would be shifted two units to the left. Dividing $(X - 2)$ by 0.5 alters the spread of the distribution so that we have a normal random variable with standard deviation 1. Therefore, if X is a normal random variable with mean 2 and standard deviation 0.5,

$$Z = \frac{X - 2}{0.5}$$

is a standard normal random variable. In general, for any arbitrary normal random variable with mean μ and standard deviation σ,

$$Z = \frac{X - \mu}{\sigma}$$

has a standard normal distribution. By transforming X into Z, we can use a table of areas computed for the standard normal curve to estimate probabilities associated with X. An outcome of the random variable Z, denoted z, is known as a *standard normal deviate* or a *z-score*.

For example, let X be a random variable that represents systolic blood pressure. For the population of 18- to 74-year-old males in the United States, systolic blood pressure is approximately normally distributed with mean 129 millimeters of mercury (mm Hg) and standard deviation 19.8 mm Hg [5]. This distribution is shown in Figure 7.13. Note that

$$Z = \frac{X - 129}{19.8}$$

is normally distributed with mean 0 and standard deviation 1.

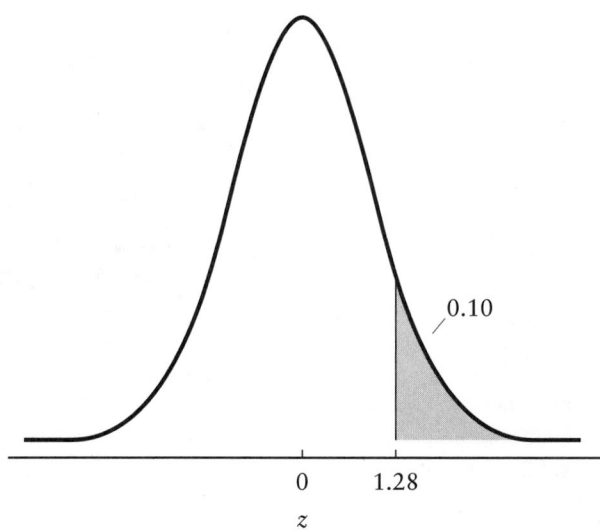

FIGURE 7.11
The standard normal curve, area to the right of $z = 1.28$

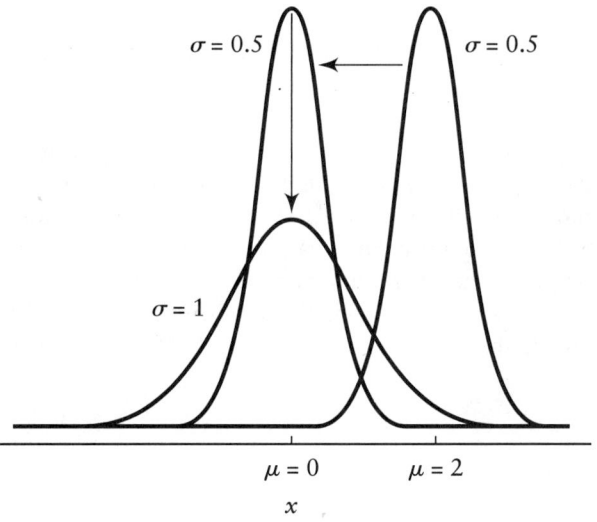

FIGURE 7.12
Transforming a normal curve with mean 2 and standard deviation 0.5 into
the standard normal curve

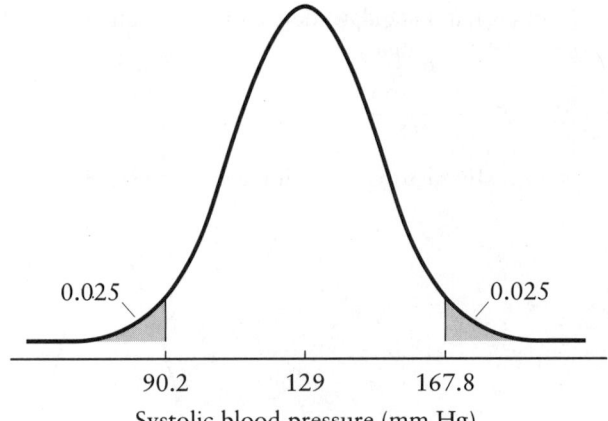

0.025   0.025

90.2   129   167.8
Systolic blood pressure (mm Hg)

FIGURE 7.13
Distribution of systolic blood pressure for males 18 to 74 years of age,
United States, 1976–1980

Suppose we wish to find the value of x that cuts off the upper 2.5% of the curve of systolic blood pressures, or, equivalently, the value of x for which $P(X > x) = 0.025$. Using Table A.3, we see that the area to the right of $z = 1.96$ is 0.025. To obtain the value of x that corresponds to this value of z, we solve the equation

$$z = 1.96$$
$$= \frac{x - 129}{19.8}$$

or

$$x = 129 + (1.96)(19.8)$$
$$= 167.8.$$

Therefore, approximately 2.5% of the men in this population—a minuscule minority—have systolic blood pressures that are greater than 167.8 mm Hg, while 97.5% have blood pressures less than 167.8 mm Hg. In other words, if we randomly select an individual from this adult male population, the probability that his systolic blood pressure is greater than 167.8 mm Hg is 0.025.

Because the standard normal curve is symmetric around $z = 0$, we know that the area to the left of $z = -1.96$ is also 0.025. By solving the equation

$$z = -1.96$$
$$= \frac{x - 129}{19.8}$$

or

$$x = 129 + (-1.96)(19.8)$$
$$= 90.2,$$

we find that 2.5% of the men have a systolic blood pressure that is less than 90.2 mm Hg. Equivalently, the probability that a randomly selected male has a systolic blood pressure less than 90.2 mm Hg is 0.025. Since 2.5% of the men in the population have systolic blood pressures greater than 167.8 mm Hg and 2.5% have values less than 90.2 mm Hg, the remaining 95% of the men must have systolic blood pressure readings that lie between 90.2 and 167.8 mm Hg.

We might also be interested in determining the proportion of men in the population who have systolic blood pressures greater than 150 mm Hg. In this case, we are given the outcome of the random variable X and must solve for the normal deviate z:

$$z = \frac{150 - 129}{19.8}$$
$$= 1.06.$$

The area to the right of $z = 1.06$ is 0.145. Therefore, approximately 14.5% of the men in this population have systolic blood pressures greater than 150 mm Hg.

Now consider the more complicated situation in which we have two normally distributed random variables. In an Australian national study of risk factor prevalence, two of the populations investigated are men whose blood pressures are within a normal or accepted range and who are not taking any corrective medication, and men who have had high blood pressure but who are at present undergoing antihypertensive drug therapy [6].

For the population of men who are not taking corrective medication, diastolic blood pressure is approximately normally distributed with mean $\mu_n = 80.7$ mm Hg and standard deviation $\sigma_n = 9.2$ mm Hg. For the men who are using antihypertensive drugs, diastolic blood pressure is also approximately normally distributed with mean $\mu_a = 94.9$ mm Hg and standard deviation $\sigma_a = 11.5$ mm Hg. These two distributions are shown in Figure 7.14. Our goal is to be able to determine whether a man has normal

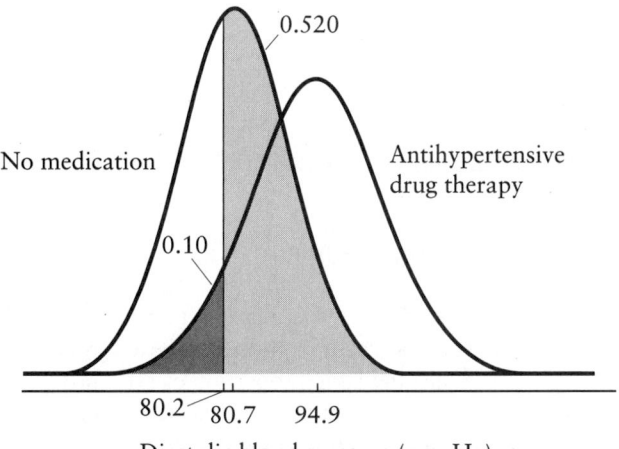

FIGURE 7.14
Distribution of systolic blood pressure for two populations of Australian males, 1980

blood pressure or whether he is taking antihypertensive medication solely on the basis of his diastolic blood pressure reading. This seemingly unimportant exercise is valuable in that it provides us with a foundation for hypothesis testing.

The first thing to notice is that because of the large amount of overlap between the two normal curves, it will be difficult to distinguish between them. Nevertheless, we will proceed: If our goal is to identify 90% of the individuals who are currently taking medication, what value of diastolic blood pressure should be designated as the lower cutoff point? Equivalently, we must find the value of diastolic blood pressure that marks off the lower 10% of this distribution. Looking at Table A.3, we find that $z = -1.28$ cuts off an area of 0.10 in the lower tail of the standard normal curve. Therefore, solving for x,

$$z = -1.28$$
$$= \frac{x - 94.9}{11.5}$$

and

$$x = 94.9 + (-1.28)(11.5)$$
$$= 80.2.$$

Approximately 90% of the men taking antihypertensive drugs have diastolic blood pressures that are greater than 80.2 mm Hg. If we use this value as our cutoff point, the other 10% of the men—those with readings below 80.2 mm Hg—represent false negatives; they are individuals currently using medication who are not identified as such.

What proportion of the men with normal blood pressures will be incorrectly labeled as antihypertensive drug users? These are the men in the drug-free population who have diastolic blood pressure readings greater than 80.2 mm Hg. Solving for z (notice that we are now using the mean and standard deviation of the population of men who are not taking corrective medication),

$$z = \frac{80.2 - 80.7}{9.2}$$
$$= -0.05.$$

An area of 0.480 lies to the left of -0.05; therefore, the area to the right of $z = -0.05$ must be

$$1.000 - 0.480 = 0.520.$$

Approximately 52% of the men with normal blood pressures would be incorrectly labeled as using medication. Note that these errors are false positive results.

To reduce the large proportion of false positive errors, the cutoff point for identifying individuals who are currently using antihypertensive drugs could be raised. If the cutoff were 90 mm Hg, for example, then

$$z = \frac{90 - 80.7}{9.2}$$
$$= 1.01,$$

and only 15.6% of the men with normal blood pressures would be wrongly classified as taking medication.

When the cutoff is raised, however, the proportion of men correctly labeled as using antihypertensive medication decreases; note that

$$z = \frac{90 - 94.9}{11.5}$$
$$= -0.43.$$

The area to the left of $z = -0.43$ is 0.334, and

$$1.000 - 0.334 = 0.666;$$

therefore, only 66.6% of the men using antihypertensive drugs would be identified. The remaining 33.4% of these men would be false negatives.

A trade-off always exists when we try to manipulate proportions of false negative and false positive outcomes. This is the same phenomenon that was observed when we were investigating the sensitivity and specificity of a diagnostic test. In general, a smaller proportion of false positive errors can be achieved only by increasing the probability of a false negative outcome, and the proportion of false negatives can be reduced only by raising the probability of a false positive. The relationship between these two types of errors in a specific application is determined by the amount of overlap in the two normal populations of interest.

7.5   Further Applications

Suppose that we are interested in investigating the probability that a patient who has been stuck with a needle infected with hepatitis B actually develops the disease. Let Y be a Bernoulli random variable that represents the disease status of a patient who has been exposed to an infected needle; Y takes the value 1 if the individual develops hepatitis and 0 if he or she does not. These two outcomes are mutually exclusive and exhaustive. If 30% of the patients who are exposed to hepatitis B become infected [7], then

$$P(Y = 1) = p$$
$$= 0.30,$$

and

$$P(Y = 0) = 1 - p$$
$$= 1 - 0.30$$
$$= 0.70.$$

If we have n independent observations of a dichotomous random variable such that each observation has a constant probability of "success" p, the total number of "successes" X follows a binomial distribution. The random variable X can assume any integer value between 0 and n; the probability that X takes on a particular value x can be expressed as

$$P(X = x) = \binom{n}{x} p^x (1 - p)^{n-x}.$$

Suppose that we select five individuals from the population of patients who have been stuck with a needle infected with hepatitis B. The number of patients in this sample who develop the disease is a binomial random variable with parameters $n = 5$ and $p = 0.30$. Its probability distribution can be represented in the following way:

$$P(X = 0) = \binom{5}{0}(0.30)^0(0.70)^{5-0}$$
$$= (1)(1)(0.70)^5$$
$$= 0.168,$$

$$P(X = 1) = \binom{5}{1}(0.30)^1(0.70)^{5-1}$$
$$= (5)(0.30)(0.70)^4$$
$$= 0.360,$$

$$P(X = 2) = \binom{5}{2}(0.30)^2(0.70)^{5-2}$$
$$= (10)(0.30)^2(0.70)^3$$
$$= 0.309,$$

$$P(X = 3) = \binom{5}{3}(0.30)^3(0.70)^{5-3}$$
$$= (10)(0.30)^3(0.70)^2$$
$$= 0.132,$$

$$P(X = 4) = \binom{5}{4}(0.30)^4(0.70)^{5-4}$$
$$= (5)(0.30)^4(0.70)$$
$$= 0.028,$$

and

$$P(X = 5) = \binom{5}{5}(0.30)^5(0.70)^{5-5}$$
$$= (1)(0.30)^5(1)$$
$$= 0.002.$$

Rather than calculate these probabilities by hand, we could have consulted Table A.1 in Appendix A. Alternatively, many statistical packages will generate probabilities associated with a binomial random variable; Table 7.2 shows the relevant output from Minitab.

TABLE 7.2
Minitab output displaying the
probability distribution of a
binomial random variable with
parameters $n = 5$ and $p = 0.30$

```
BINOMIAL WITH N=5 P=0.30
     K        P(X=K)
     0        0.1681
     1        0.3601
     2        0.3087
     3        0.1323
     4        0.0284
     5        0.0024
```

The probability that at least three individuals among the five develop hepatitis B is

$$P(X \geq 3) = P(X = 3) + P(X = 4) + P(X = 5)$$
$$= 0.132 + 0.028 + 0.003$$
$$= 0.163;$$

the probability that at most one patient develops the disease is

$$P(X \leq 1) = P(X = 0) + P(X = 1)$$
$$= 0.168 + 0.360$$
$$= 0.528.$$

In addition, the mean number of persons who would develop the disease in repeated samples of size 5 is $np = 5(0.3) = 1.5$, and the standard deviation is $\sqrt{np(1 - p)} = \sqrt{5(0.3)(0.7)} = \sqrt{1.05} = 1.03$.

If X represents the number of occurrences of some event in a specified interval of time or space such that both the mean number of occurrences and the population variance are equal to λ, X has a Poisson distribution with parameter λ. The random variable X can take on any integer value between 0 and ∞; the probability that X assumes a particular value x is

$$P(X = x) = \frac{e^{-\lambda}\lambda^x}{x!}.$$

Suppose that we are concerned with the possible spread of diphtheria and wish to know how many cases we can expect to see in a particular year. Let X represent the number of cases of diphtheria reported in the United States in a given year between 1980 and 1989. The random variable X has a Poisson distribution with parameter $\lambda = 2.5$ [8]; the probability distribution of X may be expressed as

$$P(X = x) = \frac{e^{-2.5}(2.5)^x}{x!}.$$

Therefore, the probability that no cases of diphtheria will be reported during a given year is

$$P(X = 0) = \frac{e^{-2.5}(2.5)^0}{0!}$$
$$= 0.082.$$

The probability that a single case will be reported is

$$P(X = 1) = \frac{e^{-2.5}(2.5)^1}{1!}$$
$$= 0.205;$$

similarly,

$$P(X = 2) = \frac{e^{-2.5}(2.5)^2}{2!}$$
$$= 0.257,$$

$$P(X = 3) = \frac{e^{-2.5}(2.5)^3}{3!}$$
$$= 0.214,$$

$$P(X = 4) = \frac{e^{-2.5}(2.5)^4}{4!}$$
$$= 0.134,$$

and

$$P(X = 5) = \frac{e^{-2.5}(2.5)^5}{5!}$$
$$= 0.067.$$

We could have consulted Table A.2 in Appendix A to determine these probabilities, or we could have used a statistical package.

Since the outcomes of X are mutually exclusive and exhaustive,

$$P(X \geq 4) = 1 - P(X < 4)$$
$$= 1 - (0.082 + 0.205 + 0.257 + 0.214)$$
$$= 0.242.$$

There is a 24.2% chance that we will observe four or more cases of diptheria in a given year. Similarly, the probability that we will observe six or more cases is

$$P(X \geq 6) = 1 - P(X < 6)$$
$$= 1 - (0.082 + 0.205 + 0.257 + 0.214 + 0.134 + 0.067)$$
$$= 0.041.$$

The mean number of cases per year is $\lambda = 2.5$, and the standard deviation is $\sqrt{\lambda} = \sqrt{2.5} = 1.58$.

If X can assume any value within a specified interval rather than being restricted to integers only, X is a continuous random variable. The most common continuous distribution is the normal distribution. The normal distribution is defined by two parameters—its mean μ and standard deviation σ. The mean specifies the center of the distribution; the standard deviation quantifies the amount of spread or dispersion around the mean. The shape of the normal distribution indicates that outcomes of the random variable X that are close to the mean are more likely to occur than values that are far from it.

The normal distribution with mean $\mu = 0$ and standard deviation $\sigma = 1$ is known as the standard normal distribution. Because its area has been tabulated, it is used to obtain probabilities associated with normal random variables. For example, suppose that we wish to know the area under the standard normal curve that lies between $z = -3.00$ and $z = 3.00$; equivalently, this is the area in the interval $\mu \pm 3\sigma$, shown in Figure 7.15. Looking at Table A.3, we find the area to the right of $z = 3.00$ to be 0.001. Since the standard normal curve is symmetric, the area to the left of $z = -3.00$ must be 0.001 as well. Therefore, the area between -3.00 and 3.00 is

$$P(-3 \leq Z \leq 3) = 1 - [P(Z < -3) + P(Z > 3)]$$
$$= 1 - 0.001 - 0.001$$
$$= 0.998;$$

approximately 99.8% of the area under a standard normal curve lies within ± 3 standard deviations from the mean.

If X is an arbitrary normal random variable with mean μ and standard deviation σ, then

$$Z = \frac{X - \mu}{\sigma}$$

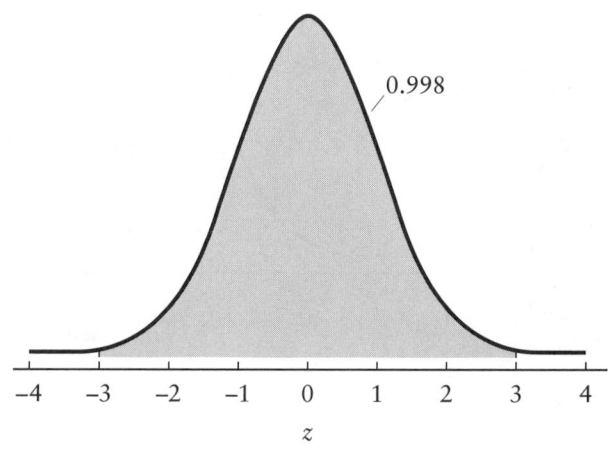

FIGURE 7.15
The standard normal curve, area between $z = -3.00$ and $z = 3.00$

is a standard normal random variable. By transforming X into Z, we can use the table of areas for the standard normal curve to estimate probabilities associated with X.

For instance, suppose that X is a random variable that represents height. For the population of 18- to 74-year-old females in the United States, height is normally distributed with mean $\mu = 63.9$ inches and standard deviation $\sigma = 2.6$ inches [9]. This distribution is illustrated in Figure 7.16. Observe that

$$Z = \frac{X - 63.9}{2.6}$$

is a standard normal random variable.

If we randomly select a woman from this population, what is the probability that she is between 60 and 68 inches tall? For $x = 60$,

$$z = \frac{60 - 63.9}{2.6}$$
$$= -1.50,$$

and for $x = 68$,

$$z = \frac{68 - 63.9}{2.6}$$
$$= 1.58.$$

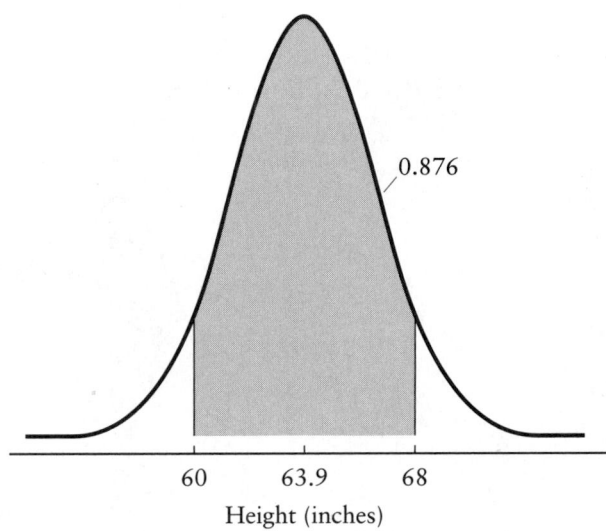

FIGURE 7.16
Distribution of height for females 18 to 74 years of age, United States, 1976–1980

As a result, the probability that x—the woman's height—lies between 60 and 68 inches is equal to the probability that z lies between -1.50 and 1.58 for the standard normal curve. The area to the left of $z = -1.50$ is 0.067, and the area to the right of $z = 1.58$ is 0.057. (Rather than Table A.3, we could use a statistical package to generate these probabilities.) Since the total area under the curve is equal to 1, the area between -1.50 and 1.58 must be

$$
\begin{aligned}
P(60 \leq X \leq 68) &= P(-1.50 \leq Z \leq 1.58) \\
&= 1 - [P(Z < -1.50) + P(Z > 1.58)] \\
&= 1 - [0.067 + 0.057] \\
&= 0.876.
\end{aligned}
$$

The probability that the woman's height is between 60 and 68 inches is 0.876.

We might also wish to know the value of height that cuts off the upper 5% of this distribution. From Table A.3, we observe that a tail area of 0.050 corresponds to $z = 1.645$. Solving for x,

$$
\begin{aligned}
z &= 1.645 \\
&= \frac{x - 63.9}{2.6}
\end{aligned}
$$

and

$$
\begin{aligned}
x &= 63.9 + (1.645)(2.6) \\
&= 68.2.
\end{aligned}
$$

Approximately 5% of the women in this population are taller than 68.2 inches.

7.6   Review Exercises

1. What is a probability distribution? What forms may a probability distribution take?

2. What are the parameters of a probability distribution?

3. What are the three properties associated with the binomial distribution?

4. What are the three properties associated with the Poisson distribution?

5. When is the binomial distribution well approximated by the Poisson?

6. What are the properties of the normal distribution?

7. Explain the importance of the standard normal distribution.

8. Let X be a discrete random variable that represents the number of diagnostic services a child receives during an office visit to a pediatric specialist; these services

include procedures such as blood tests and urinanalysis. The probability distribution for X appears below [10].

x	$P(X = x)$
0	0.671
1	0.229
2	0.053
3	0.031
4	0.010
5+	0.006
Total	1.000

(a) Construct a graph of the probability distribution of X.

(b) What is the probability that a child receives exactly three diagnostic services during an office visit to a pediatric specialist?

(c) What is the probability that he or she receives at least one service? Four or more services?

(d) What is the probability that the child receives exactly three services given that he or she receives at least one service?

9. Suppose that you are interested in monitoring air pollution in Los Angeles, California, over a one-week period. Let X be a random variable that represents the number of days out of the seven on which the concentration of carbon monoxide surpasses a specified level. Do you believe that X has a binomial distribution? Explain.

10. Consider a group of seven individuals selected from the population of 65- to 74-year-olds in the United States. The number of persons in this sample who suffer from diabetes is a binomial random variable with parameters $n = 7$ and $p = 0.125$ [11].

(a) If you wish to make a list of the seven persons chosen, in how many ways can they be ordered?

(b) Without regard to order, in how many ways can you select four individuals from this group of seven?

(c) What is the probability that exactly two of the individuals in the sample suffer from diabetes?

(d) What is the probability that four of them have diabetes?

11. According to the National Health Survey, 9.8% of the population of 18- to 24-year-olds in the United States are left-handed [9].

(a) Suppose that you select ten individuals from this population. In how many ways can the ten persons be ordered?

(b) Without regard to order, in how many ways can you select four individuals from this group of ten?

(c) What is the probability that exactly three of the ten persons are left-handed?

(d) What is the probability that at least six of the ten persons are left-handed?

(e) What is the probability that at most two individuals are left-handed?

12. According to the Behavioral Risk Factor Surveillance System, 58% of all Americans adhere to a sedentary lifestyle [12].
   (a) If you selected repeated samples of size twelve from the U.S. population, what would be the mean number of individuals per sample who do not exercise regularly? What would be the standard deviation?
   (b) Suppose that you select a sample of twelve individuals and find that ten of them do not exercise regularly. Assuming that the Surveillance System is correct, what is the probability that you would have obtained results as bad as or worse than those you observed?

13. According to the Massachusetts Department of Health, 224 women who gave birth in the state of Massachusetts in 1988 tested positive for the HIV antibody. Assume that, in time, 25% of the babies born to such mothers will also become HIV-positive.
   (a) If samples of size 224 were repeatedly selected from the population of children born to mothers with the HIV antibody, what would be the mean number of infected children per sample?
   (b) What would be the standard deviation?
   (c) Use Chebychev's inequality to describe this distribution.

14. The number of cases of tetanus reported in the United States during a single month in 1989 has a Poisson distribution with parameter $\lambda = 4.5$ [8].
   (a) What is the probability that exactly one case of tetanus will be reported during a given month?
   (b) What is the probability that at most two cases of tetanus will be reported?
   (c) What is the probability that four or more cases will be reported?
   (d) What is the mean number of cases of tetanus reported in a one-month period? What is the standard deviation?

15. In a particular county, the average number of suicides reported each month is 2.75 [13]. Assume that the number of suicides follows a Poisson distribution.
   (a) What is the probability that no suicides will be reported during a given month?
   (b) What is the probability that at most four suicides will be reported?
   (c) What is the probability that six or more suicides will be reported?

16. Let X be a random variable that represents the number of infants in a group of 2000 who die before reaching their first birthdays. In the United States, the probability that a child dies during his or her first year of life is 0.0085 [14].
   (a) What is the mean number of infants who would die in a group of this size?
   (b) What is the probability that at most five infants out of 2000 die in their first year of life?
   (c) What is the probability that between 15 and 20 infants die in their first year of life?

17. Consider the standard normal distribution with mean $\mu = 0$ and standard deviation $\sigma = 1$.
   (a) What is the probability that an outcome z is greater than 2.60?
   (b) What is the probability that z is less than 1.35?
   (c) What is the probability that z is between -1.70 and 3.10?
   (d) What value of z cuts off the upper 15% of the standard normal distribution?
   (e) What value of z marks off the lower 20% of the distribution?

18. Among females in the United States between 18 and 74 years of age, diastolic blood pressure is normally distributed with mean $\mu = 77$ mm Hg and standard deviation $\sigma = 11.6$ mm Hg [5].
    (a) What is the probability that a randomly selected woman has a diastolic blood pressure less than 60 mm Hg?
    (b) What is the probability that she has a diastolic blood pressure greater than 90 mm Hg?
    (c) What is the probability that the woman has a diastolic blood pressure between 60 and 90 mm Hg?

19. The distribution of weights for the population of males in the United States is approximately normal with mean $\mu = 172.2$ pounds and standard deviation $\sigma = 29.8$ pounds [9].
    (a) What is the probability that a randomly selected man weighs less than 130 pounds?
    (b) What is the probability that he weighs more than 210 pounds?
    (c) What is the probability that among five males selected at random from the population, at least one will have a weight outside the range 130 to 210 pounds?

20. In the Framingham Study, serum cholesterol levels were measured for a large number of healthy males. The population was then followed for 16 years. At the end of this time, the men were divided into two groups: those who had developed coronary heart disease and those who had not. The distributions of the initial serum cholesterol levels for each group were found to be approximately normal. Among individuals who eventually developed coronary heart disease, the mean serum cholesterol level was $\mu_d = 244$ mg/100 ml and the standard deviation was $\sigma_d = 51$ mg/100 ml; for those who did not develop the disease, the mean serum cholesterol level was $\mu_{nd} = 219$ mg/100 ml and the standard deviation was $\sigma_{nd} = 41$ mg/100 ml [15].
    (a) Suppose that an initial serum cholesterol level of 260 mg/100 ml or higher is used to predict coronary heart disease. What is the probability of correctly predicting heart disease for a man who will develop it?
    (b) What is the probability of predicting heart disease for a man who will not develop it?
    (c) What is the probability of failing to predict heart disease for a man who will develop it?
    (d) What would happen to the probabilities of false positive and false negative errors if the cutoff point for predicting heart disease is lowered to 250 mg/100 ml?
    (e) In this population, does initial serum cholesterol level appear to be useful for predicting coronary heart disease? Why or why not?

Bibliography

[1] National Center for Health Statistics, "Supplements to the Monthly Vital Statistics Reports: Advance Reports, 1986," *Vital and Health Statistics*, Series 24, Number 3, March 1990.

[2] Centers for Disease Control, "The Surgeon General's 1989 Report on Reducing the Health Consequences of Smoking: 25 Years of Progress," *Morbidity and Mortality Weekly Report Supplement*, Volume 38, March 24, 1989.

[3] Ross, S. M., *Introduction to Probability Models*, Orlando, Florida: Academic Press, 1985.

[4] Wilson, R., and Crouch, E. A. C., "Risk Assessment and Comparisons: An Introduction," *Science*, Volume 236, April 17, 1987, 267–270.

[5] National Center for Health Statistics, Drizd, T., Dannenberg, A. L., and Engel, A., "Blood Pressure Levels in Persons 18–74 Years of Age in 1976–1980, and Trends in Blood Pressure From 1960 to 1980 in the United States," *Vital and Health Statistics*, Series 11, Number 234, July 1986.

[6] Castelli, W. P., and Anderson, K., "Antihypertensive Treatment and Plasma Lipoprotein Levels: The Associations in Data from a Population Study," *American Journal of Medicine Supplement*, Volume 80, February 14, 1986, 23–32.

[7] Tye, L., "Many States Tackling Issue of AIDS-Infected Health Care Workers," *The Boston Globe*, May 27, 1991, 29–30.

[8] Centers for Disease Control, "Summary of Notifiable Diseases, United States, 1989," *Morbidity and Mortality Weekly Report*, Volume 39, October 5, 1990.

[9] National Center for Health Statistics, Najjar, M. F., and Rowland, M., "Anthropometric Reference Data and Prevalence of Overweight: United States, 1976–1980," *Vital and Health Statistics*, Series 11, Number 238, October 1987.

[10] National Center for Health Statistics, Woodwell, D., "Office Visits to Pediatric Specialists, 1989," *Vital and Health Statistics*, Advance Data Report Number 208, January 17, 1992.

[11] Centers for Disease Control, "Regional Variation in Diabetes Mellitus Prevalence—United States, 1988 and 1989," *Morbidity and Mortality Weekly Report*, Volume 39, November 16, 1990.

[12] Centers for Disease Control, "Coronary Heart Disease Attributable to Sedentary Lifestyle—Selected States, 1988," *Morbidity and Mortality Weekly Report*, Volume 39, August 17, 1990.

[13] Gibbons, R. D., Clark, D. C., and Fawcett, J., "A Statistical Method for Evaluating Suicide Clusters and Implementing Cluster Surveillance," *American Journal of Epidemiology Supplement*, Volume 132, July 1990, 183–191.

[14] National Center for Health Statistics, Kochanek, K. D., and Hudson, B. L., "Advance Report of Final Mortality Statistics, 1992," *Monthly Vital Statistics Report*, Volume 43, Number 6, March 22, 1995.

[15] MacMahon, S. W., and MacDonald, G. J., "A Population at Risk: Prevalence of High Cholesterol Levels in Hypertensive Patients in the Framingham Study," *American Journal of Medicine Supplement*, Volume 80, February 14, 1986, 40–47.

8

Sampling Distribution of the Mean

In the preceding chapter, we examined several theoretical probability distributions, such as the binomial distribution and the normal distribution. In all cases, the relevant population parameters were assumed to be known; this allowed us to describe the distributions completely and to calculate the probabilities associated with various outcomes. In most practical applications, however, we are not given the values of these parameters. Instead, we must attempt to describe or estimate some characteristic of a population—such as its mean or standard deviation—using the information contained in a sample of observations. The process of drawing conclusions about an entire population based on the information in a sample is known as *statistical inference*.

8.1 Sampling Distributions

Suppose that our focus is on estimating the mean value of some continuous random variable of interest. For example, we might wish to make a statement about the mean serum cholesterol level of all men residing in the United States, based on a sample drawn from this population. The obvious approach would be to use the mean of the sample as an estimate of the unknown population mean μ. The quantity $\overline{X}$ is called an *estimator* of the parameter μ. Philosophically, there are many different approaches to the process of estimation; in this case, because the population is assumed to be normally distributed, the sample mean $\overline{X}$ is a *maximum likelihood estimator* [1]. The method of maximum likelihood finds the value of the parameter that is most likely to have produced the observed sample data. This method can usually be relied on to yield reasonable estimators. Keep in mind, however, that two different samples are likely to yield different sample means; consequently, some degree of uncertainty is involved. Before we apply this estimation procedure, therefore, we first examine some of the properties of the sample mean and the ways in which it can vary.

The population under investigation can be any group that we choose. In general, we are able to estimate a population mean μ with greater precision when the group is relatively homogeneous. If there is only a small amount of variation among individuals, we can be more certain that the observations in any given sample are representative of the entire group.

It is very important that a sample provide an accurate representation of the population from which it is selected. If it does not, the conclusions drawn about the population may be distorted or biased. For instance, if we intend to make a statement about the mean serum cholesterol level of all 20- to 74-year-old males in the United States but sample only men over the age of 60, our estimate of the population mean is likely to be too high. It is crucial that the sample drawn be *random*; each individual in the population should have an equal chance of being selected. This point is discussed further in Chapter 22. In addition, we would expect that the larger the sample, the more reliable our estimate of the population mean.

Suppose that in a specified population, the mean of the continuous random variable serum cholesterol level is μ and the standard deviation is σ. We randomly select a sample of n observations from the population and compute the mean of this sample; call the sample mean $\bar{x}_1$. We then obtain a second random sample of n observations and calculate the mean of the new sample. Label this second sample mean $\bar{x}_2$. Unless everyone in the population has the same serum cholesterol level, it is very unlikely that $\bar{x}_1$ will equal $\bar{x}_2$. If we were to continue this procedure indefinitely—selecting all possible samples of size n and computing their means—we would end up with a set of values consisting entirely of sample means. Another way to think about this is to note that the estimator $\overline{X}$ is actually a random variable with outcomes $\bar{x}_1, \bar{x}_2, \bar{x}_3$, and so on.

If each mean in this series is treated as a unique observation, their collective probability distribution—the probability distribution of $\overline{X}$—is known as a *sampling distribution* of means of samples of size n. For example, if we were to select repeated samples of size 25 from the population of men residing in the United States and calculate the mean serum cholesterol level for each sample, we would end up with the sampling distribution of mean serum cholesterol levels of samples of size 25. In practice, it is not common to select repeated samples of size n from a given population; understanding the properties of the theoretical distribution of their means, however, allows us to make inference based on a **single** sample of size n.

8.2   *The Central Limit Theorem*

Given that the distribution of serum cholesterol levels in the underlying population has mean μ and standard deviation σ, the distribution of sample means computed for samples of size n has three important properties:

1. The mean of the sampling distribution is identical to the population mean μ.
2. The standard deviation of the distribution of sample means is equal to $\sigma/\sqrt{n}$. This quantity is known as the *standard error* of the mean.
3. Provided that n is large enough, the shape of the sampling distribution is approximately normal.

Intuitively, we would expect the means of all our samples to cluster around the mean of the population from which they were drawn. Although the standard deviation of the sampling distribution is related to the population standard deviation σ, there is less variability among the sample means than there is among individual observations. Even if a particular sample contains one or two extreme values, it is likely that these values will be offset by the other measurements in the group. Thus, as long as n is greater than 1, the standard error of the mean is always smaller than the standard deviation of the population. In addition, as n increases, the amount of sampling variation decreases. Finally, if n is large enough, the distribution of sample means is approximately normal. This remarkable result is known as the *central limit theorem*; it applies to any population with a finite standard deviation, regardless of the shape of the underlying distribution [2]. The farther the underlying population departs from being normally distributed, however, the larger the value of n that is necessary to ensure the normality of the sampling distribution. If the underlying population is itself normal, samples of size 1 are large enough. Even if the population is bimodal or noticeably skewed, a sample of size 30 is often sufficient.

The central limit theorem is very powerful. It holds true not only for serum cholesterol levels, but for almost any other type of measurement as well. It even applies to discrete random variables. The central limit theorem allows us to quantify the uncertainty inherent in statistical inference without having to make a great many assumptions that cannot be verified. Regardless of the distribution of X, because the distribution of the sample means is approximately normal with mean μ and standard deviation $\sigma/\sqrt{n}$, we know that if n is large enough,

$$Z = \frac{\overline{X} - \mu}{\sigma/\sqrt{n}}$$

is normally distributed with mean 0 and standard deviation 1. We have simply standardized the normal random variable $\overline{X}$ in the usual way. As a result, we can use tables of the standard normal distribution—such as Table A.3 in Appendix A—to make inference about the value of a population mean.

8.3   Applications of the Central Limit Theorem

Consider the distribution of serum cholesterol levels for all 20- to 74-year-old males living in the United States. The mean of this population is $\mu = 211$ mg/100 ml, and the standard deviation is $\sigma = 46$ mg/100 ml [3]. If we select repeated samples of size 25 from the population, what proportion of the samples will have a mean value of 230 mg/100 ml or above?

Assuming that a sample of size 25 is large enough, the central limit theorem states that the distribution of means of samples of size 25 is approximately normal with mean $\mu = 211$ mg/100 ml and standard error $\sigma/\sqrt{n} = 46/\sqrt{25} = 9.2$ mg/100 ml. This sampling distribution and the underlying population distribution are shown in Figure 8.1. Note that

$$Z = \frac{\overline{X} - 211}{9.2}$$

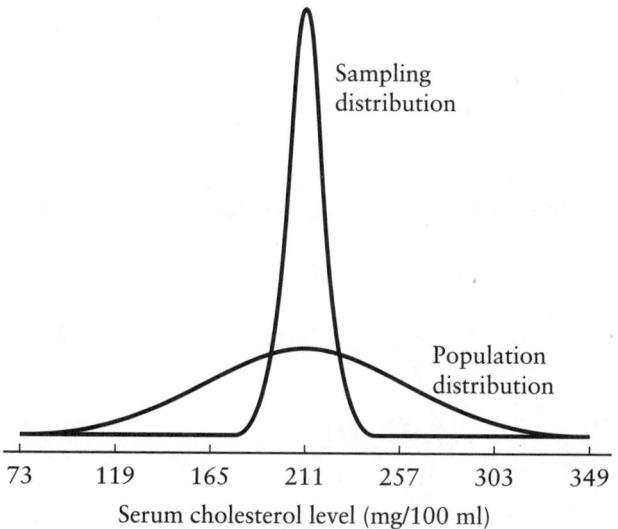

FIGURE 8.1

Distributions of individual values and means of samples of size 25 for the serum cholesterol levels of 20- to 74-year-old males, United States, 1976–1980

is a standard normal random variable. If $\bar{x} = 230$, then

$$z = \frac{230 - 211}{9.2}$$

$$= 2.07.$$

Consulting Table A.3, we find that the area to the right of $z = 2.07$ is 0.019. Only about 1.9% of the samples will have a mean greater than 230 mg/100 ml. Equivalently, if we select a single sample of size 25 from the population of 20- to 74-year-old males, the probability that the mean serum cholesterol level for this sample is 230 mg/100 ml or higher is 0.019.

What mean value of serum cholesterol level cuts off the lower 10% of the sampling distribution of means? Locating 0.100 in the body of Table A.3, we see that it corresponds to the value $z = -1.28$. Solving for $\bar{x}$,

$$z = -1.28$$

$$= \frac{\bar{x} - 211}{9.2}$$

and

$$\bar{x} = 211 + (-1.28)(9.2)$$

$$= 199.2.$$

Therefore, approximately 10% of the samples of size 25 have means that are less than or equal to 199.2 mg/100 ml.

Let us now calculate the upper and lower limits that enclose 95% of the means of samples of size 25 drawn from the population. Since 2.5% of the area under the standard normal curve lies above $z = 1.96$ and another 2.5% lies below $z = -1.96$,

$$P(-1.96 \leq Z \leq 1.96) = 0.95.$$

Thus, we are interested in outcomes of Z for which

$$-1.96 \leq Z \leq 1.96.$$

We would like to transform this inequality into a statement about $\overline{X}$. Substituting $(\overline{X} - 211)/9.2$ for Z,

$$-1.96 \leq \frac{\overline{X} - 211}{9.2} \leq 1.96.$$

Multiplying all three terms of the inequality by 9.2 and adding 211 results in

$$211 - 1.96(9.2) \leq \overline{X} \leq 211 + 1.96(9.2),$$

or

$$193.0 \leq \overline{X} \leq 229.0.$$

This tells us that approximately 95% of the means of samples of size 25 lie between 193.0 mg/100 ml and 229.0 mg/100 ml. Consequently, if we select a random sample of size 25 that is reported to be from the population of serum cholesterol levels for all 20- to 74-year-old males, and the sample has a mean that is either greater than 229.0 or less than 193.0 mg/100 ml, we should be suspicious of this claim. Either the random sample was actually drawn from a different population or a rare event has taken place. For the purposes of this discussion, a "rare event" is defined as an outcome that occurs less than 5% of the time.

Suppose we had selected samples of size 10 from the population rather than samples of size 25. In this case, the standard error of $\overline{X}$ would be $46/\sqrt{10} = 14.5$ mg/100 ml, and we would construct the inequality

$$-1.96 \leq \frac{\overline{X} - 211}{14.5} \leq 1.96.$$

The upper and lower limits that enclose 95% of the means would be

$$182.5 \leq \overline{X} \leq 239.5.$$

Note that this interval is wider than the one calculated for samples of size 25. We expect the amount of sampling variation to increase as the sample size decreases. Drawing samples of size 50 would result in upper and lower limits

$$198.2 \leq \overline{X} \leq 223.8;$$

not surprisingly, this interval is narrower than the one constructed for samples of size 25. Samples of size 100 produce the limits

$$202.0 \leq \overline{X} \leq 220.0.$$

In summary, if we include the case for which $n = 1$, we have the following results:

n	$\sigma/\sqrt{n}$	Interval Enclosing 95% of the Means	Length of Interval
1	46.0	$120.8 \leq \overline{X} \leq 301.2$	180.4
10	14.5	$182.5 \leq \overline{X} \leq 239.5$	57.0
25	9.2	$193.0 \leq \overline{X} \leq 229.0$	36.0
50	6.5	$198.2 \leq \overline{X} \leq 223.8$	25.6
100	4.6	$202.0 \leq \overline{X} \leq 220.0$	18.0

As the size of the samples increases, the amount of variability among the sample means—quantified by the standard error $\sigma/\sqrt{n}$—decreases; consequently, the limits encompassing 95% of these means move closer together. The length of an interval is simply the upper limit minus the lower limit.

Note that all the intervals we have constructed have been symmetric about the population mean 211 mg/100 ml. Clearly, there are other intervals that would also capture the appropriate proportion of the sample means. Suppose that we again wish to construct an interval that contains 95% of the means of samples of size 25. Since 1% of the area under the standard normal curve lies above $z = 2.32$ and 4% lies below $z = -1.75$, we know that

$$P(-1.75 \leq Z \leq 2.32) = 0.95.$$

As a result, we are interested in the outcomes of Z for which

$$-1.75 \leq Z \leq 2.32.$$

Substituting $(\overline{X} - 211)/9.2$ for Z, we find the interval to be

$$194.9 \leq \overline{X} \leq 232.3.$$

Therefore, we are able to say that approximately 95% of the means of samples of size 25 lie between 194.9 mg/100 ml and 232.3 mg/100 ml. It is usually preferable to construct a symmetric interval, however, primarily because it is the shortest interval that captures the appropriate proportion of the means. (An exception to this rule is the one-sided interval; we return to this special case below.) In this example, the asymmetrical interval has length $232.3 - 194.9 = 37.4$ mg/100 ml; the length of the symmetric interval is $229.0 - 193.0 = 36.0$ mg/100 ml.

We now move on to a slightly more complicated question: How large would the samples need to be for 95% of their means to lie within ± 5 mg/100 ml of the population mean μ? To answer this, it is not necessary to know the value of the parameter μ. We simply find the sample size n for which

$$P(\mu - 5 \leq \overline{X} \leq \mu + 5) = 0.95,$$

or

$$P(-5 \leq \overline{X} - \mu \leq 5) = 0.95.$$

To begin, we divide all three terms of the inequality by the standard error $\sigma/\sqrt{n} = 46/\sqrt{n}$; this results in

$$P\left(\frac{-5}{46/\sqrt{n}} \leq \frac{\overline{X} - \mu}{46/\sqrt{n}} \leq \frac{5}{46/\sqrt{n}}\right) = 0.95.$$

Since Z is equal to $(\overline{X} - \mu)/(46/\sqrt{n})$,

$$P\left(\frac{-5}{46/\sqrt{n}} \leq Z \leq \frac{5}{46/\sqrt{n}}\right) = 0.95.$$

Recall that 95% of the area under the standard normal curve lies between $z = -1.96$ and $z = 1.96$. Therefore, to find the sample size n, we could use the upper bound of the interval and solve the equation

$$z = 1.96$$
$$= \frac{5}{46/\sqrt{n}};$$

equivalently, we could use the lower bound and solve

$$z = -1.96$$
$$= \frac{-5}{46/\sqrt{n}}.$$

Taking

$$1.96 = \frac{5\sqrt{n}}{46}$$

and multiplying both sides of the equation by $46/5$, we find that

$$\sqrt{n} = \frac{1.96(46)}{5}$$

and

$$n = \left[\frac{1.96(46)}{5}\right]^2$$
$$= 325.2.$$

When we deal with sample sizes, it is conventional to round up. Therefore, samples of size 326 would be required for 95% of the sample means to lie within ± 5 mg/100 ml of the population mean μ. Another way to state this is that if we select a sample of size 326 from the population and calculate its mean, the probability that the sample mean is within ± 5 mg/100 ml of the true population mean μ is 0.95.

Up to this point, we have focused on two-sided intervals: we have found the upper and lower limits that enclose a specified proportion of the sample means. More specifically, we have focused on symmetric intervals. In some situations, however, we are interested in a one-sided interval instead. For example, we might wish to find the upper bound for 95% of the mean serum cholesterol levels of samples of size 25. Since 95% of the area under the standard normal curve lies below $z = 1.645$,

$$P(Z \leq 1.645) = 0.95.$$

Consequently, we are interested in outcomes of Z for which

$$Z \leq 1.645.$$

Substituting $(\overline{X} - 211)/9.2$ for Z produces

$$\frac{\overline{X} - 211}{9.2} \leq 1.645,$$

or

$$\overline{X} \leq 226.1.$$

Approximately 95% of the means of samples of size 25 lie below 226.1 mg/100 ml.

If we want to construct a lower bound for 95% of the mean serum cholesterol levels, we focus on values of Z that lie above -1.645; in this case, we solve

$$\frac{\overline{X} - 211}{9.2} \geq -1.645$$

to find

$$\overline{X} \geq 195.9.$$

Approximately 95% of the means of samples of size 25 lie above 195.9 mg/100 ml.

Always keep in mind that we must be cautious when making multiple statements about the sampling distribution of the means. For samples of serum cholesterol levels of size 25, we found that the probability is 0.95 that a sample mean lies within the interval

$$(193.0, 229.0).$$

We also said that the probability is 0.95 that the mean lies below 226.1 mg/100 ml, and 0.95 that it is above 195.9 mg/100 ml. Although these three statements are correct individually, they are not true simultaneously. The three events are not independent. For all of them to occur at the same time, the sample mean would have to lie in the interval

$$(195.9, 226.1);$$

the probability that this happens is not equal to 0.95.

8.4   *Further Applications*

Consider the distribution of age at the time of death for the United States population in 1979–1981. This distribution is shown in Figure 8.2; it has mean $\mu = 73.9$ years and standard deviation $\sigma = 18.1$ years, and is far from normally distributed [4]. What do we expect to happen when we sample from this population of ages?

Rather than draw samples from the population physically, we can generate a computer program to simulate this process. To conduct a *simulation*, the computer is used to model an experiment or procedure according to a specified probability distribution; in our example, the procedure would consist of selecting an individual observation from the distribution shown in Figure 8.2. The computer is then instructed to repeat the process a given number of times, keeping track of the results.

To illustrate this technique, we can use the computer to simulate the selection of four random samples of size 25 from the population of ages at the time of death for the U.S. population. Histograms of these samples are shown in Figure 8.3; their means and standard deviations are as follows:

Sample of Size 25	$\bar{x}$	s
1	71.3	18.1
2	69.2	25.6
3	74.0	14.0
4	76.8	15.0

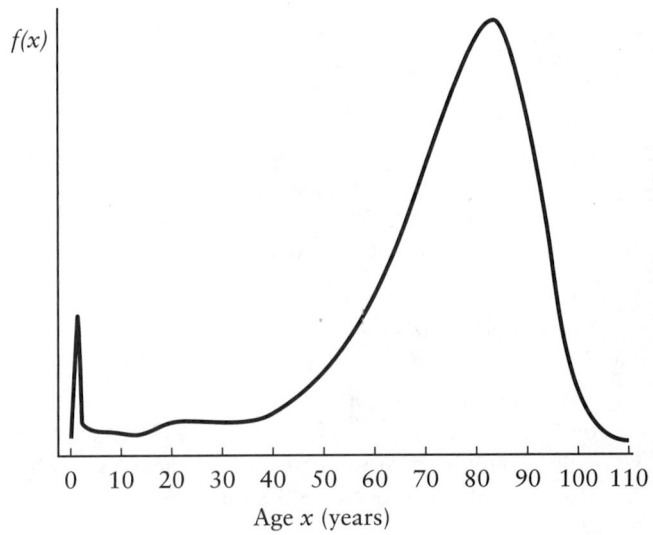

FIGURE 8.2
Distribution of age at the time of death, United States, 1979–1981

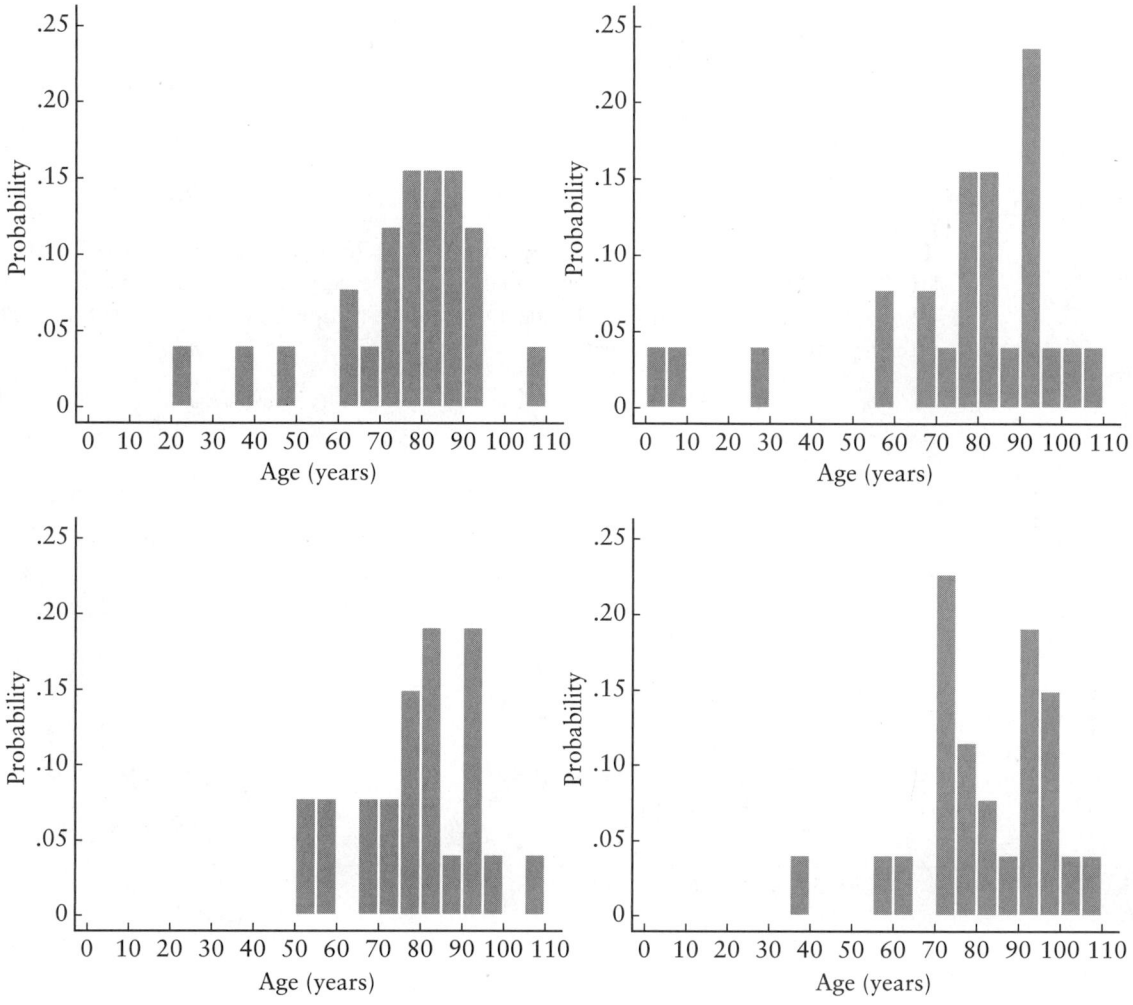

FIGURE 8.3
Histograms of four samples of size 25

Note that the four random samples are not identical. Each time we select a set of 25 measurements from the population, the observations included in the sample change. As a result, the values of $\bar{x}$ and s—our estimates of the population mean μ and standard deviation σ—differ from sample to sample. This random variation is known as *sampling variability*. In the four samples of size 25 selected above, the estimates of μ range from 69.2 years to 76.8 years. Similarly, the estimates of σ range from 14.0 to 25.6 years.

Suppose that now, instead of selecting samples of size 25, we choose four random samples of size 100 from the population of ages at the time of death. Again we use the computer to simulate this process. Histograms of the samples are displayed in Figure 8.4, and their means and standard deviations are given on the following page.

Sample of Size 100	$\bar{x}$	s
1	75.4	16.5
2	75.0	19.9
3	73.5	18.1
4	72.1	20.2

For these samples, estimates of μ range from 72.1 to 75.4 years and estimates of σ from 16.5 to 20.2 years. These ranges are smaller than the corresponding intervals for samples of size 25. We would in fact expect this; as the sample size increases, the amount of sampling variability decreases.

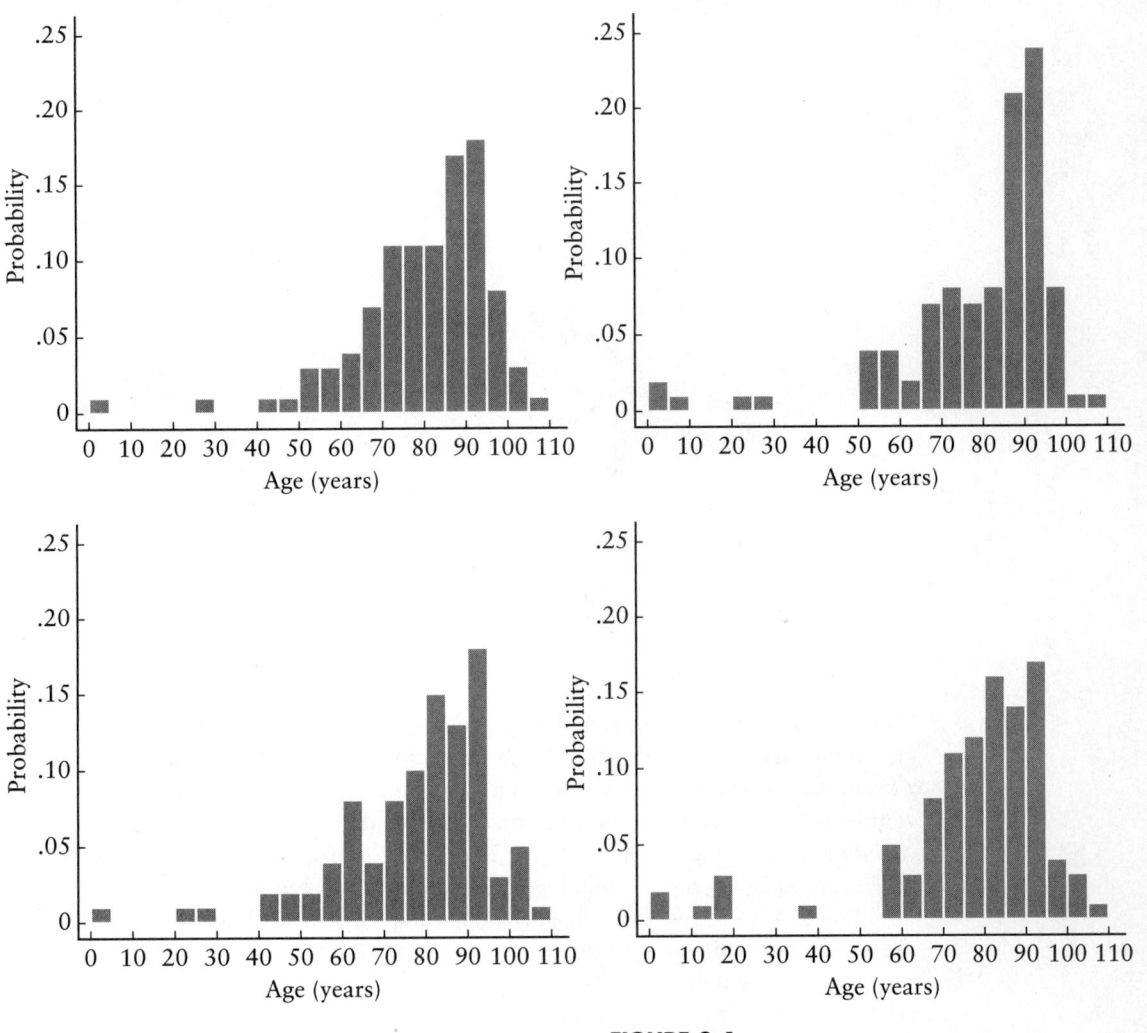

FIGURE 8.4
Histograms of four samples of size 100

We next select four random samples of size 500 from the population of ages at the time of death. Histograms appear in Figure 8.5, and the means and standard deviations are given below.

Sample of Size 500	$\bar{x}$	s
1	74.3	17.1
2	73.4	18.1
3	73.5	18.6
4	74.2	17.8

Again, the ranges of the estimates for both μ and σ decrease.

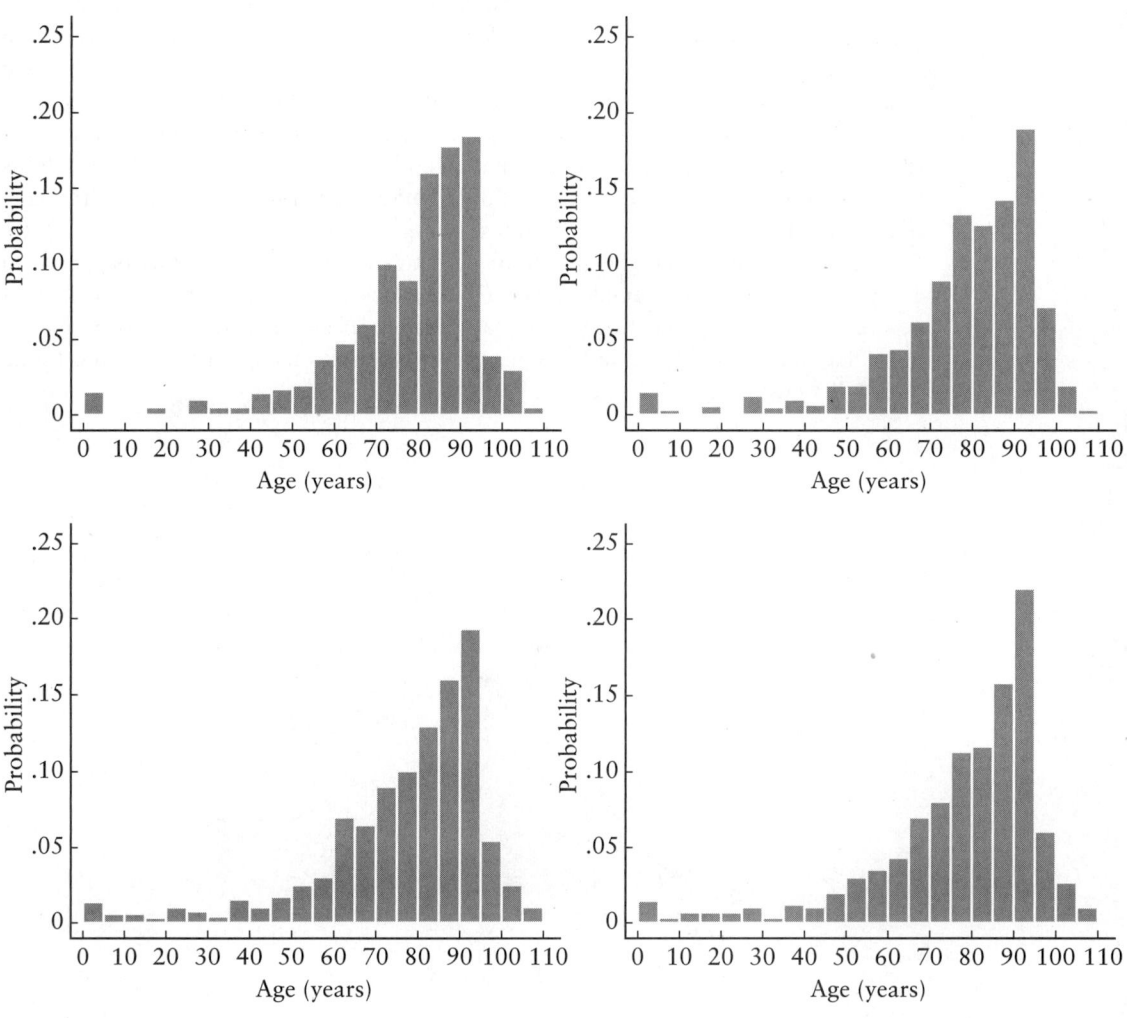

FIGURE 8.5
Histograms of four samples of size 500

Looking at Figures 8.3 through 8.5, we see that as the size of the samples increases, their distributions approach the shape of the population distribution shown in Figure 8.2. Although there are still differences among the samples, the amount of variability in the estimates $\bar{x}$ and s decreases. This property is known as *consistency*; as the samples that we select become larger and larger, the estimates of the population parameters approach their target values.

The population of ages at the time of death can also be used to demonstrate an application of the central limit theorem. To do this, we must select repeated samples of size n from the population with mean $\mu = 73.9$ years and standard deviation $\sigma = 18.1$ years and examine the distribution of the means of these samples. Theoretically, we must enumerate all possible random samples; for now, however, we select 100 samples of size 25. A histogram of the 100 sample means is displayed in Figure 8.6.

According to the central limit theorem, the distribution of the sample means possesses three properties. First, its mean should be equal to the population mean $\mu = 73.9$ years. In fact, the mean of the 100 sample means is 74.1 years. Second, we expect the standard error of the sample means to be $\sigma/\sqrt{n} = 18.1/\sqrt{25} = 3.6$ years. In reality, the standard error is 3.7 years. Finally, the distribution of sample means should be approximately normal. The shape of the histogram in Figure 8.6 and the theoretical normal distribution superimposed over it suggest that this third property holds true. Note that this is a large departure from the population distribution shown in Figure 8.2, or from any of the individual samples of size 25 shown in Figure 8.3.

Based on the sampling distribution, we can calculate probabilities associated with various outcomes of the sample mean. For instance, among samples of size 25 that are drawn from the population of ages at the time of death, what proportion have a mean that lies between 70 and 78 years? To answer this question, we must find $P(70 \leq \overline{X} \leq 78)$.

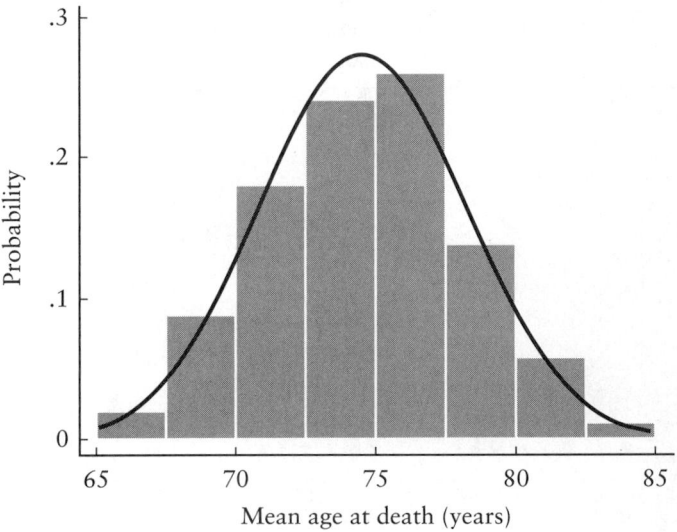

FIGURE 8.6

Histogram of 100 sample means from samples of size 25

As we just saw, the central limit theorem states that the distribution of sample means of samples of size 25 is approximately normal with mean $\mu = 73.9$ years and standard error $\sigma/\sqrt{n} = 18.1/\sqrt{25} = 3.62$ years. Therefore,

$$
Z = \frac{\overline{X} - \mu}{\sigma/\sqrt{n}}
$$
$$
= \frac{\overline{X} - 73.9}{3.62}
$$

is a standard normal random variable. If we represent the inequality in the expression

$$
P(70 \le \overline{X} \le 78)
$$

in terms of Z rather than $\overline{X}$, we can use Table A.3 to find the proportion of samples that have a mean value in this range.

We begin by subtracting 73.9 from each term in the inequality and dividing by 3.62; thus, we can express

$$
P(70 \le \overline{X} \le 78)
$$

as

$$
P\!\left(\frac{70 - 73.9}{3.62} \le \frac{\overline{X} - 73.9}{3.62} \le \frac{78 - 73.9}{3.62} \right),
$$

or

$$
P(-1.08 \le Z \le 1.13).
$$

We know that the total area underneath the standard normal curve is equal to 1. According to Table A.3, the area to the right of $z = 1.13$ is 0.129 and the area to the left of $z = -1.08$ is 0.140. Therefore,

$$
P(-1.08 \le Z \le 1.13) = 1 - 0.129 - 0.140
$$
$$
= 0.731.
$$

Approximately 73.1% of the samples of size 25 have a mean that lies between 70 and 78 years.

Now, what proportion of the means of samples of size 100 lie between 70 and 78 years? Again we must find $P(70 \le \overline{X} \le 78)$. This time, however, $\overline{X}$ has a normal distribution with mean $\mu = 73.9$ years and standard error $\sigma/\sqrt{n} = 18.1/\sqrt{100} = 1.81$ years. Therefore, we construct the inequality

$$
P\!\left(\frac{70 - 73.9}{1.81} \le \frac{\overline{X} - 73.9}{1.81} \le \frac{78 - 73.9}{1.81} \right),
$$

or

$$
P(-2.15 \le Z \le 2.27).
$$

According to Table A.3, the area to the right of $z = 2.27$ is 0.012 and the area to the left of $z = -2.15$ is 0.016. Therefore,

$$P(-2.15 \leq Z \leq 2.27) = 1 - 0.012 - 0.016$$
$$= 0.972.$$

About 97.2% of the samples of size 100 have a mean that lies between 70 and 78 years. If we were to select a single random sample of size 100 and find that its sample mean is $\bar{x} = 80$ years, either the sample actually came from a population with a different underlying mean—something higher than $\mu = 73.9$ years—or a rare event has occurred.

To address a different type of question, we might wish to find the upper and lower limits that enclose 80% of the means of samples of size 100. Consulting Table A.3, we find that 10% of the area under a standard normal curve lies above $z = 1.28$ and another 10% lies below $z = -1.28$. Since 80% of the area lies between -1.28 and 1.28, we are interested in values of Z for which

$$-1.28 \leq Z \leq 1.28,$$

and values of $\overline{X}$ for which

$$-1.28 \leq \frac{\overline{X} - 73.9}{1.81} \leq 1.28.$$

Multiplying all three terms of the inequality by 1.81 and adding 73.9 results in

$$73.9 + (-1.28)(1.81) \leq \overline{X} \leq 73.9 + (1.28)(1.81),$$

or, equivalently,

$$71.6 \leq \overline{X} \leq 76.2.$$

Therefore, 80% of the means of samples of size 100 lie between 71.6 years and 76.2 years.

8.5 Review Exercises

1. What is statistical inference?

2. Why is it important that a sample drawn from a population be random?

3. Why is it necessary to understand the properties of a theoretical distribution of means of samples of size n when in practice you will only select a single such sample?

4. What is the standard error of a sample mean? How does the standard error compare to the standard deviation of the population?

5. Explain the central limit theorem.

6. What happens to the amount of sampling variability among a set of sample means $\bar{x}_1, \bar{x}_2, \bar{x}_3, \ldots$ as the size of the samples increases?

7. What is consistency?

8. Among adults in the United States, the distribution of albumin levels (albumin is a type of protein) in cerebrospinal fluid is roughly symmetric with mean $\mu = 29.5$ mg/100 ml and standard deviation $\sigma = 9.25$ mg/100 ml [5]. Suppose that you select repeated samples of size 20 from this population and calculate the mean for each sample.
   (a) If you were to select a large number of random samples of size 20, what would be the mean of the sample means?
   (b) What would be their standard deviation? What is another name for this standard deviation of the sample means?
   (c) How does the standard deviation of the sample means compare with the standard deviation of the albumin levels themselves?
   (d) If you were to take all the different sample means and use them to construct a histogram, what would be the shape of their distribution?
   (e) What proportion of the means of samples of size 20 are larger than 33 mg/100 ml?
   (f) What proportion of the means are less than 28 mg/100 ml?
   (g) What proportion of the means are between 29 and 31 mg/100 ml?

9. Consider a random variable X that has a standard normal distribution with mean $\mu = 0$ and standard deviation $\sigma = 1$.
   (a) What can you say about the distribution of means of samples of size 10 that are drawn from this population? List three properties.
   (b) What proportion of the means of samples of size 10 are greater than 0.60?
   (c) What proportion of the means are less than -0.75?
   (d) What value cuts off the upper 20% of the distribution of means of samples of size 10?
   (e) What value cuts off the lower 10% of the distribution of means?

10. In Denver, Colorado, the distribution of daily measures of ambient nitric acid—a corrosive liquid—is skewed to the right; it has mean $\mu = 1.81 \mu$g/m^3 and standard deviation $\sigma = 2.25 \mu$g/m^3 [6]. Describe the distribution of means of samples of size 40 selected from this population.

11. In Norway, the distribution of birth weights for infants whose gestational age is 40 weeks is approximately normal with mean $\mu = 3500$ grams and standard deviation $\sigma = 430$ grams [7].
   (a) Given a newborn whose gestational age is 40 weeks, what is the probability that his or her birth weight is less than 2500 grams?
   (b) What value cuts off the lower 5% of the distribution of birth weights?
   (c) Describe the distribution of means of samples of size 5 drawn from this population. List three properties.
   (d) What value cuts off the lower 5% of the distribution of samples of size 5?

(e) Given a sample of five newborns all with gestational age 40 weeks, what is the probability that their mean birth weight is less than 2500 grams?

(f) What is the probability that only one of the five newborns has a birth weight less than 2500 grams?

12. For the population of females between the ages of 3 and 74 who participated in the National Health Interview Survey, the distribution of hemoglobin levels has mean $\mu = 13.3$ g/100 ml and standard deviation $\sigma = 1.12$ g/100 ml [8].

(a) If repeated samples of size 15 are selected from this population, what proportion of the samples will have a mean hemoglobin level between 13.0 and 13.6 g/100 ml?

(b) If the repeated samples are of size 30, what proportion will have a mean between 13.0 and 13.6 g/100 ml?

(c) How large must the samples be for 95% of their means to lie within ± 0.2 g/100 ml of the population mean μ?

(d) How large must the samples be for 95% of their means to lie within ± 0.1 g/100 ml of the population mean?

13. In the Netherlands, healthy males between the ages of 65 and 79 have a distribution of serum uric acid levels that is approximately normal with mean $\mu = 341 \mu$mol/l and standard deviation $\sigma = 79 \mu$mol/l [9].

(a) What proportion of the males have a serum uric acid level between 300 and 400 μmol/l?

(b) What proportion of samples of size 5 have a mean serum uric acid level between 300 and 400 μmol/l?

(c) What proportion of samples of size 10 have a mean serum uric acid level between 300 and 400 μmol/l?

(d) Construct an interval that encloses 95% of the means of samples of size 10. Which would be shorter, a symmetric interval or an asymmetric one?

14. For the population of adult males in the United States, the distribution of weights is approximately normal with mean $\mu = 172.2$ pounds and standard deviation $\sigma = 29.8$ pounds [10].

(a) Describe the distribution of means of samples of size 25 that are drawn from this population.

(b) What is the upper bound for 90% of the mean weights of samples of size 25?

(c) What is the lower bound for 80% of the mean weights?

(d) Suppose that you select a single random sample of size 25 and find that the mean weight for the men in the sample is $\bar{x} = 190$ pounds. How likely is this result? What would you conclude?

15. At the end of Section 8.3, it was noted that for samples of serum cholesterol levels of size 25—drawn from a population with mean $\mu = 211$ mg/100 ml and standard deviation $\sigma = 46$ mg/100 ml—the probability that a sample mean $\bar{x}$ lies within the interval (193.0, 229.0) is 0.95. Furthermore, the probability that the mean lies below 226.1 mg/100 ml is 0.95, and the probability that it is above 195.9 mg/100 ml is 0.95. For all three of these events to happen simultaneously, the sample mean $\bar{x}$ would have to lie in the interval (195.9, 226.1). What is the probability that this occurs?

Bibliography

[1] Lindgren, B. W., *Statistical Theory*, New York: Macmillan, 1976.

[2] Snedecor, G. W., and Cochran, W. G., *Statistical Methods*, Ames, IA: Iowa State University Press, 1980.

[3] National Center for Health Statistics, Fulwood, R., Kalsbeek, W., Rifkind, B., Russell-Briefel, R., Muesing, R., LaRosa, J., and Lippel, K., "Total Serum Cholesterol Levels of Adults 20–74 Years of Age: United States, 1976–1980," *Vital and Health Statistics*, Series 11, Number 236, May 1986.

[4] National Center for Health Statistics, *United States Decennial Life Tables for 1979–1981*, Volume I, Number 1, August 1985.

[5] Scully, R. E., McNeely, B. U., and Mark, E. J., "Case Record of the Massachusetts General Hospital: Weekly Clinicopathological Exercises," *The New England Journal of Medicine*, Volume 314, January 2, 1986, 39–49.

[6] Ostro, B. D., Lipsett, M. J., Wiener, M. B., and Selner, J. C., "Asthmatic Responses to Airborne Acid Aerosols," *American Journal of Public Health*, Volume 81, June 1991, 694–702.

[7] Wilcox, A. J., and Skjærven, R., "Birth Weight and Perinatal Mortality: The Effects of Gestational Age," *American Journal of Public Health*, Volume 82, March 1992, 378–382.

[8] National Center for Health Statistics, Fulwood, R., Johnson, C. L., Bryner, J. D., Gunter, E. W., and McGrath, C. R., "Hematological and Nutritional Biochemistry Reference Data for Persons 6 Months–74 Years of Age: United States, 1976–1980," *Vital and Health Statistics*, Series 11, Number 232, December 1982.

[9] Loenen, H. M. J. A., Eshuis, H., Lowik, M. R. H., Schouten, E. G., Hulshof, K. F. A. M., Odink, J., and Kok, F. J., "Serum Uric Acid Correlates in Elderly Men and Women with Special Reference to Body Composition and Dietary Intake (Dutch Nutrition Surveillance System)," *Journal of Clinical Epidemiology*, Volume 43, Number 12, 1990, 1297–1303.

[10] National Center for Health Statistics, Najjar, M. F., and Rowland, M., "Anthropometric Reference Data and Prevalence of Overweight: United States, 1976–1980," *Vital and Health Statistics*, Series 11, Number 238, October 1987.

9

Confidence Intervals

Now that we have investigated the theoretical properties of a distribution of sample means, we are ready to take the next step and apply this knowledge to the process of statistical inference. Recall that our goal is to describe or estimate some characteristic of a continuous random variable—such as its mean—using the information contained in a sample of observations.

Two methods of estimation are commonly used. The first is called *point estimation*; it involves using the sample data to calculate a single number to estimate the parameter of interest. For instance, we might use the sample mean $\bar{x}$ to estimate the population mean μ. The problem is that two different samples are very likely to result in different sample means, and thus there is some degree of uncertainty involved. A point estimate does not provide any information about the inherent variability of the estimator; we do not know how close $\bar{x}$ is to μ in any given situation. While $\bar{x}$ is more likely to be near the true population mean if the sample on which it is based is large—recall the property of consistency—a point estimate provides no information about the size of this sample. Consequently, a second method of estimation, known as *interval estimation,* is often preferred. This technique provides a range of reasonable values that are intended to contain the parameter of interest—the population mean μ, in this case—with a certain degree of confidence. This range of values is called a *confidence interval.*

9.1 Two-Sided Confidence Intervals

To construct a confidence interval for μ, we draw on our knowledge of the sampling distribution of the mean from the preceding chapter. Given a random variable X that has mean μ and standard deviation σ, the central limit theorem states that

$$Z = \frac{\overline{X} - \mu}{\sigma/\sqrt{n}}$$

has a standard normal distribution if X is itself normally distributed and an approximate standard normal distribution if it is not but n is sufficiently large. For a standard normal random variable, 95% of the observations lie between -1.96 and 1.96. In other words, the probability that Z assumes a value between -1.96 and 1.96 is

$$P(-1.96 \leq Z \leq 1.96) = 0.95.$$

Equivalently, we could substitute the quantity $(\overline{X} - \mu)/(\sigma/\sqrt{n})$ for Z and write

$$P\left(-1.96 \leq \frac{\overline{X} - \mu}{\sigma/\sqrt{n}} \leq 1.96\right) = 0.95.$$

Given this expression, we are able to manipulate the inequality inside the parentheses without altering the probability statement. We begin by multiplying all three terms of the inequality by the standard error $\sigma/\sqrt{n}$; therefore,

$$P\left(-1.96 \frac{\sigma}{\sqrt{n}} \leq \overline{X} - \mu \leq 1.96 \frac{\sigma}{\sqrt{n}}\right) = 0.95.$$

We next subtract $\overline{X}$ from each term so that

$$P\left(-1.96 \frac{\sigma}{\sqrt{n}} - \overline{X} \leq -\mu \leq 1.96 \frac{\sigma}{\sqrt{n}} - \overline{X}\right) = 0.95.$$

Finally, we multiply through by -1. Bear in mind that multiplying an inequality by a negative number reverses the direction of the inequality. Consequently,

$$P\left(1.96 \frac{\sigma}{\sqrt{n}} + \overline{X} \geq \mu \geq -1.96 \frac{\sigma}{\sqrt{n}} + \overline{X}\right) = 0.95$$

and, rearranging the terms,

$$P\left(\overline{X} - 1.96 \frac{\sigma}{\sqrt{n}} \leq \mu \leq \overline{X} + 1.96 \frac{\sigma}{\sqrt{n}}\right) = 0.95.$$

Note that $\overline{X}$ is no longer in the center of the inequality; instead, the probability statement says something about μ. The quantities $\overline{X} - 1.96(\sigma/\sqrt{n})$ and $\overline{X} + 1.96(\sigma/\sqrt{n})$ are 95% confidence limits for the population mean; we are 95% confident that the interval

$$\left(\overline{X} - 1.96 \frac{\sigma}{\sqrt{n}}, \overline{X} + 1.96 \frac{\sigma}{\sqrt{n}}\right)$$

will cover μ. This statement does **not** imply that μ is a random variable that assumes a value within the interval 95% of the time, nor that 95% of the population values lie between these limits; rather, it means that if we were to select 100 random samples from the population and use these samples to calculate 100 different confidence intervals for μ, approximately 95 of the intervals would cover the true population mean and 5 would not.

Keep in mind that the estimator $\overline{X}$ is a random variable, whereas the parameter μ is a constant. Therefore, the interval

$$\left(\overline{X} - 1.96\frac{\sigma}{\sqrt{n}}, \overline{X} + 1.96\frac{\sigma}{\sqrt{n}}\right)$$

is random and has a 95% chance of covering μ **before** a sample is selected. Since μ has a fixed value, once a sample has been drawn and the confidence limits

$$\left(\overline{x} - 1.96\frac{\sigma}{\sqrt{n}}, \overline{x} + 1.96\frac{\sigma}{\sqrt{n}}\right)$$

have been calculated, either μ is within the interval or it is not. There is no longer any probability involved.

Although a 95% confidence interval is used most often in practice, we are not restricted to this choice. We might prefer to have a greater degree of certainty regarding the value of the population mean; in this case, we could choose to construct a 99% confidence interval instead of a 95% interval. Since 99% of the observations in a standard normal distribution lie between -2.58 and 2.58, a 99% confidence interval for μ is

$$\left(\overline{X} - 2.58\frac{\sigma}{\sqrt{n}}, \overline{X} + 2.58\frac{\sigma}{\sqrt{n}}\right).$$

Approximately 99 out of 100 confidence intervals obtained from 100 independent random samples of size n drawn from this population would cover the true mean μ. As we would expect, the 99% confidence interval is wider than the 95% interval; the smaller the range of values we consider, the less confident we are that the interval covers μ.

A generic confidence interval for μ can be obtained by introducing some new notation. Let $z_{\alpha/2}$ be the value that cuts off an area of $\alpha/2$ in the upper tail of the standard normal distribution, and $-z_{\alpha/2}$ the value that cuts off an area of $\alpha/2$ in the lower tail of the distribution. If $\alpha = 0.05$, for instance, then $z_{0.05/2} = 1.96$ and $-z_{0.05/2} = -1.96$. Therefore, the general form for a $100\% \times (1 - \alpha)$ confidence interval for μ—a 95% confidence interval if $\alpha = 0.05$—is

$$\left(\overline{X} - z_{\alpha/2}\frac{\sigma}{\sqrt{n}}, \overline{X} + z_{\alpha/2}\frac{\sigma}{\sqrt{n}}\right).$$

This interval has a $100\% \times (1 - \alpha)$ chance of covering μ before a random sample is selected.

If we wish to make an interval tighter without reducing the level of confidence, we need more information about the population mean; thus, we must select a larger sample. As the sample size n increases, the standard error $\sigma/\sqrt{n}$ decreases; this results in a more narrow confidence interval. Consider the 95% confidence limits $\overline{X} \pm 1.96(\sigma/\sqrt{n})$. If we choose a sample of size 10, the confidence limits are $\overline{X} \pm 1.96(\sigma/\sqrt{10})$. If the selected sample is of size 100, then the limits are $\overline{X} \pm 1.96(\sigma/\sqrt{100})$. For an even larger sample of size 1,000, the 95% confidence limits would be $\overline{X} \pm 1.96(\sigma/\sqrt{1000})$. Summarizing these calculations, we have:

n	95% Confidence Limits for μ	Length of Interval
10	$\overline{X} \pm 0.620\sigma$	1.240σ
100	$\overline{X} \pm 0.196\sigma$	0.392σ
1000	$\overline{X} \pm 0.062\sigma$	0.124σ

As we select larger and larger random samples, the variability of $\overline{X}$—our estimator of the population mean μ—becomes smaller. The inherent variability of the underlying population, measured by σ, is always present, however.

Consider the distribution of serum cholesterol levels for all males in the United States who are hypertensive and who smoke. This distribution is approximately normal with an unknown mean μ and standard deviation $\sigma = 46$ mg/100 ml. (Even though the mean may be different, we assume for the moment that σ is the same as it was for the general population of adult males living in the United States.) We are interested in estimating the mean serum cholesterol level of this population. Before we go out and select a random sample, the probability that the interval

$$\left(\overline{X} - 1.96\frac{46}{\sqrt{n}}, \overline{X} + 1.96\frac{46}{\sqrt{n}}\right)$$

covers the true population mean μ is 0.95.

Suppose that we draw a sample of size 12 from the population of hypertensive smokers and that these men have a mean serum cholesterol level of $\overline{x} = 217$ mg/100 ml [1]. Based on this sample, a 95% confidence interval for μ is

$$\left(217 - 1.96\frac{46}{\sqrt{12}}, 217 + 1.96\frac{46}{\sqrt{12}}\right)$$

or

(191, 243).

While 217 mg/100 ml is our best guess for the mean serum cholesterol level of the population of male hypertensive smokers, the interval from 191 to 243 provides a range of reasonable values for μ. Note that this interval contains the value 211 mg/100 ml, the mean cholesterol level for all 20- to 74-year-old males in the U.S. regardless of hypertension or smoking status [2]. We are 95% confident that the limits 191 and 243 cover the true mean μ. We do **not** say that there is a 95% probability that μ lies between these values; μ is fixed and either it is between 191 and 243 or it is not.

As we previously mentioned, this confidence interval also has a frequency interpretation. Suppose that the true mean serum cholesterol level of the population of male hypertensive smokers is equal to 211 mg/100 ml, the mean level for adult males in the United States. If we were to draw 100 random samples of size 12 from this population and use each one to construct a 95% confidence interval, we would expect that, on average, 95 of the intervals would cover the true population mean $\mu = 211$ and 5 would not. This procedure was simulated and the results are illustrated in Figure 9.1. The only quantity that varies from sample to sample is $\overline{X}$. Although the centers of the intervals differ, they all

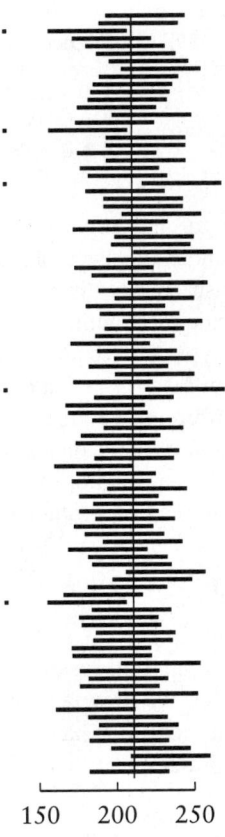

FIGURE 9.1
Set of 95% confidence intervals constructed from samples of size 12 drawn
from a normal population with mean 211 (marked by the vertical line) and
standard deviation 46

have the same length. Each of the confidence intervals that does not contain the true value
of μ is marked by a dot; note that exactly five intervals fall into this category.

Instead of generating a 95% confidence interval for the mean serum cholesterol
level, we might prefer to calculate a 99% confidence interval for μ. Using the same sample of 12 hypertensive smokers, we find the limits to be

$$\left(217 - 2.58\,\frac{46}{\sqrt{12}},\ 217 + 2.58\,\frac{46}{\sqrt{12}}\right),$$

or

$$(183, 251).$$

We are 99% confident that this interval covers the true mean serum cholesterol level of
the population. As previously noted, this interval is wider than the corresponding 95%
confidence interval.

In the example above, the length of the 99% confidence interval is $251 - 183 = 68$ mg/100 ml. How large a sample would we need to reduce the length of this interval to only 20 mg/100 ml? Since the interval is centered around the sample mean $\bar{x} = 217$ mg/100 ml, we are interested in the sample size necessary to produce the interval

$$(217 - 10, 217 + 10),$$

or

$$(207, 227).$$

Recall that the 99% confidence interval is of the form

$$\left(217 - 2.58 \frac{46}{\sqrt{n}}, \; 217 + 2.58 \frac{46}{\sqrt{n}}\right).$$

Therefore, to find the required sample size n, we must solve the equation

$$10 = \frac{2.58(46)}{\sqrt{n}}.$$

Multiplying both sides of the equality by $\sqrt{n}$ and dividing by 10, we find that

$$\sqrt{n} = \frac{2.58(46)}{10}$$

and

$$n = 140.8.$$

We would need a sample of 141 men to reduce the length of the 99% confidence interval to 20 mg/100 ml. Although the sample mean 217 mg/100 ml lies at the center of the interval, it does not play any part in determining its length; the length is a function of σ, n, and the level of confidence.

9.2   One-Sided Confidence Intervals

In some situations, we are concerned with either an upper limit for the population mean μ or a lower limit for μ, but not both. Consider the distribution of hemoglobin levels—hemoglobin is an oxygen-bearing protein found in red blood cells—for the population of children under the age of 6 who have been exposed to high levels of lead. This distribution has an unknown mean μ and standard deviation $\sigma = 0.85$ g/100 ml [3]. We know that children who have lead poisoning tend to have much lower levels of hemoglobin than children who do not. Therefore, we might be interested in finding an upper bound for μ.

To construct a one-sided confidence interval, we consider the area in one tail of the standard normal distribution only. Consulting Table A.3, we find that 95% of the observations for a standard normal random variable lie above $z = -1.645$. Therefore,

$$P(Z \geq -1.645) = 0.95.$$

Substituting $(\overline{X} - \mu)/(\sigma/\sqrt{n})$ for Z,

$$P\left(\frac{\overline{X} - \mu}{\sigma/\sqrt{n}} \geq -1.645\right) = 0.95.$$

Multiplying both sides of the inequality inside the probability statement by $\sigma/\sqrt{n}$ and then subtracting $\overline{X}$, we find that

$$P\left(-\mu \geq -\overline{X} - 1.645\,\frac{\sigma}{\sqrt{n}}\right) = 0.95$$

and

$$P\left(\mu \leq \overline{X} + 1.645\,\frac{\sigma}{\sqrt{n}}\right) = 0.95.$$

Therefore, $\overline{X} + 1.645(\sigma/\sqrt{n})$ is an upper 95% confidence bound for μ. Similarly, we could show that $\overline{X} - 1.645(\sigma/\sqrt{n})$ is the corresponding lower 95% confidence bound.

Suppose that we select a sample of 74 children who have been exposed to high levels of lead; these children have a mean hemoglobin level of $\overline{x} = 10.6$ g/100 ml [4]. Based on this sample, a one-sided 95% confidence interval for μ—the upper bound only—is

$$\mu \leq 10.6 + 1.645\left(\frac{0.85}{\sqrt{74}}\right)$$
$$\leq 10.8.$$

We are 95% confident that the true mean hemoglobin level for this population of children is at most 10.8 g/100 ml. In reality, since the value of μ is fixed, either the true mean is less than 10.8 or it is not. However, if we were to select 100 random samples of size 74 and use each one to construct a one-sided 95% confidence interval, approximately 95 of the intervals would cover the true mean μ.

9.3   Student's t Distribution

When computing confidence intervals for an unknown population mean μ, we have up to this point assumed that σ, the population standard deviation, was known. In reality, this is unlikely to be the case. If μ is unknown, σ is probably unknown as well. In this situation, confidence intervals are calculated in much the same way as we have already seen. Instead of using the standard normal distribution, however, the analysis depends on a probability distribution known as Student's t distribution. The name Student is the pseudonym of the statistician who originally discovered this distribution.

To construct a two-sided confidence interval for a population mean μ, we began by noting that

$$Z = \frac{\overline{X} - \mu}{\sigma/\sqrt{n}}$$

has an approximate standard normal distribution if n is sufficiently large. When the population standard deviation is not known, it may seem logical to substitute s, the standard deviation of a sample drawn from the population, for σ. This is, in fact, what is done. However, the ratio

$$t = \frac{\overline{X} - \mu}{s/\sqrt{n}}$$

does not have a standard normal distribution. In addition to the sampling variability inherent in $\overline{X}$—which we are using as an estimator of the population mean μ—there is also variation in s. The value of s is likely to change from sample to sample. Therefore, we must account for the fact that s may not be a reliable estimate of σ, especially when the sample with which we are working is small.

If X is normally distributed and a sample of size n is randomly chosen from this underlying population, the probability distribution of the random variable

$$t = \frac{\overline{X} - \mu}{s/\sqrt{n}}$$

is known as *Student's t distribution* with $n - 1$ degrees of freedom. We represent this using the notation t_{n-1}. Like the standard normal distribution, the t distribution is unimodal and symmetric around its mean of 0. The total area under the curve is equal to 1. However, it has somewhat thicker tails than the normal distribution; extreme values are more likely to occur with the t distribution than with the standard normal. This difference is illustrated in Figure 9.2. The shape of the t distribution reflects the extra variability introduced by the estimate s. In addition, the t distribution has a property called the *degrees of freedom*, abbreviated df. The degrees of freedom measure the amount of information available in the data that can be used to estimate σ^2; hence, they measure the reliability of s^2 as an estimate of σ^2. (The degrees of freedom are $n - 1$ rather than n because we lose 1 df by estimating the sample mean $\overline{x}$.) Recall that df $= n - 1$ is the quantity by which we divided the sum of the squared deviations around the mean, $\sum_{i=1}^{n}(x_i - \overline{x})^2$, in order to obtain the sample variance.

For each possible value of the degrees of freedom, there is a different t distribution. The distributions with smaller degrees of freedom are more spread out; as df increases, the t distribution approaches the standard normal. This occurs because as the sample size increases, s becomes a more reliable estimate of σ; if n is very large, knowing the value of s is nearly equivalent to knowing σ.

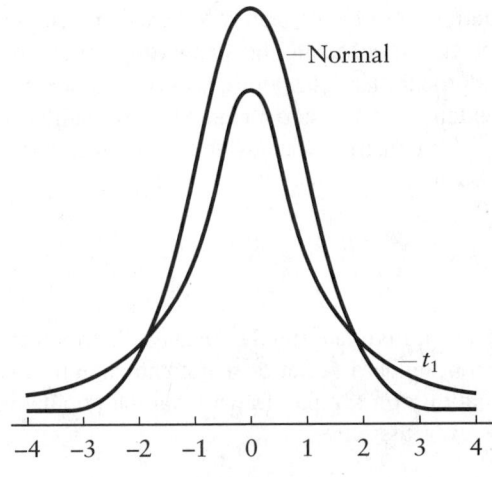

FIGURE 9.2
The standard normal distribution and Student's *t* distribution with 1 degree of freedom

Since there is a different *t* distribution for every value of the degrees of freedom, it would be quite cumbersome to work with a complete table of areas corresponding to each one. As a result, we typically rely on either a computer program or a condensed table that lists the areas under the curve for selected percentiles of the distribution only; for example, it might contain the upper 5.0, 2.5, 1.0, 0.5, and 0.05% of the distributions. When a computer is not available, condensed tables are sufficient for most applications involving the construction of confidence intervals.

Table A.4 in Appendix A is a condensed table of areas computed for the family of *t* distributions. For a particular value of df, the entry in the table represents the value of t_{n-1} that cuts off the specified area in the upper tail of the distribution. Given a *t* distribution with 10 degrees of freedom, for instance, $t_{10} = 2.228$ cuts off the upper 2.5% of the area under the curve. Since the distribution is symmetric, $t_{10} = -2.228$ marks off the lower 2.5%. The values of t_{n-1} that cut off the upper 2.5% of the distributions with various degrees of freedom are listed below.

df ($n-1$)	t_{n-1}
1	12.706
2	4.303
5	2.571
10	2.228
20	2.086
30	2.042
40	2.021
60	2.000
120	1.980
∞	1.960

For the standard normal curve, $z = 1.96$ marks the upper 2.5% of the distribution. Observe that as n increases, t_{n-1} approaches this value. In fact, when we have more than 30 degrees of freedom, we are able to substitute the standard normal distribution for the t and be off in our calculations by less than 5%.

Consider a random sample of ten children selected from the population of infants receiving antacids that contain aluminum. These antacids are often used to treat peptic or digestive disorders. The distribution of plasma aluminum levels is known to be approximately normal; however, its mean μ and standard deviation σ are not known. The mean aluminum level for the sample of ten infants is $\bar{x} = 37.2$ $\mu g/l$ and the sample standard deviation is $s = 7.13$ $\mu g/l$ [5].

Since the population standard deviation σ is not known, we must use the t distribution to find 95% confidence limits for μ. For a t distribution with $10 - 1 = 9$ degrees of freedom, 95% of the observations lie between -2.262 and 2.262. Therefore, replacing σ with s, a 95% confidence interval for the population mean μ is

$$\left(\bar{X} - 2.262\,\frac{s}{\sqrt{10}}, \bar{X} + 2.262\,\frac{s}{\sqrt{10}} \right).$$

Substituting in the values of $\bar{x}$ and s, the interval becomes

$$\left(37.2 - 2.262\,\frac{7.13}{\sqrt{10}}, 37.2 + 2.262\,\frac{7.13}{\sqrt{10}} \right),$$

or

$$(32.1, 42.3).$$

We are 95% confident that these limits cover the true mean plasma aluminum level for the population of infants receiving antacids. If we are given the additional information that the mean plasma aluminum level for the population of infants not receiving antacids is 4.13 $\mu g/l$—not a plausible value of μ for the infants who do receive them, according to the 95% confidence interval—this suggests that being given antacids greatly increases the plasma aluminum levels of children.

If the population standard deviation σ had been known and had been equal to the sample value of 7.13 $\mu g/l$, the 95% confidence interval for μ would have been

$$\left(37.2 - 1.96\,\frac{7.13}{\sqrt{10}}, 37.2 + 1.96\,\frac{7.13}{\sqrt{10}} \right),$$

or

$$(32.8, 41.6).$$

In this case, the confidence interval is slightly shorter. Most of the time, confidence intervals based on the t distribution are longer than the corresponding intervals based on the standard normal distribution. This generalization does not always apply, however;

because of the nature of sampling variability, it is possible that the value of the estimate *s* will be considerably smaller than σ for a given sample.

In a previous example, we examined the distribution of serum cholesterol levels for all males in the United States who are hypertensive and who smoke. Recall that the standard deviation of this population was assumed to be 46 mg/100 ml. On the left-hand side, Figure 9.3 shows the 95% confidence intervals for μ that were calculated from 100 random samples and were displayed in Figure 9.1. The right-hand side of the figure shows 100 additional intervals that were computed using the same samples; in each case, however, the standard deviation was not assumed to be known. Once again, 95 of the intervals contain the true mean μ, and the other 5 do not. Note that this time, however, the intervals vary in length.

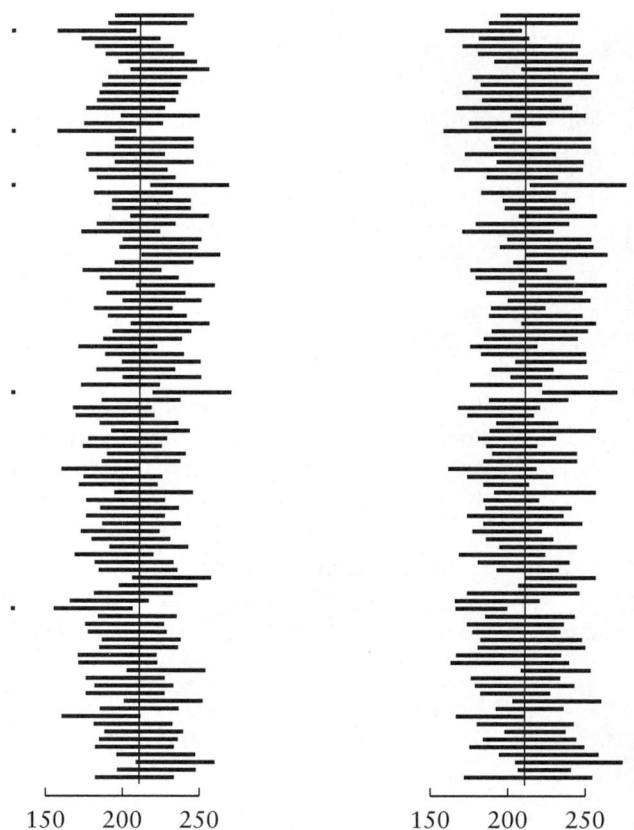

FIGURE 9.3
Two sets of 95% confidence intervals constructed from samples of size 12 drawn from normal populations with mean 211 (marked by the vertical lines), one with standard deviation 46 and the other with standard deviation unknown

9.4   Further Applications

Consider the distribution of heights for the population of individuals between the ages of 12 and 40 who suffer from fetal alcohol syndrome. Fetal alcohol syndrome is the severe end of the spectrum of disabilities caused by maternal alcohol use during pregnancy. The distribution of heights is approximately normal with unknown mean μ. We wish to find both a point estimate and a confidence interval for μ; the confidence interval provides a range of reasonable values for the parameter of interest.

When we construct a confidence interval for the mean of a continuous random variable, the technique used differs depending on whether the standard deviation of the underlying population is known or not known. For the height data, the standard deviation is assumed to be $\sigma = 6$ centimeters [6]. Therefore, we use the standard normal distribution to help us construct a 95% confidence interval. Before a sample is drawn from this population, the interval

$$\left(\overline{X} - 1.96 \frac{6}{\sqrt{n}}, \ \overline{X} + 1.96 \frac{6}{\sqrt{n}} \right)$$

has a 95% chance of covering the true population mean μ.

A random sample of 31 patients is selected from the underlying population; the mean height for these individuals is $\overline{x} = 147.4$ cm. This is the point estimate—our best guess—for the population mean μ. A 95% confidence interval based on the sample is

$$\left(147.4 - 1.96 \frac{6}{\sqrt{31}}, \ 147.4 + 1.96 \frac{6}{\sqrt{31}} \right),$$

or

$$(145.3, \ 149.5).$$

We are 95% confident that these limits cover the true mean height for the population of individuals between the ages of 12 and 40 who suffer from fetal alcohol syndrome. In reality, however, either the fixed value of μ is between 145.3 cm and 149.5 cm or it is not.

Rather than generate the confidence interval by hand, we could have used a computer to do the calculations for us. Table 9.1 shows the relevant output from Minitab. In addition to the sample size, the table displays the sample mean, the assumed standard deviation, the standard error of the mean, and the 95% confidence interval. It is also possible to calculate intervals with different levels of confidence. Table 9.2 shows a 90% confidence interval for μ. The 90% confidence interval is a bit shorter than the corresponding 95% interval; we have less confidence that this interval contains the true mean μ.

As a second example, methylphenidate is a drug that is widely used in the treatment of attention-deficit disorder. As part of a *crossover study*, ten children between the ages of 7 and 12 who suffered from this disorder were assigned to receive the drug and ten were given a placebo [7]. After a fixed period of time, treatment was withdrawn from

TABLE 9.1

Minitab output displaying a 95% confidence interval, standard deviation known

```
THE ASSUMED SIGMA = 6.000

             N      MEAN    STDEV    SE MEAN    95.0 PERCENT C.I.
HEIGHT      31     147.4    6.000     1.078    (145.288, 149.512)
```

TABLE 9.2

Minitab output displaying a 90% confidence interval, standard deviation known

```
THE ASSUMED SIGMA = 6.000

             N      MEAN    STDEV    SE MEAN    90.0 PERCENT C.I.
HEIGHT      31     147.4    6.000     1.078    (145.627, 149.173)
```

all 20 children. Subsequently, the children who had received methylphenidate were given the placebo, and those who had received the placebo now got the drug. (This is what is meant by a *crossover study*.) Measures of each child's attention and behavioral status, both on the drug and on the placebo, were obtained using an instrument called the Parent Rating Scale. Distributions of these scores are approximately normal with unknown means and standard deviations. In general, lower scores indicate an increase in attention. We wish to estimate the mean attention rating scores for children taking methylphenidate and for those taking the placebo.

Since the standard deviation is not known for either population, we use the t distribution to help us construct 95% confidence intervals. For a t distribution with $20 - 1 = 19$ degrees of freedom, 95% of the observations lie between -2.093 and 2.093. Therefore, before a sample of size 20 is drawn from the population, the interval

$$\left(\overline{X} - 2.093 \, \frac{s}{\sqrt{20}}, \overline{X} + 2.093 \, \frac{s}{\sqrt{20}} \right)$$

has a 95% chance of covering the true mean μ.

The random sample of 20 children enrolled in the study has mean attention rating score $\overline{x}_M = 10.8$ and standard deviation $s_M = 2.9$ when taking methylphenidate and mean rating score $\overline{x}_P = 14.0$ and standard deviation $s_P = 4.8$ when taking the placebo. Therefore, a 95% confidence interval for μ_M, the mean attention rating score for children taking the drug, is

$$\left(10.8 - 2.093 \, \frac{2.9}{\sqrt{20}}, 10.8 + 2.093 \, \frac{2.9}{\sqrt{20}} \right),$$

or

$$(9.44, 12.16),$$

TABLE 9.3

Stata output displaying 95% confidence intervals, standard deviation unknown

Variable	Obs	Mean	Std. Err.	[95% Conf.	Interval]
rating	20	10.8	.6484597	9.442758	12.15724

Variable	Obs	Mean	Std. Err.	[95% Conf.	Interval]
rating	20	14	1.073313	11.75353	16.24647

and a 95% confidence interval for μ_P, the mean rating score for children taking the placebo, is

$$\left(14.0 - 2.093\frac{4.8}{\sqrt{20}}, \ 14.0 + 2.093\frac{4.8}{\sqrt{20}}\right),$$

or

$$(11.75, 16.25).$$

The relevant output from Stata for both of these intervals is displayed in Table 9.3. When we look at the intervals, it appears that the mean attention rating score is likely to be lower when children with attention-deficit disorder are taking methylphenidate, implying improved attention. However, there is some overlap between the two intervals.

9.5   Review Exercises

1. Explain the difference between point and interval estimation.

2. Describe the 95% confidence interval for a population mean μ. How is the interval interpreted?

3. What are the factors that affect the length of a confidence interval for a mean? Explain briefly.

4. Describe the similarities and differences between the t distribution and the standard normal distribution. If you were trying to construct a confidence interval, when would you use one rather than the other?

5. The distributions of systolic and diastolic blood pressures for female diabetics between the ages of 30 and 34 have unknown means. However, their standard deviations are $\sigma_s = 11.8$ mm Hg and $\sigma_d = 9.1$ mm Hg, respectively [8].

(a) A random sample of ten women is selected from this population. The mean systolic blood pressure for the sample is $\bar{x}_s = 130$ mm Hg. Calculate a two-sided 95% confidence interval for μ_s, the true mean systolic blood pressure.

(b) Interpret this confidence interval.

(c) The mean diastolic blood pressure for the sample of size 10 is $\bar{x}_d = 84$ mm Hg. Find a two-sided 90% confidence interval for μ_d, the true mean diastolic blood pressure of the population.

(d) Calculate a two-sided 99% confidence interval for μ_d.

(e) How does the 99% confidence interval compare to the 90% interval?

6. Consider the t distribution with 5 degrees of freedom.
   (a) What proportion of the area under the curve lies to the right of $t = 2.015$?
   (b) What proportion of the area lies to the left of $t = -3.365$?
   (c) What proportion of the area lies between $t = -4.032$ and $t = 4.032$?
   (d) What value of t cuts off the upper 2.5% of the distribution?

7. Consider the t distribution with 21 degrees of freedom.
   (a) What proportion of the area under the curve lies to the left of $t = -2.518$?
   (b) What proportion of the area lies to the right of $t = 1.323$?
   (c) What proportion of the area lies between $t = -1.721$ and $t = 2.831$?
   (d) What value of t cuts off the lower 2.5% of the distribution?

8. Before beginning a study investigating the ability of the drug heparin to prevent bronchoconstriction, baseline values of pulmonary function were measured for a sample of 12 individuals with a history of exercise-induced asthma [9]. The mean value of forced vital capacity (FVC) for the sample is $\bar{x}_1 = 4.49$ liters and the standard deviation is $s_1 = 0.83$ liters; the mean forced expiratory volume in 1 second (FEV$_1$) is $\bar{x}_2 = 3.71$ liters and the standard deviation is $s_2 = 0.62$ liters.
   (a) Compute a two-sided 95% confidence interval for μ_1, the true population mean FVC.
   (b) Rather than a 95% interval, construct a 90% confidence interval for the true mean FVC. How does the length of the interval change?
   (c) Compute a 95% confidence interval for μ_2, the true population mean FEV$_1$.
   (d) In order to construct these confidence intervals, what assumption is made about the underlying distributions of FVC and FEV$_1$?

9. For the population of infants subjected to fetal surgery for congenital anomalies, the distribution of gestational ages at birth is approximately normal with unknown mean μ and standard deviation σ. A random sample of 14 such infants has mean gestational age $\bar{x} = 29.6$ weeks and standard deviation $s = 3.6$ weeks [10].
   (a) Construct a 95% confidence interval for the true population mean μ.
   (b) What is the length of this interval?
   (c) How large a sample would be required for the 95% confidence interval to have length 3 weeks? Assume that the population standard deviation σ is known and that $\sigma = 3.6$ weeks.
   (d) How large a sample would be needed for the 95% confidence interval to have length 2 weeks?

10. Percentages of ideal body weight were determined for 18 randomly selected insulin-dependent diabetics and are shown below [11]. A percentage of 120 means that an individual weighs 20% more than his or her ideal body weight; a percentage of 95 means that the individual weighs 5% less than the ideal.

107   119   99   114   120   104   88   114   124
116   101   121   152   100   125   114   95   117   (%)

(a) Compute a two-sided 95% confidence interval for the true mean percentage of ideal body weight for the population of insulin-dependent diabetics.
(b) Does this confidence interval contain the value 100%? What does the answer to this question tell you?

11. When eight persons in Massachusetts experienced an unexplained episode of vitamin D intoxication that required hospitalization, it was suggested that these unusual occurrences might be the result of excessive supplementation of dairy milk [12]. Blood levels of calcium and albumin for each individual at the time of hospital admission are shown below.

Calcium (mmol/l)	Albumin (g/l)
2.92	43
3.84	42
2.37	42
2.99	40
2.67	42
3.17	38
3.74	34
3.44	42

(a) Construct a one-sided 95% confidence interval—a lower bound—for the true mean calcium level of individuals who experience vitamin D intoxication.
(b) Compute a 95% lower confidence bound for the true mean albumin level of this group.
(c) For healthy individuals, the normal range of calcium values is 2.12 to 2.74 mmol/l and the range of albumin levels is 32 to 55 g/l. Do you believe that patients suffering from vitamin D intoxication have normal blood levels of calcium and albumin?

12. Serum zinc levels for 462 males between the ages of 15 and 17 are saved under the variable name zinc in the data set serzinc [13] (Appendix B, Table B.1). The units of measurement for serum zinc level are micrograms per deciliter.
(a) Find a two-sided 95% confidence interval for μ, the true mean serum zinc level for this population of males.
(b) Interpret this confidence interval.
(c) Calculate a 90% confidence interval for μ.
(d) How does the 90% confidence interval compare to the 95% interval?

13. The data set `lowbwt` contains information recorded for a sample of 100 low birth weight infants born in two teaching hospitals in Boston, Massachusetts [14] (Appendix B, Table B.7). Measurements of systolic blood pressure are saved under the variable name `sbp`, while indicators of gender—where 1 represents a male and 0 a female—are saved under the name `sex`.

(a) Compute a 95% confidence interval for the true mean systolic blood pressure of male low birth weight infants.

(b) Calculate a 95% confidence interval for the true mean systolic blood pressure of female low birth weight infants.

(c) Do you think it is possible that males and females have the same mean systolic blood pressure? Explain briefly.

Bibliography

[1] Kaplan, N. M., "Strategies to Reduce Risk Factors in Hypertensive Patients Who Smoke," *American Heart Journal*, Volume 115, January 1988, 288–294.

[2] National Center for Health Statistics, Fulwood, R., Kalsbeek, W., Rifkind, B., Russell-Briefel, R., Muesing, R., LaRosa, J., and Lippel, K., "Total Serum Cholesterol Levels of Adults 20–74 Years of Age: United States, 1976–1980," *Vital and Health Statistics*, Series 11, Number 236, May 1986.

[3] National Center for Health Statistics, Fulwood, R., Johnson, C. L., Bryner, J. D., Gunter, E. W., and McGrath, C. R., "Hematological and Nutritional Biochemistry Reference Data for Persons 6 Months–74 Years of Age: United States, 1976–1980," *Vital and Health Statistics*, Series 11, Number 232, December 1982.

[4] Clark, M., Royal, J., and Seeler, R., "Interaction of Iron Deficiency and Lead and the Hematologic Findings in Children with Severe Lead Poisoning," *Pediatrics*, Volume 81, February 1988, 247–253.

[5] Tsou, V. M., Young, R. M., Hart, M. H., and Vanderhoof, J. A., "Elevated Plasma Aluminum Levels in Normal Infants Receiving Antacids Containing Aluminum," *Pediatrics*, Volume 87, February 1991, 148–151.

[6] Streissguth, A. P., Aase, J. M., Clarren, S. K., Randels, S. P., LaDue, R. A., and Smith, D. F., "Fetal Alcohol Syndrome in Adolescents and Adults," *Journal of the American Medical Association*, Volume 265, April 17, 1991, 1961–1967.

[7] Tirosh, E., Elhasid, R., Kamah, S. C. B., and Cohen, A., "Predictive Value of Placebo Methylphenidate," *Pediatric Neurology*, Volume 9, Number 2, 1993, 131–133.

[8] Klein, B. E. K., Klein, R., and Moss, S. E., "Blood Pressure in a Population of Diabetic Persons Diagnosed After 30 Years of Age," *American Journal of Public Health*, Volume 74, April 1984, 336–339.

[9] Ahmed, T., Garrigo, J., and Danta, I., "Preventing Bronchoconstriction in Exercise-Induced Asthma with Inhaled Heparin," *The New England Journal of Medicine*, Volume 329, July 8, 1993, 90–95.

[10] Longaker, M. T., Golbus, M. S., Filly, R. A., Rosen, M. A., Chang, S. W., and Harrison, M. R., "Maternal Outcome After Open Fetal Surgery," *Journal of the American Medical Association*, Volume 265, February 13, 1991, 737–741.

[11] Saudek, C. D., Selam, J. L., Pitt, H. A., Waxman, K., Rubio, M., Jeandidier, N., Turner, D., Fischell, R. E., and Charles, M. A., "A Preliminary Trial of the Programmable Implantable Medication System for Insulin Delivery," *The New England Journal of Medicine*, Volume 321, August 31, 1989, 574–579.

[12] Jacobus, C. H., Holick, M. F., Shao, Q., Chen, T. C., Holm, I. A., Kolodny, J. M., Fuleihan, G. E. H., and Seely, E. W., "Hypervitaminosis D Associated with Drinking Milk," *The New England Journal of Medicine*, Volume 326, April 30, 1992, 1173–1177.

[13] National Center for Health Statistics, Fulwood, R., Johnson, C. L., Bryner, J. D., Gunter, E. W., and McGrath, C. R., "Hematological and Nutritional Biochemistry Reference Data for Persons 6 Months–74 Years of Age: United States, 1976–1980," *Vital and Health Statistics*, Series 11, Number 232, December 1982.

[14] Leviton, A., Fenton, T., Kuban, K. C. K., and Pagano, M., "Labor and Delivery Characteristics and the Risk of Germinal Matrix Hemorrhage in Low Birth Weight Infants," *Journal of Child Neurology*, Volume 6, October 1991, 35–40.

10

Hypothesis Testing

In our study of confidence intervals, we encountered the distribution of serum cholesterol levels for the population of males in the United States who are hypertensive and who smoke. This distribution is approximately normal with an unknown mean μ. However, we do know that the mean serum cholesterol level for the general population of all 20- to 74-year-old males is 211 mg/100 ml [1]. Therefore, we might wonder whether the mean cholesterol level of the subpopulation of men who are hypertensive smokers is 211 mg/100 ml as well. If we select a random sample of 25 men from this group and their mean serum cholesterol level is $\bar{x} = 220$ mg/100 ml, is this sample mean compatible with a hypothesized mean of 211 mg/100 ml? We know that some amount of sampling variability is to be expected. What if the sample mean is 230 mg/100 ml, or 250 mg/100 ml? How far from 211 must $\bar{x}$ be before we can conclude that μ is really equal to some other value?

10.1 General Concepts

We again concentrate on drawing some conclusion about a population parameter—the mean of a continuous random variable, in this case—using the information contained in a sample of observations. As we saw in the preceding chapter, one approach is to construct a confidence interval for μ; another is to conduct a *statistical hypothesis test*.

To perform such a test, we begin by claiming that the mean of the population is equal to some postulated value μ_0. This statement about the value of the population parameter is called the *null hypothesis*, or H_0. If we wanted to test whether the mean serum cholesterol level of the subpopulation of hypertensive smokers is equal to the mean of the general population of 20- to 74-year-old males, for instance, the null hypothesis would be

$$H_0: \mu = \mu_0 = 211 \text{ mg/100 ml.}$$

The *alternative hypothesis*, represented by H_A, is a second statement that contradicts H_0. In this case, we have

$$H_A: \mu \neq 211 \text{ mg/100 ml.}$$

Together, the null and the alternative hypotheses cover all possible values of the population mean μ; consequently, one of the two statements must be true.

After formulating the hypotheses, we next draw a random sample of size n from the population of interest. In the case of the hypertensive smokers, we selected a sample of size 12. We compare the mean of this sample, $\bar{x}$, to the postulated mean μ_0; specifically, we want to know whether the difference between the sample mean and the hypothesized mean is too large to be attributed to chance alone.

If there is evidence that the sample could **not** have come from a population with mean μ_0, we reject the null hypothesis. This occurs when, given that H_0 is true, the probability of obtaining a sample mean as extreme as or more extreme than the observed value $\bar{x}$—more extreme meaning farther away from the value μ_0—is sufficiently small. In this case, the data are not compatible with the null hypothesis; they are more supportive of the alternative. We therefore conclude that the population mean could not be μ_0. In keeping with popular phraseology, such a test result is said to be *statistically significant*. Note that statistical significance does not imply clinical or scientific significance; the test result could actually have little practical consequence.

If there is not sufficient evidence to doubt the validity of the null hypothesis, we cannot reject this claim. Instead, we concede that the population mean may be equal to μ_0. However, we do not say that we accept H_0; the test does not prove the null hypothesis. It is still possible that the population mean is some value other than μ_0, but that the random sample selected does not confirm this. Such an event can occur, for instance, when the sample chosen is too small. This point is discussed further later in this chapter.

We have stated above that if the probability of obtaining a sample mean as extreme as or more extreme than the observed $\bar{x}$ is sufficiently small, we reject the null hypothesis. But what is a "sufficiently small" probability? In most applications, 0.05 is chosen [2]. Thus, we reject H_0 when the chance that the sample could have come from a population with mean μ_0 is less than or equal to 5%. This implies that we reject incorrectly 5% of the time; given many repeated tests of significance, 5 times out of 100 we will erroneously reject the null hypothesis when it is true. To be more conservative, a probability of 0.01 is sometimes chosen. In this case, we mistakenly reject H_0 when it is true only 1% of the time. If we are willing to be less conservative, a probability of 0.10 might be used. The probability that we choose—whether 0.05, 0.01, or some other value—is known as the *significance level* of the hypothesis test. The significance level is denoted by the Greek letter α and must be specified **before** the test is actually carried out.

In many ways, a test of hypothesis can be compared to a criminal trial by jury in the United States. The individual on trial is either innocent or guilty, but is assumed to be innocent by law. After evidence pertaining to the case has been presented, the jury finds the defendant either guilty or not guilty. If the defendant is innocent and the decision of the jury is that he or she is not guilty, then the right verdict has been reached. The verdict is also correct if the defendant is guilty and is convicted of the crime.

Verdict of Jury	Defendant	
	Innocent	**Guilty**
Not Guilty	Correct	Incorrect
Guilty	Incorrect	Correct

Analogously, the true population mean is either μ_0 or not μ_0. We begin by assuming that the null hypothesis

$$H_0: \mu = \mu_0$$

is correct, and we consider the "evidence" that is presented in the form of a sample of size n. Based on our findings, the null hypothesis is either rejected or not rejected. Again there are two situations in which the conclusion drawn is correct—when the population mean is μ_0 and the null hypothesis is not rejected, and when the population mean is not μ_0 and H_0 is rejected.

Result of Test	Population	
	$\mu = \mu_0$	$\mu \neq \mu_0$
Do Not Reject	Correct	Incorrect
Reject	Incorrect	Correct

Like our legal system, the process of hypothesis testing is not perfect; there are two kinds of errors that can be made. In particular, we could either reject the null hypothesis when μ is equal to μ_0, or fail to reject it when μ is not equal to μ_0. These two types of errors—which have much in common with the false positive and false negative results that occur in diagnostic testing—are discussed in more detail in Section 10.4.

The probability of obtaining a mean as extreme as or more extreme than the observed sample mean $\bar{x}$, given that the null hypothesis

$$H_0: \mu = \mu_0$$

is true, is called the *p-value* of the test, or simply p. The p-value is compared to the predetermined significance level α to decide whether the null hypothesis should be rejected. If p is less than or equal to α, we reject H_0. If p is greater than α, we do not reject H_0. In addition to the conclusion of the test, the p-value itself is often reported in the literature.

10.2   Two-Sided Tests of Hypotheses

To conduct a test of hypothesis, we again draw on our knowledge of the sampling distribution of the mean. Assume that the continuous random variable X has mean μ_0 and known standard deviation σ. Thus, according to the central limit theorem,

$$Z = \frac{\overline{X} - \mu_0}{\sigma/\sqrt{n}}$$

has an approximate standard normal distribution if the value of n is sufficiently large. For a given sample with mean $\overline{x}$, we can calculate the corresponding outcome of Z, called the *test statistic*. We then use either a computer program or a table of the standard normal curve—such as Table A.3 in Appendix A—to determine the probability of obtaining a value of Z that is as extreme as or more extreme than the one observed. By more extreme, we mean farther away from μ_0 in the direction of the alternative hypothesis. Because it relies on the standard normal distribution, a test of this kind is called a *z-test*.

When the population standard deviation is not known, we substitute the sample value s for σ. If the underlying population is normally distributed, the random variable

$$t = \frac{\overline{X} - \mu_0}{s/\sqrt{n}}$$

has a t distribution with $n - 1$ degrees of freedom. In this case, we can calculate the outcome of t corresponding to a given $\overline{x}$ and consult our computer program or Table A.4 to find the probability of obtaining a sample mean that is more extreme than the one observed. This procedure is known as a *t-test*.

To illustrate the process of hypothesis testing, consider again the distribution of serum cholesterol levels for adult males in the United States who are hypertensive and who smoke. The standard deviation of this distribution is assumed to be $\sigma = 46$ mg/100 ml; the null hypothesis to be tested is

$$H_0: \mu = 211 \text{ mg/100 ml,}$$

where $\mu_0 = 211$ mg/100 ml is the mean serum cholesterol level for all 20- to 74-year-old males. Since the mean of the subpopulation of hypertensive smokers could be either larger than μ_0 or smaller than μ_0, we are concerned with deviations that occur in either direction. As a result, we conduct what is called a two-sided test at the $\alpha = 0.05$ level of significance. The alternative hypothesis for the two-sided test is

$$H_A: \mu \neq 211 \text{ mg/100 ml.}$$

The previously mentioned random sample of 12 hypertensive smokers has mean serum cholesterol level $\overline{x} = 217$ mg/100 ml [3]. Is it likely that this sample comes from a population with mean 211 mg/100 ml? To answer this question, we compute the test statistic

$$z = \frac{\bar{x} - \mu_0}{\sigma/\sqrt{n}}$$

$$= \frac{217 - 211}{46/\sqrt{12}}$$

$$= 0.45.$$

If the null hypothesis is true, this statistic is the outcome of a standard normal random variable. According to Table A.3, the area to the right of $z = 0.45$—which is the probability of observing $Z = 0.45$ or anything larger, given that H_0 is true—is 0.326. The area to the left of $z = -0.45$ is 0.326 as well. Thus, the area in the two tails of the standard normal distribution sums to 0.652; this is the p-value of the test. Since $p > 0.05$, we do not reject the null hypothesis. Based on this sample, the evidence is insufficient to conclude that the mean serum cholesterol level of the population of hypertensive smokers is different from 211 mg/100 ml.

Although it may not be immediately obvious, there is actually a mathematical equivalence between confidence intervals and tests of hypothesis. Because we conducted a two-sided test, any value of z that is between -1.96 and 1.96 would result in a p-value greater than 0.05. (The outcome 0.45 is just one such value.) In each case, the null hypothesis would not be rejected. On the other hand, H_0 would be rejected for any value of z that is either less than -1.96 or greater than 1.96. Because they indicate when we reject and when we do not, the numbers -1.96 and 1.96 are called the *critical values* of the test statistic.

Another way to look at this is to note that the null hypothesis will fail to be rejected when μ_0 is any value that lies within the 95% confidence interval for μ. Recall that in Chapter 9 we found a 95% confidence interval for the mean serum cholesterol level of hypertensive smokers to be

(191, 243).

Any value of μ_0 that lies in this interval would result in a test statistic that is between -1.96 and 1.96. Therefore, if the null hypothesis had been

$H_0: \mu = 240$ mg/100 ml,

H_0 would not have been rejected. Similarly, the null hypothesis would not be rejected for $\mu_0 = 195$ mg/100 ml. In contrast, any value of μ_0 that lies outside of the 95% confidence interval for μ—such as $\mu_0 = 260$ mg/100 ml—would result in a rejection of the null hypothesis at the $\alpha = 0.05$ level. These values produce test statistics either less than -1.96 or greater than 1.96.

Although confidence intervals and tests of hypotheses lead us to the same conclusions, the information provided by each is somewhat different. The confidence interval supplies a range of reasonable values for the parameter μ and tells us something about the uncertainty in our point estimate $\bar{x}$. The hypothesis test helps us to decide whether the postulated value of the mean is likely to be correct or not, and provides a specific *p*-value.

Returning to the test itself, the value $\mu_0 = 211$ mg/100 ml was selected for the null hypothesis because it is the mean serum cholesterol level of the population of all 20- to 74-year-old males. Consequently, H_0 claims that the mean serum cholesterol level of males who are hypertensive smokers is identical to the mean cholesterol level of the general population of males. The hypothesis was established with an interest in obtaining evidence that would cause it to be rejected in favor of the alternative; a rejection would have implied that the mean serum cholesterol level of male hypertensive smokers is **not** equal to the mean of the population as a whole.

As a second example, consider the random sample of ten children selected from the population of infants receiving antacids that contain aluminum. The underlying distribution of plasma aluminum levels for this population is approximately normal with an unknown mean μ and standard deviation σ. However, we do know that the mean plasma aluminum level for the sample of size 10 is $\bar{x} = 37.20$ μg/l and that its standard deviation is $s = 7.13$ μg/l [4]. Furthermore, the mean plasma aluminum level for the population of infants not receiving antacids is 4.13 μg/l. Is it likely that the data in our sample could have come from a population with mean $\mu_0 = 4.13$ μg/l? To find out, we conduct a test of hypothesis; the null hypothesis is

$$H_0: \mu = 4.13 \ \mu g/l,$$

and the alternative is

$$H_A: \mu \neq 4.13 \ \mu g/l.$$

We are interested in deviations from the mean that could occur in either direction; we would want to know if μ is actually larger than 4.13 or if it is smaller. Therefore, we conduct a two-sided test at the $\alpha = 0.05$ level of significance.

Because we do not know the population standard deviation σ, we use a *t*-test rather than a *z*-test. The test statistic is

$$t = \frac{\bar{x} - \mu_0}{s/\sqrt{n}},$$

or

$$t = \frac{37.20 - 4.13}{7.13/\sqrt{10}}$$

$$= 14.67.$$

If the null hypothesis is true, this outcome has a *t* distribution with $10 - 1 = 9$ df. Consulting Table A.4, we observe that the total area to the right of $t_9 = 14.67$ and to the left of $t_9 = -14.67$ is less than $2(0.0005) = 0.001$. Therefore, $p < 0.05$, and we reject the null hypothesis

$$H_0: \mu = 4.13 \, \mu g/l.$$

This sample of infants provides evidence that the mean plasma aluminum level of children receiving antacids is not equal to the mean aluminum level of children who do not receive them. In fact, since the sample mean $\bar{x}$ is larger than μ_0, the true mean aluminum level is higher than $4.13 \, \mu g/l$.

10.3   One-Sided Tests of Hypotheses

Before we conduct a test of hypothesis, we must decide whether we are concerned with deviations from μ_0 that could occur in both directions—meaning either higher or lower than μ_0—or in one direction only. This choice determines whether we consider the area in two tails of the appropriate curve when calculating a p-value or the area in a single tail. The decision must be made **before** a random sample is selected; it should not be influenced by the outcome of the sample. If prior knowledge indicates that μ cannot be less than μ_0, the only values of $\bar{x}$ that will provide evidence against the null hypothesis

$$H_0: \mu = \mu_0$$

are those that are much larger than μ_0. In a situation such as this, the null hypothesis is more properly stated as

$$H_0: \mu \leq \mu_0$$

and the alternative hypothesis as

$$H_A: \mu > \mu_0.$$

For example, most people would agree that it is unreasonable to believe that exposure to a toxic substance—such as ambient carbon monoxide or sulphur dioxide—could possibly be beneficial to humans. Therefore, we anticipate only harmful effects and conduct a one-sided test. A two-sided test is always the more conservative choice; in general, the p-value of a two-sided test is twice as large as that of a one-sided test.

Consider the distribution of hemoglobin levels for the population of children under the age of 6 who have been exposed to high levels of lead. This distribution has an unknown mean μ; its standard deviation is assumed to be $\sigma = 0.85$ g/100 ml [5]. We might wish to know whether the mean hemoglobin level of this population is equal to the mean of the general population of children under the age of 6, $\mu = 12.29$ g/100 ml. We believe that if the hemoglobin levels of exposed children differ from those of unexposed children, they must on average be lower; therefore, we are concerned only with deviations from the mean that are below μ_0. The null hypothesis for the test is

$$H_0: \mu \geq 12.29 \text{ g/100 ml}$$

and the one-sided alternative is

$$H_A: \mu < 12.29 \text{ g/100 ml.}$$

H_0 would be rejected for values of $\bar{x}$ that are lower than 12.29, but not for those that are higher. We conduct the one-sided test at the $\alpha = 0.05$ level of significance; since σ is known, we use the normal distribution rather than the t.

A random sample of 74 children who have been exposed to high levels of lead has a mean hemoglobin level of $\bar{x} = 10.6$ g/100 ml [6]. Therefore, the appropriate test statistic is

$$z = \frac{\bar{x} - \mu_0}{\sigma/\sqrt{n}}$$

$$= \frac{10.6 - 12.29}{0.85/\sqrt{74}}$$

$$= -17.10.$$

According to Table A.3, the area to the left of $z = -17.10$ is less than 0.001. Since this p-value is smaller than $\alpha = 0.05$, we reject the null hypothesis

$$H_0: \mu \geq 12.29 \text{ g/100 ml}$$

in favor of the alternative. Because this is a one-sided test, any value of z that is less than or equal to the critical value -1.645 would have led us to reject the null hypothesis. (Also note that 12.29 lies above 10.8, the upper one-sided 95% confidence bound for μ calculated in Chapter 9.)

In this example, H_0 was chosen to test the statement that the mean hemoglobin level of the population of children who have been exposed to lead is the same as that of the general population, 12.29 g/100 ml. By rejecting H_0, we conclude that this is not the case; the mean hemoglobin level for children who have been exposed to lead is in fact lower than the mean for children who have not been exposed.

The choice between a one-sided and a two-sided test can be extremely controversial. Not infrequently, a one-sided test achieves significance when a two-sided test does not. Consequently, the decision is often made on nonscientific grounds. In response to this practice, some journal editors are reluctant to publish studies that employ one-sided tests. This may be an overreaction to something that the intelligent reader is able to discern. In any event, we will avoid further discussion of this debate.

10.4   Types of Error

As noted in Section 10.1, two kinds of errors can be made when we conduct a test of hypothesis. The first is called a *type I error*; it is also known as a *rejection error* or an *α error*. A type I error is made if we reject the null hypothesis

$$H_0: \mu = \mu_0$$

when H_0 is true. The probability of making a type I error is determined by the significance level of the test; recall that

$$\alpha = P(\text{reject } H_0 \mid H_0 \text{ is true}).$$

If we were to conduct repeated, independent tests of hypotheses setting the significance level at 0.05, we would erroneously reject a true null hypothesis 5% of the time.

Consider the case of a drug that has been deemed effective for reducing high blood pressure. After being treated with this drug for a given period of time, a population of individuals suffering from hypertension has mean diastolic blood pressure μ_d, a value that is clinically lower than the mean diastolic blood pressure of untreated hypertensives. Now suppose that another company produces a generic version of the same drug. We would like to know whether the generic drug is as effective at reducing high blood pressure as the brand-name version. To determine this, we examine the distribution of diastolic blood pressures for a sample of individuals who have been treated with the generic drug; if μ is the mean of this population, we use the sample to test the null hypothesis

$$H_0: \mu = \mu_d.$$

What if the manufacturer of the generic drug actually submits the brand-name product for testing in place of its own version? Vitarine Pharmaceuticals, a New York–based drug company, has reportedly done just that [7]. The company has made similar substitutions on four different occasions. In a situation such as this, we know that the null hypothesis must be true—we are testing the drug that itself set the standard. Therefore, if the test of hypothesis leads us to reject H_0 and pronounce the "generic" drug to be either more or less efficacious than the brand-name version, a type I error has been made.

The second kind of error that can be made during a hypothesis test is a *type II error*, also known as an *acceptance error* or a *β error*. A type II error is made if we fail to reject the null hypothesis

$$H_0: \mu = \mu_0$$

when H_0 is false. The probability of committing a type II error is represented by the Greek letter β, where

$$\beta = P(\text{do not reject } H_0 \mid H_0 \text{ is false}).$$

If $\beta = 0.10$, for instance, the probability that we do not reject the null hypothesis when $\mu \neq \mu_0$ is 0.10, or 10%. The two types of errors that can be made are summarized below.

	Population	
Result of Test	$\mu = \mu_0$	$\mu \neq \mu_0$
Do Not Reject	Correct	Type II Error
Reject	Type I Error	Correct

Recall the distribution of serum cholesterol levels for all 20- to 74-year-old males in the United States. The mean of this population is $\mu = 211$ mg/100 ml, and the standard deviation is $\sigma = 46$ mg/100 ml. Suppose that we do not know the true mean of this population; however, we do know that the mean serum cholesterol level for the subpopulation of 20- to 24-year-old males is 180 mg/100 ml. Since older men tend to have higher cholesterol levels than do younger men on average, we would expect the mean cholesterol level of the population of 20- to 74-year-olds to be higher than 180 mg/100 ml. (And indeed it is, although we are pretending not to know this.) Therefore, if we were to conduct a one-sided test of the null hypothesis

$$H_0: \mu \leq 180 \text{ mg/100 ml}$$

against the alternative hypothesis

$$H_A: \mu > 180 \text{ mg/100 ml},$$

we would expect H_0 to be rejected. It is possible, however, that it would not be. The probability of reaching this incorrect conclusion—a type II error—is β.

What is the value of β associated with a test of the null hypothesis

$$H_0: \mu \leq 180 \text{ mg/100 ml},$$

assuming that we select a sample of size 25? To determine this, we first find the mean serum cholesterol level our sample must have for H_0 to be rejected. Since we are conducting a one-sided test at the $\alpha = 0.05$ level of significance, H_0 would be rejected for $z \geq 1.645$; this is the critical value of the test. Writing out the test statistic

$$z = \frac{\bar{x} - \mu_0}{\sigma/\sqrt{n}},$$

we have

$$1.645 = \frac{\bar{x} - 180}{46/\sqrt{25}},$$

and, solving for $\bar{x}$,

$$\bar{x} = 180 + \frac{1.645(46)}{\sqrt{25}}$$

$$= 195.1.$$

As shown in Figure 10.1, the area to the right of $\bar{x} = 195.1$ corresponds to the upper 5% of the sampling distribution of means of samples of size 25 when $\mu = 180$. Therefore, the null hypothesis

$$H_0: \mu \leq 180 \text{ mg/100 ml}$$

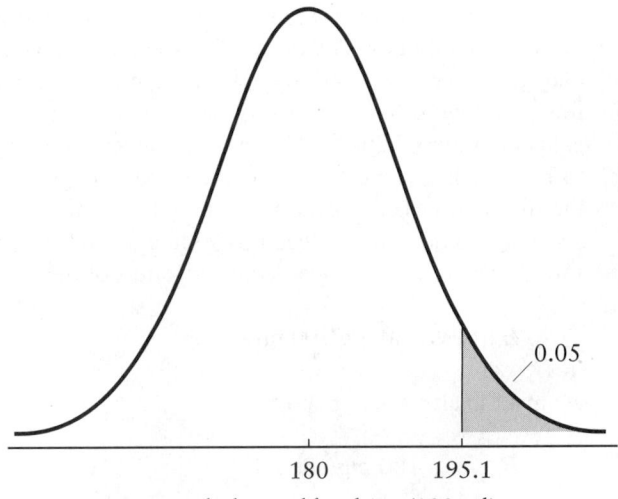

FIGURE 10.1

Distribution of means of samples of size 25 for the serum cholesterol levels
of males 20 to 74 years of age, $\mu = 180$ mg/100 ml

would be rejected if our sample has a mean $\bar{x}$ that is greater than or equal to
195.1 mg/100 ml. A sample with a smaller mean would not provide sufficient evidence
to reject H_0 in favor of H_A at the 0.05 level of significance.

Recall that the probability of making a type II error, or β, is the probability of not
rejecting the null hypothesis given that H_0 is false. Therefore, it is the chance of obtaining a sample mean that is less than 195.1 mg/100 ml given that the true population
mean is not 180 but is instead $\mu_1 = 211$ mg/100 ml. To find the value of β, we again
consider the sampling distribution of means of samples of size 25; this time, however,
we let $\mu = 211$. This distribution is shown on the right side of Figure 10.2. Since a sample mean less than $\bar{x} = 195.1$ mg/100 ml implies that we do not reject H_0, we would
like to know what proportion of this new distribution centered at 211 mg/100 ml lies
below 195.1. Observe that

$$z = \frac{195.1 - 211.0}{46/\sqrt{25}}$$

$$= -1.73.$$

The area under the standard normal curve that lies to the left of $z = -1.73$ is 0.042.
Therefore, β—the probability of failing to reject

$$H_0: \mu \leq 180 \text{ mg/100 ml}$$

when the true population mean is $\mu_1 = 211$ mg/100 ml—is equal to 0.042.

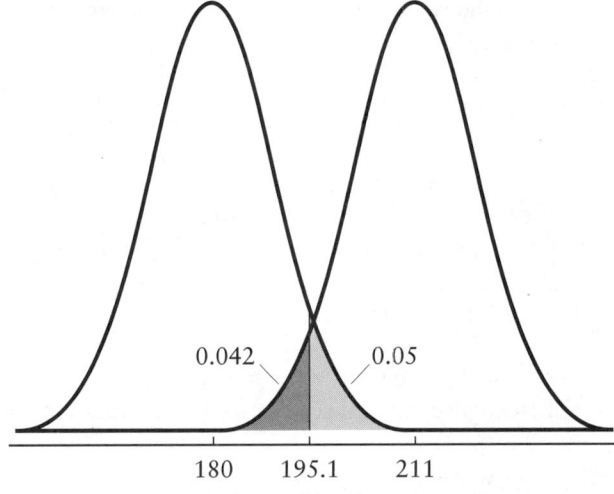

FIGURE 10.2
Distribution of means of samples of size 25 for the serum cholesterol
levels of males 20 to 74 years of age, $\mu = 180$ mg/100 ml versus
$\mu = 211$ mg/100 ml

While α, the probability of committing a type I error, is determined by looking at the case in which H_0 is true and μ is equal to μ_0, β is found by considering the situation in which H_0 is false and μ does not equal μ_0. If μ is not equal to μ_0, however, there are an infinite number of possible values that μ could assume. The type II error is calculated for a **single** such value, μ_1; in the previous example, μ_1 was chosen to be 211 mg/100 ml. (We selected 211 because in this unusual example, we knew it to be the true population mean.) If we had chosen a different alternative population mean, we would have computed a different value for β. The closer μ_1 is to μ_0, the more difficult it is to reject the null hypothesis.

10.5   Power

If β is the probability of making a type II error, $1 - \beta$ is called the *power* of the test of hypothesis. The power is the probability of rejecting the null hypothesis when H_0 is false. In other words, it is the probability of avoiding a type II error:

power = P(reject $H_0 \mid H_0$ is false).

The power may also be thought of as the likelihood that a particular study will detect a deviation from the null hypothesis given that one exists. Like β, the power must be computed for a particular alternative population mean μ_1.

In the serum cholesterol example above, the power of the one-sided test of hypothesis is

$$1 - \beta = 1 - 0.042$$
$$= 0.958.$$

Consequently, for a test conducted at the 0.05 level of significance and using a sample of size 25, there is a 95.8% chance of rejecting the null hypothesis

$$H_0: \mu \leq 180 \text{ mg/100 ml}$$

given that H_0 is false and the true population mean is $\mu_1 = 211$ mg/100 ml. Note that this could also have been expressed in the following way:

$$\text{power} = P(\text{reject } \mu \leq 180 \mid \mu = 211)$$
$$= P(\overline{X} \geq 195.1 \mid \mu = 211)$$
$$= P(Z \geq -1.73)$$
$$= 1 - P(Z < -1.73)$$
$$= 1 - 0.042$$
$$= 0.958.$$

The quantity $1 - \beta$ would have assumed a different value if we had set μ_1 equal to 200 mg/100 ml, and yet another value if we had let μ_1 be 220 mg/100 ml. If we were to plot the values of $1 - \beta$ against all possible alternative population means, we would end up with what is known as a *power curve*. A power curve for the test of the null hypothesis

$$H_0: \mu \leq 180 \text{ mg/100 ml}$$

is shown in Figure 10.3. Note that when $\mu_1 = 180$,

$$\text{power} = P(\text{reject } \mu \leq 180 \mid \mu = 180)$$
$$= P(\text{reject } \mu \leq 180 \mid H_0 \text{ is true})$$
$$= \alpha$$
$$= 0.05.$$

The power of the test approaches 1 as the alternative mean moves farther and farther away from the null value of 180 mg/100 ml.

Investigators generally try to design tests of hypotheses so that they have high power. It is not enough to know that we have a small probability of rejecting H_0 when it is true; we would also like to have a large probability of rejecting the null hypothesis when it is false. In most practical applications, a power less than 80% is considered insufficient. One way to increase the power of a test is to raise the significance level α. If we increase α, we cut off a smaller portion of the tail of the sampling distribution cen-

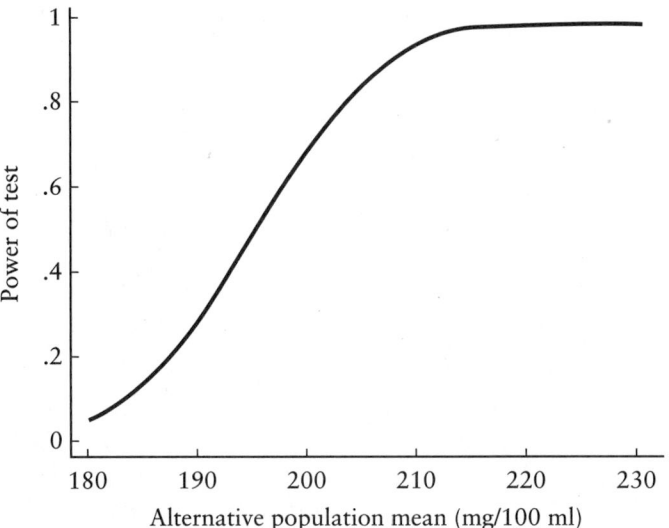

FIGURE 10.3
Power curve for $\mu_0 = 180$, $\alpha = 0.05$, and $n = 25$

tered at μ_1. Correspondingly, β becomes smaller and the power, $1 - \beta$, increases. If α had been equal to 0.10 for the test of the null hypothesis

$$H_0: \mu \leq 180 \text{ mg/100 ml,}$$

for instance, β would have been 0.018 and the power 0.982. This situation is illustrated in Figure 10.4; compare it to Figure 10.2, where α was equal to 0.05. Bear in mind, however, that by raising α we are increasing the probability of making a type I error.

This trade-off between α and β is similar to that observed to exist between the sensitivity and the specificity of a diagnostic test. Recall that by increasing the sensitivity of a test, we automatically decrease its specificity; alternatively, increasing the specificity lowers the sensitivity. The same is true for α and β. The balance between the two types of error is a delicate one, and their relative importance varies depending on the time and the situation. In 1692, during the Salem witch trials, Increase Mather published a sermon, signed by himself and 14 other parsons, wherein he stated [8],

> It were better that ten suspected witches should escape, than that one innocent person should be condemned.

In the eighteenth century, Benjamin Franklin said,

> It is better that 100 guilty persons should escape than that one innocent person should suffer.

More recently, however, an editorial on child abuse claimed that it is "equally important" to identify and punish child molesters and to exonerate those who are falsely accused [9].

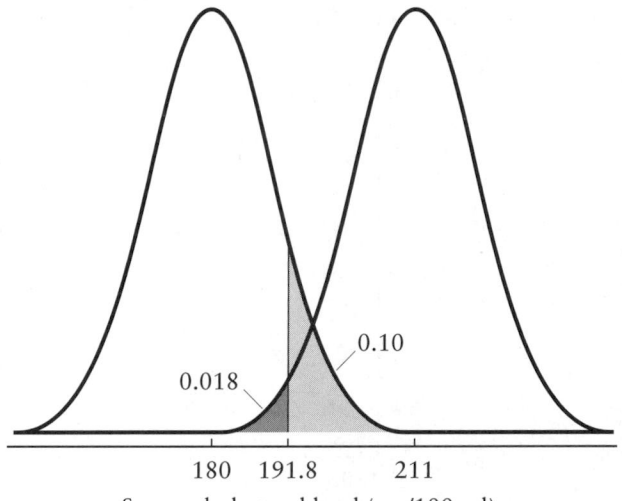

0.10

0.018

180   191.8   211

Serum cholesterol level (mg/100 ml)

FIGURE 10.4
Distributions of means of samples of size 25 for the serum cholesterol
levels of males 20 to 74 years of age, $\mu = 180$ mg/100 ml versus
$\mu = 211$ mg/100 ml

The more information we have—meaning the larger our sample—the less likely we are to commit an error of either type. Regardless of our decision, however, the possibility that we have made a mistake always exists.

The only way to diminish α and β simultaneously is to reduce the amount of overlap in the two normal distributions—the one centered at μ_0 and the one centered at μ_1. One way that this can be accomplished is by considering only large deviations from μ_0. The farther apart the values of μ_0 and μ_1, the greater the power of the test. An alternative is to increase the sample size n. By increasing n, we decrease the standard error $\sigma/\sqrt{n}$; this causes the two sampling distributions to become more narrow, which, in turn, lessens the amount of overlap. The standard error also decreases if we reduce the underlying population standard deviation σ, but this is usually not possible. Another option that we have not yet mentioned is to find a "more powerful" test statistic. This topic is discussed further in Chapter 13.

10.6   Sample Size Estimation

In the preceding section, we outlined a method for calculating the power of a test conducted at the α level using a sample of size n. In the early stages of planning a study, however, investigators usually want to reverse the situation and determine the sample size that will be necessary to provide a specified power. For example, suppose that we wish to test the null hypothesis

$$H_0: \mu \leq 180 \text{ mg/100 ml}$$

at the $\alpha = 0.01$ level of significance. Once again, μ is the mean serum cholesterol level of the population of 20- to 74-year-old males in the United States; the standard deviation is $\sigma = 46$ mg/100 ml. If the true population mean is as large as 211 mg/100 ml, we want to risk only a 5% chance of failing to reject the null hypothesis; consequently, we set β equal to 0.05 and the power of the test to 0.95. Under these circumstances, how large a sample would we require?

Since $\alpha = 0.01$ instead of 0.05, we begin by noting that H_0 would be rejected for $z \geq 2.32$. Substituting $(\bar{x} - 180)/(46/\sqrt{n})$ for the normal deviate z, we let

$$z = 2.32$$
$$= \frac{\bar{x} - 180}{46/\sqrt{n}}.$$

Solving for $\bar{x}$,

$$\bar{x} = 180 + 2.32\left(\frac{46}{\sqrt{n}}\right).$$

Therefore, we would reject the null hypothesis if the sample mean $\bar{x}$ takes any value greater than or equal to $180 + 2.32(46/\sqrt{n})$.

Now consider the desired power of the test. If the true mean serum cholesterol level were actually $\mu_1 = 211$ mg/100 ml—so that the normal deviate z could be expressed as $(\bar{x} - 211)/(46/\sqrt{n})$—we would want to reject the null hypothesis with probability $1 - \beta = 1 - 0.05 = 0.95$. The value of z that corresponds to $\beta = 0.05$ is $z = -1.645$; therefore,

$$z = -1.645$$
$$= \frac{\bar{x} - 211}{46/\sqrt{n}}$$

and

$$\bar{x} = 211 - 1.645\left(\frac{46}{\sqrt{n}}\right).$$

Setting the two expressions for the sample mean $\bar{x}$ equal to each other,

$$180 + 2.32\left(\frac{46}{\sqrt{n}}\right) = 211 - 1.645\left(\frac{46}{\sqrt{n}}\right).$$

Multiplying both sides of the equality by $\sqrt{n}$ and collecting terms,

$$\sqrt{n}(211 - 180) = [2.32 - (-1.645)](46)$$

and

$$n = \left[\frac{(2.32 + 1.645)(46)}{(211 - 180)}\right]^2$$
$$= 34.6.$$

By convention, we always round up when calculating a sample size. Therefore, a sample of 35 men would be required.

Using notation introduced in Chapter 9, it is possible to write a more general formula for calculating sample size. Recall that z_α represents the value that cuts off an area of α in the upper tail of the standard normal distribution, while $-z_\alpha$ is the value that cuts off an area of α in the lower tail of the distribution. If we conduct a one-sided test of the null hypothesis

$$H_0: \mu \leq \mu_0$$

against the alternative

$$H_0: \mu > \mu_0$$

at the α level of significance, H_0 would be rejected for any test statistic that takes a value $z \geq z_\alpha$. Similarly, considering the desired power of the test $1 - \beta$, the generic value of z that corresponds to a probability β is $z = -z_\beta$. The two different expressions for $\bar{x}$ are

$$\bar{x} = \mu_0 + z_\alpha\left(\frac{\sigma}{\sqrt{n}}\right)$$

and

$$\bar{x} = \mu_1 - z_\beta\left(\frac{\sigma}{\sqrt{n}}\right),$$

and setting them equal to each other gives us

$$n = \left[\frac{[z_\alpha - (-z_\beta)](\sigma)}{(\mu_1 - \mu_0)}\right]^2$$
$$= \left[\frac{(z_\alpha + z_\beta)(\sigma)}{(\mu_1 - \mu_0)}\right]^2.$$

This is the sample size necessary to achieve a power of $1 - \beta$ when we conduct a one-sided test at the α level.

Several factors influence the size of n. If we reduce the type I error α, then z_α—the cutoff point for rejecting H_0—would increase in value; this would result in a larger sample size. Similarly, if we lower the type II error β or increase the power, then $-z_\beta$ gets smaller or more negative. Again, this would produce a larger value of n. If we consider an alternative population mean that is closer to the hypothesized value, the difference $\mu_1 - \mu_0$ would decrease and the sample size increase. It makes sense that we would need a bigger sample size to detect a smaller difference. Finally, the larger the variability of the underlying population σ, the larger the sample size required.

In the serum cholesterol level example, we knew that the hypothesized population mean μ_0 had to be smaller than the alternative μ_1; consequently, we conducted a one-sided test. If it is not known whether μ_0 is larger or smaller than μ_1, a two-sided test is appropriate. In this case, we must modify the critical value of z that would cause the null hypothesis to be rejected. When $\alpha = 0.01$, for instance,

$$H_0 : \mu = 180 \text{ mg/100 ml}$$

would be rejected for $z \geq 2.58$, not $z \geq 2.32$. Substituting this value into the equation above,

$$n = \left[\frac{(2.58 + 1.645)(46)}{(211 - 180)} \right]^2$$
$$= 39.3,$$

and a sample of size 40 would be required. More generally, H_0 will be rejected at the α level for $z \geq z_{\alpha/2}$ (and also for $z \leq -z_{\alpha/2}$), and the sample size formula becomes

$$n = \left[\frac{(z_{\alpha/2} + z_\beta)(\sigma)}{(\mu_1 - \mu_0)} \right]^2.$$

Note that the sample size for a two-sided test is always larger than the sample size for the corresponding one-sided test.

10.7   Further Applications

Consider once again the distribution of heights for the population of 12- to 40-year-olds who suffer from fetal alcohol syndrome. This distribution is approximately normal with unknown mean μ; its standard deviation is $\sigma = 6$ centimeters [10]. We might wish to know whether the mean height for this population is equal to the mean height for individuals in the same age group who do not have fetal alcohol syndrome.

The first step in conducting a test of hypothesis is to make a formal claim about the value of μ_0. Since the mean height of 12- to 40-year-olds who do not suffer from fetal alcohol syndrome is 160.0 centimeters, the null hypothesis is

$$H_0 : \mu = 160.0 \text{ cm}.$$

We are concerned with deviations from μ_0 that could occur in either direction; thus, we conduct a two-sided test at the $\alpha = 0.05$ level of significance. The alternative hypothesis is

$$H_A : \mu \neq 160.0 \text{ cm}.$$

For a random sample of size 31 selected from the population of 12- to 40-year-olds who suffer from fetal alcohol syndrome, the mean height is $\bar{x} = 147.4$ cm. If the true mean height of this group is $\mu = 160.0$ cm, what is the probability of selecting a sample with a mean as low as 147.4? To answer this question, we calculate the test statistic

$$z = \frac{\bar{x} - \mu_0}{\sigma/\sqrt{n}}$$

$$= \frac{147.4 - 160.0}{6/\sqrt{31}}$$

$$= -11.69.$$

We use a z-test rather than a t-test because the value of σ is known. Since z is the outcome of a standard normal random variable, we consult Table A.3 and find that the area to the left of $z = -11.69$ and to the right of $z = 11.69$ is much less than 0.001. Therefore, since $p < 0.05$, we reject the null hypothesis at the 0.05 level of significance. The random sample provides evidence that the mean height for the population of individuals who have fetal alcohol syndrome is different from the mean height for those who do not; the persons who suffer from this condition tend to be shorter, on average.

Rather than work through the calculations ourselves, we could have let a computer perform the test of hypothesis for us. Table 10.1 shows the appropriate output from Minitab. The output restates the null and alternative hypotheses and the assumed population standard deviation σ; it then gives various summary measures, the test statistic z, and the p-value of the test. Minitab does not provide us with a conclusion, however; that we must do for ourselves.

Another way to approach this problem would have been to construct a confidence interval for μ, the true mean height for the population of 12- to 40-year-olds who suffer from fetal alcohol syndrome. In Chapter 9, we found a two-sided 95% confidence interval for μ to be

(145.3, 149.5).

Since this interval does not contain the value 160.0, we know that the null hypothesis

$$H_0: \mu = 160.0 \text{ cm}$$

would be rejected in favor of H_A at the 0.05 level of significance.

When the standard deviation of a population is not known, we use the sample standard deviation s in place of σ to conduct a test of hypothesis. Consider the distribution of the concentration of benzene—a chemical believed to be harmful to humans—

TABLE 10.1
Minitab output for the z-test

```
TEST OF MU = 160.0 VS MU N.E. 160.0
THE ASSUMED SIGMA = 6.0
```

	N	MEAN	STDEV	SE MEAN	Z	P VALUE
HEIGHT	31	147.4	6.000	1.078	−11.69	0.000

in a specified brand of cigar. This distribution is approximately normal with an unknown mean μ and standard deviation σ. We are told that the mean concentration of benzene in a brand of cigarette that is being used as the standard is $81\mu g/g$ tobacco [11], and would like to know whether the mean concentration of benzene in the cigars is equal to that in the cigarettes. To determine this, we test the null hypothesis

$$H_0: \mu = 81\mu g/g.$$

We are interested in deviations from the mean that could occur in either direction, and thus conduct a two-sided test at the $\alpha = 0.05$ level of significance. The alternative hypothesis is

$$H_A: \mu \neq 81\mu g/g.$$

A random sample of seven cigars has mean benzene concentration $\bar{x} = 151\mu g/g$ and standard deviation $s = 9\mu g/g$. Is it possible that these observations could have come from a population with mean $\mu = 81\mu g/g$? To determine this, we calculate the test statistic

$$
\begin{aligned}
t &= \frac{\bar{x} - \mu_0}{s/\sqrt{n}} \\
&= \frac{151 - 81}{9/\sqrt{7}} \\
&= 20.6.
\end{aligned}
$$

The statistic t is the outcome of a random variable that has a t distribution with $7 - 1 = 6$ degrees of freedom. Consulting Table A.4, we observe that the total area under the curve to the left of -20.6 and to the right of 20.6 is less than 0.001. Since $p < 0.05$, we reject the null hypothesis. The random sample of size 7 suggests that the cigars contain a higher concentration of benzene than do the cigarettes.

Table 10.2 displays the relevant output from Stata. Note that the lower portion of the output provides the test statistics and p-values for three different alternative hypotheses. The information in the center is for the two-sided test, while the information on either side is for the two possible one-sided tests. (Keep in mind that we should have determined in advance which particular test we are interested in.) In addition, Stata displays the 95% confidence interval for the true population mean μ.

Now recall the distribution of hemoglobin levels for the population of children under the age of 6 who have been exposed to high levels of lead. The mean of this population is $\mu = 10.60$ g/100 ml, and its standard deviation is $\sigma = 0.85$ g/100 ml. Suppose that we do not know the true population mean μ; however, we do know that the mean hemoglobin level for the general population of children under the age of 6 is 12.29 g/100 ml. If we were to conduct a test of the null hypothesis

$$H_0: \mu = 12.29 \text{ g/100 ml},$$

we would expect that this false hypothesis would be rejected. Assuming that we select a very small random sample of size 5 from the population of children who have been

TABLE 10.2

Stata output for the t-test

One-sample t test						Number of obs = 7	
Variable	Mean	Std. Err.	t	P > |t|	[95% Conf. Interval]		
benzene	151	3.40168	44.3898	0.0000	142.6764	159.3236	

Degrees of freedom: 6

Ho: mean(benzene) = 81

Ha: mean < 81	Ha: mean ~= 81	Ha: mean > 81
t = 20.5781	t = 20.5781	t = 20.5781
P < t = 1.0000	P > |t| = 0.0000	P > t = 0.0000

exposed to lead, what is the probability that we will make a type II error—or fail to reject H_0 when it is false—given that the true population mean is $\mu_1 = 10.60$ g/100 ml?

To answer this question, we begin by finding the mean hemoglobin level that the sample must have for H_0 to be rejected. We believe that the hemoglobin levels of children who have been exposed to lead must on average be lower than those of unexposed children. If we conduct a one-sided test at the $\alpha = 0.05$ level of significance, the null hypothesis would be rejected for $z \leq -1.645$. Since

$$z = \frac{\bar{x} - \mu_0}{\sigma/\sqrt{n}},$$

we have

$$z = -1.645$$
$$= \frac{\bar{x} - 12.29}{0.85/\sqrt{5}}$$

and

$$\bar{x} = 12.29 - \frac{1.645(0.85)}{\sqrt{5}}$$
$$= 11.66.$$

Therefore, the null hypothesis

$$H_0: \mu \geq 12.29 \text{ g/100 ml}$$

would be rejected in favor of the alternative

$$H_A: \mu < 12.29 \text{ g/100 ml}$$

if the sample of size 5 has a mean $\bar{x}$ that is less than or equal to 11.66 g/100 ml. This area corresponds to the lower 5% of the sampling distribution of means of samples of size 5 when $\mu = 12.29$ g/100 ml; it is illustrated in Figure 10.5.

The quantity β is the probability of making a type II error, or failing to reject H_0 given that it is false and the true population mean is $\mu_1 = 10.60$ g/100 ml. To find β, we consider the sampling distribution of means of samples of size 5 when $\mu = 10.60$ g/100 ml. Since a sample mean greater than 11.66 g/100 ml implies that we do not reject H_0, we must determine what proportion of the distribution centered at 10.60 g/100 ml lies to the right of 11.66. Observe that

$$z = \frac{11.66 - 10.60}{0.85/\sqrt{5}}$$
$$= 2.79.$$

Consulting Table A.3, we find that the area under the standard normal curve that lies to the right of $z = 2.79$ is 0.003. Therefore, β is equal to 0.003.

The power of the test—or the probability of rejecting the null hypothesis given that H_0 is false and the true population mean is $\mu_1 = 10.60$ g/100 ml—is

$$1 - \beta = 1 - 0.003$$
$$= 0.997.$$

Even with a sample size of only 5, we are almost certain to reject H_0. This is in part because the standard deviation of the underlying population is quite small relative to the difference in means $\mu_1 - \mu_0$.

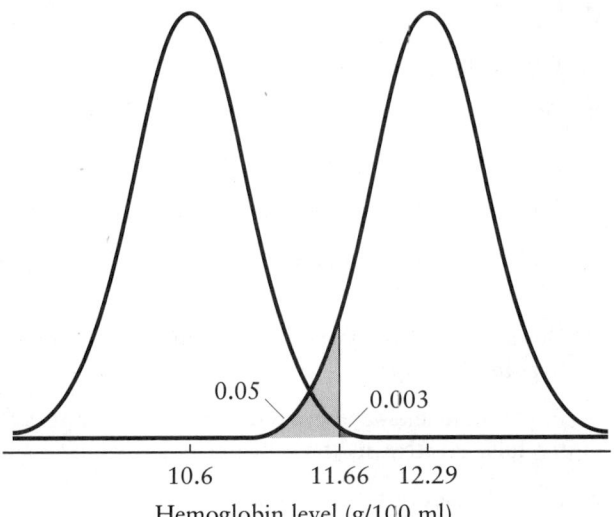

FIGURE 10.5

Distribution of means of samples of size 5 for the hemoglobin levels of children under the age of 6, $\mu = 10.60$ g/100 ml versus $\mu = 12.29$ g/100 ml

Now suppose that we are planning a new study to attempt to determine the mean hemoglobin level for the population of children under the age of 6 who have been exposed to high levels of lead. We are told that the general population of children in this age group has mean hemoglobin level $\mu = 12.29$ g/100 ml and standard deviation $\sigma = 0.85$ g/100 ml. If the true mean hemoglobin level for exposed children is 0.5 g/100 ml lower than that for unexposed children, we want the test to have 80% power to detect this difference. We plan to use a one-sided test conducted at the 0.05 level of significance. What sample size will we need for this study?

We begin by assembling all the pieces necessary to carry out a sample size calculation. Since a one-sided test will be performed at the $\alpha = 0.05$ level, $z_\alpha = 1.645$. We want a power of 0.80; therefore, $\beta = 0.20$ and $z_\beta = 0.84$. The hypothesized mean of the population is $\mu_0 = 12.29$ g/100 ml, and the alternative mean is 0.5 units less than this, or $\mu_1 = 11.79$ g/100 ml. We do not know the standard deviation of the hemoglobin levels for exposed children; however, we are willing to assume that it is the same as that for unexposed children. Therefore, $\sigma = 0.85$ g/100 ml. Putting these pieces together,

$$n = \left[\frac{(z_\alpha + z_\beta)(\sigma)}{(\mu_1 - \mu_0)} \right]^2$$
$$= \left[\frac{(1.645 + 0.84)(0.85)}{(11.79 - 12.29)} \right]^2$$
$$= 17.8.$$

Thus, a sample of size 18 would be required.

10.8 Review Exercises

1. What is the purpose of a test of hypothesis?

2. Does a hypothesis test ever prove the null hypothesis? Explain.

3. What is a p-value? What does the p-value mean in words?

4. Briefly explain the relationship between confidence intervals and hypothesis testing.

5. Under what circumstances might you use a one-sided test of hypothesis rather than a two-sided test?

6. Describe the two types of errors that can be made when you conduct a test of hypothesis.

7. Explain the analogy between type I and type II errors in a test of hypothesis and the false positive and false negative results that occur in diagnostic testing.

8. List four factors that affect the power of a test.

9. The distribution of diastolic blood pressures for the population of female diabetics between the ages of 30 and 34 has an unknown mean μ_d and standard deviation $\sigma_d = 9.1$ mm Hg. It may be useful to physicians to know whether the mean of this

population is equal to the mean diastolic blood pressure of the general population of females in this age group, 74.4 mm Hg [12].

(a) What is the null hypothesis of the appropriate test?

(b) What is the alternative hypothesis?

(c) A sample of ten diabetic women is selected; their mean diastolic blood pressure is $\bar{x}_d = 84$ mm Hg. Using this information, conduct a two-sided test at the $\alpha = 0.05$ level of significance. What is the p-value of the test?

(d) What conclusion do you draw from the results of the test?

(e) Would your conclusion have been different if you had chosen $\alpha = 0.01$ instead of $\alpha = 0.05$?

10. *E. canis* infection is a tick-borne disease of dogs that is sometimes contracted by humans. Among infected humans, the distribution of white blood cell counts has an unknown mean μ and a standard deviation σ. In the general population, the mean white blood cell count is 7250/mm^3 [13]. It is believed that persons infected with *E. canis* must on average have lower white blood cell counts.

(a) What are the null and alternative hypotheses for a one-sided test?

(b) For a sample of 15 infected persons, the mean white blood cell count is $\bar{x} = 4767/\text{mm}^3$ and the standard deviation is $s = 3204/\text{mm}^3$ [14]. Conduct the test at the $\alpha = 0.05$ level.

(c) What do you conclude?

11. Body mass index is calculated by dividing a person's weight by the square of his or her height; it is a measure of the extent to which the individual is overweight. For the population of middle-aged men who later develop diabetes mellitus, the distribution of baseline body mass indices is approximately normal with an unknown mean μ and standard deviation σ. A sample of 58 men selected from this group has mean $\bar{x} = 25.0$ kg/m^2 and standard deviation $s = 2.7$ kg/m^2 [15].

(a) Construct a 95% confidence interval for the population mean μ.

(b) At the 0.05 level of significance, test whether the mean baseline body mass index for the population of middle-aged men who do develop diabetes is equal to 24.0 kg/m^2, the mean for the population of men who do not. What is the p-value of the test?

(c) What do you conclude?

(d) Based on the 95% confidence interval, would you have expected to reject or not to reject the null hypothesis? Why?

12. The population of male industrial workers in London who have never experienced a major coronary event has mean systolic blood pressure 136 mm Hg and mean diastolic blood pressure 84 mm Hg [16]. You might be interested in determining whether these values are the same as those for the population of industrial workers who **have** suffered a coronary event.

(a) A sample of 86 workers who have experienced a major coronary event has mean systolic blood pressure $\bar{x}_s = 143$ mm Hg and standard deviation $s_s = 24.4$ mm Hg. Test the null hypothesis that the mean systolic blood pressure for the population of industrial workers who have experienced such an event is identical to the mean for the workers who have not, using a two-sided test at the $\alpha = 0.10$ level.

(b) The same sample of men has mean diastolic blood pressure $\bar{x}_d = 87$ mm Hg and standard deviation $s_d = 16.0$ mm Hg. Test the null hypothesis that the mean diastolic blood pressure for the population of workers who have experienced a major coronary event is identical to the mean for the workers who have not.

(c) How do the two groups of workers compare?

13. Over the years, the Food and Drug Administration of the United States (FDA) has worked very hard to avoid making type II errors. A type II error occurs when the FDA approves a drug that is not both safe and effective. Despite the agency's efforts, however, bad drugs do on occasion get through to the public. For example, Omniflox, an antibiotic, had to be recalled less than six months after its approval due to reports of severe adverse reactions, which included a number of deaths. Similarly, Fenoterol, an inhaled drug intended to relieve asthma attacks, was found to increase the risk of death rather than decrease it [17]. Is there any way for the FDA to completely eliminate the occurrence of type II errors? Explain.

14. Data from the Framingham Study allow us to compare the distributions of initial serum cholesterol levels for two populations of males: those who go on to develop coronary heart disease and those who do not. The mean serum cholesterol level of the population of men who do not develop heart disease is $\mu = 219$ mg/100 ml and the standard deviation is $\sigma = 41$ mg/100 ml [18]. Suppose, however, that you do not know the true population mean; instead, you hypothesize that μ is equal to 244 mg/100 ml. This is the mean initial serum cholesterol level of men who eventually develop the disease. Since it is believed that the mean serum cholesterol level for the men who do not develop heart disease cannot be higher than the mean level for men who do, a one-sided test conducted at the $\alpha = 0.05$ level of significance is appropriate.

(a) What is the probability of making a type I error?

(b) If a sample of size 25 is selected from the population of men who do not go on to develop coronary heart disease, what is the probability of making a type II error?

(c) What is the power of the test?

(d) How could you increase the power?

(e) You wish to test the null hypothesis

$$H_0: \mu \geq 244 \text{ mg/100 ml}$$

against the alternative

$$H_A: \mu < 244 \text{ mg/100 ml}$$

at the $\alpha = 0.05$ level of significance. If the true population mean is as low as 219 mg/100 ml, you want to risk only a 5% chance of failing to reject H_0. How large a sample would be required?

(f) How would the sample size change if you were willing to risk a 10% chance of failing to reject a false null hypothesis?

15. In Norway, the distribution of birth weights for full-term infants whose gestational age is 40 weeks is approximately normal with mean $\mu = 3500$ grams and standard deviation $\sigma = 430$ grams [19]. An investigator plans to conduct a study to deter-

mine whether the birth weights of full-term babies whose mothers smoked throughout pregnancy have the same mean. If the true mean birth weight for the infants whose mothers smoked is as low as 3200 grams (or as high as 3800 grams), the investigator wants to risk only a 10% chance of failing to detect this difference. A two-sided test conducted at the 0.05 level of significance will be used. What sample size is needed for this study?

16. The Bayley Scales of Infant Development yield scores on two indices—the Psychomotor Development Index (PDI) and the Mental Development Index (MDI)—which can be used to assess a child's level of functioning in each of these areas at approximately one year of age. Among normal healthy infants, both indices have a mean value of 100. As part of a study assessing the development and neurologic status of children who have undergone reparative heart surgery during the first three months of life, the Bayley Scales were administered to a sample of one-year-old infants born with congenital heart disease. The data are contained in the data set `heart` [20] (Appendix B, Table B.12); PDI scores are saved under the variable name `pdi`, while MDI scores are saved under `mdi`.

   (a) At the 0.05 level of significance, test the null hypothesis that the mean PDI score for children born with congenital heart disease who undergo reparative heart surgery during the first three months of life is equal to 100, the mean score for healthy children. Use a two-sided test. What is the *p*-value? What do you conclude?

   (b) Conduct the analogous test of hypothesis for the mean MDI score. What do you conclude?

   (c) Construct 95% confidence intervals for the true mean PDI score and the true mean MDI score for this population of children with congenital heart disease. Does either of these intervals contain the value 100? Would you have expected that they would?

Bibliography

[1] National Center for Health Statistics, Fulwood, R., Kalsbeek, W., Rifkind, B., Russell-Briefel, R., Muesing, R., LaRosa, J., and Lippel, K., "Total Serum Cholesterol Levels of Adults 20–74 Years of Age: United States, 1976–1980," *Vital and Health Statistics*, Series 11, Number 236, May 1986.

[2] Gauvreau, K., and Pagano, M., "Why 5%?," *Nutrition*, Volume 10, 1994, 93–94.

[3] Kaplan, N. M., "Strategies to Reduce Risk Factors in Hypertensive Patients Who Smoke," *American Heart Journal*, Volume 115, January 1988, 288–294.

[4] Tsou, V. M., Young, R. M., Hart, M. H., and Vanderhoof, J. A., "Elevated Plasma Aluminum Levels in Normal Infants Receiving Antacids Containing Aluminum," *Pediatrics*, Volume 87, February 1991, 148–151.

[5] National Center for Health Statistics, Fulwood, R., Johnson, C. L., Bryner, J. D., Gunter, E. W., and McGrath, C. R., "Hematological and Nutritional Biochemistry Reference Data for Persons 6 Months–74 Years of Age: United States, 1976–1980," *Vital and Health Statistics*, Series 11, Number 232, December 1982.

[6] Clark, M., Royal, J., and Seeler, R., "Interaction of Iron Deficiency and Lead and the Hematologic Findings in Children with Severe Lead Poisoning," *Pediatrics*, Volume 81, February 1988, 247–253.

[7] "Firm Admits Using Rival's Drug in Tests," *The Boston Globe*, July 1, 1989, 41.

[8] Davidson, J.W., and Lytle, M.H., *After the Fact: The Art of Historical Detection*, Third Edition, Volume 1, New York: McGraw-Hill, Inc., 1992, 26.

[9] "Child Abuse—and Trial Abuse," *The New York Times*, January 20, 1990, 24.

[10] Streissguth, A. P., Aase, J. M., Clarren, S. K., Randels, S. P., LaDue, R. A., and Smith, D. F., "Fetal Alcohol Syndrome in Adolescents and Adults," *Journal of the American Medical Association*, Volume 265, April 17, 1991, 1961–1967.

[11] Appel, B. R., Guirguis, G., Kim, I., Garbin, O., Fracchia, M., Flessel, C. P., Kizer, K. W., Book, S. A., and Warriner, T. E., "Benzene, Benzo(a)Pyrene, and Lead in Smoke from Tobacco Products Other Than Cigarettes," *American Journal of Public Health*, Volume 80, May 1990, 560–564.

[12] Klein, B. E. K., Klein, R., and Moss, S. E., "Blood Pressure in a Population of Diabetic Persons Diagnosed After 30 Years of Age," *American Journal of Public Health*, Volume 74, April 1984, 336–339.

[13] National Center for Health Statistics, Fulwood, R., Johnson, C. L., Bryner, J. D., Gunter, E. W., and McGrath, C. R., "Hematological and Nutritional Biochemistry Reference Data for Persons 6 Months–74 Years of Age: United States, 1976–1980," *Vital and Health Statistics*, Series 11, Number 232, December 1982.

[14] Rohrbach, B. W., Harkess, J. R., Ewing, S. A., Kudlac, J., McKee, G. L., and Istre, G. R., "Epidemiologic and Clinical Characteristics of Persons with Serologic Evidence of *E. canis* Infection," *American Journal of Public Health*, Volume 80, April 1990, 442–445.

[15] Feskens, E. J. M., and Kromhout, D., "Cardiovascular Risk Factors and the 25 Year Incidence of Diabetes Mellitus in Middle-Aged Men," *American Journal of Epidemiology*, Volume 130, December 1989, 1101–1108.

[16] Meade, T. W., Cooper, J. A., and Peart, W. S., "Plasma Renin Activity and Ischemic Heart Disease," *The New England Journal of Medicine*, Volume 329, August 26, 1993, 616-619.

[17] Burkholz, H., *The FDA Follies*, New York: Basic Books, 1994, 107–113.

[18] MacMahon, S. W., and MacDonald, G. J., "A Population at Risk: Prevalence of High Cholesterol Levels in Hypertensive Patients in the Framingham Study," *American Journal of Medicine Supplement*, Volume 80, February 14, 1986, 40–47.

[19] Wilcox, A. J., and Skjærven, R., "Birth Weight and Perinatal Mortality: The Effect of Gestational Age," *American Journal of Public Health*, Volume 82, March 1992, 378–382.

[20] Bellinger, D. C., Jonas, R. A., Rappaport, L. A., Wypij, D., Wernovsky, G., Kuban, K. C. K., Barnes, P. D., Holmes, G. L., Hickey, P. R., Strand, R. D., Walsh, A. Z., Helmers, S. L., Constantinou, J. E., Carrazana, E. J., Mayer, J. E., Hanley, F. L., Castaneda, A. R., Ware, J. H., and Newburger, J. W., "Developmental and Neurologic Status of Children After Heart Surgery with Hypothermic Circulatory Arrest or Low-Flow Cardiopulmonary Bypass," *The New England Journal of Medicine*, Volume 332, March 2, 1995, 549–555.

11

Comparison of Two Means

In the preceding chapter, we used a statistical test of hypothesis to compare the unknown mean of a single population to some fixed, known value μ_0. In practical applications, however, it is far more common to compare the means of two different populations, where both means are unknown. Often, the two groups have received different treatments or undergone different exposures.

The idea of comparing populations to draw a conclusion about their similarities or differences has been around for hundreds of years. In the sixteenth century, for example, it was believed that gunshot wounds were susceptible to infection and therefore needed to be cauterized. One of the earliest large-scale uses of gunpowder occurred during the French invasion of Italy in 1537. This expedition was the first for a French army surgeon named Ambroise Paré; he reports on an attack on Turin [1]:

> Now all the said soldiers at the Château, seeing our men coming with great fury, did all they could to defend themselves and killed and wounded a great number of our soldiers with pikes, arquebuses, and stones, where the surgeons had much work cut out for them. Now I was at that time a freshwater soldier, I had not yet seen wounds made by gunshot at the first dressing. It is true that I had read Jean de Vigo, first book, "Of Wounds in General," chapter eight, that wounds made by firearms participate of venenosity, because of the powder, and for their cure he commands to cauterize them with oil of elder, scalding hot, in which should be mixed a little theriac, and in order not to err before using the said oil, knowing that such a thing could bring great pain to the patient, I wished to know first, how the other surgeons did for the first dressing which was to apply the said oil as hot as possible, into the wound with tents and setons, of whom I took courage to do as they did. At last my oil lacked and I was constrained to apply in its place a digestive made of the yolks of eggs, oil of roses and turpentine. That night I could not sleep at my ease, fearing by lack of cauterization that I should find the wounded on whom I had failed to put the said oil dead or empoisoned, which made me rise very early to visit them, where beyond my hope, I found those upon whom I had put the digestive medicament feeling little pain, and their wounds without inflammation or swelling having rested fairly well throughout the

> night; the others to whom I had applied the said boiling oil, I had found feverish, with great pain and swelling about their wounds. Then I resolved with myself never more to burn thus cruelly poor men wounded with gunshot.

The results of this comparison—one of the first documented clinical trials—were totally convincing. The same could be said of studies investigating the use of penicillin in the treatment of bacterial diseases. Unfortunately, such trials are the exception rather than the rule; progress is usually measured much more slowly.

This chapter introduces a procedure for deciding whether the differences we observe between sample means are too large to be attributed to chance alone. A test of hypothesis involving two samples is similar in many respects to a test conducted for a single sample. We begin by specifying a null hypothesis; in most cases, we are interested in testing whether the two population means are equal. We then calculate the probability of obtaining a pair of sample means as discrepant as or more discrepant than the observed means given that the null hypothesis is true. If this probability is sufficiently small, we reject the null hypothesis and conclude that the two population means are different. As before, we must specify a level of significance α and state whether we are interested in a one-sided or a two-sided test. The specific form of the analysis depends on the nature of the two sets of observations involved; in particular, we must determine whether the data come from paired or independent samples.

11.1   Paired Samples

The distinguishing characteristic of paired samples is that for each observation in the first group, there is a corresponding observation in the second group. In the technique known as *self-pairing*, measurements are taken on a single subject at two distinct points in time. One common example of self-pairing is the "before and after" experiment, in which each individual is examined before a certain treatment has been applied and then again after the treatment has been completed. A second type of pairing occurs when an investigator matches the subjects in one group with those in a second group so that the members of a pair are as much alike as possible with respect to important characteristics such as age and gender.

Pairing is frequently employed in an attempt to control for extraneous sources of variation that might otherwise influence the results of the comparison. If measurements are made on the same subject rather than on two different individuals, a certain amount of biological variability is eliminated. We do not have to worry about the fact that one subject is older than the other, or that one is male and the other female. The intent of pairing, therefore, is to make a comparison more precise.

Consider the data taken from a study in which each of 63 adult males with coronary artery disease is subjected to a series of exercise tests on a number of different occasions. On one day, a patient first undergoes an exercise test on a treadmill; the length of time from the start of the test until the patient experiences angina—pain or spasms in the chest—is recorded. He is then exposed to plain room air for approximately one hour. At the end of this time, he performs a second exercise test; time until the onset of angina is again recorded. The observation of interest is the percent decrease in time to angina between the first and the second tests. If during the first test, for example, a man has an attack of angina after 983 seconds, and during the second test he has an attack after 957 seconds, his percent decrease in time to angina is

$$\frac{983 - 957}{983} = 0.026$$
$$= 2.6\%.$$

The unknown population mean of this distribution of percent decreases is μ_1; for the 63 patients in the sample, the observed mean percent decrease is $\bar{x}_1 = 0.96\%$ [2].

On another day, the same patient undergoes a similar series of tests. This time, however, he is exposed to a mixture of air and carbon monoxide during the interval between the tests. The amount of carbon monoxide added to the air is intended to increase the patient's carboxyhemoglobin level—a biological measure of exposure—to 4%; this level is lower than that typically endured by smokers, but is about what might be experienced by an individual sitting in heavy traffic in an area with poor ventilation. Again the observation of interest is the percent decrease in time to angina between the first and second tests. The unknown mean of this distribution is μ_2; the sample mean for the group of 63 subjects is $\bar{x}_2 = 7.59\%$.

For the first ten patients in the study, the percent decreases in time to angina for each of the two occasions are displayed in Figure 11.1. Note that for eight of the men, the measure increases; for the other two, it decreases. We would like to determine whether there is any evidence of a difference in the percent decrease in time to angina between the trial in which the subjects are exposed to carbon monoxide and the trial in which they are not. Since we believe that excessive exposure to carbon monoxide could

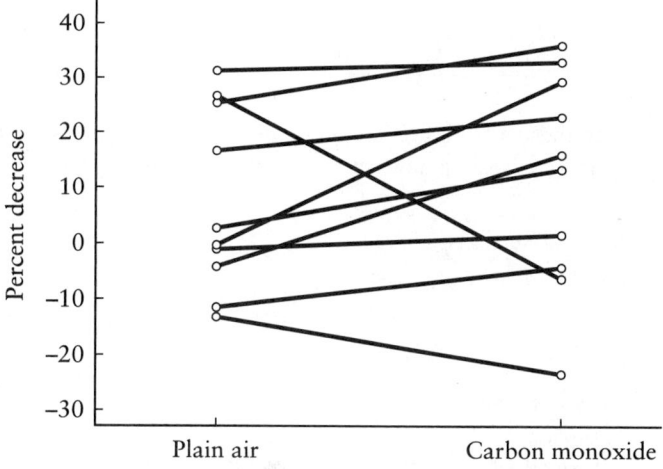

FIGURE 11.1
Percent decrease in time to angina on two different occasions for each of ten men with coronary artery disease

not possibly be beneficial to an individual's health, we consider deviations that occur in one direction only; we therefore conduct a one-sided test at the $\alpha = 0.05$ level of significance. The null hypothesis is

$$H_0: \mu_1 \geq \mu_2,$$

or

$$H_0: \mu_1 - \mu_2 \geq 0,$$

and the alternative is

$$H_A: \mu_1 - \mu_2 < 0.$$

In this study, each patient is subjected to the same series of tests both with and without exposure to carbon monoxide. This self-pairing eliminates any distortions or biases that might be introduced by comparing patients who differ with respect to age, weight, or severity of coronary artery disease. Because the data consist of paired samples, the *paired t-test* is the appropriate method of analysis.

Rather than consider the two sets of observations to be distinct samples, we focus on the **difference** in measurements within each pair. Suppose that our two groups of observations are as follows:

Sample 1	Sample 2
x_{11}	x_{12}
x_{21}	x_{22}
x_{31}	x_{32}
.	.
.	.
x_{n1}	x_{n2}

In these samples, x_{11} and x_{12} are a pair, x_{21} and x_{22} are a pair, and so on. We use these data to create a new set of observations that represent the differences within each pair:

$$d_1 = x_{11} - x_{12}$$
$$d_2 = x_{21} - x_{22}$$
$$d_3 = x_{31} - x_{32}$$
$$\vdots$$
$$d_n = x_{n1} - x_{n2}$$

Instead of analyzing the individual observations, we use the difference between the members of each pair as the variable of interest. Since the difference is a single measurement, our analysis reduces to the one-sample problem and we apply the hypothesis testing procedure introduced in Chapter 10.

To do this, we first note that the mean of the set of differences is

$$\bar{d} = \frac{\sum_{i=1}^{n} d_i}{n};$$

this sample mean provides a point estimate for the true difference in population means $\mu_1 - \mu_2$. The standard deviation of the differences is

$$s_d = \sqrt{\frac{\sum_{i=1}^{n} (d_i - \bar{d})^2}{n - 1}}.$$

If we denote the true difference in population means by

$$\delta = \mu_1 - \mu_2$$

and wish to test whether these two means are equal, we can write the null hypothesis as

$$H_0: \delta = 0$$

and the alternative as

$$H_A: \delta \neq 0.$$

Assuming that the population of differences is normally distributed, H_0 can be tested by computing the statistic

$$t = \frac{\bar{d} - \delta}{s_d/\sqrt{n}};$$

note that $s_d/\sqrt{n}$ is the standard error of $\bar{d}$. If the null hypothesis is true, this quantity has a t distribution with $n - 1$ degrees of freedom. We compare the outcome of t to the values in Table A.4 in Appendix A to find p, the probability of observing a mean difference as large as or larger than $\bar{d}$ given that $\delta = 0$. (Or, as always, we can use a computer program to do the calculations for us.) If $p \leq \alpha$, we reject H_0. If $p > \alpha$, we do not reject the null hypothesis.

Returning to the study of males with coronary artery disease, we focus on the difference in measurements for a given individual. For each of the 63 men in the study, we calculate the percent decrease in time to angina when he is exposed to carbon monoxide minus the percent decrease when he is exposed to unadulterated air. The mean of these differences is

$$\bar{d} = \frac{\sum_{i=1}^{63} d_i}{63}$$

$$= -6.63,$$

and the standard deviation is

$$s_d = \sqrt{\frac{\sum_{i=1}^{63}(d_i - \overline{d})^2}{63 - 1}}$$
$$= 20.29.$$

As can be seen from Figure 11.2, the differences are fairly symmetric and can be considered to be approximately normally distributed. Therefore, if we rewrite the null hypothesis of the test as

$$H_0: \delta \geq 0,$$

we can evaluate H_0 using the statistic

$$t = \frac{\overline{d} - \delta}{s_d/\sqrt{n}}$$

or

$$t = \frac{-6.63 - 0}{20.29/\sqrt{63}}$$
$$= -2.59.$$

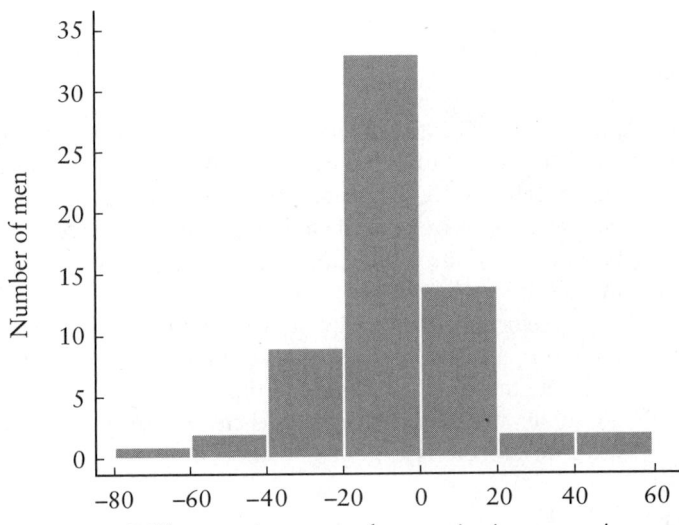

Differences in percent decrease in time to angina

FIGURE 11.2

Differences in the percent decrease in time to angina for a sample of 63
men with coronary artery disease

Consulting Table A.4, we observe that for a t distribution with $63 - 1 = 62$ degrees of freedom, the area under the curve to the left of $t_{62} = -2.59$ is between 0.005 and 0.01. Therefore, $0.005 < p < 0.01$. Rejecting the null hypothesis at the 0.05 level, we conclude that there is a significant difference between the mean percent decrease in time to angina when the patients are exposed to carbon monoxide and the decrease when they are exposed to air alone. Exposure to carbon monoxide increases the percent decrease in time to angina; in other words, exposed patients tend to develop angina more quickly.

As we have mentioned, the sample mean $\overline{d}$ provides a point estimate of the true difference in population means $\delta = \mu_1 - \mu_2$. However, we might also be interested in calculating an upper confidence bound for δ. Again we rely on the one-sample technique. For a t distribution with 62 degrees of freedom, 95% of the observations lie above -1.671. Therefore,

$$P(t \geq -1.671) = P\left(\frac{\overline{d} - \delta}{s_d/\sqrt{n}} \geq -1.671\right)$$
$$= 0.95.$$

A one-sided 95% confidence interval for δ is

$$\delta \leq \overline{d} + 1.671 \frac{s_d}{\sqrt{n}}$$
$$= -6.63 + 1.671 \frac{20.29}{\sqrt{63}}$$
$$= -2.36.$$

We are 95% confident that the true difference in population means is less than or equal to -2.36%. Stated in another way, we are 95% confident that the **decrease** in time to angina after exposure to carbon monoxide is at least 2.36%.

11.2 Independent Samples

Now suppose that we are given the measurements of serum iron level for two samples of children: one group healthy and the other group suffering from cystic fibrosis, a congenital disease of the mucous glands. The two underlying populations are independent and normally distributed. If the population of diseased children has mean serum iron level μ_1 and the population of healthy children has mean μ_2, we might again be interested in testing the null hypothesis that the two population means are identical. This can be expressed as either

$$H_0: \mu_1 - \mu_2 = 0$$

or

$$H_0: \mu_1 = \mu_2.$$

The alternative hypothesis is

$$H_A: \mu_1 \neq \mu_2.$$

From the normal population with mean μ_1 and standard deviation σ_1, we draw a random sample of size n_1. The mean of this sample is denoted by $\bar{x}_1$ and its standard deviation by s_1. Similarly, we select a random sample of size n_2 from the normal population with mean μ_2 and standard deviation σ_2. The mean of this sample is represented by $\bar{x}_2$ and its standard deviation by s_2. Note that the numbers of observations in the two samples—n_1 and n_2—need not be the same.

		Group 1	Group 2
Population	Mean	μ_1	μ_2
	Standard Deviation	σ_1	σ_2
Sample	Mean	$\bar{x}_1$	$\bar{x}_2$
	Standard Deviation	s_1	s_2
	Sample Size	n_1	n_2

Two different situations arise in the comparison of independent samples. In the first, the variances of the underlying populations either are known to be equal to each other or are assumed to be equal. This leads to the *two-sample t-test*, which is omnipresent in the literature. In the second, the variances are not assumed to be the same; in this case, the standard t-test is no longer valid. Before a test of means is carried out, many individuals believe that a preliminary test of variances should be conducted to distinguish between these two alternatives. Others object to the test on philosophical grounds: It is highly sensitive to the assumption of normality and has low power in many situations where the t-test should be avoided [3]. In addition, a modification of the two-sample test performed without this initial check has been shown to have high power in settings where it is not known whether the underlying population variances are equal [4]. Since it is generally either unnecessary or ineffective, we do not recommend the use of a preliminary test of variances in this text.

11.2.1 Equal Variances

We first consider the situation in which it either is known or is reasonable to assume that the two population variances are identical. Recall that for a single normal population with mean μ and standard deviation σ, the central limit theorem states that the sample mean $\overline{X}$ is approximately normally distributed—assuming that n is large enough—with mean μ and standard error $\sqrt{\sigma^2/n} = \sigma/\sqrt{n}$. As a result,

$$z = \frac{\bar{x} - \mu}{\sigma/\sqrt{n}}$$

is the outcome of a standard normal random variable. When we are dealing with samples from two independent normal populations, an extension of the central limit theorem says that the difference in sample means $\overline{X}_1 - \overline{X}_2$ is approximately normal with mean

$\mu_1 - \mu_2$ and standard error $\sqrt{\sigma_1^2/n_1 + \sigma_2^2/n_2}$. Since it is assumed that the population variances are equal, we substitute the common value σ^2 for both σ_1^2 and σ_2^2. Consequently, we know that

$$z = \frac{(\bar{x}_1 - \bar{x}_2) - (\mu_1 - \mu_2)}{\sqrt{\sigma^2/n_1 + \sigma^2/n_2}}$$

$$= \frac{(\bar{x}_1 - \bar{x}_2) - (\mu_1 - \mu_2)}{\sqrt{\sigma^2[(1/n_1) + (1/n_2)]}}$$

is the outcome of a standard normal random variable. If the value of the population variance σ^2 is known, this statistic can be used to test the null hypothesis

$$H_0: \mu_1 = \mu_2.$$

As we have previously noted, it is much more common that the true value of σ^2 is not known. In this case, we use the test statistic

$$t = \frac{(\bar{x}_1 - \bar{x}_2) - (\mu_1 - \mu_2)}{\sqrt{s_p^2\,[(1/n_1) + (1/n_2)]}}$$

instead. The quantity s_p^2 is a pooled estimate of the common variance σ^2. Under the null hypothesis that the population means are identical, $\mu_1 - \mu_2$ is equal to 0, and the test statistic t has a t distribution with $(n_1 - 1) + (n_2 - 1) = n_1 + n_2 - 2$ degrees of freedom. We can compare the value of this statistic to the critical values in Table A.4 to find p, the probability of observing a discrepancy as large as $\bar{x}_1 - \bar{x}_2$ given that μ_1 is equal to μ_2. If $p \leq \alpha$, we reject the null hypothesis. If $p > \alpha$, we do not reject H_0.

The pooled estimate of the variance, s_p^2, combines information from both of the samples to produce a more reliable estimate of σ^2. It can be calculated in two different ways. If we know the values of all the observations in the samples, we apply the formula

$$s_p^2 = \frac{\sum_{i=1}^{n_1} (x_{i1} - \bar{x}_1)^2 + \sum_{j=1}^{n_2} (x_{j2} - \bar{x}_2)^2}{n_1 + n_2 - 2}.$$

If we are given s_1 and s_2 only, we must use

$$s_p^2 = \frac{(n_1 - 1)s_1^2 + (n_2 - 1)s_2^2}{n_1 + n_2 - 2}.$$

This second formula demonstrates that s_p^2 is actually a weighted average of the two sample variances s_1^2 and s_2^2, where each variance is weighted by the degrees of freedom associated with it. If n_1 is equal to n_2, s_p^2 is the simple arithmetic average; otherwise, more weight is given to the variance of the larger sample. Recalling that

$$s_1^2 = \frac{\sum_{i=1}^{n_1} (x_{i1} - \bar{x}_1)^2}{n_1 - 1}$$

and

$$s_2^2 = \frac{\sum_{j=1}^{n_2} (x_{j2} - \bar{x}_2)^2}{n_2 - 1},$$

we can see that the two formulas for computing s_p^2 are mathematically equivalent.

As an illustration of the two-sample t-test, consider the distributions of serum iron levels for the population of healthy children and the population of children with cystic fibrosis. The distributions are both approximately normal; denote the mean serum iron level of the healthy children by μ_1 and the mean iron level of the children with the disease by μ_2. The standard deviations of the two populations—σ_1 and σ_2—are unknown but are assumed from previous work to be equal. We would like to determine whether children with cystic fibrosis have a normal level of iron in their blood on average; therefore, we test the null hypothesis that the two population means are identical,

$$H_0: \mu_1 = \mu_2.$$

A random sample is selected from each population. The sample of $n_1 = 9$ healthy children has mean serum iron level $\bar{x}_1 = 18.9\ \mu\text{mol/l}$ and standard deviation $s_1 = 5.9\ \mu\text{mol/l}$; the sample of $n_2 = 13$ children with cystic fibrosis has mean iron level $\bar{x}_2 = 11.9\ \mu\text{mol/l}$ and standard deviation $s_2 = 6.3\ \mu\text{mol/l}$ [5]. Is it likely that the observed difference in sample means—18.9 versus 11.9 $\mu\text{mol/l}$—is the result of chance variation, or should we conclude that the discrepancy is due to a true difference in population means?

In some cases, an investigator will begin an analysis by constructing a separate confidence interval for the mean of each individual population. As an illustration, 95% confidence intervals for the mean serum iron levels of children with and without cystic fibrosis are displayed in Figure 11.3. In general, if the two intervals do not overlap, this

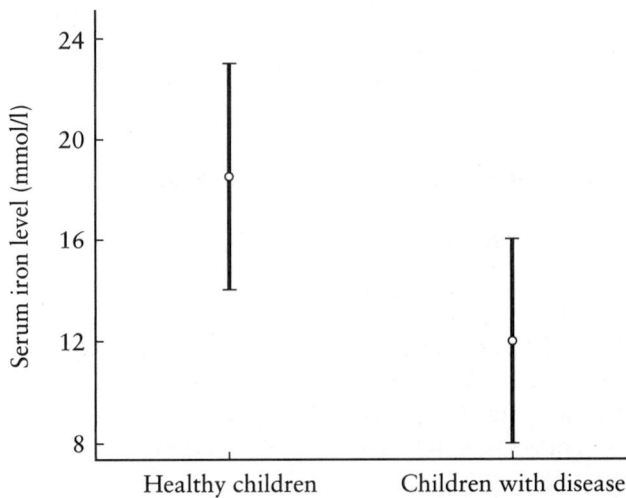

FIGURE 11.3

95% confidence intervals for the mean serum iron levels of healthy children and children with cystic fibrosis

suggests that the population means are in fact different. It should be kept in mind, however, that this technique is not a formal test of hypothesis. In our example, there is a small amount of overlap between the intervals; consequently, it is not possible to draw any type of meaningful conclusion.

Note that the two samples of children were randomly selected from distinct normal populations; in addition, the population variances are assumed to be equal. The two-sample t-test is thus the appropriate technique to apply. The null hypothesis states that there is no difference in the underlying population mean iron levels for the two groups of children: Since we wish to detect a difference that could occur in either direction—we would want to know if the children with cystic fibrosis have either a higher mean or a lower mean than the children without this disease—we conduct a two-sided test and set the significance level at $\alpha = 0.05$. The alternative hypothesis is

$$H_A: \mu_1 \neq \mu_2.$$

To carry out the test, we begin by calculating the pooled estimate of the variance

$$s_p^2 = \frac{(n_1 - 1)s_1^2 + (n_2 - 1)s_2^2}{n_1 + n_2 - 2}$$

$$= \frac{(9 - 1)(5.9)^2 + (13 - 1)(6.3)^2}{9 + 13 - 2}$$

$$= \frac{(8)(34.81) + (12)(39.69)}{20}$$

$$= 37.74.$$

We next calculate the test statistic

$$t = \frac{(\bar{x}_1 - \bar{x}_2) - (\mu_1 - \mu_2)}{\sqrt{s_p^2 \left[(1/n_1) + (1/n_2)\right]}}$$

$$= \frac{(18.9 - 11.9) - 0}{\sqrt{(37.74)\left[(1/9) + (1/13)\right]}}$$

$$= 2.63.$$

Consulting Table A.4, we observe that for a t distribution with $n_1 + n_2 - 2 = 9 + 13 - 2 = 20$ degrees of freedom, the total area under the curve to the right of $t_{20} = 2.63$ is between 0.005 and 0.01. Therefore, the sum of the areas to the right of $t_{20} = 2.63$ and to the left of $t_{20} = -2.63$ must be between 0.01 and 0.02. Since p is less than 0.05, we reject the null hypothesis

$$H_0: \mu_1 = \mu_2$$

at the 0.05 level of significance. The difference between the mean serum iron level of healthy children and the mean level of children with cystic fibrosis is statistically significant; based on these samples, it appears that children with cystic fibrosis suffer from an iron deficiency.

The quantity $\bar{x}_1 - \bar{x}_2$ provides a point estimate for the true difference in population means $\mu_1 - \mu_2$; once again, however, we may also wish to construct a confidence interval. Note that for a t distribution with 20 degrees of freedom, 95% of the observations lie between -2.086 and 2.086. As a result,

$$P\left(-2.086 \leq \frac{(\bar{X}_1 - \bar{X}_2) - (\mu_1 - \mu_2)}{\sqrt{s_p^2\left[(1/n_1) + (1/n_2)\right]}} \leq 2.086\right) = 0.95.$$

Rearranging terms, we find that the 95% confidence limits for $\mu_1 - \mu_2$ are

$$(\bar{x}_1 - \bar{x}_2) \pm (2.086)\sqrt{s_p^2\left[\frac{1}{n_1} + \frac{1}{n_2}\right]},$$

or

$$(18.9 - 11.9) \pm (2.086)\sqrt{(37.74)\left[\frac{1}{9} + \frac{1}{13}\right]}.$$

Therefore, we are 95% confident that the interval

$$(1.4, 12.6)$$

covers $\mu_1 - \mu_2$, the true difference in mean serum iron levels for the two populations of children. Unlike the separate intervals displayed in Figure 11.3, this confidence interval for the difference in means is mathematically equivalent to the two-sample test of hypothesis conducted at the 0.05 level. Notice that the interval does not contain the value 0.

11.2.2 Unequal Variances

We now turn to the situation in which the variances of the two populations are not assumed to be equal. In this case, a modification of the two-sample t-test must be applied. Instead of using s_p^2 as an estimate of the common variance σ^2, we substitute s_1^2 for σ_1^2 and s_2^2 for σ_2^2. Therefore, the appropriate test statistic is

$$t = \frac{(\bar{x}_1 - \bar{x}_2) - (\mu_1 - \mu_2)}{\sqrt{(s_1^2/n_1) + (s_2^2/n_2)}}.$$

This case is unlike the one in which we had equal variances, in that the exact distribution of t is difficult to derive. Therefore, it is necessary to use an approximation [6]. We begin by calculating the quantity

$$v = \frac{[(s_1^2/n_1) + (s_2^2/n_2)]^2}{[(s_1^2/n_1)^2/(n_1 - 1) + (s_2^2/n_2)^2/(n_2 - 1)]};$$

the value of v is then rounded down to the nearest integer. Under the null hypothesis, we can approximate the distribution of t by a t distribution with v degrees of freedom. As usual, we compare the value of the statistic to the critical values in Table A.4—or use a computer program—to decide whether or not we should reject H_0.

Suppose that we are interested in investigating the effects of an antihypertensive drug treatment on persons over the age of 60 who suffer from isolated systolic hypertension. By definition, individuals with this condition have a systolic blood pressure greater than 160 mm Hg while their diastolic blood pressure is below 90 mm Hg. Before the beginning of the study, subjects who had been randomly selected to take the active drug and those chosen to receive a placebo were comparable with respect to systolic blood pressure. After one year of participating in the study, the mean systolic blood pressure for patients receiving the drug is denoted by μ_1 and the mean for those receiving the placebo by μ_2. The standard deviations of the two populations are unknown, and we do not feel justified in assuming that they are equal. Since we would like to determine whether the mean systolic blood pressures of the patients in these two different groups remain the same, we test the null hypothesis

$$H_0: \mu_1 = \mu_2.$$

We begin by selecting a random sample from each of the two groups. The sample of $n_1 = 2308$ individuals receiving the active drug treatment has mean systolic blood pressure $\bar{x}_1 = 142.5$ mm Hg and standard deviation $s_1 = 15.7$ mm Hg; the sample of $n_2 = 2293$ persons receiving the placebo has mean $\bar{x}_2 = 156.5$ mm Hg and standard deviation $s_2 = 17.3$ mm Hg [7]. We are interested in detecting differences that could occur in either direction, and thus conduct a two-sided test at the $\alpha = 0.05$ level of significance. The alternative hypothesis is

$$H_0: \mu_1 \neq \mu_2.$$

Since the two groups of patients were selected from independent normal populations and the variances are not assumed to be equal, the modified two-sample test should be applied. (Note that the modified test does not assume that the variances are unequal; it just does not require that they be the same.) In this case, the test statistic is

$$
\begin{aligned}
t &= \frac{(\bar{x}_1 - \bar{x}_2) - (\mu_1 - \mu_2)}{\sqrt{(s_1^2/n_1) + (s_2^2/n_2)}} \\
&= \frac{(142.5 - 156.5) - 0}{\sqrt{[(15.7)^2/2308] + [(17.3)^2/2293]}} \\
&= -28.74.
\end{aligned}
$$

We next calculate the approximate degrees of freedom; since $s_1^2 = (15.7)^2 = 246.49$ and $s_2^2 = (17.3)^2 = 299.29$,

$$
\begin{aligned}
v &= \frac{[(s_1^2/n_1) + (s_2^2/n_2)]^2}{[(s_1^2/n_1)^2/(n_1 - 1) + (s_2^2/n_2)^2/(n_2 - 1)]} \\
&= \frac{[(246.49/2308) + (299.29/2293)]^2}{[(246.49/2308)^2/(2308 - 1) + (299.29/2293)^2/(2293 - 1)]} \\
&= 4550.5.
\end{aligned}
$$

Rounding down to the nearest integer, $v = 4550$. Since a t distribution with 4550 degrees of freedom is for all practical purposes identical to the standard normal distribution, we may consult either Table A.3 or Table A.4. In both cases, we find that p is less than 0.001. As a result, we reject the null hypothesis

$$H_0: \mu_1 = \mu_2$$

at the 0.05 level of significance. After one year, individuals who receive the active drug treatment have a lower mean systolic blood pressure than those who receive the placebo.

Once again, we might wish to construct a confidence interval for the true difference in population means $\mu_1 - \mu_2$. For a t distribution with 4550 degrees of freedom—or the standard normal distribution—95% of the observations lie between -1.96 and 1.96. Consequently,

$$P\left(-1.96 \leq \frac{(\overline{X}_1 - \overline{X}_2) - (\mu_1 - \mu_2)}{\sqrt{(s_1^2/n_1) + (s_2^2/n_2)}} \leq 1.96\right) = 0.95.$$

If we rearrange terms, the 95% confidence limits for $\mu_1 - \mu_2$ are

$$(\overline{x}_1 - \overline{x}_2) \pm (1.96)\sqrt{\frac{s_1^2}{n_1} + \frac{s_2^2}{n_2}},$$

or

$$(142.5 - 156.5) \pm (1.96)\sqrt{\frac{(15.7)^2}{2308} + \frac{(17.3)^2}{2293}}.$$

Therefore, we are 95% confident that the interval

$$(-15.0, -13.0)$$

covers $\mu_1 - \mu_2$, the true difference in mean systolic blood pressures for the two populations. Note that this interval does not contain the value 0, and is therefore consistent with the results of the modified two-sample test.

11.3   Further Applications

We now return to the study that investigated the effects of carbon monoxide exposure on patients with coronary artery disease. Previously we compared the mean percent decrease in time to angina when a group of 63 adult males were exposed to a level of carbon monoxide intended to raise their carboxyhemoglobin levels to 4% to the mean percent decrease when they were exposed to uncontaminated air. We now wish to make a similar comparison, but this time the same patients are exposed to carbon monoxide targeted to increase their carboxyhemoglobin levels to only 2%. In this part of the study, therefore, each patient is exposed to a lower concentration of carbon monoxide. The population means for the percent decreases associated with exposure to air and exposure to carbon monoxide are represented by μ_1 and μ_2, respectively.

Again, we would like to know whether the two population means μ_1 and μ_2 are identical. Since we still believe that exposure to carbon monoxide cannot be beneficial to an individual's health, we are concerned with deviations that occur in one direction only. Therefore, we conduct a one-sided test at the $\alpha = 0.05$ level of significance; the null hypothesis is

$$H_0: \mu_1 \geq \mu_2,$$

or

$$H_0: \delta \geq 0,$$

where $\delta = \mu_1 - \mu_2$, and the alternative hypothesis is

$$H_A: \delta < 0.$$

Rather than working with the two individual sets of observations, we instead focus on the difference in the percent decreases in time to angina calculated for each individual. In this way, we are able to perform a one-sample test. The mean of these differences—a point estimate for the true difference in population means δ—is

$$\overline{d} = \frac{\sum_{i=1}^{62} d_i}{62}$$
$$= -4.95,$$

and their standard deviation is

$$s_d = \sqrt{\frac{\sum_{i=1}^{62} (d_i - \overline{d})^2}{62 - 1}}$$
$$= 19.05.$$

(One man was unable to participate on the day when he would have been exposed to the lower level of carbon monoxide; therefore, the sample size is 62 rather than 63.) The test statistic for the paired t-test is

$$t = \frac{\overline{d} - \delta}{s_d/\sqrt{n}},$$

or

$$t = \frac{-4.95 - 0}{19.05/\sqrt{62}}$$
$$= -2.05.$$

For a t distribution with 61 degrees of freedom, $0.01 < p < 0.025$. Therefore, we reject the null hypothesis at the 0.05 level. The paired samples suggest that the mean percent decrease in time to angina when the patient is exposed to a low level of carbon monoxide is larger than the mean percent decrease when the patient is not exposed; once again, exposed patients tend to develop angina more quickly.

Rather than work through the paired t-test ourselves, we could have used a computer to perform the calculations for us. In most statistical packages, there are two ways to do this: We can either perform the test on the original sets of observations, allowing the computer to generate the differences, or compute the differences ourselves and carry out a one-sample test. The appropriate output from Stata, using the original measurements, is shown in Table 11.1. In addition to summary statistics for the percent decreases associated with exposure to carbon monoxide and exposure to plain air, summaries for the differences are provided as well. The output also reports the null hypothesis and the three possible alternatives, along with a test statistic and a p-value for each one. In this case, we are interested in the alternative hypothesis on the left; thus, $p = 0.0226$. Note that the computer provides us with a more precise p-value than did Table A.4.

Now consider a different study, which was designed to investigate the effects of lactose consumption on carbohydrate energy absorption among premature infants. In particular, we are interested in determining whether a reduction in the intake of lactose—a sugar found in milk—increases or decreases energy absorption. In this study, one group of newborns was fed their mothers' breast milk; another group received a for-

TABLE 11.1

Stata output for the paired t-test

Paired t test					Number of obs = 62	
Variable	Mean	Std.Err.	t	P > |t|	[95% Conf.	Interval]
carbon2	.9254365	2.232755	.414482	0.6800	-3.539232	5.390105
air	5.873768	1.801627	3.26026	0.0018	2.271192	9.476344
diff	-4.948331	2.418982	-2.04563	0.0451	-9.785384	-.111278

Degrees of freedom: 61

Ho: mean diff = 0

Ha: diff < 0	Ha: diff ~= 0	Ha: diff > 0
t = -2.046	t = -2.046	t = -2.046
P < t = 0.0226	P > |t| = 0.0451	P > t = 0.9774

mula that contained only half as much lactose. The distributions of carbohydrate energy absorption for the two populations are approximately normal. We believe it is reasonable to assume that they have equal variances, and would like to know whether they also have identical means. Since we are concerned with deviations that could occur in either direction, we test the null hypothesis

$$H_0: \mu_1 = \mu_2$$

against the two-sided alternative

$$H_A: \mu_1 \neq \mu_2.$$

A random sample of $n_1 = 8$ infants who were fed their mothers' breast milk has mean energy absorption $\bar{x}_1 = 87.38\%$ and standard deviation $s_1 = 4.56\%$; a sample of $n_2 = 10$ newborns who were given the formula has mean $\bar{x}_2 = 90.14\%$ and standard deviation $s_2 = 4.58\%$ [8]. Since the samples are independent and the underlying population variances are assumed to be equal—an assumption that seems reasonable based on the values of s_1 and s_2—we apply the two-sample t-test.

We begin by calculating the pooled estimate of the variance,

$$s_p^2 = \frac{(n_1 - 1)s_1^2 + (n_2 - 1)s_2^2}{n_1 + n_2 - 2}$$

$$= \frac{(8 - 1)(4.56)^2 + (10 - 1)(4.58)^2}{8 + 10 - 2}$$

$$= 20.90.$$

The value s_p^2 combines information from both samples of children to produce a more reliable estimate of the common variance σ^2. The test statistic is

$$t = \frac{(\bar{x}_1 - \bar{x}_2) - (\mu_1 - \mu_2)}{\sqrt{s_p^2[(1/n_1) + (1/n_2)]}}$$

$$= \frac{(87.38 - 90.14) - 0}{\sqrt{(20.90)[(1/8) + (1/10)]}}$$

$$= -1.27.$$

For a t distribution with $8 + 10 - 2 = 16$ degrees of freedom, the total area under the curve to the left of -1.27 and to the right of 1.27 is greater than $2(0.10) = 0.20$. Therefore, we do not reject the null hypothesis. Based on these samples, lactose intake in newborns does not appear to have an effect on carbohydrate energy absorption.

Once again, we could have used a computer to conduct the test of hypothesis for us. Sample output from Stata is shown in Table 11.2. Since we are interested in a two-sided test, we focus on the information in the center of the lower part of the output. While Table A.4 allowed us to say that $p > 0.20$, the computer tells us that $p = 0.2213$.

TABLE 11.2

Stata output displaying the two-sample t-test, assuming equal variances

Two-sample t test with equal variances				M: Number of obs = 8 F: Number of obs = 10		
Variable	Mean	Std.Err.	t	P > |t|	[95% Conf.	Interval]
milk	87.38	1.612203	54.1991	0.0000	83.56774	91.19226
formula	90.14	1.448323	62.2375	0.0000	86.86367	93.41633
diff	-2.76	2.168339	-1.27286	0.2213	-7.356674	1.836674

Degrees of freedom: 16

$$\text{Ho: mean(x) } - \text{ mean(y) } = \text{ diff } = 0$$

Ha: diff < 0	Ha: diff ~= 0	Ha: diff > 0
t = -1.2729	t = -1.2729	t = -1.2729
P < t = 0.1106	P > |t| = 0.2213	P > t = 0.8894

In a study conducted to investigate the risk factors for heart disease among patients suffering from diabetes, one of the characteristics examined was body mass index. Body mass index is a measure of the extent to which an individual is overweight. We wish to determine whether the mean body mass index for male diabetics is equal to the mean for female diabetics. In each group, the distribution of indices is approximately normal; we have no reason to believe that the variances must be equal, and do not wish to make this assumption. Consequently, we test the null hypothesis

$$H_0: \mu_1 = \mu_2$$

against the two-sided alternative

$$H_A: \mu_1 \neq \mu_2,$$

using the modified version of the two-sample test.

A random sample is selected from each population. The $n_1 = 207$ male diabetics have mean body mass index $\bar{x}_1 = 26.4 \text{ kg/m}^2$ and standard deviation $s_1 = 3.3 \text{ kg/m}^2$; the $n_2 = 127$ female diabetics have mean body mass index $\bar{x}_2 = 25.4 \text{ kg/m}^2$ and standard deviation $s_2 = 5.2 \text{ kg/m}^2$ [9]. The test statistic is

$$t = \frac{(\bar{x}_1 - \bar{x}_2) - (\mu_1 - \mu_2)}{\sqrt{(s_1^2/n_1) + (s_2^2/n_2)}}$$

$$= \frac{(26.4 - 25.4) - 0}{\sqrt{[(3.3)^2/207] + [(5.2)^2/127]}}$$

$$= 1.94.$$

Since $s_1^2 = (3.3)^2 = 10.89$ and $s_2^2 = (5.2)^2 = 27.04$, we find the approximate degrees of freedom to be

$$v = \frac{[(s_1^2/n_1) + (s_2^2/n_2)]^2}{[(s_1^2/n_1)^2/(n_1 - 1) + (s_2^2/n_2)^2/(n_2 - 1)]}$$

$$= \frac{[(10.89/207) + (27.04/127)]^2}{[(10.89/207)^2/(207 - 1) + (27.04/127)^2/(127 - 1)]}$$

$$= 188.9.$$

Rounding down to the nearest integer, $v = 188$. For a t distribution with 188 degrees of freedom, $0.05 < p < 0.10$. The results of this test are borderline; although we would reject the null hypothesis

$$H_0: \mu_1 = \mu_2$$

at the 0.10 level of significance, we would not reject it at the 0.05 level. It appears that males who have diabetes may tend to have a slightly higher body mass index—and therefore to be more overweight—than females suffering from this disease.

Again, we could have used the computer to perform these calculations. The appropriate output from SAS is shown in Table 11.3. In addition to summary statistics for each of the independent groups, the output displays the test statistic, degrees of freedom, and p-value for the test that assumes equal variances as well as the one that does not. (It also includes a preliminary test of variances, even if we do not ask for it.) The p-value for the modified test is approximately equal to 0.05 if we round down to two decimal places. Again we conclude that male diabetics have a tendency to be more overweight than female diabetics, but we now know that p is a lot closer to 0.05 than it is to

TABLE 11.3

SAS output displaying the two-sample t-test, allowing either equal or unequal variances

			TTEST PROCEDURE			

Variable: BMI

GROUP	N	Mean	Std Dev	Std Error	Minimum	Maximum
M	207	26.4	3.3	0.229366	19.7	32.8
F	127	25.4	5.2	0.461425	17.5	35.2

Variances	T	DF	Prob> |T|
Unequal	1.9407	188.9	0.0538
Equal	2.1505	332.0	0.0322

For H0: Variances are equal, F = 0.403   DF = (126,206)

Prob > F = 0.000

0.10. Note that p for the test that assumes equal variances is actually a little less than 0.05. However, since we had no reason to believe that the variances should be equal—and, in fact, the sample standard deviations s_1 and s_2 suggest that they are unlikely to be the same—it is safer to use the modified test. Although this test is less precise than the traditional two-sample t-test if the variances truly are identical, it is more reliable if they are not.

11.4   Review Exercises

1. What is the main difference between paired and independent samples?

2. Explain the purpose of paired data. In certain situations, what might be the advantage of using paired samples rather than independent ones?

3. When should you use the two-sample t-test? When must the modified version of the test be applied?

4. What is the rationale for using a pooled estimate of the variance in the two-sample t-test?

5. A crossover study was conducted to investigate whether oat bran cereal helps to lower serum cholesterol levels in hypercholesterolemic males. Fourteen such individuals were randomly placed on a diet that included either oat bran or corn flakes; after two weeks, their low-density lipoprotein (LDL) cholesterol levels were recorded. Each man was then switched to the alternative diet. After a second two-week period, the LDL cholesterol level of each individual was again recorded. The data from this study are shown below [10].

| Subject | LDL (mmol/l) | |
	Corn Flakes	Oat Bran
1	4.61	3.84
2	6.42	5.57
3	5.40	5.85
4	4.54	4.80
5	3.98	3.68
6	3.82	2.96
7	5.01	4.41
8	4.34	3.72
9	3.80	3.49
10	4.56	3.84
11	5.35	5.26
12	3.89	3.73
13	2.25	1.84
14	4.24	4.14

(a) Are the two samples of data paired or independent?
(b) What are the appropriate null and alternative hypotheses for a two-sided test?
(c) Conduct the test at the 0.05 level of significance. What is the p-value?
(d) What do you conclude?

6. Suppose that you are interested in determining whether exposure to the organochlorine DDT, which has been used extensively as an insecticide for many years, is associated with breast cancer in women. As part of a study that investigated this issue, blood was drawn from a sample of women diagnosed with breast cancer over a six-year period and from a sample of healthy control subjects matched to the cancer patients on age, menopausal status, and date of blood donation. Each woman's blood level of DDE—an important byproduct of DDT in the human body—was measured, and the difference in levels for each patient and her matched control calculated. A sample of 171 such differences has mean $\bar{d} = 2.7$ ng/ml and standard deviation $s_d = 15.9$ ng/ml [11].
   (a) Test the null hypothesis that the mean blood levels of DDE are identical for women with breast cancer and for healthy control subjects. What do you conclude?
   (b) Would you expect a 95% confidence interval for the true difference in population mean DDE levels to contain the value 0? Explain.

7. The following data come from a study that examines the efficacy of saliva cotinine as an indicator for exposure to tobacco smoke. In one part of the study, seven subjects—none of whom were heavy smokers and all of whom had abstained from smoking for at least one week prior to the study—were each required to smoke a single cigarette. Samples of saliva were taken from all individuals 2, 12, 24, and 48 hours after smoking the cigarette. The cotinine levels at 12 hours and at 24 hours are shown below [12].

| Subject | Cotinine Levels (nmol/l) ||
	After 12 Hours	After 24 Hours
1	73	24
2	58	27
3	67	49
4	93	59
5	33	0
6	18	11
7	147	43

Let μ_{12} represent the population mean cotinine level 12 hours after smoking the cigarette and μ_{24} the mean cotinine level 24 hours after smoking. It is believed that μ_{24} must be lower than μ_{12}.
(a) Construct a one-sided 95% confidence interval for the true difference in population means $\mu_{12} - \mu_{24}$.
(b) Test the null hypothesis that the population means are identical at the $\alpha = 0.05$ level of significance. What do you conclude?

8. A study was conducted to determine whether an expectant mother's cigarette smoking has any effect on the bone mineral content of her otherwise healthy child. A sample of 77 newborns whose mothers smoked during pregnancy has mean bone mineral content $\bar{x}_1 = 0.098$ g/cm and standard deviation $s_1 = 0.026$ g/cm; a sample of 161 infants whose mothers did not smoke has mean $\bar{x}_2 = 0.095$ g/cm and standard deviation $s_2 = 0.025$ g/cm [13]. Assume that the underlying population variances are equal.

   (a) Are the two samples of data paired or independent?

   (b) State the null and alternative hypotheses of the two-sided test.

   (c) Conduct the test at the 0.05 level of significance. What do you conclude?

9. In an investigation of pregnancy-induced hypertension, one group of women with this disorder was treated with low-dose aspirin, and a second group was given a placebo. A sample consisting of 23 women who received aspirin has mean arterial blood pressure 111 mm Hg and standard deviation 8 mm Hg; a sample of 24 women who were given the placebo has mean blood pressure 109 mm Hg and standard deviation 8 mm Hg [14].

   (a) At the 0.01 level of significance, test the null hypothesis that the two populations of women have the same mean arterial blood pressure.

   (b) Construct a 99% confidence interval for the true difference in population means. Does this interval contain the value 0?

10. As part of the Women's Health Trial, one group of women were encouraged to adopt a low-fat diet while a second group received no dietary counseling. A year later, the women in the intervention group had successfully maintained their diets. At that time, a study was undertaken to determine whether their husbands also had a reduced level of fat intake [15].

    (a) In the intervention group, a sample of 156 husbands has mean daily fat intake $\bar{x}_1 = 54.8$ grams and standard deviation $s_1 = 28.1$ grams. In the control group, a sample of 148 husbands has mean intake $\bar{x}_2 = 69.5$ grams and standard deviation $s_2 = 34.7$ grams. Calculate separate 95% confidence intervals for the true mean fat intakes of men in each group. Use these intervals to construct a graph like Figure 11.3. Does the graph suggest that the two population means are likely to be equal to each other?

    (b) Formally test the null hypothesis that the two groups of men have the same mean dietary fat intake using a two-sided test. What do you conclude?

    (c) Construct a 95% confidence interval for the true difference in population means.

    (d) A researcher might also be interested in knowing whether the men differ with respect to the intake of other types of food, such as protein or carbohydrates. In the intervention group, the husbands have mean daily carbohydrate intake $\bar{x}_1 = 172.5$ grams and standard deviation $s_1 = 68.8$ grams. In the control group, the men have mean carbohydrate intake $\bar{x}_2 = 185.5$ grams and standard deviation $s_2 = 69.0$ grams. Test the null hypothesis that the two populations have the same mean carbohydrate intake. What do you conclude?

11. The table below compares the levels of carboxyhemoglobin for a group of non-smokers and a group of cigarette smokers. Sample means and standard deviations

are shown [16]. It is believed that the mean carboxyhemoglobin level of the smokers must be higher than the mean level of the nonsmokers. There is no reason to assume that the underlying population variances are identical.

Group	n	Carboxyhemoglobin (%)
Nonsmokers	121	$\bar{x} = 1.3, s = 1.3$
Smokers	75	$\bar{x} = 4.1, s = 2.0$

(a) What are the null and alternative hypotheses of the one-sided test?

(b) Conduct the test at the 0.05 level of significance. What do you conclude?

12. Suppose that you wish to compare the characteristics of tuberculosis meningitis in patients infected with HIV and those who are not infected. In particular, you would like to determine whether the two populations have the same mean age. A sample of 37 infected patients has mean age $\bar{x}_1 = 27.9$ years and standard deviation $s_1 = 5.6$ years; a sample of 19 patients who are not infected has mean age $\bar{x}_2 = 38.8$ years and standard deviation $s_2 = 21.7$ years [17].

(a) Test the null hypothesis that the two populations of patients have the same mean age at the 0.05 level of significance.

(b) Do you expect that a 95% confidence interval for the true difference in population means would contain the value 0? Why or why not?

13. Consider the numbers of community hospital beds per 1000 population that are available in each state in the United States and in the District of Columbia. The data for both 1980 and 1986 are on your disk in a data set called bed [18] (Appendix B, Table B.13). The values for 1980 are saved under the variable name bed80; those for 1986 are saved under bed86. A second data set, called bed2, contains the same information in a different format. The numbers of beds per 1000 population for both calendar years are saved under the variable name bed, and an indicator of year under the name year.

(a) Generate descriptive statistics for the numbers of hospital beds in each year.

(b) Since there are two observations for each state—one for 1980 and one for 1986—the data are actually paired. A common error in analyzing this type of data is to ignore the pairing and assume that the samples are independent. Compare the mean number of community hospital beds per 1000 population in 1980 to the mean number of beds in 1986 using the two-sample t-test. What do you conclude?

(c) Now compare the mean number of beds in 1980 to the mean number in 1986 using the paired t-test.

(d) Comment on the differences between the two tests. Do you reach the same conclusion in each case?

(e) Generate a 95% confidence interval for the true difference in the mean number of hospital beds in 1980 and 1986.

14. The data set lowbwt contains information for a sample of 100 low birth weight infants born in two teaching hospitals in Boston, Massachusetts [19] (Appendix B,

Table B.7). Measurements of systolic blood pressure are saved under the variable name `sbp` and indicators of gender—with 1 representing a male and 0 a female—under the name `sex`.

(a) Construct a histogram of systolic blood pressure measurements for this sample. Based on the graph, do you believe that blood pressure is approximately normally distributed?

(b) Test the null hypothesis that among low birth weight infants, the mean systolic blood pressure for boys is equal to the mean for girls. Use a two-sided test at the 0.05 level of significance. What do you conclude?

15. The Bayley Scales of Infant Development yield scores on two indices—the Psychomotor Development Index (PDI) and the Mental Development Index (MDI)—that can be used to assess a child's level of functioning at approximately one year of age. As part of a study investigating the development and neurologic status of children who had undergone reparative heart surgery during the first three months of life, the Bayley Scales were administered to a sample of one-year-old infants born with congenital heart disease. The children had been randomized to one of two different treatment groups, known as "circulatory arrest" and "low-flow bypass." The groups differed in the specific way in which the reparative surgery was performed. Unlike circulatory arrest, low-flow bypass maintains continuous circulation through the brain; although it is felt to be preferable by some physicians, it also has its own associated risk of brain injury. The data for this study are saved in the data set `heart` [20] (Appendix B, Table B.12). PDI scores are saved under the variable name `pdi`, MDI scores under `mdi`, and indicators of treatment group under `trtment`. For this variable, 0 represents circulatory arrest and 1 is for low-flow bypass.

(a) At the 0.05 level of significance, test the null hypothesis that the mean PDI score at one year of age for the circulatory arrest treatment group is equal to the mean PDI score for the low-flow group. What is the *p*-value for this test?

(b) Test the null hypothesis that the mean MDI scores are identical for the two treatment groups. What is the *p*-value?

(c) What do these tests suggest about the relationship between a child's surgical treatment group during the first three months of life and his or her subsequent developmental status at one year of age?

Bibliography

[1] Packard, F. R., *The Life and Times of Ambroise Paré*, 1510–1590, New York: Paul B. Hoeber, 1921.

[2] Allred, E. N., Bleecker, E. R., Chaitman, B. R., Dahms, T. E., Gottlieb, S. O., Hackney, J. D., Hayes, D., Pagano, M., Selvester, R. H., Walden, S. M., and Warren, J., "Acute Effects of Carbon Monoxide Exposure on Individuals with Coronary Artery Disease," *Health Effects Institute Research Report Number 25*, November 1989.

[3] Markowski, C. A., and Markowski, E. P., "Conditions for the Effectiveness of a Preliminary Test of Variance," *The American Statistician*, Volume 44, November 1990, 322–326.

[4] Moser, B. K., and Stevens, G. R., "Homogeneity of Variance in the Two-Sample Means Test," *The American Statistician*, Volume 46, February 1992, 19–21.

[5] Zempsky, W. T., Rosenstein, B. J., Carroll, J. A., and Oski, F. A., "Effect of Pancreatic Enzyme Supplements on Iron Absorption," *American Journal of Diseases of Children*, Volume 143, August 1989, 966–972.

[6] Satterthwaite, F. W., "An Approximate Distribution of Estimates of Variance Components," *Biometrics Bulletin*, Volume 2, December 1946, 110–114.

[7] SHEP Cooperative Research Group, "Prevention of Stroke by Antihypertensive Drug Treatment in Older Persons with Isolated Systolic Hypertension: Final Results of the Systolic Hypertension in the Elderly Program (SHEP)," *Journal of the American Medical Association*, Volume 265, June 26, 1991, 3255–3264.

[8] Kien, C. L., Liechty, E. A., and Mullett, M. D., "Effects of Lactose Intake on Nutritional Status in Premature Infants," *Journal of Pediatrics*, Volume 116, March 1990, 446–449.

[9] Barrett-Connor, E. L., Cohn, B. A., Wingard, D. L., and Edelstein, S. L., "Why Is Diabetes Mellitus a Stronger Risk Factor for Fatal Ischemic Heart Disease in Women Than in Men?," *Journal of the American Medical Association*, Volume 265, February 6, 1991, 627–631.

[10] Anderson, J. W., Spencer, D. B., Hamilton, C. C., Smith, S. F., Tietyen, J., Bryant, C. A., and Oeltgen, P., "Oat-Bran Cereal Lowers Serum Total and LDL Cholesterol in Hypercholesterolemic Men," *American Journal of Clinical Nutrition*, Volume 52, September 1990, 495–499.

[11] Wolff, M. S., Toniolo, P. G., Lee, E. W., Rivera, M., and Dubin, N., "Blood Levels of Organochlorine Residues and Risk of Breast Cancer," *Journal of the National Cancer Institute*, Volume 85, April 21, 1993, 648–652.

[12] DiGiusto, E., and Eckhard, I., "Some Properties of Saliva Cotinine Measurements in Indicating Exposure to Tobacco Smoking," *American Journal of Public Health*, Volume 76, October 1986, 1245–1246.

[13] Venkataraman, P. S., and Duke, J. C., "Bone Mineral Content of Healthy, Full-term Neonates: Effect of Race, Gender, and Maternal Cigarette Smoking," *American Journal of Diseases of Children*, Volume 145, November 1991, 1310–1312.

[14] Schiff, E., Barkai, G., Ben-Baruch, G., and Mashiach, S., "Low-Dose Aspirin Does Not Influence the Clinical Course of Women with Mild Pregnancy-Induced Hypertension," *Obstetrics and Gynecology*, Volume 76, November 1990, 742–744.

[15] Shattuck, A. L., White, E., and Kristal, A. R., "How Women's Adopted Low-Fat Diets Affect Their Husbands," *American Journal of Public Health*, Volume 82, September 1992, 1244–1250.

[16] Jarvis, M. J., Tunstall-Pedoe, H., Feyerabend, C., Vesey, C., and Saloojee, Y., "Comparison of Tests Used to Distinguish Smokers from Nonsmokers," *American Journal of Public Health*, Volume 77, November 1987, 1435–1438.

[17] Berenguer, J., Moreno, S., Laguna, F., Vicente, T., Adrados, M., Ortega, A., González-LaHoz, J., and Bouza, E., "Tuberculosis Meningitis in Patients Infected with the Human Immunodeficiency Virus," *The New England Journal of Medicine*, Volume 326, March 5, 1992, 668–672.

[18] National Center for Health Statistics, *Health United States 1988*, Public Health Service, Hyattsville, MD: March 1989.

[19] Leviton, A., Fenton, T., Kuban, K. C. K., and Pagano, M., "Labor and Delivery Characteristics and the Risk of Germinal Matrix Hemorrhage in Low Birth Weight Infants," *Journal of Child Neurology*, Volume 6, October 1991, 35-40.

[20] Bellinger, D. C., Jonas, R. A., Rappaport, L. A., Wypij, D., Wernovsky, G., Kuban, K. C. K., Barnes, P. D., Holmes, G. L., Hickey, P. R., Strand, R. D., Walsh, A. Z., Helmers, S. L., Constantinou, J. E., Carrazana, E. J., Mayer, J. E., Hanley, F. L., Castaneda, A. R., Ware, J. H., and Newburger, J. W., "Developmental and Neurologic Status of Children After Heart Surgery with Hypothermic Circulatory Arrest or Low-Flow Cardiopulmonary Bypass," *The New England Journal of Medicine*, Volume 332, March 2, 1995, 549–555.

12

Analysis of Variance

In the preceding chapter, we covered techniques for determining whether a difference exists between the means of two independent populations. It is not unusual, however, to encounter situations in which we wish to test for differences among three or more independent means rather than just two. The extension of the two-sample t-test to three or more samples is known as the *analysis of variance*.

12.1 One-Way Analysis of Variance

12.1.1 The Problem

When we discussed the paired t-test in Chapter 11, we examined data from a study that investigated the effects of carbon monoxide exposure on patients with coronary artery disease by subjecting them to a series of exercise tests. The men involved in the study were recruited from three different medical centers—the Johns Hopkins University School of Medicine, the Rancho Los Amigos Medical Center, and the St. Louis University School of Medicine. Before combining the subjects into one large group to conduct the analysis, we should have first examined some baseline characteristics to ensure that the patients from different centers were in fact comparable.

One characteristic that we might wish to consider is pulmonary function at the start of the study. If the patients from one medical center begin with measures of forced expiratory volume in 1 second that are much larger—or much smaller—than those from the other centers, the results of the analysis may be affected. Therefore, given that the populations of patients in the three centers have mean baseline FEV_1 measurements μ_1, μ_2, and μ_3 respectively, we would like to test the null hypothesis that the population means are identical. This may be expressed as

$$H_0: \mu_1 = \mu_2 = \mu_3.$$

The alternative hypothesis is that at least one of the population means differs from one of the others.

In general, we are interested in comparing the means of k different populations. Suppose that the k populations are independent and normally distributed. We begin by drawing a random sample of size n_1 from the normal population with mean μ_1 and standard deviation σ_1. The mean of this sample is denoted by $\bar{x}_1$ and its standard deviation by s_1. Similarly, we select a random sample of size n_2 from the normal population with mean μ_2 and standard deviation σ_2, and so on for the remaining populations. The numbers of observations in each sample need not be the same.

		Group 1	**Group 2**	**...**	**Group k**
Population	Mean	μ_1	μ_2	...	μ_k
	Standard Deviation	σ_1	σ_2	...	σ_k
Sample	Mean	$\bar{x}_1$	$\bar{x}_2$	...	$\bar{x}_k$
	Standard Deviation	s_1	s_2	...	s_k
	Sample Size	n_1	n_2	...	n_k

For the study investigating the effects of carbon monoxide exposure on individuals with coronary artery disease, the FEV_1 distributions of patients associated with each of the three medical centers make up distinct populations. From the population of FEV_1 measurements for the patients at Johns Hopkins University, we select a sample of size $n_1 = 21$. From the population at Rancho Los Amigos we draw a sample of size $n_2 = 16$, and from the one at St. Louis University we select a sample of size $n_3 = 23$. The data, along with their sample means and standard deviations, are shown in Table 12.1 [1]. A 95% confidence interval for the true mean FEV_1 of men at each medical center is shown in Figure 12.1. Based on this graph, the mean FEV_1 for the patients at Johns Hopkins may be a little lower than the means for the other two groups; however, all three intervals overlap. We would like to conduct a more formal analysis.

Presented with these data, we might attempt to compare the three population means by evaluating all possible pairs of sample means using the two-sample t-test. For a total of three groups, the number of tests required is $\binom{3}{2} = 3$. We would compare group 1 to group 2, group 1 to group 3, and group 2 to group 3. We assume that the variances of the underlying populations are all equal, or

$$\sigma_1^2 = \sigma_2^2 = \sigma_3^2 = \sigma^2.$$

The pooled estimate of the common variance, which we denote by s_W^2, combines information from all three samples; in particular,

$$s_W^2 = \frac{(n_1 - 1)s_1^2 + (n_2 - 1)s_2^2 + (n_3 - 1)s_3^2}{n_1 + n_2 + n_3 - 3}.$$

This quantity is simply an extension of s_p^2, the pooled estimate of the variance used for the two-sample t-test.

TABLE 12.1

Forced expiratory volume in one second for patients with coronary artery disease sampled at three different medical centers

Johns Hopkins	Rancho Los Amigos	St. Louis
3.23	3.22	2.79
3.47	2.88	3.22
1.86	1.71	2.25
2.47	2.89	2.98
3.01	3.77	2.47
1.69	3.29	2.77
2.10	3.39	2.95
2.81	3.86	3.56
3.28	2.64	2.88
3.36	2.71	2.63
2.61	2.71	3.38
2.91	3.41	3.07
1.98	2.87	2.81
2.57	2.61	3.17
2.08	3.39	2.23
2.47	3.17	2.19
2.47		4.06
2.74		1.98
2.88		2.81
2.63		2.85
2.53		2.43
		3.20
		3.53
$n_1 = 21$	$n_2 = 16$	$n_3 = 23$
$\bar{x}_1 = 2.63$ liters	$\bar{x}_2 = 3.03$ liters	$\bar{x}_3 = 2.88$ liters
$s_1 = 0.496$ liters	$s_2 = 0.523$ liters	$s_3 = 0.498$ liters

Performing all possible pairs of tests is not a problem if the number of populations is relatively small. In the instance where $k = 3$, there are only three such tests. If $k = 10$, however, the process becomes much more complicated. In this case, we would have to perform $\binom{10}{2} = 45$ different pairwise tests.

Even more important, another problem that arises when all possible two-sample t-tests are conducted is that this procedure is likely to lead to an incorrect conclusion. Suppose that the three population means are in fact equal and that we conduct all three pairwise tests. We assume, for argument's sake, that the tests are independent and set the significance level for each one at 0.05. By the multiplicative rule, the probability of failing to reject a null hypothesis of no difference in all instances—and thereby drawing the correct conclusion in each of the three tests—would be

$$P(\text{fail to reject in all three tests}) = (1 - 0.05)^3$$
$$= (0.95)^3$$
$$= 0.857.$$

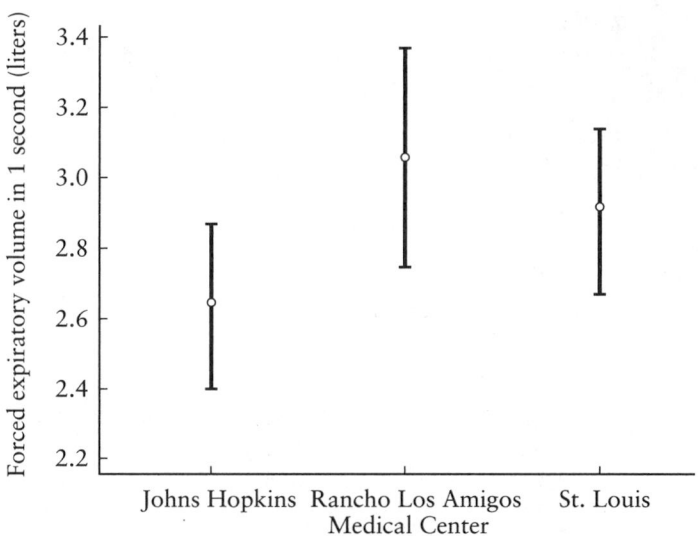

FIGURE 12.1
95% confidence intervals for the true mean forced expiratory volumes
in one second at three medical centers

Consequently, the probability of rejecting the null hypothesis in at least one of the tests would be

$$P(\text{reject in at least one test}) = 1 - 0.857$$
$$= 0.143.$$

Since we know that the null hypothesis is true in each case, 0.143 is the overall probability of making a type I error. As we can see, the combined probability of a type I error for the set of three tests is much larger than 0.05. In reality, the problem is even more complex; since each of the *t*-tests is conducted using the same set of data, we cannot assume that they are all independent. We would like to be able to use a testing procedure in which the overall probability of making a type I error is equal to some predetermined level α. The one-way analysis of variance is such a technique.

12.1.2   Sources of Variation

As its name implies, the one-way analysis of variance procedure is dependent on estimates of spread or dispersion. The term *one-way* indicates that there is a single factor or characteristic that distinguishes the various populations from each other; in the study of carbon monoxide exposure, that characteristic is the medical center at which a subject was recruited. When we work with several different populations with a common variance σ^2, two measures of variability can be computed: the variation of the individual values around their population means and the variation of the population means around the overall mean. If the variability within the k different populations is small relative to the variability among their respective means, this suggests that the population means are in fact different.

To test the null hypothesis

$$H_0: \mu_1 = \mu_2 = \cdots = \mu_k$$

for a set of k populations, we first need to find a measure of the variability of the individual observations around their population means. The pooled estimate of the common variance σ^2 is one such measure; if we let $n = n_1 + n_2 + \cdots + n_k$, then

$$s_W^2 = \frac{(n_1 - 1)s_1^2 + (n_2 - 1)s_2^2 + \cdots + (n_k - 1)s_k^2}{n_1 + n_2 + \cdots + n_k - k}$$

$$= \frac{(n_1 - 1)s_1^2 + (n_2 - 1)s_2^2 + \cdots + (n_k - 1)s_k^2}{n - k}.$$

This quantity is simply a weighted average of the k individual sample variances. The subscript W refers to the "within-groups" variability.

We next need an expression that estimates the extent to which the population means vary around the overall mean. If the null hypothesis is true and the means are identical, the amount of variability expected will be the same as that for an individual population; thus, this quantity also estimates the common variance σ^2. In particular,

$$s_B^2 = \frac{n_1(\bar{x}_1 - \bar{x})^2 + n_2(\bar{x}_2 - \bar{x})^2 + \cdots + n_k(\bar{x}_k - \bar{x})^2}{k - 1}.$$

The terms $(\bar{x}_i - \bar{x})^2$ are the squared deviations of the sample means $\bar{x}_i$ from the grand mean $\bar{x}$. The *grand mean* is defined as the overall average of the n observations that make up the k different samples; consequently,

$$\bar{x} = \frac{n_1\bar{x}_1 + n_2\bar{x}_2 + \cdots + n_k\bar{x}_k}{n_1 + n_2 + \cdots + n_k}$$

$$= \frac{n_1\bar{x}_1 + n_2\bar{x}_2 + \cdots + n_k\bar{x}_k}{n}.$$

The subscript B denotes the "between-groups" variability.

Now that we have these two different estimates of the variance, we would like to be able to answer the following question: Do the sample means vary around the grand mean more than the individual observations vary around the sample means? If they do, this implies that the corresponding population means are in fact different. To test the null hypothesis that the population means are identical, we use the test statistic

$$F = \frac{s_B^2}{s_W^2}.$$

Under the null hypothesis, both s_W^2 and s_B^2 estimate the common variance σ^2, and F is close to 1. If there is a difference among populations, the between-groups variance exceeds the within-groups variance and F is greater than 1. Under H_0, the ratio F has an F distribution with $k - 1$ and $n - k$ degrees of freedom; the degrees of freedom correspond to the numerator and the denominator, respectively. We represent this distribution using the notation $F_{k-1, n-k}$, or, more generically, F_{df_1, df_2}. If we have only two independent samples, the *F-test* reduces to the two-sample *t*-test.

The F distribution is similar to the t in that it is not unique; there is a different F distribution for each possible pair of values df_1 and df_2. Unlike the t distribution, however, the F distribution cannot assume negative values. In addition, it is skewed to the right. The extent to which it is skewed is determined by the values of the degrees of freedom. The F distribution with 4 and 2 degrees of freedom is shown in Figure 12.2.

Table A.5 in Appendix A is a table of critical values computed for the family of F distributions. Only selected percentiles are included—in this case, the upper 10.0, 5.0, 2.5, 1.0, 0.05, and 0.01% of the distributions. The degrees of freedom for the numerator are displayed across the top of the table, and the degrees of freedom for the denominator are listed down the left-hand side. For any given combination, the corresponding entry in the table represents the value of F_{df_1, df_2} that cuts off the specified area in the upper tail of the distribution. Given an F distribution with 4 and 2 degrees of freedom, for example, the table shows that $F_{4, 2} = 19.25$ cuts off the upper 5% of the curve.

Again examining the FEV_1 data collected for patients from three different medical centers, we are interested in testing

$$H_0: \mu_1 = \mu_2 = \mu_3,$$

the null hypothesis that the mean forced expiratory volumes in one second for the three centers are identical. To begin, we verify that the FEV_1 measurements are approximately normally distributed. Based on the histograms shown in Figure 12.3, this appears to be a reasonable assumption. Next, since we feel it is fair to assume that the popula-

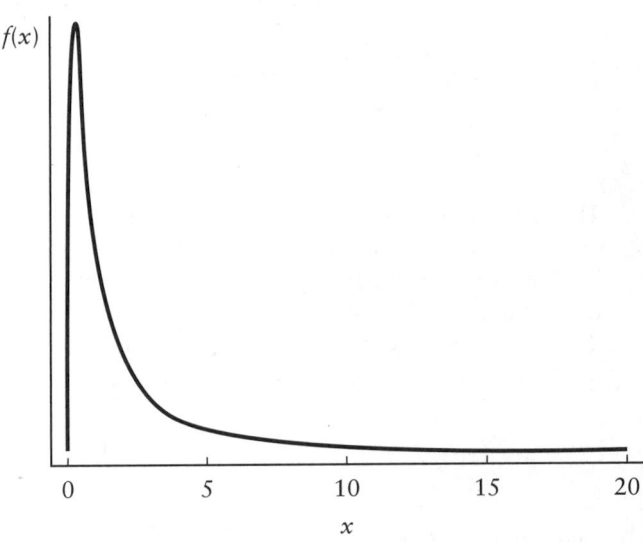

FIGURE 12.2
The F distribution with 4 and 2 degrees of freedom

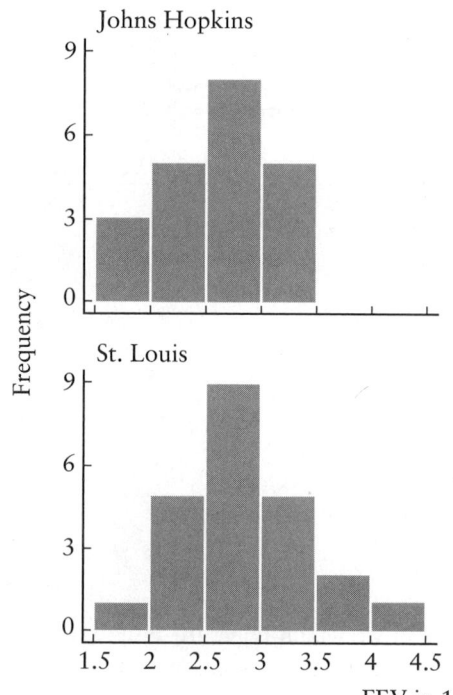

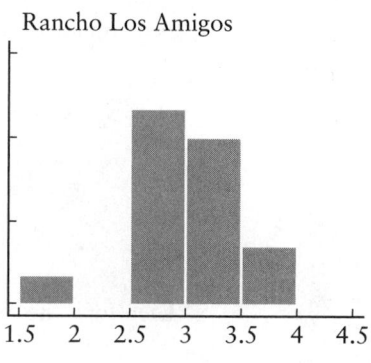

FIGURE 12.3
Histograms for measurements of forced expiratory volume in one second for subjects at three medical centers

tion variances are identical (note that the sample standard deviations are all quite close), we compute the estimate of the within-groups variance:

$$s_W^2 = \frac{(n_1 - 1)s_1^2 + (n_2 - 1)s_2^2 + (n_3 - 1)s_3^2}{n_1 + n_2 + n_3 - 3}$$

$$= \frac{(21 - 1)(0.496)^2 + (16 - 1)(0.523)^2 + (23 - 1)(0.498)^2}{21 + 16 + 23 - 3}$$

$$= 0.254 \text{ liters}^2.$$

Since

$$\bar{x} = \frac{n_1 \bar{x}_1 + n_2 \bar{x}_2 + n_3 \bar{x}_3}{n_1 + n_2 + n_3}$$

$$= \frac{21(2.63) + 16(3.03) + 23(2.88)}{21 + 16 + 23}$$

$$= 2.83 \text{ liters,}$$

the estimate of the between-groups variance is

$$s_B^2 = \frac{n_1(\bar{x}_1 - \bar{x})^2 + n_2(\bar{x}_2 - \bar{x})^2 + n_3(\bar{x}_3 - \bar{x})^2}{3 - 1}$$

$$= \frac{21(2.63 - 2.83)^2 + 16(3.03 - 2.83)^2 + 23(2.88 - 2.83)^2}{3 - 1}$$

$$= 0.769 \text{ liters}^2.$$

Therefore, the test statistic is

$$\text{F} = \frac{s_B^2}{s_W^2}$$

$$= \frac{0.769}{0.254}$$

$$= 3.028.$$

For an F distribution with $k - 1 = 3 - 1 = 2$ and $n - k = 60 - 3 = 57$ degrees of freedom, $0.05 < p < 0.10$. Although we would reject the null hypothesis at the 0.10 level, we do not reject it at the 0.05 level. There may possibly be some difference among the mean FEV_1 measurements for these three populations.

12.2 Multiple Comparisons Procedures

As we have seen, one-way analysis of variance may be used to test the null hypothesis that k population means are identical,

$$H_0: \mu_1 = \mu_2 = \cdots = \mu_k.$$

What happens, however, if we reject H_0? Although we can conclude that the population means are not all equal, we cannot be more specific than this. We do not know whether all the means are different from one another or only some of them are different. Once we reject the null hypothesis, therefore, we often want to conduct additional tests to find out where the differences lie.

Many different techniques for conducting multiple comparisons exist; they typically involve testing each pair of means individually. In the preceding section, we mentioned that one possible approach is to perform a series of $\binom{k}{2}$ two-sample t-tests. As we noted, however, performing multiple tests increases the probability of making a type I error. We can avoid this problem by being more conservative in our individual comparisons; by reducing the individual α levels, we ensure that the overall level of significance is kept at a predetermined level.

The significance level for each of the individual comparisons depends on the number of tests being conducted. The greater the number of tests, the smaller it must be. To set the overall probability of making a type I error at 0.05, for example, we should use

$$\alpha^* = \frac{0.05}{\binom{k}{2}}$$

as the significance level for an individual comparison. This modification is known as the *Bonferroni correction*.

For the case in which we have $k = 3$ populations, a total of $\binom{3}{2} = 3$ tests are required. Consequently, if we wish to set the overall level of significance at 0.10, we must use

$$\alpha^* = \frac{0.10}{3}$$
$$= 0.033$$

as the level for each individual test.

To conduct a test of the null hypothesis

$$H_0: \mu_i = \mu_j,$$

we calculate

$$t_{ij} = \frac{\bar{x}_i - \bar{x}_j}{\sqrt{s_w^2[(1/n_i) + (1/n_j)]}},$$

the test statistic for a two-sample t-test. Note that instead of using the data from only two samples to estimate the common variance σ^2, we take advantage of the additional information that is available and use all k samples. Under the null hypothesis, t_{ij} has a t distribution with $n - k$ degrees of freedom.

For the comparison of baseline FEV_1 values among the three medical centers, we begin by considering populations 1 and 2, the patients at Johns Hopkins and those at Rancho Los Amigos. In this case,

$$t_{12} = \frac{\bar{x}_1 - \bar{x}_2}{\sqrt{s_w^2[(1/n_1) + (1/n_2)]}}$$
$$= \frac{2.63 - 3.03}{\sqrt{0.254[(1/21) + (1/16)]}}$$
$$= -2.39.$$

For a t distribution with $n - k = 60 - 3 = 57$ degrees of freedom, $p = 0.02$. Therefore, we reject the null hypothesis at the 0.033 level and conclude that μ_1 is not equal to μ_2. Looking at the sample means, the mean baseline FEV_1 for patients at Johns Hopkins is lower than the mean for men at Rancho Los Amigos.

We next compare populations 1 and 3, the group at Johns Hopkins and the group at St. Louis. The test statistic is

$$t_{13} = \frac{\bar{x}_1 - \bar{x}_3}{\sqrt{s_w^2[(1/n_1) + (1/n_3)]}}$$

$$= \frac{2.63 - 2.88}{\sqrt{0.254[(1/21) + (1/23)]}}$$

$$= -1.64.$$

Since $p > 0.10$, we do not have sufficient evidence to reject the null hypothesis that μ_1 equals μ_3.

Finally, we compare the patients at Rancho Los Amigos and those at St. Louis:

$$t_{23} = \frac{\bar{x}_2 - \bar{x}_3}{\sqrt{s_w^2[(1/n_2) + (1/n_3)]}}$$

$$= \frac{3.03 - 2.88}{\sqrt{0.254[(1/16) + (1/23)]}}$$

$$= 0.91.$$

This time $p > 0.20$, and we are unable to reject the null hypothesis that μ_2 is equal to μ_3. In summary, therefore, we find that the mean baseline FEV_1 measurement for patients at Johns Hopkins is somewhat lower than the mean for the patients at Rancho Los Amigos; we cannot make any further distinctions among the medical centers.

One disadvantage of the Bonferroni multiple comparisons procedure is that it can suffer from a lack of power. It is highly conservative and may fail to detect a difference in means that actually exists. However, there are many competing multiple-comparisons procedures that could be used instead [2]. The appropriate technique to apply in a given situation depends on a variety of factors, including the types of comparisons to be made (are all possible pairwise tests being performed as in the example above, for instance, or are two or more treatment groups being compared to a single control group?), whether the comparisons were specified before collecting and summarizing the data or after, and whether all the samples contain equal numbers of observations.

12.3 Further Applications

A study was conducted to follow three groups of overweight males for a period of one year. The first group decreased their energy intake by dieting but did not participate in an exercise program. The second group exercised regularly but did not alter their eating habits. The third group changed neither their diet nor their level of physical activity. At the end of one year, total change in body weight was measured for each individual.

Among these three populations, is there any evidence of a difference in mean change in body weight?

We begin by noting that changes in weight tend to be normally distributed, and then select a random sample from each population. The sample of 42 men who dieted has mean change in body weight $\bar{x}_1 = -7.2$ kg and standard deviation $s_1 = 3.7$ kg. The sample of 47 men who participated in an exercise program has mean $\bar{x}_2 = -4.0$ kg and standard deviation $s_2 = 3.9$ kg. Finally, the sample of 42 men who neither dieted nor exercised has mean $\bar{x}_3 = 0.6$ kg and standard deviation $s_3 = 3.7$ kg [3].

We are interested in testing the null hypothesis that the mean changes in total body weight are identical for the three normal populations, or

$$H_0: \mu_1 = \mu_2 = \mu_3.$$

The alternative hypothesis is that at least one of the population means differs from one of the others. We assume that the underlying population variances are all equal, a reasonable assumption given the sample standard deviations above. It should be noted, however, that the one-way analysis of variance is relatively insensitive to departures from this assumption; even if the variances are not quite identical, the technique still works pretty well.

To conduct the test, we begin by computing an estimate of the within-groups variance:

$$
\begin{aligned}
s_W^2 &= \frac{(n_1 - 1)s_1^2 + (n_2 - 1)s_2^2 + (n_3 - 1)s_3^2}{n_1 + n_2 + n_3 - 3} \\
&= \frac{(42 - 1)(3.7)^2 + (47 - 1)(3.9)^2 + (42 - 1)(3.7)^2}{42 + 47 + 42 - 3} \\
&= 14.24 \text{ kg}^2.
\end{aligned}
$$

Since the grand mean of the data is

$$
\begin{aligned}
\bar{x} &= \frac{n_1 \bar{x}_1 + n_2 \bar{x}_2 + n_3 \bar{x}_3}{n_1 + n_2 + n_3} \\
&= \frac{42(-7.2) + 47(-4.0) + 42(0.6)}{42 + 47 + 42} \\
&= -3.55 \text{ kg},
\end{aligned}
$$

the estimate of the between-groups variance is

$$
\begin{aligned}
s_B^2 &= \frac{n_1(\bar{x}_1 - \bar{x})^2 + n_2(\bar{x}_2 - \bar{x})^2 + n_3(\bar{x}_3 - \bar{x})^2}{3 - 1} \\
&= \frac{42(-7.2 + 3.55)^2 + 47(-4.0 + 3.55)^2 + 42(0.6 + 3.55)^2}{3 - 1} \\
&= 646.20 \text{ kg}^2.
\end{aligned}
$$

Consequently, the test statistic is

$$F = \frac{s_B^2}{s_W^2}$$

$$= \frac{646.20}{14.24}$$

$$= 45.38.$$

For an F distribution with $k - 1 = 3 - 1 = 2$ and $n - k = 131 - 3 = 128$ degrees of freedom, $p < 0.001$. Therefore, we reject the null hypothesis and conclude that the mean changes in total body weight are not identical for the three populations.

Now that we have determined that the population means are not all the same, we would like to find out where the specific differences lie. One way to do this is to apply the Bonferroni multiple comparisons procedure. If we plan to conduct all possible pairwise tests and we wish to set the overall probability of making a type I error at 0.05, the significance level for an individual comparison is

$$\alpha* = \frac{0.05}{\binom{3}{2}}$$

$$= \frac{0.05}{3}$$

$$= 0.0167.$$

We begin by considering the group of men who dieted and those who were involved in an exercise program. To test the null hypothesis that the mean changes in total body weight are identical for these two populations,

$$H_0 : \mu_1 = \mu_2,$$

we calculate the test statistic

$$t_{12} = \frac{\bar{x}_1 - \bar{x}_2}{\sqrt{s_W^2[(1/n_1) + (1/n_2)]}}$$

$$= \frac{-7.2 - (-4.0)}{\sqrt{14.24[(1/42) + (1/47)]}}$$

$$= -3.99.$$

For a t distribution with $n - k = 131 - 3 = 128$ degrees of freedom, $p < 0.001$. Therefore, we reject the null hypothesis at the 0.0167 level of significance and conclude that μ_1 differs from μ_2. Looking at the sample means, we see that the mean decrease in total body weight is greater for the men who dieted.

We next compare means for the men who dieted and for those who neither dieted nor exercised. In this case, the test statistic is

$$t_{13} = \frac{\bar{x}_1 - \bar{x}_3}{\sqrt{s_w^2[(1/n_1) + (1/n_3)]}}$$

$$= \frac{-7.2 - 0.6}{\sqrt{14.24[(1/42) + (1/42)]}}$$

$$= -9.47.$$

Since $p < 0.001$, we again reject the null hypothesis of equal means and conclude that μ_1 differs from μ_3. The mean decrease in body weight is larger for the group of men who dieted.

Finally, we compare means for the men who exercised and those who did not change their lifestyles in any way, calculating

$$t_{23} = \frac{\bar{x}_2 - \bar{x}_3}{\sqrt{s_w^2[(1/n_2) + (1/n_3)]}}$$

$$= \frac{-4.0 - 0.6}{\sqrt{14.24[(1/47) + (1/42)]}}$$

$$= -5.74.$$

We find that $p < 0.001$; as a result, we conclude that μ_2 is not equal to μ_3. The mean decrease in total body weight is greater for the men who participated in an exercise program. In summary, all three of the population means are different from each other. The mean change in total body weight is lowest for the men who dieted—meaning that they lost the most weight—followed by that for the men who exercised. The mean is highest for those who made no changes in their lifestyles.

Rather than work through all these computations ourselves, we could have used a computer to carry out the analysis for us. The appropriate Stata output is displayed in Table 12.2. The top portion of the output is called an analysis of variance (ANOVA) table. On the far right, the ANOVA table lists the test statistic F and its associated p-value. A p-value displayed as 0.0000 means that $p < 0.0001$. The column labeled MS, or mean squares, contains the between- and within-groups estimates of the variance. The columns labeled SS and df contain the numerators and denominators of these estimates, respectively. Note that each of the numerators is actually a sum of squared deviations from the mean; with s_B^2 we are concerned with the deviations of the sample means from the grand mean, and with s_w^2 we use the deviations of the individual observations from the sample means. Consequently, the between- and within-groups estimates of the variance may be thought of as averages of squared deviations.

The bottom portion of Table 12.2 shows the results of the Bonferroni multiple comparisons procedure. For each of the three possible pairwise t-tests, the table lists the difference in sample means $\bar{x}_i - \bar{x}_j$ and the corresponding p-value. In this case, all three p-values are less than 0.001.

TABLE 12.2
Stata output illustrating the one-way analysis of variance and the Bonferroni multiple comparisons procedure

Source	SS	df	MS	F	Prob > F
	Analysis of Variance				
Between groups	1292.40	2	646.20	45.38	0.0000
Within groups	1822.72	128	14.24		
Total	3115.12	130	23.96		

Comparison of wtchange by group
(Bonferroni)

Row Mean– Col Mean	1	2
2	-3.20 0.000	
3	-7.80 0.000	-4.60 0.000

12.4 Review Exercises

1. When you are testing the equality of several population means, what problems arise if you attempt to perform all possible pairwise t-tests?

2. What is the idea behind the one-way analysis of variance? What two measures of variation are being compared?

3. What are the properties of the F distribution?

4. Describe the purpose of the Bonferroni correction for multiple comparisons.

5. Consider the F distribution with 8 and 16 degrees of freedom.
   (a) What proportion of the area under the curve lies to the right of $F = 2.09$?
   (b) What value of F cuts off the upper 1% of the distribution?
   (c) What proportion of the area under the curve lies to the left of $F = 4.52$?

6. Consider the F distribution with 3 and 30 degrees of freedom.
   (a) What proportion of the area under the curve lies to the right of $F = 5.24$?
   (b) What proportion of the area under the curve lies to the left of $F = 2.92$?
   (c) What value of F cuts off the upper 2.5% of the distribution?
   (d) What value of F cuts off the upper 0.1%?

7. A study of patients with insulin-dependent diabetes was conducted to investigate the effects of cigarette smoking on renal and retinal complications. Before examining the results of the study, you wish to compare the baseline measures of sys-

tolic blood pressure across four different subgroups: nonsmokers, current smokers, ex-smokers, and tobacco chewers. A sample is selected from each subgroup; the relevant data are shown below [4]. Means and standard deviations are expressed in mm Hg. Assume that systolic blood pressure is normally distributed.

	n	$\bar{x}$	s
Nonsmokers	269	115	13.4
Current Smokers	53	114	10.1
Ex-smokers	28	118	11.6
Tobacco Chewers	9	126	12.2

(a) Calculate the estimate of the within-groups variance.
(b) Calculate the estimate of the between-groups variance.
(c) At the 0.05 level of significance, test the null hypothesis that the mean systolic blood pressures of the four groups are identical. What do you conclude?
(d) If you find that the population means are not all equal, use the Bonferroni multiple comparisons procedure to determine where the differences lie. What is the significance level of each individual test?

8. One of the goals of the Edinburgh Artery Study was to investigate the risk factors for peripheral arterial disease among persons 55 to 74 years of age. You wish to compare mean LDL cholesterol levels, measured in mmol/liter, among four different populations of subjects: patients with intermittent claudication or interruptions in movement, those with major asymptomatic disease, those with minor asymptomatic disease, and those with no evidence of disease at all. Samples are selected from each population; summary statistics are shown below [5].

	n	$\bar{x}$	s
Intermittent Claudication	73	6.22	1.62
Major Asymptomatic Disease	105	5.81	1.43
Minor Asymptomatic Disease	240	5.77	1.24
No Disease	1080	5.47	1.31

(a) Test the null hypothesis that the mean LDL cholesterol levels are the same for each of the four populations. What are the degrees of freedom associated with this test?
(b) What do you conclude?
(c) What assumptions about the data must be true for you to use the one-way analysis of variance technique?
(d) Is it necessary to take an additional step in this analysis? If so, what is it? Explain.

9. A study was conducted to assess the performance of outpatient substance abuse treatment centers. Three different types of units were evaluated: private for-profit (FP), private not-for-profit (NFP), and public. Among the performance measures considered were minutes of individual therapy per session and minutes of group

therapy per session. Samples were selected from each type of treatment center; summary statistics are shown in the table [6].

Treatment Centers	Individual Therapy			Group Therapy		
	n	$\bar{x}$	s	n	$\bar{x}$	s
FP	37	49.46	15.47	30	105.83	42.91
NFP	312	54.76	11.41	296	98.68	31.27
Public	169	53.25	11.08	165	94.17	27.12

(a) Given these numbers, how do the different types of treatment centers compare with respect to average minutes of therapy per session?

(b) Construct 95% confidence intervals for the mean minutes of individual therapy per session in each type of treatment center. Do the same for the mean minutes of group therapy. In either case, do you notice anything that suggests that the population means may not be the same?

(c) Test the null hypothesis that the mean minutes of individual therapy per session are identical for each type of center. If necessary, carry out a multiple-comparisons procedure.

(d) Test the null hypothesis that the mean minutes of group therapy per session are the same for each type of center. Again, use a multiple comparisons procedure if necessary.

(e) How do the different treatment centers compare to each other?

10. For the study discussed in the text that investigates the effect of carbon monoxide exposure on patients with coronary artery disease, baseline measures of pulmonary function were examined across medical centers. Another characteristic that you might wish to investigate is age. The relevant measurements are saved on your disk in a data set called cad (Appendix B, Table B.14). Values of age are saved under the variable name age, and indicators of center are saved under center.

(a) Do you feel that it is appropriate to use one-way analysis of variance to evaluate these data? Explain.

(b) For each medical center, what are the sample mean and standard deviation for the values of age?

(c) What is the between-groups estimate of the variance?

(d) What is the within-groups estimate of the variance?

(e) Test the null hypothesis that at the time when the study was conducted, the mean ages for men at all three centers were identical. What do you conclude?

11. The data set lowbwt contains information for a sample of 100 low birth weight infants born in two teaching hospitals in Boston, Massachusetts [7] (Appendix B, Table B.7). Systolic blood pressure measurements are saved under the variable name sbp and indicators of gender—where 1 represents a male and 0 a female—under the name sex.

(a) Assuming equal variances for males and females, use the two-sample t-test to evaluate the null hypothesis that among low birth weight infants, the mean systolic blood pressure for girls is identical to that for boys.

(b) Even though there are only two populations instead of three or more, test the same null hypothesis using the one-way analysis of variance.

(c) In this text, the statement has been made that in the case of two independent samples, the F-test used in the one-way analysis of variance reduces to the two-sample t-test. Do you believe this to be true? Explain briefly.

Bibliography

[1] Allred, E. N., Bleecker, E. R., Chaitman, B. R., Dahms, T. E., Gottlieb, S. O., Hackney, J. D., Hayes, D., Pagano, M., Selvester, R. H., Walden, S. M., and Warren, J., "Acute Effects of Carbon Monoxide Exposure on Individuals with Coronary Artery Disease," *Health Effects Institute Research Report Number 25*, November 1989.

[2] Ott, L., *An Introduction to Statistical Methods and Data Analysis*, North Scituate, MA: Duxbury, 1977.

[3] Wood, P. D., Stefanick, M. L., Dreon, D. M., Frey-Hewitt, B., Garay, S. C., Williams, P. T., Superko, H. R., Fortmann, S. P., Albers, J. J., Vranizan, K. M., Ellsworth, N. M., Terry, R. B., and Haskell, W. L., "Changes in Plasma Lipids and Lipoproteins in Overweight Men During Weight Loss Through Dieting as Compared with Exercise," *The New England Journal of Medicine*, Volume 319, November 3, 1988, 1173–1179.

[4] Chase, H. P., Garg, S. K., Marshall, G., Berg, C. L., Harris, S., Jackson, W. E., and Hamman, R. E., "Cigarette Smoking Increases the Risk of Albuminuria Among Subjects with Type I Diabetes," *Journal of the American Medical Association*, Volume 265, February 6, 1991, 614–617.

[5] Fowkes, F. G. R., Housley, E., Riemersma, R. A., Macintyre, C. C. A., Cawood, E. H. H., Prescott, R. J., and Ruckley, C. V., "Smoking, Lipids, Glucose Intolerance, and Blood Pressure as Risk Factors for Peripheral Atherosclerosis Compared with Ischemic Heart Disease in the Edinburgh Artery Study," *American Journal of Epidemiology*, Volume 135, February 15, 1992, 331–340.

[6] Wheeler, J. R. C., Fadel, H., and D'Aunno, T. A., "Ownership and Performance of Outpatient Substance Abuse Treatment Centers," *American Journal of Public Health*, Volume 82, May 1992, 711–718.

[7] Leviton, A., Fenton, T., Kuban, K. C. K., and Pagano, M., "Labor and Delivery Characteristics and the Risk of Germinal Matrix Hemorrhage in Low Birth Weight Infants," *Journal of Child Neurology*, Volume 6, October 1991, 35–40.

13

Nonparametric Methods

For all the statistical tests that we have studied up to this point, the populations from which the data were sampled were assumed to be either normally distributed or approximately so. In fact, this property is necessary for the tests to be valid. Since the forms of the underlying distributions are assumed to be known and only the values of certain parameters—such as the means and the standard deviations—are not, these tests are said to be *parametric*. If the data do not conform to the assumptions made by such traditional techniques, nonparametric methods of statistical inference should be used instead. *Nonparametric techniques* make fewer assumptions about the nature of the underlying distributions. As a result, they are sometimes called *distribution-free methods*.

Nonparametric tests of hypotheses follow the same general procedure as the parametric tests that we have already studied. We begin by making some claim about the underlying populations in the form of a null hypothesis; we then calculate the value of a test statistic using the data contained in a random sample of observations. Depending on the magnitude of this statistic, we either reject or do not reject the null hypothesis.

13.1 The Sign Test

The *sign test* may be used to compare two samples of observations when the populations from which the values are drawn are not independent. In this respect, it is similar to the paired t-test. Like the t-test, it does not examine the two groups individually; instead, it focuses on the difference in values for each pair. However, it does not require that the population of differences be normally distributed. The sign test is used to evaluate the null hypothesis that in the underlying population of differences among pairs, the median difference is equal to 0.

Consider a study designed to investigate the amount of energy expended by patients with the congenital disease cystic fibrosis. We would like to compare energy expenditure while at rest for persons suffering from this disease and for healthy individuals matched to the patients on certain important characteristics. However, we do not feel comfortable in assuming that either resting energy expenditure or the differences between measurements are normally distributed. Therefore, we use the sign test to evaluate the null hypothesis that the median difference is equal to 0. For a two-sided test, the alternative hypothesis is that the median difference is not 0.

To carry out the sign test, we begin by selecting a random sample of pairs of observations. Table 13.1 shows the measurements of resting energy expenditure for samples of 13 patients with cystic fibrosis and 13 healthy individuals matched to the patients on age, sex, height, and weight [1]. Using these values, we calculate the difference for each pair of observations. If the difference is greater than 0, the pair is assigned a plus sign; if it is less than 0, it receives a minus sign. Differences of exactly 0 provide no information about which individual in the pair has a higher resting energy expenditure, and are thus excluded from the analysis. When differences are excluded, the sample size n must be reduced accordingly.

We next count the number of plus signs in the sample; this total is denoted by D. Under the null hypothesis that the median difference is equal to 0, we would expect to have approximately equal numbers of plus and minus signs. Equivalently, the probability that a particular difference is positive is $1/2$, and the probability that it is negative is also $1/2$. If a plus sign is considered to be a "success," the n plus and minus signs can be thought of as the outcomes of a Bernoulli random variable with probability of success $p = 0.5$. The total number of plus signs D is a binomial random variable with parameters n and p. Therefore, the mean number of plus signs in a sample of size n is $np = n/2$, and the standard deviation is $\sqrt{np(1-p)} = \sqrt{n/4}$.

TABLE 13.1

Resting energy expenditure (REE) for patients with cystic fibrosis and healthy individuals matched on age, sex, height, and weight

Pair	REE (kcal/day)		Difference	Sign
	CF	**Healthy**		
1	1153	996	157	+
2	1132	1080	52	+
3	1165	1182	−17	−
4	1460	1452	8	+
5	1634	1162	472	+
6	1493	1619	−126	−
7	1358	1140	218	+
8	1453	1123	330	+
9	1185	1113	72	+
10	1824	1463	361	+
11	1793	1632	161	+
12	1930	1614	316	+
13	2075	1836	239	+

If D is either much larger or much smaller than $n/2$, we would want to reject H_0. We evaluate the null hypothesis by considering the test statistic

$$z_+ = \frac{D - (n/2)}{\sqrt{n/4}}.$$

If the null hypothesis is true and the sample size n is large, z_+ follows an approximate normal distribution with mean 0 and standard deviation 1. This test is called the sign test because it depends only on the signs of the calculated differences, not on their actual magnitudes.

For the data in Table 13.1, there are $D = 11$ plus signs. In addition,

$$\frac{n}{2} = \frac{13}{2}$$
$$= 6.5,$$

and

$$\sqrt{\frac{n}{4}} = \sqrt{\frac{13}{4}}$$
$$= \sqrt{3.25}$$
$$= 1.80.$$

Therefore,

$$z_+ = \frac{D - (n/2)}{\sqrt{n/4}}$$
$$= \frac{11 - 6.5}{1.80}$$
$$= 2.50.$$

The area under the standard normal curve to the right of $z = 2.50$ and to the left of $z = -2.50$ is $p = 2(0.006) = 0.012$; since p is less than 0.05, we reject the null hypothesis and conclude that the median difference among pairs is not equal to 0. Because most of the differences are positive, we can infer that resting energy expenditure is higher among persons with cystic fibrosis than it is among healthy individuals. This could be due to a number of factors, including differences in metabolism and the increased effort required to breathe.

If the sample size n is small, less than about 20, the test statistic z_+ cannot always be assumed to have a standard normal distribution. In this case, we use a different procedure to evaluate H_0. Recall that under the null hypothesis, D is a binomial random variable with parameters n and $p = 1/2$. Therefore, we can use the binomial distribution itself to calculate the probability of observing D positive differences—or some number more extreme—given that H_0 is true.

For the resting energy expenditure data in Table 13.1, we found $D = 11$ plus signs. Under the null hypothesis that the median difference is equal to 0, we would expect only $13/2$, or 6.5. The probability of observing 11 or more plus signs is

$$P(D \geq 11) = P(D = 11) + P(D = 12) + P(D = 13)$$

$$= \binom{13}{11}(0.5)^{11}(0.5)^{13-11} + \binom{13}{12}(0.5)^{12}(0.5)^{13-12}$$

$$+ \binom{13}{13}(0.5)^{13}(0.5)^{13-13}$$

$$= 0.0095 + 0.0016 + 0.0001$$

$$= 0.0112.$$

This is the p-value of the one-sided test; we are considering only the case in which D is larger than 6.5. The p-value of the corresponding two-sided test is approximately $2(0.0112) = 0.0224$. Once again, we would reject the null hypothesis at the 0.05 level of significance and conclude that resting energy expenditure is higher among patients with cystic fibrosis.

13.2   The Wilcoxon Signed-Rank Test

Although the sign test frees us from having to make any assumptions about the underlying distribution of differences, it also ignores some potentially useful information: the magnitude of these differences. As a result, the sign test is not often used in practice. Instead, the *Wilcoxon signed-rank test* can be used to compare two samples from populations that are not independent. Like the sign test, it does not examine the two groups individually; instead, it focuses on the difference in values for each pair of observations. It does not require that the population of differences be normally distributed. However, it does take into account the magnitudes of the differences as well as their signs. The Wilcoxon signed-rank test is used to evaluate the null hypothesis that in the underlying population of differences among pairs, the median difference is equal to 0.

Suppose that we would now like to investigate the use of the drug amiloride as a therapy for patients with cystic fibrosis. It is believed that this drug may help to improve air flow in the lungs and thereby delay the loss of pulmonary function often associated with the disease. Forced vital capacity (FVC) is the volume of air that a person can expel from the lungs in 6 seconds; we would like to compare the reduction in FVC that occurs over a 25-week period of treatment with the drug to the reduction that occurs in the same patients over a similar period of time during treatment with a placebo. However, we are not willing to assume that differences in reduction in FVC are normally distributed.

To conduct the Wilcoxon signed-rank test, we proceed as follows. We begin by selecting a random sample of n pairs of observations. Table 13.2 shows the measurements of FVC reduction for a sample of 14 patients with cystic fibrosis [2]. We next calculate the difference for each pair of observations and, ignoring the signs of these differences, rank their absolute values from smallest to largest. A difference of 0 is not ranked; it is eliminated from the analysis, and the sample size is reduced by 1 for each pair eliminated. Tied observations are assigned an average rank. If the two smallest differences had both taken the value 11, for instance, then each would have received a rank of $(1 + 2)/2 = 1.5$. Finally, we assign each rank either a plus or a minus sign

TABLE 13.2
Reduction in forced vital capacity (FVC) for a sample of patients with cystic fibrosis

Subject	Reduction in FVC (ml)		Difference	Rank	Signed Rank	
	Placebo	**Drug**				
1	224	213	11	1	1	
2	80	95	−15	2		−2
3	75	33	42	3	3	
4	541	440	101	4	4	
5	74	−32	106	5	5	
6	85	−28	113	6	6	
7	293	445	−152	7		−7
8	−23	−178	155	8	8	
9	525	367	158	9	9	
10	−38	140	−178	10		−10
11	508	323	185	11	11	
12	255	10	245	12	12	
13	525	65	460	13	13	
14	1023	343	680	14	14	
					86	−19

depending on the sign of the difference. For example, the difference in Table 13.2 that has the second smallest absolute value is −15; therefore, this observation receives rank 2. Because the difference itself is negative, however, the observation's signed rank is −2.

The next step in the test is to compute the sum of the positive ranks and the sum of the negative ranks. Ignoring the signs, we denote the smaller sum by T. Under the null hypothesis that the median of the underlying population of differences is equal to 0, we would expect the sample to have approximately equal numbers of positive and negative ranks. Furthermore, the sum of the positive ranks should be comparable in magnitude to the sum of the negative ranks. We evaluate this hypothesis by considering the statistic

$$z_T = \frac{T - \mu_T}{\sigma_T},$$

where

$$\mu_T = \frac{n(n + 1)}{4}$$

is the mean sum of the ranks and

$$\sigma_T = \sqrt{\frac{n(n+1)(2n+1)}{24}}$$

is the standard deviation [3]. If H_0 is true and the sample size n is large enough,

$$z_T = \frac{T - \mu_T}{\sigma_T}$$

follows an approximate normal distribution with mean 0 and standard deviation 1.

The differences in Table 13.2 are displayed in Figure 13.1 in the form of a histogram. The graph confirms that differences in reduction in FVC may not be normally distributed (although we cannot say for sure, since the sample size is small); therefore, the paired t-test would not be appropriate. Proceeding with the Wilcoxon signed-rank test, the sum of the positive ranks is 86 and the sum of the negative ranks is -19. Ignoring the signs, $T = 19$. Also,

$$\mu_T = \frac{n(n+1)}{4}$$
$$= \frac{14(14+1)}{4}$$
$$= 52.5,$$

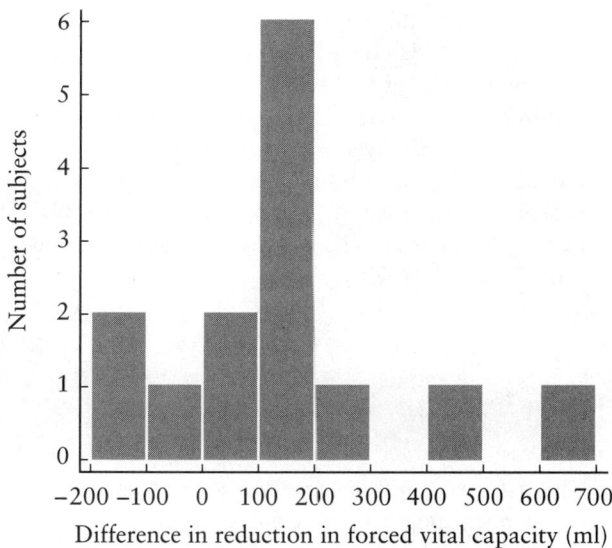

FIGURE 13.1

Differences in reduction in forced vital capacity for a sample of patients
with cystic fibrosis

and

$$\sigma_T = \sqrt{\frac{n(n+1)(2n+1)}{24}}$$

$$= \sqrt{\frac{14(14+1)[2(14)+1]}{24}}$$

$$= 15.93.$$

Solving for z_T, we find that

$$z_T = \frac{T - \mu_T}{\sigma_T}$$

$$= \frac{19 - 52.5}{15.93}$$

$$= -2.10.$$

The area under the standard normal curve to the left of $z = -2.10$ and to the right of $z = 2.10$ is $p = 2(0.018) = 0.036$. Since p is less than $\alpha = 0.05$, we reject the null hypothesis and conclude that the median difference is not equal to 0. Most of the differences are positive; this suggests that the reduction in forced vital capacity is greater during treatment with the placebo than it is during treatment with the drug. In other words, use of the drug does diminish the loss of pulmonary function.

If n is small, the test statistic z_T cannot be assumed to follow a standard normal distribution. In this case, tables are available to help us determine whether we should reject the null hypothesis. Table A.6 in Appendix A displays the distribution function of the smaller sum of ranks T for samples of size n less than or equal to 12. The possible values of T, represented by T_0, are listed down the left-hand side of the table; the sample sizes are displayed across the top. For each combination of T_0 and n, the entry in the table is the probability that T is less than or equal to T_0. If $n = 8$, for instance, the probability that T is less than or equal to 5 is 0.0391. This is the p-value of the one-sided test of hypothesis. The p-value of the appropriate two-sided test is approximately $2(0.0391) = 0.0782$.

13.3   The Wilcoxon Rank Sum Test

The *Wilcoxon rank sum test* is used to compare two samples that have been drawn from independent populations. Consequently, it is a nonparametric counterpart of the two-sample t-test. Unlike the t-test, it does not require that the underlying populations be normally distributed or that their variances be equal. It does, however, assume that the distributions have the same general shape. The Wilcoxon rank sum test evaluates the null hypothesis that the medians of the two populations are identical.

Consider the distributions of normalized mental age scores for two populations of children suffering from phenylketonuria (PKU). Individuals with this disorder are unable to metabolize the protein phenylalanine. It has been suggested that an elevated level of serum phenylalanine increases a child's likelihood of mental deficiency. The members of the first group have average daily serum phenylalanine levels below 10.0 mg/dl; those in the second group have average levels above 10.0 mg/dl. We would like to compare normalized mental age scores for these two populations of children. However, we are not willing to assume that normalized mental age scores are normally distributed in patients with this disorder.

To carry out the Wilcoxon rank sum test, we begin by selecting an independent random sample from each of the populations of interest. Table 13.3 displays samples taken from the two populations of children with PKU; there are 21 children with low exposure and 18 children with high exposure [4]. We next combine the two samples into one large group, order the observations from smallest to largest, and assign a rank to each one. If there are tied observations, we assign an average rank to all measurements with the same value. Note, for instance, that two of the children in the sample have a

TABLE 13.3

Normalized mental age scores (nMA) for two samples of children suffering from phenylketonuria

Low Exposure (< 10.0 mg/dl)		High Exposure (≥10.0 mg/dl)	
nMA (mos)	Rank	nMA (mos)	Rank
34.5	2.0	28.0	1.0
37.5	6.0	35.0	3.0
39.5	7.0	37.0	4.5
40.0	8.0	37.0	4.5
45.5	11.5	43.5	9.0
47.0	14.5	44.0	10.0
47.0	14.5	45.5	11.5
47.5	16.0	46.0	13.0
48.7	19.5	48.0	17.0
49.0	21.0	48.3	18.0
51.0	23.0	48.7	19.5
51.0	23.0	51.0	23.0
52.0	25.5	52.0	25.5
53.0	28.0	53.0	28.0
54.0	31.5	53.0	28.0
54.0	31.5	54.0	31.5
55.0	34.5	54.0	31.5
56.5	36.0	55.0	34.5
57.0	37.0		313.0
58.5	38.5		
58.5	38.5		
	467.0		

normalized mental age score of 37.0 months. Since these observations are fourth and fifth in the ordered list of 39 measurements, we assign an average rank of $(4 + 5)/2 = 4.5$ to each one. Similarly, three subjects have a normalized mental age score of 51.0 months; these observations each receive a rank of $(22 + 23 + 24)/3 = 23$.

The next step in the test is to find the sum of the ranks corresponding to each of the original samples. The smaller of the two sums is denoted by W. Under the null hypothesis that the underlying populations have identical medians, we would expect the ranks to be distributed randomly between the two groups. Therefore, the average ranks for each of the samples should be approximately equal. We test this hypothesis by calculating the statistic

$$z_W = \frac{W - \mu_W}{\sigma_W},$$

where

$$\mu_W = \frac{n_S(n_S + n_L + 1)}{2}$$

is the mean sum of the ranks and

$$\sigma_W = \sqrt{\frac{n_S n_L (n_S + n_L + 1)}{12}}$$

is the standard deviation of W [3]. In these equations, n_S represents the number of observations in the sample that has the smaller sum of ranks and n_L the number of observations in the sample with the larger sum. For large values of n_S and n_L,

$$z_W = \frac{W - \mu_W}{\sigma_W}$$

follows an approximate standard normal distribution, assuming that the null hypothesis is true.

Using the data in Table 13.3, the values of normalized mental age score for each phenylalanine exposure group are shown in Figure 13.2. While the scores may not be normally distributed—they appear to be skewed to the left—the histograms do have the same basic shape. The sum of the ranks in the low exposure group is 467, and the sum in the high exposure group is 313; therefore, $W = 313$. In addition,

$$\mu_W = \frac{n_S(n_S + n_L + 1)}{2}$$

$$= \frac{18(18 + 21 + 1)}{2}$$

$$= 360,$$

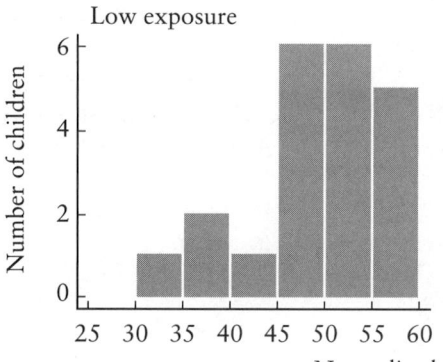 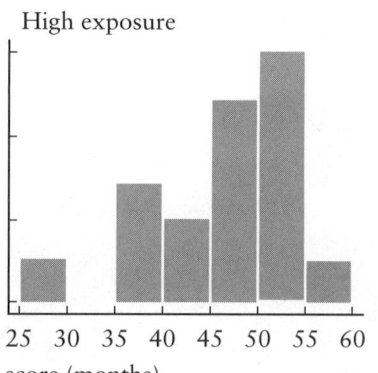

FIGURE 13.2
Normalized mental age scores for two samples of children suffering from phenylalaninemia

and

$$\sigma_W = \sqrt{\frac{n_S n_L (n_S + n_L + 1)}{12}}$$

$$= \sqrt{\frac{18(21)(18 + 21 + 1)}{12}}$$

$$= 35.5.$$

Substituting these values into the equation for the test statistic, we have

$$z_W = \frac{W - \mu_W}{\sigma_W}$$

$$= \frac{313 - 360}{35.5}$$

$$= -1.32.$$

Since $p = 2(0.093) = 0.186$ is greater than 0.05, we do not reject the null hypothesis. The samples do not provide evidence of a difference in median normalized mental age scores for the two populations; children with high serum phenylalanine exposure have achieved a level of mental functioning that is comparable to the level for children with low exposure.

If n_S and n_L are very small, z_W cannot always be assumed to follow a standard normal distribution. Table A.7 in Appendix A shows the distribution functions of the smaller sum of ranks W for sample sizes n_S and n_L that are each less than or equal to 10. The possible values of W, denoted by W_0, are displayed on the left-hand sides of the tables; values of n_1 and n_2 appear across the top. In this case, n_2 is the larger of the two sample sizes n_S and n_L, and n_1 is the smaller. For each combination of W_0, n_1, and n_2, the entry in the table corresponds to the probability that W is less than or equal to W_0. Suppose that $n_S = 5$ and $n_L = 4$. Therefore, $n_1 = 4$ and $n_2 = 5$. The probability that W is less than or equal to 13 is 0.0556. This is the p-value of the one-sided test of hypothesis; the p-value of the corresponding two-sided test is $2(0.0556) = 0.1112$.

13.4 Advantages and Disadvantages of Nonparametric Methods

Nonparametric techniques have several advantages over traditional methods of statistical inference. One advantage is that they do not incorporate all the restrictive assumptions characteristic of parametric tests. They do not require that the underlying populations be normally distributed, for instance. At most, the populations should have the same basic shape. In addition, because nonparametric tests deal with ranks rather than with the actual values of the observations, they can be performed relatively quickly for small samples. Their use of ranks makes nonparametric techniques less sensitive to measurement error than traditional tests, and also permits the use of ordinal rather than continuous data. Since it does not make sense to calculate either a mean or a standard deviation for ordinal values, parametric tests are usually not appropriate.

Nonparametric methods also have a number of disadvantages. If the assumptions underlying a parametric test are satisfied, the nonparametric test is less powerful than the comparable parametric technique. This means that if the null hypothesis is false, the nonparametric test would require a larger sample to provide sufficient evidence to reject it. If the data do come from an underlying normal population, the power of the Wilcoxon tests is approximately 95% of that of the t-tests; if the t-test requires 19 observations to achieve a particular power, the Wilcoxon test requires 20 observations to achieve the same power. In addition, the hypotheses tested by nonparametric techniques tend to be less specific than those tested by traditional methods. Because they rely on ranks rather than on the actual values of the observations, nonparametric tests do not use everything that is known about a distribution. This, of course, presumes that our information about the underlying population is correct. Finally, if a large proportion of the observations are tied, σ_T and σ_W overestimate the standard deviations of T and W, respectively. To compensate for this, a correction term must be added to the calculations [3].

13.5 Further Applications

A study was conducted to investigate the use of extracorporeal membrane oxygenation (ECMO)—a mechanical system for oxygenating the blood—in the treatment of newborns with neonatal respiratory failure. It is thought that the use of this procedure may reduce the output of an infant's left ventricle, thereby decreasing the amount of blood pumped to the body. Thus, we would like to compare left ventricular dimension before and during the use of ECMO. We are not willing to assume that the population of differences in left ventricular dimension is normally distributed; therefore, we use the Wilcoxon signed-rank test to evaluate the null hypothesis that the median difference is equal to 0.

Table 13.4 shows the relevant data for a sample of 15 infants suffering from respiratory distress [5]. We begin by calculating the difference for each pair of observations and then, ignoring the signs of these differences, we rank their absolute values from smallest to largest. Tied observations are assigned an average rank. Since a difference of 0 is not ranked and there are four of these in the data set, the sample size is reduced to $n = 11$. Each rank is then assigned either a plus or a minus sign depending on the sign of the difference itself.

If we ignore the signs, the smaller sum of ranks is $T = 18.5$. In addition, the mean sum of the ranks is

$$\mu_T = \frac{n(n + 1)}{4}$$
$$= \frac{11(11 + 1)}{4}$$
$$= 33,$$

TABLE 13.4

Left ventricular dimension (LVD) for a sample of infants suffering from neonatal respiratory failure

Subject	Before ECMO	During ECMO	Difference	Rank	Signed Rank	
1	1.6	1.6	0.0	—		
2	2.0	2.0	0.0	—		
3	1.2	1.2	0.0	—		
4	1.6	1.6	0.0	—		
5	1.6	1.5	0.1	2.5	2.5	
6	1.7	1.6	0.1	2.5	2.5	
7	1.6	1.5	0.1	2.5	2.5	
8	1.6	1.7	−0.1	2.5		−2.5
9	1.6	1.4	0.2	5.5	5.5	
10	1.7	1.5	0.2	5.5	5.5	
11	1.0	1.3	−0.3	8.0		−8.0
12	1.5	1.8	−0.3	8.0		−8.0
13	1.5	1.8	−0.3	8.0		−8.0
14	1.4	1.8	−0.4	10.0		−10.0
15	1.5	2.0	−0.5	11.0	18.5	−11.0
						−47.5

LVD (centimeters): Before ECMO, During ECMO columns.

and the standard deviation is

$$\sigma_T = \sqrt{\frac{n(n + 1)(2n + 1)}{24}}$$

$$= \sqrt{\frac{11(11 + 1)(22 + 1)}{24}}$$

$$= 11.25.$$

Therefore, we find the test statistic to be

$$z_T = \frac{T - \mu_T}{\sigma_T}$$

$$= \frac{18.5 - 33}{11.25}$$

$$= -1.29.$$

The p-value of the test is 2(0.099), or 0.198. Since p is greater than 0.05, we do not reject the null hypothesis; the sample fails to provide evidence that the median difference is not equal to 0. Treatment with ECMO appears to have no effect on the output of the infants' left ventricles.

For the data in Table 13.4, we end up with a sample of size 11. Since n is small, we cannot be sure that z_T has a standard normal distribution. Recall that $T = 18.5$; to be conservative, we round T up to 19. Consulting Table A.6, we see that the p-value of the two-sided test is 2(0.1202) = 0.2404. Again, we would fail to reject the null hypothesis.

Rather than work through the calculations ourselves, we could use a computer to conduct the signed-rank test for us. The relevant Minitab output is shown in Table 13.5. The output includes the null and alternative hypotheses for the two-sided test, the sample size, the sum of the positive ranks, and the corresponding p-value (calculated by incorporating a small correction factor in the test statistic z_T). Note that the estimated median of the differences is only slightly below 0.

Emphysema is a swelling of the air sacs in the lungs that is characterized by labored breathing and an increased susceptibility to infection. Carbon monoxide diffusing capacity, denoted Dl_{CO}, is a measure of lung function that has been tested as a possible diagnostic tool for detecting emphysema. Consider the distributions of CO-diffusing capacity for the population of healthy individuals and for the population of

TABLE 13.5

Minitab output displaying the Wilcoxon signed-rank test

TEST OF MEDIAN = 0.0000 VERSUS MEDIAN N.E. 0.0000

N	N FOR TEST	WILCOXON STATISTIC	P-VALUE	ESTIMATED MEDIAN
15	11	18.5	0.213	−0.07500

patients with emphysema. We are not willing to assume that these distributions are necessarily normal. Therefore, using a two-sided test conducted at the $\alpha = 0.05$ level of significance, we evaluate the null hypothesis that the two populations have the same median Dl_{CO} by means of the Wilcoxon rank sum test.

Table 13.6 shows the data for random samples of 13 individuals who have emphysema and 23 who do not [6]. The observations are combined into one large group and ordered from smallest to largest; a rank is assigned to each one. Separating the ranks according to the original samples, the smaller sum of ranks corresponds to the sample of persons who have emphysema. Therefore, $W = 168$. In addition,

$$\mu_W = \frac{n_S(n_S + n_L + 1)}{2}$$

$$= \frac{13(13 + 23 + 1)}{2}$$

$$= 240.5$$

TABLE 13.6

Carbon monoxide diffusing capacity for samples of individuals with and without emphysema

Emphysema		No Emphysema	
Dl_{CO}	Rank	Dl_{CO}	Rank
7.51	2	6.19	1
10.81	3	12.11	5
11.75	4	14.12	8
12.59	6	15.50	11
13.47	7	15.52	12
14.18	9	16.56	13
15.25	10	17.06	14
17.40	15	19.59	18
17.75	16	20.21	19
19.13	17	20.35	20
20.93	21	21.05	22
25.73	28	21.41	23
26.16	30	23.39	24
	168	23.60	25
		24.05	26
		25.59	27
		25.79	29
		26.29	31
		29.60	32
		30.88	33
		31.42	34
		32.66	35
		36.16	36
			498

is the mean sum of the ranks, and

$$\sigma_W = \sqrt{\frac{n_S n_L (n_S + n_L + 1)}{12}}$$

$$= \sqrt{\frac{13(23)(13 + 23 + 1)}{12}}$$

$$= 30.36$$

is the standard deviation of W. Under the null hypothesis that the two underlying populations have identical medians, the test statistic

$$z_W = \frac{W - \mu_W}{\sigma_W}$$

$$= \frac{168 - 240.5}{30.36}$$

$$= -2.39$$

follows an approximate standard normal distribution. The p-value of the test is $2(0.008) = 0.016$. Therefore, we reject the null hypothesis at the 0.05 level of significance. The samples do provide evidence that the median CO-diffusing capacity of the population of individuals who have emphysema is different from the median of the population of those who do not. In general, people suffering from emphysema have lower CO-diffusing capacities.

Again, we can perform the calculations using a computer. Sample output from Stata appears in Table 13.7. The top portion of the output displays the sum of the ranks for each of the two samples, as well as the mean (expected) sum of ranks. The bottom part lists the null hypothesis, the test statistic z_W, and the p-value of the two-sided test. Stata also provides the variance of the smaller sum of ranks, and includes an adjustment for ties if necessary.

TABLE 13.7

Stata output illustrating the Wilcoxon rank sum test

```
Two-sample Wilcoxon rank-sum (Mann-Whitney) test
```

group	obs	rank sum	expected
1	13	168	240.5
2	23	498	425.5
combined	36	666	666

```
         unadjusted variance        921.92
         adjustment for ties          0.00
           adjusted variance        921.91

   H₀:   dlco(emph==0)   =   dlco(emph==1)
                     z   =            -2.388
           Prob > |z|    =            0.0170
```

13.6   Review Exercises

1. How do nonparametric tests differ from parametric ones?

2. What are the advantages and disadvantages of using the sign test to analyze paired observations?

3. How does the Wilcoxon signed-rank test improve on the sign test?

4. How do the assumptions of the Wilcoxon rank sum test differ from those underlying the two-sample *t*-test?

5. What are the advantages and disadvantages of using ranks rather than continuous measurements to conduct tests of hypotheses?

6. Consult the resting energy expenditure data for patients with cystic fibrosis and healthy individuals matched to the patients on age, sex, height, and weight that were presented in Table 13.1.
   (a) Using the Wilcoxon signed-rank test, evaluate the null hypothesis that the median of the population of differences is equal to 0. What do you conclude?
   (b) Compare the results of the signed-rank test to those obtained when the sign test was used.

7. Suppose that you are interested in examining the effects of the transition from fetal to postnatal circulation among premature infants. For each of 14 healthy newborns, respiratory rate is measured at two different times—once when the infant is less than 15 days old, and again when he or she is more than 25 days old [7].

Subject	Respiratory Rate (breaths/minute)	
	Time 1	Time 2
1	62	46
2	35	42
3	38	40
4	80	42
5	48	36
6	48	46
7	68	45
8	26	40
9	48	42
10	27	40
11	43	46
12	67	31
13	52	44
14	88	48

(a) Using the sign test, evaluate the null hypothesis that the median difference in respiratory rates for the two times is equal to 0.

(b) Evaluate the same hypothesis using the Wilcoxon signed-rank test.

(c) Do you reach the same conclusion in each case?

8. Nineteen individuals with asthma were enrolled in a study investigating the respiratory effects of sulphur dioxide. During the study, two measurements were obtained for each subject. The first is the increase in specific airway resistance (SAR)—a measure of bronchoconstriction—from the time when the individual is at rest until after he or she has been exercising for 5 minutes; the second is the increase in SAR for the same subject after he or she has undergone a similar exercise test conducted in an atmosphere of 0.25 ppm sulphur dioxide [8].

	Increase in SAR (cm $H_2O \times$ sec)	
Subject	Air	SO_2
1	0.82	0.72
2	0.86	1.05
3	1.86	1.40
4	1.64	2.30
5	12.57	13.49
6	1.56	0.62
7	1.28	2.41
8	1.08	2.32
9	4.29	8.19
10	1.37	6.33
11	14.68	19.88
12	3.64	8.87
13	3.89	9.25
14	0.58	6.59
15	9.50	2.17
16	0.93	9.93
17	0.49	13.44
18	31.04	16.25
19	1.66	19.89

(a) At the $\alpha = 0.05$ level of significance, test the null hypothesis that the median difference in increase in specific airway resistance for the two occasions is equal to 0. What do you conclude?

(b) Do you feel that it would have been appropriate to use the paired t-test to evaluate these data? Why or why not?

9. Suppose that you have a set of 12 measurements, 6 drawn from one population and 6 from another. The two populations are not independent. You wish to use the signed-rank test to evaluate the null hypothesis that in the underlying population of differences, the median difference is equal to 0.

(a) If $T = 2$, what is the p-value of the two-sided test?

(b) Suppose that the sum of the positive ranks is 15 and the sum of the negative ranks is -6. What is the p-value of the two-sided test?

10. The following data are taken from a study that compares adolescents who have bulimia to healthy adolescents with similar body compositions and levels of physical activity. The data consist of measures of daily caloric intake for random samples of 23 bulimic adolescents and 15 healthy ones [9].

Daily Caloric Intake (kcal/kg)				
Bulimic			**Healthy**	
15.9	18.9	25.1	20.7	30.6
16.0	19.6	25.2	22.4	33.2
16.5	21.5	25.6	23.1	33.7
17.0	21.6	28.0	23.8	36.6
17.6	22.9	28.7	24.5	37.1
18.1	23.6	29.2	25.3	37.4
18.4	24.1	30.9	25.7	40.8
18.9	24.5		30.6	

(a) Test the null hypothesis that the median daily caloric intake of the population of individuals suffering from bulimia is equal to the median caloric intake of the healthy population. Conduct a two-sided test at the 0.05 level of significance.

(b) Do you believe that adolescents with bulimia require a lower daily caloric intake than do healthy adolescents?

11. The characteristics of low birth weight children dying of sudden infant death syndrome were examined for both females and males. The ages at time of death for samples of 11 girls and 16 boys are shown below [10].

Age (days)		
Females	**Males**	
53	46	115
56	52	133
60	58	134
60	59	167
78	77	175
87	78	
102	80	
117	81	
134	84	
160	103	
277	114	

(a) Test the null hypothesis that the median ages at death are identical for the two populations. What do you conclude?

(b) Do you feel that it would have been appropriate to use the two-sample t-test to analyze these data? Why or why not?

12. Suppose that you have two independent populations and wish to use the rank sum test to evaluate the null hypothesis that their medians are identical. You select a sample of size 4 from the first population and a sample of size 5 from the second.
    (a) If $W = 16$, what is the p-value of the two-sided test?
    (b) If the sum of the ranks in the first sample is 12 and the sum in the second sample is 33, what is the p-value of the two-sided test?

13. The numbers of community hospital beds per 1000 population that are available in each state in the United States and in the District of Columbia are saved in the data set bed [11] (Appendix B, Table B.13). The values for 1980 are saved under the variable name bed80, and those for 1986 are saved under bed86. The data set bed2 contains the same information in a different format. The numbers of beds per 1000 population for both calendar years are saved under the variable name bed, and an indicator of year under the name year. In Chapter 11, these data were analyzed by means of the t-test. The same information will now be used in a nonparametric analysis.
    (a) Construct a pair of box plots for the numbers of community hospital beds per 1000 population in 1980 and in 1986.
    (b) Use the Wilcoxon signed-rank test to determine whether the median difference in the number of beds is equal to 0.
    (c) Compare the median number of hospital beds in 1980 to the median number in 1986 using the Wilcoxon rank sum test.
    (d) Comment on the differences between the two Wilcoxon tests. Do you reach the same conclusion in each case? Which was the appropriate test to use in this situation?
    (e) Compare the results obtained using the nonparametric techniques to those obtained with the parametric t-tests in Chapter 11. Consider the underlying assumptions of the tests.

14. A study was conducted to evaluate the effectiveness of a work site health promotion program in reducing the prevalence of cigarette smoking. Thirty-two work sites were randomly assigned either to implement the health program or to make no changes for a period of two years. The promotion program consisted of health education classes combined with a payroll-based incentive system. The data collected during the study are saved in the data set program [12] (Appendix B, Table B.15). For each work site, smoking prevalence at the start of the study is saved under the variable name baseline, and smoking prevalence at the end of the two-year period under the name followup. The variable group contains the value 1 for work sites that implemented the health program and 2 for the sites that did not.
    (a) For the work sites that implemented the health promotion program, test the null hypothesis that the median difference in smoking prevalence over the two-year period is equal to 0.
    (b) Test the same null hypothesis for the sites that did not make any changes.
    (c) Evaluate the null hypothesis that the median difference in smoking prevalence over the two-year period for work sites that implemented the program is equal to the median difference for sites that did not.
    (d) Do you believe that the health promotion program was effective in reducing the prevalence of smoking? Explain.

15. The data set `lowbwt` contains measurements for a sample of 100 low birth weight infants born in two teaching hospitals in Boston, Massachusetts [13] (Appendix B, Table B.7). The values of Apgar score—an index of neonatal asphyxia, or oxygen deprivation—recorded five minutes after birth are saved under the variable name `apgar5`. The Apgar score is an ordinal random variable that takes values between 0 and 10. Indicators of gender, where 1 represents a male and 0 a female, are saved under the name `sex`.

(a) Construct a pair of box plots of five-minute Apgar score for males and females.

(b) Test the null hypothesis that among low birth weight infants, the median Apgar score for boys is equal to the median score for girls. What do you conclude?

16. A study was conducted to determine whether women who do not have health insurance coverage are less likely to be screened for breast cancer, and whether their disease is more advanced at the time of diagnosis [14]. The medical records for a sample of women who were privately insured and for a sample who were uninsured were examined. The stage of breast cancer at diagnosis was assigned a number between 1 and 5, where 1 denotes the least advanced disease and 5 the most advanced. The relevant observations are saved in a data set called `insure` [14] (Appendix B, Table B.16); the stage of disease is saved under the variable name `stage`, and an indicator of group status—which takes the value 0 for women who were uninsured and 1 for those who were privately insured—under the name `group`.

(a) Could the two-sample *t*-test be used to analyze these data? Why or why not?

(b) Test the null hypothesis that the median stage of cancer for the privately insured women is identical to the median stage of cancer for the uninsured women.

(c) Do these data suggest that uninsured women have more advanced disease than insured women at the time of diagnosis?

Bibliography

[1] Spicher, V., Roulet, M., and Schutz, Y., "Assessment of Total Energy Expenditure in Free-Living Patients with Cystic Fibrosis," *Journal of Pediatrics*, Volume 118, June 1991, 865–872.

[2] Knowles, M. R., Church, N. L., Waltner, W. E., Yankaskas, J. R., Gilligan, P., King, M., Edwards, L. J., Helms, R. W., and Boucher, R. C., "A Pilot Study of Aerosolized Amiloride for the Treatment of Lung Disease in Cystic Fibrosis," *The New England Journal of Medicine*, Volume 322, April 26, 1990, 1189–1194.

[3] Hollander, M., and Wolfe, D. A., *Nonparametric Statistical Methods*, New York: Wiley, 1973.

[4] Wrona, R. M., "A Clinical Epidemiologic Study of Hyperphenylalaninemia," *American Journal of Public Health*, Volume 69, July 1979, 673–679.

[5] Burch, K. D., Covitz, W., Lovett, E. J., Howell, C., and Kanto, W. P., "The Significance of Ductal Shunting During Extracorporeal Membrane Oxygenation," *Journal of Pediatric Surgery*, Volume 24, September 1989, 855–859.

[6] Morrison, N. J., Abboud, R. T., Ramadan, F., Miller, R. R., Gibson, N. N., Evans, K. G., Nelems, B., and Müller, N. L., "Comparison of Single Breath Carbon Monoxide Diffusing

Capacity and Pressure-Volume Curves in Detecting Emphysema," *American Review of Respiratory Disease*, Volume 139, May 1989, 1179–1187.

[7] Lee, L. A., Kimball, T. R., Daniels, S. R., Khoury, P., and Meyer, R. A., "Left Ventricular Mechanics in the Preterm Infant and their Effect on the Measurement of Cardiac Performance," *The Journal of Pediatrics*, Volume 120, January 1992, 114–119.

[8] Bethel, R. A., Sheppard, D., Geffroy, B., Tam, E., Nadel, J. A., and Boushey, H. A., "Effect of 0.25 ppm Sulphur Dioxide on Airway Resistance in Freely Breathing, Heavily Exercising, Asthmatic Subjects," *American Review of Respiratory Disease*, Volume 31, April 1985, 659–661.

[9] Gwirtsman, H. E., Kaye, W. H., Obarzanek, E., George, D. T., Jimerson, D. C., and Ebert, M. H., "Decreased Caloric Intake in Normal-Weight Patients with Bulimia: Comparison with Female Volunteers," *American Journal of Clinical Nutrition*, Volume 49, January 1989, 86–92.

[10] Walker, A. M., Jick, H., Perera, D. R., Thompson, R. S., and Knauss, T. A., "Diphtheria–Tetanus–Pertussis Immunization and Sudden Infant Death Syndrome," *American Journal of Public Health*, Volume 77, August 1987, 945–951.

[11] National Center for Health Statistics, *Health United States 1988*, Hyattsville, MD: Public Health Service, March 1989.

[12] Jeffery, R. W., Forster, J. L., French, S. A., Kelder, S. H., Lando, H. A., McGovern, P. G., Jacobs, D. R., and Baxter, J. E., "The Healthy Worker Project: A Work-Site Intervention for Weight Control and Smoking Cessation," *American Journal of Public Health*, Volume 83, March 1993, 395–401.

[13] Leviton, A., Fenton, T., Kuban, K. C. K., and Pagano, M., "Labor and Delivery Characteristics and the Risk of Germinal Matrix Hemorrhage in Low Birth Weight Infants," *Journal of Child Neurology*, Volume 6, October 1991, 35–40.

[14] Ayanian, J. Z., Kohler, B. A., Abe, T., and Epstein, A. M., "The Relation Between Health Insurance Coverage and Clinical Outcomes Among Women with Breast Cancer," *The New England Journal of Medicine*, Volume 329, July 29, 1993, 326–331.

14
Inference on Proportions

In previous chapters, we applied the techniques of statistical inference to continuous or measured data. In particular, we investigated the properties of population means and, in the case of nonparametric methods, medians. We now extend the methodology of statistical inference to include enumerated data, or counts. The basic underlying principles remain the same, and the normal distribution again plays a key role.

When studying counts, we are usually concerned with the proportion of times that an event occurs rather than the number of times. In the mid-nineteenth century, for example, the Vienna General Hospital—the Allgemeines Krankenhaus of the University of Vienna—had two obstetrical divisions [1]. Every year, approximately 3500 babies were delivered in each division. There were two major differences between them, however. In the first division, all deliveries were supervised by obstetricians and medical students; in the second, they were overseen by midwives. Furthermore, the proportion of women who died of puerperal fever—an infection developing during childbirth—was between 0.17 and 0.23 in the first division, whereas the proportion of women who died was about 0.017 in the second division.

Ignac Semmelweiss, the assistant to the professor of obstetrics, was convinced that this tenfold difference in proportions was not due to chance alone. His research led him to conclude that the discrepancy existed because, in addition to delivering babies, the obstetricians and students dissected several cadavers per day. Since the germ theory of disease had not yet been proposed, proper hygiene was not practiced; individuals went freely from dissections to deliveries without taking sanitary precautions. Believing that this practice was the root of the problem, Semmelweiss changed the procedure. He insisted that obstetricians wash their hands in a chlorine solution before being allowed to attend a delivery. In the subsequent year, the proportions of women who died were 0.012 in the first division and 0.013 in the second. Unfortunately, Semmelweiss was ahead of his time. His conclusions were not generally accepted; in fact, his discovery caused him to lose his position. Would such a discrepancy in proportions be ignored today, or would we accept that the two divisions are in fact different? To address this issue, we investigate some aspects of the variability of proportions.

14.1   Normal Approximation to the Binomial Distrib... *n*

The binomial distribution provides a foundation for the analysis of propor... that if we have a sequence of n independent Bernoulli trials—each of which... one of two mutually exclusive outcomes, often designated "success" and "failur... each trial has a constant probability of success p, the total number of successe... binomial random variable. The probability distribution of X, represented by the for...

$$P(X = x) = \binom{n}{x} p^x (1 - p)^{n-x},$$

can be used to make statements about the possible outcomes of the random variable. In particular, we can use this expression to compute the probabilities associated with specified outcomes x.

Suppose that we select a random sample of 30 individuals from the population of adults in the United States. As we learned in Chapter 7, the probability that a member of this population currently smokes cigarettes, cigars, or a pipe is equal to 0.29 [2]; therefore, the total number of smokers in the sample is a binomial random variable with parameters $n = 30$ and $p = 0.29$. For a given sample of size 30, what is the probability that six or fewer of its members smoke? Applying the additive rule of probability, we find that

$$
\begin{aligned}
P(X \le 6) &= P(X = 0) + P(X = 1) + P(X = 2) + P(X = 3) \\
&\quad + P(X = 4) + P(X = 5) + P(X = 6) \\
&= (1)(0.29)^0(0.71)^{30} + (30)(0.29)^1(0.71)^{29} \\
&\quad + (435)(0.29)^2(0.71)^{28} + (4060)(0.29)^3(0.71)^{27} \\
&\quad + (27{,}405)(0.29)^4(0.71)^{26} + (142{,}506)(0.29)^5(0.71)^{25} \\
&\quad + (593{,}775)(0.29)^6(0.71)^{24} \\
&= 0.190.
\end{aligned}
$$

We can also calculate the probabilities associated with the outcomes of a binomial random variable X using an approximate procedure based on the normal distribution. When the sample size n is large, the binomial distribution is cumbersome to work with, and this alternative procedure can be considerably more convenient. In Chapter 7 we saw that as the sample size grows larger, the shape of a binomial distribution increasingly resembles that of a normal distribution. Furthermore, the mean number of successes per sample is np, and the variance is $np(1 - p)$. If n is sufficiently large, therefore, we can approximate the distribution of X by a normal distribution with the same mean and variance as the binomial. A widely used criterion states that n is "sufficiently large" if both np and $n(1 - p)$ are greater than or equal to 5. (Some people believe that this condition is not conservative enough, and prefer that both np and $n(1 - p)$ be greater than or equal to 10.) In this case,

$$Z = \frac{X - np}{\sqrt{np(1 - p)}}$$

is approximately normal with mean 0 and standard deviation 1.

Using the normal approximation to the binomial distribution, we would like to find the proportion of samples of size 30 in which at most six individuals are current smokers. Since $np = 30(0.29) = 8.7$ and $n(1 - p) = 30(0.71) = 21.3$ are both greater than 5, we note that

$$P(X \leq 6) = P\left(\frac{X - np}{\sqrt{np(1 - p)}} \leq \frac{6 - np}{\sqrt{np(1 - p)}}\right)$$

$$= P\left(Z \leq \frac{6 - (30)(0.29)}{\sqrt{30(0.29)(0.71)}}\right)$$

$$= P(Z \leq -1.09).$$

The area under the standard normal curve that lies to the left of $z = -1.09$ is 0.138; this is equal to the probability that at most six of the individuals smoke. Note that in this instance the normal approximation provides only a rough estimate of the exact binomial probability 0.190.

It has been shown that a better approximation to the binomial distribution can be obtained by adding 0.5 to the specified outcome x if we are interested in the probability that X is less than x, and subtracting 0.5 if we are calculating the probability that X is greater than x. If we wish to find $P(X \leq 6)$, for instance, we would replace this quantity by $P(X \leq 6 + 0.5) = P(X \leq 6.5)$; similarly, to compute $P(X \geq 6)$, we would replace it by $P(X \geq 6 - 0.5) = P(X \geq 5.5)$. The term 0.5 that appears in these expressions is known as a *continuity correction.* The continuity correction is used to compensate for the fact that the discrete binomial distribution is being approximated by a continuous normal distribution.

Return to the problem of finding the proportion of samples of size 30 in which at most six individuals are current smokers. Applying the continuity correction, we find that

$$P(X \leq 6.5) = P\left(Z \leq \frac{6.5 - (30)(0.29)}{\sqrt{30(0.29)(0.71)}}\right)$$

$$= P(Z \leq -0.89).$$

The area under the standard normal curve to the left of $z = -0.89$ is 0.187. In this example, the use of a continuity correction results in a much better approximation of the exact binomial probability 0.190.

14.2   *Sampling Distribution of a Proportion*

As we have noted, we are usually interested in estimating the proportion of times that a particular event occurs in a given population rather than the number of times. For example, we might wish to make a statement about the proportion of patients who survive five years after being diagnosed with lung cancer based on a sample drawn from this group. If the sample is of size n, and x of its members are alive five years after diagnosis, we could estimate the population proportion p by

$$\hat{p} = \frac{x}{n}.$$

The sample proportion $\hat{p}$, or p-hat, is the maximum likelihood estimator of p. Recall that this is the value of the parameter that is most likely to have produced the observed sample data. Before we use $\hat{p}$ to make inference, however, we should first investigate some of its properties.

In the population of patients who have been diagnosed with lung cancer, we could represent five-year survival by a 1 and death by a 0. The mean of this set of values is equal to the proportion of 1s in the population, or p. The standard deviation is $\sqrt{p(1-p)}$. Suppose that we randomly select a sample of size n from the population and denote the proportion of 1s in the sample by $\hat{p}_1$. Similarly, we could select a second sample of size n and denote the proportion of 1s in this new sample by $\hat{p}_2$. If we were to continue this procedure indefinitely, we would end up with a set of values consisting entirely of sample proportions. Treating each proportion in the series as a unique observation, their collective probability distribution is a sampling distribution of proportions for samples of size n. According to the central limit theorem, the distribution of sample proportions has the following properties:

1. The mean of the sampling distribution is the population mean p.

2. The standard deviation of the distribution of sample proportions is equal to $\sqrt{p(1-p)/n}$. As in the case of the mean, this quantity is known as the *standard error*.

3. The shape of the sampling distribution is approximately normal provided that n is sufficiently large.

Because the distribution of sample proportions is approximately normal with mean p and standard deviation $\sqrt{p(1-p)/n}$, we know that

$$Z = \frac{\hat{p} - p}{\sqrt{p(1-p)/n}}$$

is normally distributed with mean 0 and standard deviation 1. As a result, we can use tables of the standard normal distribution to make inference about the value of a population proportion.

For example, consider five-year survival among patients who have been diagnosed with lung cancer. The mean proportion of individuals surviving is $p = 0.10$; the standard deviation is $\sqrt{p(1-p)} = \sqrt{0.10(1-0.10)} = 0.30$ [3]. If we select repeated samples of size 50 from this population, what fraction will have a sample proportion of 0.20 or higher? It is again recommended that np and $n(1-p)$ be greater than or equal to 5. Since $np = 50(0.10) = 5$ and $n(1 - p) = 50(0.90) = 45$, the central limit theorem states that the distribution of sample proportions is approximately normal with mean $p = 0.10$ and standard error $\sqrt{p(1 - p)/n} = \sqrt{0.10(1 - 0.10)/50} = 0.0424$. Therefore,

$$P(\hat{p} \geq 0.20) = P\left(\frac{\hat{p} - p}{\sqrt{p(1 - p)/n}} \geq \frac{0.20 - p}{\sqrt{p(1 - p)/n}}\right)$$

$$= P\left(Z \geq \frac{0.20 - 0.10}{0.0424}\right)$$

$$= P(Z \geq 2.36).$$

Consulting Table A.3, we observe that the area under the standard normal curve that lies to the right of $z = 2.36$ is 0.009. Only about 0.9% of the samples will have a sample proportion of 0.20 or higher.

A continuity correction could have been applied in this example, just as it was used in the preceding section. As long as n is reasonably large, however, the effect of the correction is negligible. Therefore, we omit it here and in the remainder of this chapter.

14.3   Confidence Intervals

To construct a confidence interval for a population proportion, we follow the same procedure that we used for a population mean. We begin by drawing a sample of size n and using these observations to compute the sample proportion $\hat{p}$; $\hat{p}$ is a point estimate of p. Applying the results of the preceding section, we then note that

$$Z = \frac{\hat{p} - p}{\sqrt{p(1 - p)/n}}$$

is a normal random variable with mean 0 and standard deviation 1, assuming that n is sufficiently large. We know that for a standard normal distribution, 95% of the possible outcomes lie between -1.96 and 1.96. Therefore,

$$P\left(-1.96 \leq \frac{\hat{p} - p}{\sqrt{p(1 - p)/n}} \leq 1.96\right) = 0.95$$

and

$$P\left(\hat{p} - 1.96\sqrt{\frac{p(1 - p)}{n}} \leq p \leq \hat{p} + 1.96\sqrt{\frac{p(1 - p)}{n}}\right) = 0.95.$$

The values $\hat{p} - 1.96\sqrt{p(1 - p)/n}$ and $\hat{p} + 1.96\sqrt{p(1 - p)/n}$ are 95% confidence limits for the population proportion p. Note, however, that these quantities depend on the value of p. Since p is unknown, we must estimate it using the sample proportion $\hat{p}$. Consequently,

$$\left(\hat{p} - 1.96 \sqrt{\frac{\hat{p}(1 - \hat{p})}{n}}, \hat{p} + 1.96 \sqrt{\frac{\hat{p}(1 - \hat{p})}{n}} \right)$$

is an approximate 95% confidence interval for p. This interval has approximately a 95% chance of covering the true population proportion before a sample is selected.

Using similar reasoning, we could also construct an approximate one-sided confidence interval for p, or intervals with different levels of confidence. A generic, approximate two-sided $100\% \times (1 - \alpha)$ confidence interval for p takes the form

$$\left(\hat{p} - z_{\alpha/2} \sqrt{\frac{\hat{p}(1 - \hat{p})}{n}}, \hat{p} + z_{\alpha/2} \sqrt{\frac{\hat{p}(1 - \hat{p})}{n}} \right),$$

where $z_{\alpha/2}$ is the value that cuts off an area of $\alpha/2$ in the upper tail of the standard normal distribution.

Consider the distribution of five-year survival for individuals under 40 who have been diagnosed with lung cancer. This distribution has an unknown population mean p. In a randomly selected sample of 52 patients, only six survive five years [4]. Therefore,

$$\hat{p} = \frac{x}{n}$$

$$= \frac{6}{52}$$

$$= 0.115$$

is a point estimate for p. Since $n\hat{p} = 52(0.115) = 6.0$ and $n(1 - \hat{p}) = 52(0.885) = 46.0$, the sample size is large enough to justify the use of the normal approximation; an approximate 95% confidence interval for p is

$$\left(0.115 - 1.96 \sqrt{\frac{0.115(1 - 0.115)}{52}}, 0.115 + 1.96 \sqrt{\frac{0.115(1 - 0.115)}{52}} \right),$$

or

$(0.028, 0.202).$

While 0.115 is our best guess for the population proportion, the interval provides a range of reasonable values for p. We are 95% confident that these limits cover the true proportion of individuals under 40 who survive five years.

Although this interval was constructed using the normal approximation to the binomial distribution, we could actually have generated a confidence interval for p using the binomial distribution itself. This method is especially valuable for small samples in which the use of the normal approximation cannot be justified, and provides an exact rather than an approximate interval. Because the computations involved are considerably more complex than those for the approximate interval, we will not present the procedure here [5]; however, a number of statistical software packages use the exact method to generate confidence intervals for proportions.

14.4   Hypothesis Testing

As mentioned above, the distribution of five-year survival for individuals under 40 who have been diagnosed with lung cancer has an unknown population proportion p. We do know, however, that the proportion of patients surviving five years among those who are over 40 at the time of diagnosis is 8.2% [4]. Is it possible that the proportion surviving in the under-40 population is 0.082 as well? To determine whether this is the case, we conduct a statistical test of hypothesis.

We begin by making a claim about the value of the population proportion p. If we wish to test whether the fraction of lung cancer patients surviving at least five years after diagnosis is the same among persons under 40 as it is among those over 40, the null hypothesis is

$$H_0: p = 0.082.$$

For a two-sided test conducted at the 0.05 level of significance, the alternative hypothesis would be

$$H_A: p \neq 0.082.$$

We next draw a random sample of dichotomous observations from the underlying population and find the probability of obtaining a sample proportion as extreme or more extreme than $\hat{p}$, given that the true population mean is p. We do this by calculating the test statistic

$$z = \frac{\hat{p} - p_0}{\sqrt{p_0(1 - p_0)/n}}.$$

If n is sufficiently large and the null hypothesis is true, this ratio is normally distributed with mean 0 and standard deviation 1. Depending on its magnitude and the resulting p-value, therefore, we either reject or do not reject H_0.

For the sample of 52 persons under 40 who have been diagnosed with lung cancer, $\hat{p} = 0.115$. Therefore, the appropriate test statistic is

$$z = \frac{\hat{p} - p_0}{\sqrt{p_0(1 - p_0)/n}}$$
$$= \frac{0.115 - 0.082}{\sqrt{0.082(1 - 0.082)/52}}$$
$$= 0.87.$$

According to Table A.3, the p-value of the test is 0.384. Since this is greater than the level of significance 0.05, we do not reject the null hypothesis. These samples do not provide evidence that the proportions of lung cancer patients who survive five years beyond diagnosis differ in the two age groups.

Since the approximate 95% confidence interval for p contains the value 0.082, the confidence interval would have led us to the same conclusion. When we are dealing with proportions, however, this is not always the case. Because the standard error is calculated in different ways for confidence intervals and tests of hypothesis—the population proportion is estimated by the sample proportion $\hat{p}$ for the confidence interval and by the postulated value p_0 for the hypothesis test—the mathematical equivalence between these two methods that existed for means does not apply here.

Finally, just as we were able to construct a confidence interval for a population proportion using the binomial distribution itself rather than the normal approximation, an exact test of hypothesis is possible as well [5]. Once again, however, we will not present this technique here; we simply note that many statistical packages use the exact method to conduct hypothesis tests for proportions.

14.5   Sample Size Estimation

When designing a study, investigators often wish to determine the sample size n that will be necessary to provide a specified power. Recall that the power of a test of hypothesis is the probability that we will reject the null hypothesis given that it is false. When dealing with proportions, power calculations are a little more complex than they were for tests based on means; however, the reasoning is quite similar.

For instance, suppose that we wish to test the null hypothesis

$$H_0: p \leq 0.082$$

against the alternative

$$H_A: p > 0.082$$

at the $\alpha = 0.01$ level of significance. Once again, p is the proportion of lung cancer patients under 40 at diagnosis who survive at least five years. Although we previously conducted a two-sided test, we are now concerned only with values of p that are greater than 0.082. If the true population proportion is as large as 0.200, we want to risk only a 5% chance of failing to reject the null hypothesis; therefore, β is equal to 0.05, and the power of the test is 0.95. How large a sample would be required?

Since $\alpha = 0.01$ and we are conducting a one-sided test, we begin by noting that H_0 would be rejected for $z \geq 2.32$. Therefore, we set

$$z = 2.32$$
$$= \frac{\hat{p} - 0.082}{\sqrt{0.082(1 - 0.082)/n}}.$$

Solving for $\hat{p}$,

$$\hat{p} = 0.082 + 2.32\sqrt{\frac{0.082(1 - 0.082)}{n}}.$$

We would reject the null hypothesis if the sample proportion $\hat{p}$ is greater than this value.

We now focus on the desired power of the test. If the true proportion of patients surviving for five years is 0.200, we want to reject the null hypothesis with probability $1 - \beta = 0.95$. The value of z that corresponds to $\beta = 0.05$ is $z = -1.645$; therefore,

$$z = -1.645$$
$$= \frac{\hat{p} - 0.200}{\sqrt{0.200(1 - 0.200)/n}},$$

and, solving for $\hat{p}$,

$$\hat{p} = 0.200 - 1.645\sqrt{\frac{0.200(1 - 0.200)}{n}}.$$

Setting the two expressions for $\hat{p}$ equal to each other and solving for the sample size n,

$$n = \left[\frac{2.32\sqrt{0.082(1 - 0.082)} + 1.645\sqrt{0.200(1 - 0.200)}}{0.200 - 0.082}\right]^2$$
$$= 120.4.$$

Rounding up, a sample of size 121 would be required.

In general, if the probability of making a type I error is α and the probability of making a type II error is β, the sample size n is

$$n = \left[\frac{z_\alpha\sqrt{p_0(1 - p_0)} + z_\beta\sqrt{p_1(1 - p_1)}}{p_1 - p_0}\right]^2$$

for a one-sided test of hypothesis. Note that p_0 is the hypothesized population proportion and p_1 is the alternative. The magnitudes of these proportions—along with the values of α and β—determine the necessary sample size n. If we are interested in conducting a two-sided test, we must make an adjustment to the preceding formula. In this case, the null hypothesis would be rejected for $z \geq z_{\alpha/2}$ (and also for $z \leq -z_{\alpha/2}$). As a result, the required sample size would be

$$n = \left[\frac{z_{\alpha/2}\sqrt{p_0(1 - p_0)} + z_\beta\sqrt{p_1(1 - p_1)}}{p_1 - p_0}\right]^2.$$

14.6 Comparison of Two Proportions

Just as we did for the mean, we can again generalize the procedure of hypothesis testing to accommodate the comparison of two proportions. Most often we are interested in testing the null hypothesis that the proportions from two independent populations are identical, or

$$H_0: p_1 = p_2,$$

against the alternative

$$H_A: p_1 \neq p_2.$$

To conduct the test, we draw a random sample of size n_1 from the population with mean p_1. If there are x_1 successes in the sample, then

$$\hat{p}_1 = \frac{x_1}{n_1}.$$

Similarly, we select a sample of size n_2 from the population with mean p_2; therefore,

$$\hat{p}_2 = \frac{x_2}{n_2}.$$

In order to determine whether the observed difference in sample proportions $\hat{p}_1 - \hat{p}_2$ is too large to be attributed to chance alone, we then calculate the probability of obtaining a pair of proportions as discrepant as or more discrepant than those observed, given that the null hypothesis is true. If this probability is sufficiently small, we reject H_0 and conclude that the two population proportions are different. As always, we must specify a level of significance α before conducting the test.

If the null hypothesis is true and the population proportions p_1 and p_2 are in fact equal, the data from both samples can be combined to estimate this common parameter; in particular,

$$\hat{p} = \frac{n_1 \hat{p}_1 + n_2 \hat{p}_2}{n_1 + n_2}$$

$$= \frac{x_1 + x_2}{n_1 + n_2}.$$

The quantity $\hat{p}$ is a weighted average of the two sample proportions $\hat{p}_1$ and $\hat{p}_2$.

Under the null hypothesis, the estimator of the standard error of the difference $\hat{p}_1 - \hat{p}_2$ takes the form $\sqrt{\hat{p}(1 - \hat{p})[(1/n_1) + (1/n_2)]}$. Thus, the appropriate test statistic is

$$z = \frac{(\hat{p}_1 - \hat{p}_2) - (p_1 - p_2)}{\sqrt{\hat{p}(1 - \hat{p})[(1/n_1) + (1/n_2)]}}.$$

If n_1 and n_2 are sufficiently large, this statistic has a normal distribution with mean 0 and standard deviation 1. A commonly used though conservative criterion is that each of the quantities $n_1\hat{p}, n_1(1 - \hat{p}), n_2\hat{p}$, and $n_2(1 - \hat{p})$ be greater than or equal to 5. If these conditions are satisfied, we compare the value of the statistic to the critical values in Table A.3 to find the p-value of the test; based on the magnitude of p, we either reject or do not reject the null hypothesis.

In a study investigating morbidity and mortality among pediatric victims of motor vehicle accidents, information regarding the effectiveness of seat belts was collected over an 18-month period. Two random samples were selected, one from the population of children who were wearing a seat belt at the time of the accident, and the other from the population who were not. We would like to test

$$H_0: p_1 = p_2,$$

the null hypothesis that the proportions of children who die as a result of the accident are identical in the two populations. To do this, we conduct a two-sided test at the 0.05 level of significance; the alternative hypothesis is

$$H_A: p_1 \neq p_2.$$

In the sample of 123 children who were wearing a seat belt at the time of the accident, 3 died [6]. Therefore,

$$\hat{p}_1 = \frac{x_1}{n_1}$$
$$= \frac{3}{123}$$
$$= 0.024.$$

In the sample of 290 children who were not wearing a seat belt, 13 died. Consequently,

$$\hat{p}_2 = \frac{x_2}{n_2}$$
$$= \frac{13}{290}$$
$$= 0.045.$$

Is this discrepancy in sample proportions—a difference of 0.021, or 2.1%—too large to be attributed to chance?

If the population proportions p_1 and p_2 are in fact equal, their common value p is estimated by

$$\hat{p} = \frac{x_1 + x_2}{n_1 + n_2}$$

$$= \frac{3 + 13}{123 + 290}$$

$$= 0.039.$$

Substituting the values of $\hat{p}_1$, $\hat{p}_2$, and $\hat{p}$ into the expression for the test statistic, we find that

$$z = \frac{(\hat{p}_1 - \hat{p}_2) - (p_1 - p_2)}{\sqrt{\hat{p}(1 - \hat{p})[(1/n_1) + (1/n_2)]}}$$

$$= \frac{(0.024 - 0.045) - 0}{\sqrt{(0.039)(1 - 0.039)[(1/123) + (1/290)]}}$$

$$= -1.01.$$

According to Table A.3, the p-value of the test is 0.312. Therefore, we cannot reject the null hypothesis. The samples collected in this particular study do not provide evidence that the proportions of children dying differ between those who were wearing seat belts and those who were not.

The quantity $\hat{p}_1 - \hat{p}_2$ provides a point estimate for the true difference in population proportions; we might also wish to construct a confidence interval for this difference. Using the normal approximation, the 95% confidence limits for $p_1 - p_2$ are

$$\left(\hat{p}_1 - \hat{p}_2 \pm 1.96 \sqrt{\frac{\hat{p}_1(1 - \hat{p}_1)}{n_1} + \frac{\hat{p}_2(1 - \hat{p}_2)}{n_2}} \right).$$

Note that, once again, the standard error of the difference in proportions is not the same as that used in the test of significance. In the hypothesis test, the standard error

$$\sqrt{\hat{p}(1 - \hat{p})[(1/n_1) + (1/n_2)]}$$

was employed, based on the assumption that the null hypothesis was true. This assumption is not necessary in the calculation of a confidence interval.

Since $\hat{p}_1 = 0.024$ and $\hat{p}_2 = 0.045$, an approximate 95% confidence interval for $p_1 - p_2$ is

$$\left((0.024 - 0.045) \pm 1.96 \sqrt{\frac{0.024(1 - 0.024)}{123} + \frac{0.045(1 - 0.045)}{290}} \right),$$

or

$$(-0.057, 0.015).$$

We are 95% confident that this interval covers the true difference in the proportions of children dying in each population. Note that the interval contains the value 0; thus, it is consistent with the test of hypothesis conducted at the 0.05 level.

14.7   *Further Applications*

Suppose we are interested in investigating the cognitive abilities of children weighing less than 1500 grams at birth. Although their birth weights are extremely low, many of these children exhibit normal growth patterns during the first year of life. A small group does not. These children suffer from perinatal growth failure, a condition that prevents them from developing properly. One indicator of perinatal growth failure is that during the first several months of life, the infant has a head circumference measurement that is far below normal.

We would like to examine the relationship between perinatal growth failure and subsequent cognitive ability. In particular, we wish to estimate the proportion of children suffering from this condition who, when they reach 8 years of age, have intelligence quotient (IQ) scores that are below 70. In the general population, IQ scores are scaled to have mean 100; a score less than 70 suggests a deficiency in cognitive ability. To estimate the proportion of children with IQs in this range, a random sample of 33 infants with perinatal growth failure was chosen. At the age of 8, eight children have scores below 70 [7]. Therefore,

$$\hat{p} = \frac{x}{n}$$

$$= \frac{8}{33}$$

$$= 0.242$$

is a point estimate for p.

In addition to this point estimate, we also wish to construct a 99% confidence interval for p. Since $n\hat{p} = 33(0.242) = 8.0$ and $n(1 - \hat{p}) = 33(0.758) = 25.0$, we can use the normal approximation to the binomial distribution. Consequently, we know that

$$\left(\hat{p} - 2.58 \sqrt{\frac{\hat{p}(1 - \hat{p})}{n}}, \hat{p} + 2.58 \sqrt{\frac{\hat{p}(1 - \hat{p})}{n}} \right)$$

has a 99% chance of covering p before a sample is selected, and an approximate 99% confidence interval for p is

$$\left(0.242 - 2.58 \sqrt{\frac{0.242(1 - 0.242)}{33}}, 0.242 + 2.58 \sqrt{\frac{0.242(1 - 0.242)}{33}} \right),$$

or

$$(0.050, 0.434).$$

This interval provides a range of reasonable values for p. We are 99% confident that these limits cover the true proportion of children who suffered from perinatal growth failure as infants and subsequently have IQ scores below 70.

Although we do not know the true value of p for this population, we do know that 3.2% of the children who exhibited normal growth in the perinatal period have IQ scores below 70 when they reach school age. We would like to know whether this is also true of the children who suffered from perinatal growth failure. Since we are concerned with deviations that could occur in either direction, we conduct a two-sided test of the null hypothesis

$$H_0: p = 0.032$$

at the 0.01 level of significance. The alternative hypothesis is

$$H_A: p \neq 0.032.$$

Based on the random sample of 33 infants with perinatal growth failure, $\hat{p} = 0.242$. If the true population proportion is 0.032, what is the probability of selecting a sample with a proportion as high as 0.242? To answer this question, we calculate the test statistic

$$
\begin{aligned}
z &= \frac{\hat{p} - p}{\sqrt{p(1-p)/n}} \\
&= \frac{0.242 - 0.032}{\sqrt{0.032(1 - 0.032)/33}} \\
&= 6.85.
\end{aligned}
$$

According to Table A.3, the p-value of the test is less than 0.001. We reject the null hypothesis at the 0.01 level and conclude that, among children who experienced perinatal growth failure, the proportion having IQ scores below 70 is larger than 0.032.

In a study conducted to investigate the nonclinical factors associated with the method of surgical treatment received for early-stage breast cancer, some patients underwent a modified radical mastectomy while others had a partial mastectomy accompanied by radiation therapy. We are interested in determining whether the age of the patient affects the type of treatment she receives. In particular, we want to know whether the proportions of women under 55 are identical in the two treatment groups. Consequently, we test the null hypothesis

$$H_0: p_1 = p_2$$

against the two-sided alternative

$$H_A: p_1 \neq p_2.$$

A random sample of 658 women who underwent a partial mastectomy and subsequent radiation therapy contains 292 women under 55; a sample of 1580 women who received a modified radical mastectomy contains 397 women under 55 [8]. Therefore,

$$\hat{p}_1 = \frac{x_1}{n_1}$$

$$= \frac{292}{658}$$

$$= 0.444,$$

and

$$\hat{p}_2 = \frac{x_2}{n_2}$$

$$= \frac{397}{1580}$$

$$= 0.251.$$

If the two proportions p_1 and p_2 are identical, an estimate of their common value p is

$$\hat{p} = \frac{x_1 + x_2}{n_1 + n_2}$$

$$= \frac{292 + 397}{658 + 1580}$$

$$= 0.308.$$

Since $n_1\hat{p}$, $n_1(1 - \hat{p})$, $n_2\hat{p}$, and $n_2(1 - \hat{p})$ are all greater than 5, the test statistic is

$$z = \frac{(\hat{p}_1 - \hat{p}_2) - (p_1 - p_2)}{\sqrt{\hat{p}(1 - \hat{p})[(1/n_1) + (1/n_2)]}}$$

$$= \frac{(0.444 - 0.251) - 0}{\sqrt{(0.308)(1 - 0.308)[(1/658) + (1/1580)]}}$$

$$= 9.01.$$

According to Table A.3, the p-value of the test is less than 0.001. Therefore, we reject the null hypothesis at the 0.05 level and conclude that women who undergo a modified radical mastectomy tend to be older than those who receive a partial mastectomy and radiation therapy.

Since $\hat{p}_1 = 0.444$ and $\hat{p}_2 = 0.251$, $\hat{p}_1 - \hat{p}_2 = 0.193$ is a point estimate for the true difference in population proportions. An approximate 95% confidence interval for $p_1 - p_2$ is

$$\left((0.444 - 0.251) \pm 1.96 \sqrt{\frac{0.444(1 - 0.444)}{658} + \frac{0.251(1 - 0.251)}{1580}} \right)$$

or

(0.149, 0.237).

We are 95% confident that the true difference in the proportions of women under 55 in the two treatment groups is contained between 0.149 and 0.237.

14.8 Review Exercises

1. When is it appropriate to use the normal approximation to the binomial distribution?

2. Describe the function of a continuity correction. When is it most useful?

3. What factors affect the length of a confidence interval for a proportion? Explain.

4. When you are working with a difference in proportions, why does the estimated standard error used in the construction of a confidence interval differ from that used in a hypothesis test?

5. Suppose that you select a random sample of 40 children from the population of newborn infants in Mexico. The probability that a child in this population weighs at most 2500 grams is 0.15 [9].
   (a) For the sample of size 40, what is the probability that four or fewer of the infants weigh at most 2500 grams? Compute the exact binomial probability.
   (b) Using the normal approximation to the binomial distribution, estimate the probability that four or fewer of the children weigh at most 2500 grams.
   (c) Do these two methods provide consistent results?

6. A study was conducted to investigate the relationship between maternal smoking during pregnancy and the presence of congenital malformations in the child. Among children who suffer from an abnormality other than Down's syndrome or an oral cleft, 32.8% have mothers who smoked during pregnancy [10]. This proportion is homogeneous for children with various types of defects.
   (a) If you were to select repeated samples of size 25 from this population, what could you say about the distribution of sample proportions? List three properties.
   (b) Among the samples of size 25, what fraction has a sample proportion of 0.45 or higher?
   (c) What fraction has a sample proportion of 0.20 or lower?
   (d) What value of p cuts off the lower 10% of the distribution?

7. Returning to the same study examining the relationship between maternal smoking during pregnancy and congenital malformations, consider children born with an oral cleft. In a random sample of 27 such infants, 15 have mothers who smoked during pregnancy [10].
   (a) What is a point estimate for p? Construct a 95% confidence interval for the population proportion.

(b) You would like to know whether the proportion of mothers who smoked during pregnancy for children with an oral cleft is identical to the proportion of mothers who smoked for children with other types of malformations. What is the null hypothesis of the appropriate test?

(c) What is the alternative hypothesis?

(d) Conduct the test at the 0.01 level of significance.

(e) What do you conclude?

(f) If the true population proportion of children with an oral cleft is as low as 0.25, you want to risk only a 10% chance of failing to reject the null hypothesis. If you are conducting a two-sided test at the 0.01 level of significance, how large a sample would be required?

8. As part of a recent study conducted in France investigating the effectiveness of the drug mifepristone (RU 486) for terminating early pregnancy, 488 women were administered mifepristone followed 48 hours later by a single dose of a second drug, misoprostol. In 473 of these women, the pregnancy was terminated and the conceptus completely expelled [11].

(a) Estimate the proportion of successfully terminated early pregnancies among women using the described treatment regimen.

(b) Construct a 95% confidence interval for the true population proportion p.

(c) Interpret this confidence interval.

(d) Calculate a 90% confidence interval for p.

(e) How does the 90% confidence interval compare to the 95% interval?

9. In New York City, a study was conducted to evaluate whether any information that is available at the time of birth can be used to identify children with special educational needs. In a random sample of 45 third-graders enrolled in the special education program of the public school system, 4 have mothers who have had more than 12 years of schooling [12].

(a) Construct a 90% confidence interval for the population proportion of children with special educational needs whose mothers have had more than 12 years of schooling.

(b) In 1980, 22% of all third-graders enrolled in the New York City public school system had mothers who had had more than 12 years of schooling. Suppose you wish to know whether this proportion is the same for children in the special education program. What are the null and alternative hypotheses of the appropriate test?

(c) Conduct the test at the 0.05 level of significance.

(d) What do you conclude?

(e) If the true population proportion of children with special educational needs whose mothers have had more than 12 years of schooling is as high as 0.22, you want to risk only a 5% chance of failing to reject the null hypothesis. If you are conducting a two-sided test at the 0.05 level of significance, how large a sample would be required?

10. Suppose that you are interested in determining whether the advice given by a physician during a routine physical examination is effective in encouraging patients to stop smoking. In a study of current smokers, one group of patients was given a brief

talk about the hazards of smoking and was encouraged to quit. A second group received no advice pertaining to smoking. All patients were given a follow-up exam. In the sample of 114 patients who had received the advice, 11 reported that they had quit smoking; in the sample of 96 patients who had not, 7 had quit smoking [13].

(a) Estimate the true difference in population proportions $p_1 - p_2$.

(b) Construct a 95% confidence interval for this difference.

(c) At the 0.05 level of significance, test the null hypothesis that the proportions of patients who quit smoking are identical for those who received advice and those who did not.

(d) Do you believe that the advice given by physicians is effective? Why or why not?

11. A study was conducted to investigate the use of community-based treatment programs among Medicaid beneficiaries suffering from severe mental illness. The study involved assigning a sample of 311 patients to a prepaid medical plan and a sample of 310 patients to the traditional Medicaid program. After a specified period of time, the number of persons in each group who had visited a community crisis center in the previous three months was determined. Among the individuals assigned to the prepaid plan, 13 had visited a crisis center; among those receiving traditional Medicaid, 22 had visited a center [14].

(a) For each group, estimate the proportion of patients who had visited a community crisis center in the previous three months.

(b) At the 0.10 level of significance, test the null hypothesis that the proportions are identical in the two populations.

(c) What do you conclude?

12. Suppose you are interested in investigating the factors that affect the prevalence of tuberculosis among intravenous drug users. In a group of 97 individuals who admit to sharing needles, 24.7% had a positive tuberculin skin test result; among 161 drug users who deny sharing needles, 17.4% had a positive test result [15].

(a) Assuming that the population proportions of positive skin test results are in fact equal, estimate their common value p.

(b) Test the null hypothesis that the proportions of intravenous drug users who have a positive tuberculin skin test result are identical for those who share needles and those who do not.

(c) What do you conclude?

(d) Construct a 95% confidence interval for the true difference in proportions.

13. The data set lowbwt contains information for a sample of 100 low birth weight infants born in two teaching hospitals in Boston, Massachusetts [16] (Appendix B, Table B.7). Indicators of a maternal diagnosis of toxemia during the pregnancy—a condition characterized by high blood pressure and other potentially serious complications—are saved under the variable name tox. The value 1 represents a diagnosis of toxemia and 0 means no such diagnosis.

(a) Estimate the proportion of low birth weight infants whose mothers experienced toxemia during pregnancy.

(b) Construct a 95% confidence interval for the true population proportion p.

(c) Is this interval an exact confidence interval or one that is based on the normal approximation to the binomial distribution?

Bibliography

[1] Nuland, S. B., *Doctors: The Biography of Medicine,* New York: Vintage Books, 1989.

[2] Centers for Disease Control, "The Surgeon General's 1989 Report on Reducing the Health Consequences of Smoking: 25 Years of Progress," *Morbidity and Mortality Weekly Report Supplement,* March 24, 1989.

[3] Myers, M. H., and Gloecker-Ries, L. A., "Cancer Patient Survival Rates: SEER Program Results for 10 Years of Follow-Up," *Ca—A Cancer Journal for Clinicians,* Volume 39, January-February 1989, 21–32.

[4] Jubelirer, S. J., and Wilson, R. A., "Lung Cancer in Patients Younger than 40 Years of Age," *Cancer,* Volume 67, March 1, 1991, 1436–1438.

[5] Rosner, B., *Fundamentals of Biostatistics,* Fourth Edition, Belmont, CA: Wadsworth Publishing Company, 1995.

[6] Osberg, J. S., and DiScala, C., "Morbidity Among Pediatric Motor Vehicle Crash Victims: The Effectiveness of Seat Belts," *American Journal of Public Health,* Volume 82, March 1992, 422–425.

[7] Hack, M., Breslau, N., Weissman, B., Aram, D., Klein, N., and Borawski, E., "Effect of Very Low Birth Weight and Subnormal Head Size on Cognitive Abilities at School Age," *The New England Journal of Medicine,* Volume 325, July 25, 1991, 231–237.

[8] Satariano, E. R., Swanson, G. M., and Moll, P. P., "Nonclinical Factors Associated with Surgery Received for Treatment of Early-Stage Breast Cancer," *American Journal of Public Health,* Volume 82, February 1992, 195–198.

[9] United Nations Children's Fund, *The State of the World's Children 1991,* New York: Oxford University Press.

[10] Khoury, M. J., Weinstein, A., Panny, S., Holtzman, N. A., Lindsay, P. K., Farrel, K., and Eisenberg, M., "Maternal Cigarette Smoking and Oral Clefts: A Population-Based Study," *American Journal of Public Health,* Volume 77, May 1987, 623–625.

[11] Peyron, R., Aubeny, E., Targosz, V., Silvestre, L., Renault, M., Elkik, F., Leclerc, P., Ulmann, A., and Baulieu, E., "Early Termination of Pregnancy with Mifepristone (RU 486) and the Orally Active Prostaglandin Misoprostol," *New England Journal of Medicine,* Volume 328, May 27, 1993, 1509–1513.

[12] Goldberg, D., McLaughlin, M., Grossi, M., Tytun, A., and Blum, S., "Which Newborns in New York City Are at Risk for Special Education Placement?" *American Journal of Public Health,* Volume 82, March 1992, 438–440.

[13] Folsom, A. R., and Grim, R. H., "Stop Smoking Advice by Physicians: A Feasible Approach?" *American Journal of Public Health,* Volume 77, July 1987, 849–850.

[14] Christianson, J. B., Lurie, N., Finch, M., Moscovice, I. S., and Hartley, D., "Use of Community-Based Mental Health Programs by HMOs: Evidence from a Medicaid Demonstration," *American Journal of Public Health,* Volume 82, June 1992, 790–796.

[15] Graham, N. M. H., Nelson, K. E., Solomon, L., Bonds, M., Rizzo, R. T., Scavotto, J., Astemborski, J., and Vlahov, D., "Prevalence of Tuberculin Positivity and Skin Test Anergy in HIV-1-Seropositive and -Seronegative Intravenous Drug Users," *Journal of the American Medical Association,* Volume 267, January 15, 1992, 369–373.

[16] Leviton, A., Fenton, T., Kuban, K. C. K., and Pagano, M., "Labor and Delivery Characteristics and the Risk of Germinal Matrix Hemorrhage in Low Birth Weight Infants," *Journal of Child Neurology,* Volume 6, October 1991, 35–40.

15

Contingency Tables

In the preceding chapter, we used the normal approximation to the binomial distribution to conduct tests of hypotheses for two independent proportions. However, we could have achieved the same results using an alternative technique. When working with nominal data that have been grouped into categories, we often arrange the counts in a tabular format known as a *contingency table*. In the simplest case, two dichotomous random variables are involved; the rows of the table represent the outcomes of one variable, and the columns represent the outcomes of the other. The entries in the table are the counts that correspond to a particular combination of categories.

15.1 The Chi-Square Test

15.1.1 2 × 2 Tables

Consider the 2 × 2 table below, which displays the results of a study investigating the effectiveness of bicycle safety helmets in preventing head injury [1]. The data consist of a random sample of 793 individuals who were involved in bicycle accidents during a specified one-year period.

| Head Injury | Wearing Helmet | | Total |
	Yes	No	
Yes	17	218	235
No	130	428	558
Total	147	646	793

Of the 793 individuals who were involved in bicycle accidents, 147 were wearing safety helmets at the time of the incident and 646 were not. Among those wearing helmets, 17 suffered head injuries requiring the attention of a doctor, whereas the remaining 130 did not; among the individuals not wearing safety helmets, 218 sustained serious head injuries, and 428 did not. The entries in the contingency table—17, 130, 218, and 428—are thus the observed counts within each combination of categories.

To examine the effectiveness of bicycle safety helmets, we wish to know whether there is an association between the incidence of head injury and the use of helmets among individuals who have been involved in accidents. To determine this, we test the null hypothesis

H_0: The proportion of persons suffering head injuries among the population of individuals wearing safety helmets at the time of the accident is equal to the proportion of persons sustaining head injuries among those not wearing helmets.

against the alternative

H_A: The proportions of persons suffering head injuries are not identical in the two populations.

We conduct the test at the $\alpha = 0.05$ level of significance.

The first step in carrying out the test is to calculate the expected count for each cell of the contingency table, given that H_0 is true. Under the null hypothesis, the proportions of individuals experiencing head injuries among those wearing helmets and those not wearing helmets are identical; therefore, we can ignore the two separate categories and treat all 793 individuals as a single homogeneous sample. In this sample, the overall proportion of persons sustaining head injuries is

$$\frac{235}{793} = 29.6\%,$$

and the proportion not experiencing head injuries is

$$\frac{558}{793} = 70.4\%.$$

As a result, of the 147 individuals wearing safety helmets at the time of the accident, we would expect that 29.6%, or

$$147(0.296) = 147\left(\frac{235}{793}\right)$$
$$= 43.6,$$

suffer head injuries, and 70.4%, or

$$147(0.704) = 147\left(\frac{558}{793}\right)$$
$$= 103.4,$$

do not. Similarly, among the 646 bicyclists not wearing safety helmets, we would expect that 29.6%, or

$$646(0.296) = 646\left(\frac{235}{793}\right)$$
$$= 191.4,$$

sustain head injuries, whereas 70.4%, or

$$646(0.704) = 646\left(\frac{558}{793}\right)$$
$$= 454.6,$$

do not. In general, the expected count for a given cell in the table is equal to the row total multiplied by the column total divided by the table total. To minimize roundoff error, we usually compute expected counts to a fraction of a person. Thus, for the four categories in the original table, the expected counts are

Head Injury	Wearing Helmet Yes	No	Total
Yes	43.6	191.4	235.0
No	103.4	454.6	558.0
Total	147.0	646.0	793.0

In general, if a 2 × 2 table of observed frequencies for a sample of size n can be represented as follows,

Variable 1	Variable 2 Yes	No	Total
Yes	a	b	$a + b$
No	c	d	$c + d$
Total	$a + c$	$b + d$	n

the corresponding table of expected counts is

Variable 1	Variable 2 Yes	No	Total
Yes	$(a + b)(a + c)/n$	$(a + b)(b + d)/n$	$a + b$
No	$(c + d)(a + c)/n$	$(c + d)(b + d)/n$	$c + d$
Total	$a + c$	$b + d$	n

The row and column totals in the table of expected counts are identical to those in the observed table. These marginal totals have been held fixed by design; we calculate the cell entries that would have been expected given that there is no association between the row and column classifications **and** that the number of individuals within each group remains constant.

The *chi-square test* compares the observed frequencies in each category of the contingency table (represented by O) with the expected frequencies given that the null hypothesis is true (denoted by E). It is used to determine whether the deviations between the observed and the expected counts, $O - E$, are too large to be attributed to chance. Since there is more than one cell in the table, these deviations must be combined in some way. To perform the test for the counts in a contingency table with r rows and c columns, we calculate the sum

$$X^2 = \sum_{i=1}^{rc} \frac{(O_i - E_i)^2}{E_i},$$

where rc is the number of cells in the table. The probability distribution of this sum is approximated by a *chi-square (χ^2) distribution* with $(r - 1)(c - 1)$ degrees of freedom. A 2×2 table has $(2 - 1)(2 - 1) = 1$ degree of freedom; a 3×4 table has $(3 - 1)(4 - 1) = 6$ degrees of freedom. To ensure that the sample size is large enough to make this approximation valid, no cell in the table should have an expected count less than 1, and no more than 20% of the cells should have an expected count less than 5 [2].

Like the F distribution, the chi-square distribution is not symmetric. A chi-square random variable cannot be negative; it assumes values from zero to infinity and is skewed to the right. As is true for all probability distributions, however, the total area beneath the curve is equal to 1. As with the t and F distributions, there is a different chi-square distribution for each possible value of the degrees of freedom. The distributions with small degrees of freedom are highly skewed; as the number of degrees of freedom increases, the distributions become less skewed and more symmetric. This is illustrated in Figure 15.1.

Table A.8 in Appendix A is a condensed table of areas for the chi-square distribution with various degrees of freedom. For a particular value of df, the entry in the table is the outcome of χ^2_{df} that cuts off the specified area in the upper tail of the distribution. Given a chi-square distribution with 1 degree of freedom, for instance, $\chi^2_1 = 3.84$ cuts off the upper 5% of the area under the curve.

Since all the expected counts for the data pertaining to bicycle safety helmets are greater than 5, we can proceed with the chi-square test. The next step is to calculate the sum

$$X^2 = \sum_{i=1}^{rc} \frac{(O_i - E_i)^2}{E_i}.$$

For a 2×2 contingency table, this test statistic has an approximate chi-square distribution with $(2 - 1)(2 - 1) = 1$ degree of freedom. Note that we are using discrete observations to estimate χ^2, a continuous distribution. The approximation is quite good for tables with many degrees of freedom, but may not be valid for 2×2 tables that have

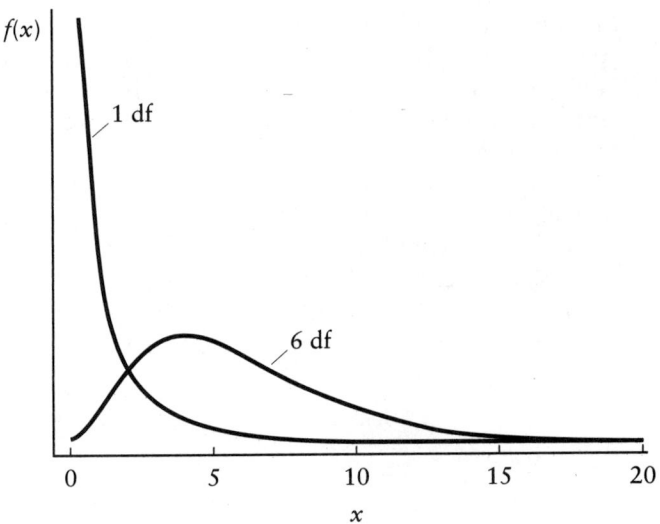

FIGURE 15.1
Chi-square distributions with 1 and 6 degrees of freedom

df $= 1$. Therefore, we often apply a continuity correction in this situation. When we are working with a 2×2 table, the corrected test statistic is

$$X^2 = \sum_{i=1}^{4} \frac{(|O_i - E_i| - 0.5)^2}{E_i},$$

where $|O_i - E_i|$ is the absolute value of the difference between O_i and E_i. The term -0.5 in the numerator is often referred to as the *Yates correction*. The effect of this term is to decrease the value of the test statistic, and thus increase the corresponding p-value. Although the Yates correction has been used extensively in the past, many investigators have begun to question its validity. Some believe that it results in an overly conservative test that may fail to reject a false null hypothesis [3]. As long as n is reasonably large, however, the effect of the correction factor is negligible.

Substituting the observed and expected counts for the bicycle safety helmet data into the corrected equation, we find that

$$X^2 = \frac{(|17 - 43.6| - 0.5)^2}{43.6} + \frac{(|130 - 103.4| - 0.5)^2}{103.4}$$
$$+ \frac{(|218 - 191.4| - 0.5)^2}{191.4} + \frac{(|428 - 454.6| - 0.5)^2}{454.6}$$
$$= 15.62 + 6.59 + 3.56 + 1.50$$
$$= 27.27.$$

For a χ^2 distribution with 1 degree of freedom, Table A.8 tells us that the p-value is less than 0.001. Although we are considering only one tail of the chi-square distribution, the test is two-sided; large outcomes of $(O_i - E_i)^2$ can result when the observed value is greater than the expected value and also when it is smaller. Since $p < \alpha$, we reject the null hypothesis and conclude that the proportions of individuals suffering head injuries are not identical in the two populations. Among persons involved in bicycle accidents, the use of a safety helmet reduces the incidence of head injury.

If we represent a 2×2 table in the general format shown below,

	Variable 2		
Variable 1	Yes	No	Total
Yes	a	b	$a + b$
No	c	d	$c + d$
Total	$a + c$	$b + d$	n

the test statistic X^2 can also be expressed as

$$X^2 = \frac{n[|ad - bc| - (n/2)]^2}{(a + c)(b + d)(a + b)(c + d)}.$$

Because there is no need to compute a table of expected counts, this form of the statistic is more convenient computationally. However, it does not have the same straightforward interpretation as a measure of the deviations between observed and expected counts.

As mentioned previously, the chi-square test is based on an approximation that works best when the sample size n is large. However, we can also conduct a test of hypothesis comparing two proportions using an alternative technique that allows us to compute the exact probability of the occurrence of the observed frequencies in the contingency table, given that there is no association between the row and column classifications and that the marginal totals remain fixed. This technique, known as *Fisher's exact test,* is especially useful when the sample size is small. Because the computations involved can be arduous, it is not presented in this text [4]. However, a number of statistical software packages perform Fisher's exact test for 2×2 tables in addition to the chi-square test.

15.1.2 $r \times c$ Tables

In the case of a 2×2 table, the chi-square test for independent proportions is equivalent to the hypothesis test that uses the normal approximation to the binomial distribution. The chi-square test is used more often in practice, however. In addition to being relatively easy to compute, it has another advantage: It can be generalized to accommodate the comparison of three or more proportions. In this situation, we arrange the data as an $r \times c$ contingency table, where r is the number of rows and c is the number of columns.

Consider the following data, taken from a study that investigates the accuracy of death certificates. In two different hospitals, the results of 575 autopsies were compared to the causes of death listed on the certificates. One of the hospitals that participated in the study was a community hospital, labeled A; the other was a university hospital, labeled B. The data are displayed in the form of a 2×3 contingency table [5].

	Death Certificate Status			
Hospital	Confirmed Accurate	Inaccurate No Change	Incorrect Recoding	Total
A	157	18	54	229
B	268	44	34	346
Total	425	62	88	575

Of the 575 death certificates considered, 425 were confirmed to be accurate, 62 either lacked information or contained inaccuracies but did not require recoding of the underlying cause of death, and 88 were incorrect and required recoding. We would like to determine whether the results of the study suggest different practices in completing death certificates at the two hospitals. To do this, we test the null hypothesis,

H_0: Within each category of certificate status, the proportions of death certificates in Hospital A are identical,

against the alternative hypothesis that the proportions are not the same. Another way to formulate the null hypothesis is to say that there is no association between hospital and death certificate status; the alternative hypothesis would be that there is an association. In either case, we use the chi-square test and set the significance level at $\alpha = 0.05$.

The first step in carrying out the test is to calculate the expected count for each cell in the contingency table. Recall that the expected count is equal to the row total multiplied by the column total divided by the table total. For instance, among death certificates confirmed to be accurate, we expect that

$$\frac{229 \times 425}{575} = 169.3$$

would be in Hospital A, whereas

$$\frac{346 \times 425}{575} = 255.7$$

would be in Hospital B. For all six categories in the table, the expected counts are as follows:

Hospital	Certificate Status			Total
	Confirmed Accurate	Inaccurate No Change	Incorrect Recoding	
A	169.3	24.7	35.0	229.0
B	255.7	37.3	53.0	346.0
Total	425.0	62.0	88.0	575.0

Since all expected counts are greater than 5, we next calculate the sum

$$X^2 = \sum_{i=1}^{6} \frac{(O_i - E_i)^2}{E_i} .$$

We are working with a 2×3 contingency table, and consequently do not need to include the Yates continuity correction. Therefore,

$$
\begin{aligned}
X^2 &= \frac{(157 - 169.3)^2}{169.3} + \frac{(18 - 24.7)^2}{24.7} + \frac{(54 - 35.0)^2}{35.0} \\
&\quad + \frac{(268 - 255.7)^2}{255.7} + \frac{(44 - 37.3)^2}{37.3} + \frac{(34 - 53.0)^2}{53.0} \\
&= 0.89 + 1.82 + 10.31 + 0.59 + 6.81 + 1.20 \\
&= 21.62.
\end{aligned}
$$

For a χ^2 distribution with $(2 - 1)(3 - 1) = 2$ degrees of freedom, $p < 0.001$. Since p is less than α, we reject the null hypothesis and conclude that the proportions of death certificates in Hospital A are not identical for the three different categories of certificate status; equivalently, there is an association between hospital and death certificate status. Again consulting the original contingency table, we note that Hospital A has 36.9% of the certificates confirmed to be accurate, 29.0% of those that are inaccurate but do not need to be changed, and 61.4% of those that require recoding. It appears that Hospital A, the community hospital, contains a larger proportion of death certificates that are incorrect and require recoding than Hospital B.

15.2   McNemar's Test

Assuming that we are still interested in comparing proportions, what can we do if our data are paired rather than independent? As with continuous data, the distinguishing characteristic of paired samples for counts is that each observation in the first group has a corresponding observation in the second group. In this situation, we use *McNemar's test* to evaluate hypotheses about the data.

Consider the following information taken from a study that investigates acute myocardial infarction—or heart attack—among Navajos residing in the United States. In the study, 144 victims of acute myocardial infarction were age- and gender-matched with 144 individuals free of heart disease. The members of each pair were then asked whether they had ever been diagnosed with diabetes. The results are presented below [6].

	MI		
Diabetes	**Yes**	**No**	**Total**
Yes	46	25	71
No	98	119	217
Total	144	144	288

Among the 144 individuals who had experienced acute myocardial infarction, 46 were diagnosed with diabetes; among those who were free of heart disease, only 25 suffered from diabetes. We would like to know what these samples tell us about the proportions of diabetics in the two heart disease groups.

The preceding table has the same format as the 2×2 contingency table used in the comparison of proportions from independent samples; therefore, we might be tempted to apply the chi-square test. Note, however, that we have a total of 288 observations but only 144 pairs. Each matched pair in the study provides two responses: one for the individual who suffered a myocardial infarction, and one for the individual who did not. Since the chi-square test disregards the paired nature of the data, it is inappropriate in this situation. We must somehow take the pairing into account in our analysis.

Suppose that we take another look at the raw data from the study, but this time classify them in the following manner:

	No MI		
MI	**Diabetes**	**No Diabetes**	**Total**
Diabetes	9	37	46
No Diabetes	16	82	98
Total	25	119	144

Of the 46 Navajos who had experienced acute myocardial infarction and who were diabetic, 9 were matched with individuals who had diabetes and 37 with individuals who did not. Of the 98 infarction victims who did not suffer from diabetes, 16 were paired with diabetics and 82 were not. Each entry in the table corresponds to the combination of responses for a matched pair rather than an individual person.

We can now conduct a proper paired analysis. However, we must first change the statement of the null hypothesis to reflect the new format of the data. Instead of testing whether the proportion of diabetics among individuals who have experienced acute myocardial infarction is equal to the proportion of diabetics among those who have not, we test

H_0: There are equal numbers of pairs in which the victim of acute myocardial infarction is a diabetic and the matched individual free of heart disease is not, and in which the person without heart disease is a diabetic but the individual who has experienced an infarction is not,

or, more concisely,

H_0: There is no association between diabetes and the occurrence of acute myocardial infarction.

The alternative hypothesis is that an association exists. We conduct this test at the $\alpha = 0.05$ level of significance.

The *concordant pairs*—or the pairs of responses in which either two diabetics or two nondiabetics are matched—provide no information for testing a null hypothesis about differences in diabetic status. Therefore, we discard these data and focus only on the *discordant pairs,* or the pairs of responses in which a person who has diabetes is paired with an individual who does not.

Let r represent the number of pairs in which the victim of acute myocardial infarction suffers from diabetes and the individual free of heart disease does not, and s the number of pairs in which the person without heart disease is a diabetic but the individual who had an infarction is not. If the null hypothesis is true, r and s should be approximately equal. Therefore, if the difference between r and s is large, we would want to reject the null hypothesis of no association. To conduct McNemar's test, we calculate the statistic

$$X^2 = \frac{[|r - s| - 1]^2}{(r + s)}.$$

This ratio has an approximate chi-square distribution with 1 degree of freedom. The term -1 in the numerator is a continuity correction; again, we are using discrete counts to estimate the continuous χ^2 distribution.

For the data investigating the relationship between diabetes and myocardial infarction, $r = 37$ and $s = 16$. Therefore,

$$X^2 = \frac{[|37 - 16| - 1]^2}{(37 + 16)}$$

$$= \frac{[21 - 1]^2}{53}$$

$$= 7.55.$$

For a chi-square distribution with 1 degree of freedom, $0.001 < p < 0.01$. Since p is less than α, we reject the null hypothesis. For the given population of Navajos, we conclude that if there is a difference between individuals who experience an infarction and those who do not, victims of acute myocardial infarction are more likely to suffer from diabetes than the individuals free from heart disease who have been matched on age and gender.

15.3   The Odds Ratio

Although the chi-square test allows us to determine whether an association exists between two independent nominal random variables, and McNemar's test does the same for paired dichotomous variables, neither test provides a measure of the strength of the association. However, methods are available for estimating the magnitude of the effect using the observations in a single sample. For a 2×2 table showing information on two independent dichotomous variables, for instance, one such measure is the *odds ratio,* or *relative odds.*

In Chapter 6 we stated that if an event occurs with probability p, the *odds* in favor of the event are $p/(1 - p)$ to 1. If an event occurs with probability $1/2$, for instance, the odds in favor of the event are $(1/2)/(1 - 1/2) = (1/2)/(1/2) = 1$ to 1. Conversely, if the odds in favor of an event are a to b, the probability that the event occurs is $a/(a + b)$.

If we have two dichotomous random variables that represent a disease and an exposure, the odds ratio is defined as the odds in favor of disease among exposed individuals divided by the odds in favor of disease among the unexposed, or

$$OR = \frac{P(\text{disease} \mid \text{exposed})/[1 - P(\text{disease} \mid \text{exposed})]}{P(\text{disease} \mid \text{unexposed})/[1 - P(\text{disease} \mid \text{unexposed})]}.$$

Alternatively, the odds ratio may be defined as the odds of exposure among diseased individuals divided by the odds of exposure among the nondiseased, or

$$OR = \frac{P(\text{exposure} \mid \text{diseased})/[1 - P(\text{exposure} \mid \text{diseased})]}{P(\text{exposure} \mid \text{nondiseased})/[1 - P(\text{exposure} \mid \text{nondiseased})]}.$$

These two different expressions for the relative odds are mathematically equivalent; consequently, the odds ratio can be estimated for both cohort and case-control studies.

Suppose that our data—which consist of a sample of n individuals—are again arranged in the form of a 2×2 contingency table:

	Exposed	Unexposed	Total
Disease	a	b	$a + b$
No Disease	c	d	$c + d$
Total	$a + c$	$b + d$	n

In this case, we would estimate that

$$P(\text{disease} \mid \text{exposed}) = \frac{a}{a+c},$$

and

$$P(\text{disease} \mid \text{unexposed}) = \frac{b}{b+d};$$

therefore,

$$1 - P(\text{disease} \mid \text{exposed}) = 1 - \frac{a}{a+c}$$

$$= \frac{c}{a+c},$$

and

$$1 - P(\text{disease} \mid \text{unexposed}) = 1 - \frac{b}{b+d}$$

$$= \frac{d}{b+d}.$$

Using these results, we can express an estimator of the odds ratio as

$$\widehat{OR} = \frac{[a/(a+c)]/[c/(a+c)]}{[b/(b+d)]/[d/(b+d)]}$$

$$= \frac{a/c}{b/d}$$

$$= \frac{ad}{bc}.$$

This estimator is simply the cross-product ratio of the entries in the 2×2 table.

The odds ratio is not the same as the relative risk; *relative risk* is the probability of disease among exposed individuals divided by the probability of disease among the unexposed. The estimator of the relative risk is

$$\widehat{RR} = \frac{P(\text{disease} \mid \text{exposed})}{P(\text{disease} \mid \text{unexposed})}$$

$$= \frac{a/(a+c)}{b/(b+d)}$$

$$= \frac{a(b+d)}{b(a+c)}.$$

When we are dealing with a rare disease—or values of a and b that are small relative to the values of c and d—the relative risk can be approximated by the odds ratio:

$$\widehat{RR} = \frac{a(b + d)}{b(a + c)}$$

$$\approx \frac{ad}{bc}$$

$$= \widehat{OR}.$$

Because the sampling distribution of $\widehat{OR}$ has better statistical properties than that of $\widehat{RR}$, it is generally preferable to work with the relative odds.

The following data are taken from a study that attempts to determine whether the use of electronic fetal monitoring during labor affects the frequency of caesarean section deliveries. Caesarian delivery can be thought of as the "disease," and electronic monitoring as the "exposure." Of the 5824 infants included in the study, 2850 were electronically monitored during labor and 2974 were not [7]. The outcomes are as follows:

Caesarian Delivery	EFM Exposure		Total
	Yes	No	
Yes	358	229	587
No	2492	2745	5237
Total	2850	2974	5824

The odds of being delivered by caesarean section in the group that was monitored relative to the group that was not monitored are estimated by

$$\widehat{OR} = \frac{(358)(2745)}{(229)(2492)}$$

$$= 1.72.$$

These data suggest that the odds of being delivered by caesarean section are 1.72 times higher for fetuses that are electronically monitored during labor than for fetuses that are not monitored. Therefore, there does appear to be a moderate association between the use of electronic fetal monitoring and the eventual method of delivery. This does not imply, however, that electronic monitoring somehow **causes** a caesarean delivery; it is possible that the fetuses at higher risk are the ones that are monitored.

The cross-product ratio is simply a point estimate of the strength of association between two dichotomous random variables. To gauge the uncertainty in this estimate, we must calculate a confidence interval as well; the width of the interval reflects the amount of variability in our estimator $\widehat{OR}$. Recall that the derivation of the expression for a 95% confidence interval for a mean μ,

$$\left(\bar{x} - 1.96\frac{\sigma}{\sqrt{n}}, \bar{x} + 1.96\frac{\sigma}{\sqrt{n}}\right),$$

relied on the assumption that the underlying population values were normally distributed. When computing a confidence interval for the odds ratio, we must make the same assumption of underlying normality. A problem arises, however, in that the probability distribution of the odds ratio is skewed to the right. Although it cannot take on negative values, the relative odds can assume any positive value between 0 and infinity. In contrast, the probability distribution of the natural logarithm of the odds ratio is more symmetric and approximately normal. Therefore, when calculating a confidence interval for the odds ratio, we typically work in the log scale. To ensure that the sample size is large enough, the expected value of each entry in the contingency table should be at least 5.

The expression for a 95% confidence interval for the natural logarithm of the odds ratio is

$$(\ln (\widehat{OR}) - 1.96 \, se\,[\ln (\widehat{OR})], \, \ln (\widehat{OR}) + 1.96 \, se\,[\ln (\widehat{OR})]).$$

Therefore, to compute a confidence interval for $\ln (OR)$, we must first know the standard error of this quantity. For a 2×2 table arranged in the following manner,

	Exposed	Unexposed	Total
Disease	a	b	$a + b$
No Disease	c	d	$c + d$
Total	$a + c$	$b + d$	n

the standard error of $\ln (\widehat{OR})$ is estimated by

$$\widehat{se}\,[\ln (\widehat{OR})] = \sqrt{\frac{1}{a} + \frac{1}{b} + \frac{1}{c} + \frac{1}{d}}.$$

If any one of the entries in the contingency table is equal to 0, the standard error is undefined. In this case, adding 0.5 to each of the values a, b, c, and d will correct the situation and still provide a reasonable estimate; thus, the modified estimate of the standard error is

$$\widehat{se}\,[\ln (\widehat{OR})] = \sqrt{\frac{1}{(a + 0.5)} + \frac{1}{(b + 0.5)} + \frac{1}{(c + 0.5)} + \frac{1}{(d + 0.5)}}.$$

The appropriate estimate for the standard error can be substituted into the expression for the confidence interval above. To find a 95% confidence interval for the odds ratio itself, we take the antilogarithms of the upper and lower limits of the interval for $\ln (OR)$ to get

$$(e^{\ln(\widehat{OR}) - 1.96 \, \widehat{se}[\ln(\widehat{OR})]}, \, e^{\ln(\widehat{OR}) + 1.96 \, \widehat{se}\,[\ln(\widehat{OR})]}).$$

For the data examining the relationship between electronic fetal monitoring status during labor and the eventual method of delivery, the log of the estimated odds ratio is

$$\ln (\widehat{OR}) = \ln (1.72)$$
$$= 0.542.$$

The estimated standard error of $\ln(\widehat{OR})$ is

$$\widehat{se}\,[\ln(\widehat{OR})] = \sqrt{\frac{1}{a} + \frac{1}{b} + \frac{1}{c} + \frac{1}{d}}$$

$$= \sqrt{\frac{1}{358} + \frac{1}{229} + \frac{1}{2492} + \frac{1}{2745}}$$

$$= 0.089.$$

Therefore, a 95% confidence interval for the log of the odds ratio is

$$(0.542 - 1.96(0.089),\ 0.542 + 1.96(0.089)),$$

or

$$(0.368, 0.716),$$

and a 95% confidence interval for the odds ratio itself is

$$(e^{0.368},\ e^{0.716}),$$

or

$$(1.44, 2.05).$$

We are 95% confident that the odds of delivery by caesarean section are between 1.44 and 2.05 times higher for fetuses that are monitored during labor than for those that are not monitored. Note that this interval does not contain the value 1. An odds ratio of 1 would imply that fetuses that are monitored and those that are not monitored have identical odds of caesarean delivery.

An odds ratio can also be calculated to estimate the strength of association between two paired dichotomous random variables. Using the notation introduced in Section 15.2, r and s represent the numbers of each type of discordant pair in the study. For the investigation of the relationship between acute myocardial infarction and diabetes among Navajos residing in the United States, for example, r represented the number of pairs in which the subject who had suffered an acute myocardial infarction was diabetic and the matched individual free of heart disease was not, and s the number of pairs in which the person without heart disease was diabetic and the subject who had an infarction was not. In this case, the relative odds of suffering from diabetes for individuals who have experienced acute myocardial infarction versus those who have not are estimated by

$$\widehat{OR} = \frac{r}{s}$$

$$= \frac{37}{16}$$

$$= 2.31.$$

By noting that the estimated standard error of $\ln(\widehat{OR})$ for paired dichotomous data is equal to $\sqrt{(r+s)/rs}$, we can also calculate a confidence interval for the true population odds ratio. A 95% interval for the natural logarithm of the odds ratio is

$$\ln(\widehat{OR}) \pm 1.96 \operatorname{se}[\ln(\widehat{OR})].$$

Since

$$\ln(\widehat{OR}) = \ln(2.31)$$
$$= 0.837$$

and

$$\widehat{\operatorname{se}}[\ln(\widehat{OR})] = \sqrt{\frac{r+s}{rs}}$$
$$= \sqrt{\frac{37+16}{37(16)}}$$
$$= 0.299,$$

the 95% confidence interval for $\ln(OR)$ would be

$$(0.837 - 1.96(0.299), 0.837 + 1.96(0.299)),$$

or

$$(0.251, 1.423).$$

A 95% confidence interval for the odds ratio itself is then

$$(e^{0.251}, e^{1.423}),$$

or

$$(1.29, 4.15).$$

Note that this interval does not contain the value 1.

15.4 Berkson's Fallacy

Although the odds ratio is a useful measure of the strength of association between two dichotomous random variables, it provides a valid estimate of the magnitude of the effect only if the sample of observations on which it is based is random. This point is sometimes forgotten; a restricted sample, such as a sample of patients from a single hospital, is usually much easier to obtain. This restricted sample is then used to make inference about the population as a whole.

In one study, the investigators surveyed 2784 individuals—257 of whom were hospitalized—and determined whether each subject suffered from a disease of the circulatory system or a respiratory illness or both [8]. If they had limited their questioning to the 257 hospitalized patients only, the results would have been as follows:

Circulatory Disease	Respiratory Disease		Total
	Yes	**No**	
Yes	7	29	36
No	13	208	221
Total	20	237	257

An estimate of the relative odds of having respiratory illness among individuals who suffer from a disease of the circulatory system versus those who do not is

$$\widehat{OR} = \frac{(7)(208)}{(29)(13)}$$
$$= 3.86;$$

the chi-square test of the null hypothesis that there is no association between the two diseases yields

$$X^2 = \frac{(|7 - 2.8| - 0.5)^2}{2.8} + \frac{(|29 - 33.2| - 0.5)^2}{33.2}$$
$$+ \frac{(|13 - 17.2| - 0.5)^2}{17.2} + \frac{(|208 - 203.8| - 0.5)^2}{203.8}$$
$$= 4.89 + 0.41 + 0.80 + 0.07$$
$$= 6.17.$$

For a chi-square distribution with 1 degree of freedom, $0.01 < p < 0.025$. Therefore, we reject the null hypothesis at the 0.05 level and conclude that individuals who have a disease of the circulatory system are more likely to suffer from respiratory illness than individuals who do not.

Now consider the entire sample of 2784 individuals, which consists of both hospitalized and nonhospitalized subjects:

Circulatory Disease	Respiratory Disease		Total
	Yes	**No**	
Yes	22	171	193
No	202	2389	2591
Total	224	2560	2784

For these data, the estimate of the odds ratio is

$$\widehat{OR} = \frac{(22)(2389)}{(171)(202)}$$
$$= 1.52.$$

This value is much lower than the estimate of the relative odds calculated for the hospitalized patients only. In addition, the value of the chi-square test statistic is

$$X^2 = \frac{(|22 - 15.53| - 0.5)^2}{15.53} + \frac{(|171 - 177.47| - 0.5)^2}{177.47}$$
$$+ \frac{(|202 - 208.47| - 0.5)^2}{208.47} + \frac{(|2389 - 2382.53| - 0.5)^2}{2382.53}$$
$$= 2.29 + 0.20 + 0.17 + 0.01$$
$$= 2.67.$$

For a chi-square distribution with 1 degree of freedom, $p > 0.10$. We can no longer reject the null hypothesis that there is no association between respiratory and circulatory diseases at the 0.05 level of significance.

Why do the conclusions drawn from these two samples differ so drastically? To answer this question, we must consider the rates of hospitalization that occur within each of the four disease subgroups. Among the 22 individuals suffering from both circulatory and respiratory disease, 7 are hospitalized. Therefore, the rate of hospitalization for this subgroup is

$$\frac{7}{22} = 31.8\%.$$

The rate of hospitalization among subjects with respiratory illness alone is

$$\frac{13}{202} = 6.4\%.$$

Among individuals with circulatory disease only, the rate is

$$\frac{29}{171} = 17.0\%,$$

and among persons suffering from neither disease, the rate of hospitalization is

$$\frac{208}{2389} = 8.7\%.$$

Thus, individuals with both circulatory and respiratory diseases are much more likely to be hospitalized than individuals in any of the three other subgroups. Also, subjects with circulatory disease are more likely to be hospitalized than those with respiratory illness. If we sample only patients who are hospitalized, therefore, our conclusions about the relationship between these two diseases will be biased. We are more likely to select an individual who is suffering from both illnesses than a person in any of the other subgroups, and more likely to select a person with circulatory disease than one with respiratory problems. As a result, we observe an association that does not actually exist. This kind of spurious relationship among variables—which is evident only because of the way in which the sample was chosen—is known as *Berkson's fallacy.*

15.5   *Further Applications*

When we are presented with independent samples of nominal data that have been grouped into categories, the chi-square test can be used to determine whether the proportions of some event of interest are identical in the various groups. For example, consider the following data, taken from a study investigating an outbreak of gastroenteritis—an inflammation of the membranes of the stomach and small intestine—following a lunch served in the cafeteria of a U.S. high school. Among a sample of 263 students who bought lunch in the school cafeteria on the day in question, 225 ate prepared sandwiches and 38 did not [9]. The numbers of cases of gastroenteritis in each group are displayed below.

	Ate Sandwich		
	Yes	**No**	**Total**
Ill	109	4	113
Not Ill	116	34	150
Total	225	38	263

Among the students who ate prepared sandwiches, 48.4% became ill; among those who did not, 10.5% became ill. We would like to test the null hypothesis that there is no association between the consumption of a sandwich and the onset of gastroenteritis, or

H_0: The proportion of students becoming ill among those who ate the sandwiches is equal to the proportion becoming ill among those who did not eat the sandwiches,

at the $\alpha = 0.05$ level of significance. The alternative hypothesis is that an association does exist.

To begin, we must calculate the expected count for each cell of the 2×2 table. Under the null hypothesis, the proportions of students developing gastroenteritis are identical in the two groups. Treating all 263 students as a single sample, the overall proportion of students becoming ill is

$$\frac{113}{263} = 43.0\%,$$

and the proportion of students not becoming ill is

$$\frac{150}{263} = 57.0\%.$$

Therefore, among the 225 students who ate the sandwiches, we would expect that 43%, or

$$225\left(\frac{113}{263}\right) = 96.7,$$

become ill and that 57%, or

$$225\left(\frac{150}{263}\right) = 128.3,$$

do not. Similarly, among the 38 students who did not eat the sandwiches, we expect that

$$38\left(\frac{113}{263}\right) = 16.3$$

become ill and

$$38\left(\frac{150}{263}\right) = 21.7$$

do not. Thus, the table of expected counts is as follows:

	Ate Sandwich		
	Yes	No	Total
Ill	96.7	16.3	113.0
Not Ill	128.3	21.7	150.0
Total	225.0	38.0	263.0

Since the expected count in each cell of the table is greater than 5, we proceed with the chi-square test by calculating the statistic

$$X^2 = \sum_{i=1}^{4} \frac{(|O_i - E_i| - 0.5)^2}{E_i}$$

$$= \frac{(|109 - 96.7| - 0.5)^2}{96.7} + \frac{(|4 - 16.3| - 0.5)^2}{16.3}$$

$$+ \frac{(|116 - 128.3| - 0.5)^2}{128.3} + \frac{(|34 - 21.7| - 0.5)^2}{21.7}$$

$$= 1.44 + 8.54 + 1.09 + 6.42$$

$$= 17.49.$$

For a chi-square distribution with 1 degree of freedom, $p < 0.001$. Since p is less than α, we reject H_0 and conclude that the proportions of students developing gastroenteritis are not identical in the two groups. Among students eating lunch at the high school cafeteria on the day in question, the consumption of a prepared sandwich was associated with an increased risk of illness.

Most computer packages are able to perform the chi-square test. The relevant output from SAS is shown in Table 15.1. The output consists of the observed 2×2 contingency table, the percentage of observations in each cell of the table, and a wide variety of test statistics and their corresponding p-values. Note that the "Chi-Square" test statistic of 19.074—the first one listed in the bottom portion of the output—is somewhat different from the value we calculated by hand; this is because SAS is not applying the continuity correction factor. The "Continuity Adj. Chi-Square" statistic in the third row down is much closer to our calculated value and differs only because of rounding error.

Now consider the situation in which we have paired samples of dichotomous data. The following information comes from a study that examines changes in smoking status over a two-year period [10]. In 1980, a sample of 2110 adults over the age of 18 were asked to identify themselves as smokers or nonsmokers. In 1982, the same 2110 individuals were again asked whether they were currently smokers or nonsmokers. Of

TABLE 15.1

SAS output displaying the chi-square test

```
                    TABLE OF SANDWICH BY ILL

        SANDWICH          ILL
        Frequency
        Percent                0          1       Total
                    0         34          4          38
                             12.93       1.52       14.45
                    1        116        109         225
                             44.11      41.44       85.55
        Total                150        113         263
                             57.03      42.97      100.00

            STATISTICS FOR TABLE OF SANDWICH BY ILL
```

Statistic	DF	Value	Prob
Chi-Square	1	19.074	0.001
Likelihood Ratio Chi-Square	1	22.101	0.001
Continuity Adj.Chi-Square	1	17.558	0.001
Mantel-Haenszel Chi-Square	1	19.002	0.001
Fisher's Exact Test (Left)			1.000
(Right)			3.97E-06
(2-Tail)			5.23E-06
Phi Coefficient		0.269	
Contingency Coefficient		0.260	
Cramer's V		0.269	
Sample Size = 263			

the 717 individuals who smoked in 1980, 620 were still smoking in 1982, and 97 had stopped. Of the 1393 nonsmokers in 1980, 1317 remained nonsmokers in 1982, and 76 had begun to smoke. Each entry in the table corresponds to the paired response of a single individual.

1980	1982		Total
	Smoker	Nonsmoker	
Smoker	620	97	717
Nonsmoker	76	1317	1393
Total	696	1414	2110

We would like to test the null hypothesis that there is no association between smoking status and year, or, more formally,

H_0: Among the individuals who changed their smoking status between 1980 and 1982, equal numbers switched from being smokers to nonsmokers and from being nonsmokers to smokers.

The alternative hypothesis is that there is an association, or that there was a tendency for smoking status to change in one direction or the other. In order to reach a conclusion, we conduct McNemar's test at the $\alpha = 0.05$ level of significance.

Note that we have $r = 97$ pairs in which a smoker becomes a nonsmoker and $s = 76$ pairs in which a nonsmoker becomes a smoker. To evaluate H_0, we calculate the test statistic

$$X^2 = \frac{[|r - s| - 1]^2}{(r + s)}$$
$$= \frac{[|97 - 76| - 1]^2}{(97 + 76)}$$
$$= 2.31.$$

For a chi-square distribution with 1 degree of freedom, $p > 0.10$. Since p is greater than α, we cannot reject the null hypothesis. The sample does not provide evidence that there is an association between smoking status and year.

Stata output illustrating the results of McNemar's test is shown in Table 15.2. Note that for this particular test, Stata ignores the variable names; the labels "Cases" and "Controls" refer to each of the two paired samples—1980 and 1982 for our data— and "Exposed" and "Unexposed" classify the status of each subject. Once again, the test statistic of 2.55 differs from the one we calculated by hand because the computer package does not apply a continuity correction factor.

In both of the previous examples, we determined whether an association existed between two dichotomous random variables; we did not measure the strength of the association, however. For a 2×2 contingency table, one way in which the effect can be quantified is by means of an odds ratio. For the data from the study that examined the

TABLE 15.2

Stata output displaying McNemar's test

Cases	Controls Exposed	Unexposed	Total
Exposed	620	97	717
Unexposed	76	1317	1393
Total	696	1414	2110

McNemar's chi2(1) = 2.55   Pr>chi2=0.1104

outbreak of gastroenteritis in a U.S. high school, the odds of becoming ill among those who ate prepared sandwiches relative to those who did not are estimated by

$$\widehat{OR} = \frac{(109)(34)}{(4)(116)}$$
$$= 7.99.$$

The odds of becoming ill for students who ate prepared sandwiches are 7.99 times larger than the odds for students who did not; therefore, there does appear to be a strong association between eating one of the sandwiches and a student's health.

To measure the uncertainty in this point estimate, we can calculate a confidence interval for the true population odds ratio. Note that the logarithm of $\widehat{OR}$ is

$$\ln(\widehat{OR}) = \ln(7.99)$$
$$= 2.078,$$

and that the estimated standard error of $\ln(\widehat{OR})$ is

$$\widehat{se}[\ln(\widehat{OR})] = \sqrt{\frac{1}{109} + \frac{1}{4} + \frac{1}{116} + \frac{1}{34}}$$
$$= 0.545.$$

Therefore, a 95% confidence interval for the log of the odds ratio is

$$(2.078 - 1.96(0.545), 2.078 + 1.96(0.545))$$

or

$$(1.010, 3.146),$$

and a 95% confidence interval for the odds ratio itself is

$$(e^{1.010}, e^{3.146})$$

or

(2.75, 23.24).

We are 95% confident that the odds of becoming ill for students who ate sandwiches are between 2.75 and 23.24 times larger than the odds for students who did not. Note that this interval does not contain the value 1.

For the paired dichotomous data from the study that looked at changes in smoking status over a two-year interval, the odds of smoking in 1980 versus smoking in 1982 are estimated by

$$\widehat{OR} = \frac{97}{76}$$
$$= 1.28.$$

To calculate a 95% confidence interval for the true population odds ratio, note first that the logarithm of $\widehat{OR}$ is

$$\ln(\widehat{OR}) = \ln(1.28)$$
$$= 0.244,$$

and then that the estimated standard error of $\ln(\widehat{OR})$ is

$$\widehat{se}\,[\ln(\widehat{OR})] = \sqrt{\frac{97 + 76}{97(76)}}$$
$$= 0.153.$$

Therefore, a 95% confidence interval for the log of the odds ratio is

$$(0.244 - 1.96(0.153),\ 0.244 + 1.96(0.153))$$

or

$$(-0.056, 0.544),$$

and a 95% confidence interval for the odds ratio itself is

$$(e^{-0.056},\ e^{0.544})$$

or

$$(0.95, 1.72).$$

This interval does contain the value 1.

15.6 Review Exercises

1. How does the chi-square test statistic use the observed frequencies in a contingency table to determine whether an association exists between two nominal random variables?

2. Describe the properties of the chi-square distribution.

3. How does the null hypothesis evaluated using McNemar's test differ from that evaluated by the chi-square test?

4. Explain the difference between relative odds and relative risk.

5. How can the use of a restricted rather than a random sample affect the results of an analysis?

6. Consider the chi-square distribution with 2 degrees of freedom.
   (a) What proportion of the area under the curve lies to the right of $\chi^2 = 9.21$?
   (b) What proportion of the area lies to the right of $\chi^2 = 7.38$?
   (c) What value of χ^2 cuts off the upper 10% of the distribution?

7. Consider the chi-square distribution with 17 degrees of freedom.
   (a) What proportion of the area under the curve lies to the right of $\chi^2 = 33.41$?
   (b) What proportion of the area lies to the left of $\chi^2 = 27.59$?
   (c) What value of χ^2 cuts off the upper 10% of the distribution?

8. The following data come from a study designed to investigate drinking problems among college students. In 1983, a group of students were asked whether they had ever driven an automobile while drinking. In 1987, after the legal drinking age was raised, a different group of college students were asked the same question [11].

Drove While Drinking	Year		Total
	1983	1987	
Yes	1250	991	2241
No	1387	1666	3053
Total	2637	2657	5294

   (a) Use the chi-square test to evaluate the null hypothesis that the population proportions of students who drove while drinking are the same in the two calendar years.
   (b) What do you conclude about the behavior of college students?
   (c) Again test the null hypothesis that the proportions of students who drove while drinking are identical for the two calendar years. This time, use the method based on the normal approximation to the binomial distribution that was presented in Section 14.6. Do you reach the same conclusion?
   (d) Construct a 95% confidence interval for the true difference in population proportions.

(e) Does the 95% confidence interval contain the value 0? Would you have expected that it would?

9. A study was conducted to evaluate the relative efficacy of supplementation with calcium versus calcitriol in the treatment of postmenopausal osteoporosis [12]. Calcitriol is an agent that has the ability to increase gastrointestinal absorption of calcium. A number of patients withdrew from this study prematurely due to the adverse effects of treatment, which include thirst, skin problems, and neurologic symptoms. The relevant data appear below.

Treatment	Withdrawal Yes	Withdrawal No	Total
Calcitriol	27	287	314
Calcium	20	288	308
Total	47	575	622

(a) Compute the sample proportion of subjects who withdrew from the study in each treatment group.
(b) Test the null hypothesis that there is no association between treatment group and withdrawal from the study at the 0.05 level of significance. What do you conclude?

10. In a survey conducted in Italy, physicians with different specialties were questioned regarding the surgical treatment of early breast cancer. In particular, they were asked whether they would recommend radical surgery regardless of a patient's age (R), conservative surgery only for younger patients (CR), or conservative surgery regardless of age (C). The results of this survey are presented below [13].

Physician Specialty	Surgery R	Surgery CR	Surgery C	Total
Internal	6	22	42	70
Surgery	23	61	127	211
Radiotherapy	2	3	54	59
Oncology	1	12	43	56
Gynecology	1	12	31	44
Total	33	110	297	440

(a) At the 0.05 level of significance, test the null hypothesis that there is no association between physician specialty and recommended treatment.
(b) What do you conclude?

11. The following table compiles data from six studies designed to investigate the accuracy of death certificates. The results of 5373 autopsies were compared to the causes of death listed on the certificates. Of those considered, 3726 certificate were confirmed to be accurate, 783 either lacked information or contained inaccuracies but did not require recoding of the underlying cause of death, and 864 were incorrect and required recoding [14].

Date of Study	Certificate Status			Total
	Confirmed Accurate	Inaccurate No Change	Incorrect Recoding	
1955–1965	2040	367	327	2734
1970	149	60	48	257
1970–1971	288	25	70	383
1975–1977	703	197	252	1152
1977–1978	425	62	88	575
1980	121	72	79	272
Total	3726	783	864	5373

(a) Do you believe that the results are homogeneous or consistent across studies?

(b) It should be noted that autopsies are not performed at random; in fact, many are done because the cause of death listed on the certificate is uncertain. What problems may arise if you attempt to use the results of these studies to make inference about the population as a whole?

12. In a study of intraobserver variability in the assessment of cervical smears, 3325 slides were screened for the presence or absence of abnormal squamous cells. Each slide was screened by a particular observer and then rescreened six months later by the same observer. The results of this study are shown below [15].

First Screening	Second Screening		Total
	Present	Absent	
Present	1763	489	2252
Absent	403	670	1073
Total	2166	1159	3325

(a) Do these data support the null hypothesis that there is no association between time of screening and diagnosis?

The data could also be displayed in the following manner:

Abnormal Cells	Screening		Total
	First	Second	
Present	2252	2166	4418
Absent	1073	1159	2232
Total	3325	3325	6650

(b) Is there anything wrong with this presentation? How would you analyze these data?

13. Suppose that you are interested in investigating the association between retirement status and heart disease. One concern might be the age of the subjects: An older person is more likely to be retired, and also more likely to have heart disease. In one study, therefore, 127 victims of cardiac arrest were matched on a number of characteristics that included age with 127 healthy control subjects; retirement status was then ascertained for each subject [16].

	Cardiac Arrest		
Healthy	Retired	Not Retired	Total
Retired	27	12	39
Not Retired	20	68	88
Total	47	80	127

(a) Test the null hypothesis that there is no association between retirement status and cardiac arrest.
(b) What do you conclude?
(c) Estimate the relative odds of being retired for healthy individuals versus those who have experienced cardiac arrest.
(d) Construct a 95% confidence interval for the true population odds ratio. Does this interval contain the value 1? What does this mean?

14. In response to a study suggesting a link between lack of circumcision in males and cervical cancer in their female partners, an investigation was conducted to assess the accuracy of reported circumcision status. Before registering at a cancer institute, male patients were asked to fill out a questionnaire; the data requested included circumcision status. This information was confirmed by interview. Subsequently, each man received a complete physical examination during which the physician noted whether he was circumcised or not. The data, collected over a two-month period, are presented below [17].

	Patient Statement		
Examination	Yes	No	Total
Yes	37	47	84
No	19	89	108
Total	56	136	192

At the 0.05 level of significance, does there appear to be an association between the results of the examination and the patient's own response? If so, what is the relationship?

15. In an effort to evaluate the health effects of air pollutants containing sulphur, individuals living near a pulp mill in Finland were questioned about various symptoms following a strong emission released in September 1987 [18]. The same subjects were questioned again four months later, during a low-exposure period. A summary of the responses related to the occurrence of headaches is presented below.

Low Exposure	High Exposure		Total
	Yes	No	
Calcitriol	2	2	4
Calcium	8	33	41
Total	10	35	45

(a) Test the null hypothesis that there is no association between exposure to air pollutants containing sulphur and the occurrence of headaches.

(b) What do you conclude?

16. In a study of the risk factors for invasive cervical cancer that was conducted in Germany, the following data were collected relating smoking status to the presence or absence of cervical cancer [19].

	Smoker	Nonsmoker	Total
Cancer	108	117	225
No Cancer	163	268	431
Total	271	385	656

(a) Estimate the relative odds of invasive cervical cancer for smokers versus non-smokers.

(b) Calculate a 95% confidence interval for the population odds ratio.

(c) Test the null hypothesis that there is no association between smoking status and the presence of cervical cancer at the 0.05 level of significance.

17. In France, a study was conducted to investigate potential risk factors for ectopic pregnancy. Of the 279 women who had experienced ectopic pregnancy, 28 had suffered from pelvic inflammatory disease. Of the 279 women who had not, 6 had suffered from pelvic inflammatory disease [20].

(a) Construct a 2×2 contingency table for these data.

(b) Estimate the relative odds of suffering ectopic pregnancy for women who have had pelvic inflammatory disease versus women who have not.

(c) Find a 99% confidence interval for the population relative odds.

18. The data on the following page were taken from a study investigating the associations between spontaneous abortion and various risk factors [21].

Alcohol Use (Drinks per Week)	Number of Pregnancies	Spontaneous Abortions
0	33164	6793
1–2	9099	2068
3–6	3069	776
7–20	1527	456
21+	287	98

(a) For each level of alcohol consumption, estimate the probability that a woman who becomes pregnant will undergo a spontaneous abortion.

(b) For each category of alcohol use, estimate the relative odds of experiencing a spontaneous abortion for women who consume some amount of alcohol versus those who do not consume any.

(c) In each case, calculate a 95% confidence interval for the odds ratio.

(d) What do you conclude?

19. In a study of HIV infection among women entering the New York State prison system, 475 inmates were cross-classified with respect to HIV seropositivity and their histories of intravenous drug use. These data are stored on your disk in a data set called `prison` [22] (Appendix B, Table B.17). The indicators of seropositivity are saved under the variable name `hiv` and those of intravenous drug use under `ivdu`.

(a) Among women who have used drugs intravenously, what proportion are HIV-positive? Among women who have not used drugs intravenously, what proportion are HIV-positive?

(b) At the 0.05 level of significance, test the null hypothesis that there is no association between history of intravenous drug use and HIV seropositivity.

(c) What do you conclude?

(d) Estimate the relative odds of being HIV-positive for women who have used drugs intravenously versus those who have not.

20. A study was conducted to determine whether geographic variations in the use of medical and surgical services could be explained in part by differences in the appropriateness with which physicians use these services. One concern might be that a high rate of inappropriate use of a service is associated with high overall use within a particular region. For the procedure coronary angiography, three geographic areas were studied: a high-use site (Site 1), a low-use urban site (Site 2), and a low-use rural site (Site 3). Within each geographical region, each use of this procedure was classified as appropriate, equivocal, or inappropriate by a panel of expert physicians. These data are saved on your disk in a data set called `angio` [23] (Appendix B, Table B.18). Site number is saved under the variable name `site`, and level of appropriateness under `appropro`.

(a) At the 0.05 level of significance, test the null hypothesis that there is no association between geographic region and the appropriateness of use of coronary angiography.

(b) What do you conclude?

21. Two different questionnaire formats designed to measure alcohol consumption—one encompassing all types of food in the diet and the other specifically targeting alcohol use—were compared for men and women between 50 and 65 years of age living in a particular community. For each of the alcoholic beverages beer, liquor, red wine, and white wine, each subject was classified as either a nondrinker (never or less than one drink per month) or a drinker (one or more drinks per month) according to each of the questionnaires. The relevant information pertaining to beer consumption is saved in the data set `alcohol` [24] (Appendix B, Table B.19); categories for the generic questionnaire are saved under the name `genques` and those for the questionnaire targeting alcohol use are saved under `alcques`.

(a) Test the null hypothesis that there is no association between drinking statuses on the two different types of questionnaires.

(b) What do you conclude?

Bibliography

[1] Thompson, R. S., Rivara, F. P., and Thompson, D. C., "A Case-Control Study of the Effectiveness of Bicycle Safety Helmets," *The New England Journal of Medicine,* Volume 320, May 25, 1989, 1361–1367.

[2] Cochran, W. G., "Some Methods for Strengthening the Common χ^2 Test," *Biometrics,* Volume 10, December 1954, 417–451.

[3] Grizzle, J. E., "Continuity Correction in the χ^2 Test for 2×2 Tables," *The American Statistician,* Volume 21, October 1967, 28–32.

[4] Rosner, B., *Fundamentals of Biostatistics,* Fourth Edition, Belmont, CA: Wadsworth Publishing Company, 1995.

[5] Schottenfeld, D., Eaton, M., Sommers, S. C., Alonso, D. R., and Wilkinson, C., "The Autopsy as a Measure of Accuracy of the Death Certificate," *Bulletin of the New York Academy of Medicine,* Volume 58, December 1982, 778–794.

[6] Coulehan, J. L., Lerner, G., Helzlsouer, K., Welty, T. K., and McLaughlin, J., "Acute Myocardial Infarction Among Navajo Indians, 1976–1983," *American Journal of Public Health,* Volume 76, April 1986, 412–414.

[7] McCusker, J., Harris, D. R., and Hosmer, D. W., "Association of Electronic Fetal Monitoring During Labor with Caesarean Section Rate and with Neonatal Morbidity and Mortality," *American Journal of Public Health,* Volume 78, September 1988, 1170–1174.

[8] Roberts, R. S., Spitzer, W. O., Delmore, T., and Sackett, D. L., "An Empirical Demonstration of Berkson's Bias," *Journal of Chronic Diseases,* Volume 31, February 1978, 119–128.

[9] Gross, T. P., Conde, J. G., Gary, G. W., Harting, D., Goeller, D., and Israel, E., "An Outbreak of Acute Infectious Nonbacterial Gastroenteritis in a High School in Maryland," *Public Health Reports,* Volume 104, March–April 1989, 164–169.

[10] Kirscht, J. P., Brock, B. M., and Hawthorne, V. M., "Cigarette Smoking and Changes in Smoking Among a Cohort of Michigan Adults, 1980-82," *American Journal of Public Health,* Volume 77, April 1987, 501–502.

[11] Engs, R. C., and Hanson, D. J., "University Students' Drinking Patterns and Problems: Examining the Effects of Raising the Purchase Age," *Public Health Reports,* Volume 103, November–December 1988, 667–673.

[12] Tilyard, M. W., Spears, G. F. S., Thomson, J., and Dovey, S., "Treatment of Postmenopausal Osteoporosis with Calcitriol or Calcium," *The New England Journal of Medicine,* Volume 326, February 6, 1992, 357–362.

[13] Liberati, A., Apolone, G., Nicolucci, A., Confalonieri, C., Fossati, R., Grilli, R., Torri, V., Mosconi, P., and Alexanian, A., "The Role of Attitudes, Beliefs, and Personal Characteristics of Italian Physicians in the Surgical Treatment of Early Breast Cancer," *American Journal of Public Health,* Volume 81, January 1991, 38–42.

[14] Kircher, T., Nelson, J., and Burdo, H., "The Autopsy as a Measure of Accuracy of the Death Certificate," *The New England Journal of Medicine*, Volume 313, November 14, 1985, 1263–1269.

[15] Klinkhamer, P. J. J. M., Vooijs, G. P., and de Haan, A. F. J., "Intraobserver and Interobserver Variability in the Quality Assessment of Cervical Smears," *Acta Cytologica,* Volume 33, March–April 1989, 215–218.

[16] Siscovick, D. S., Strogatz, D. S., Weiss, N. S., and Rennert, G., "Retirement and Primary Cardiac Arrest in Males," *American Journal of Public Health,* Volume 80, February 1990, 207–208.

[17] Lilienfeld, A. M., and Graham, S., "Validity of Determining Circumcision Status by Questionnaire as Related to Epidemiological Studies of Cancer of the Cervix," *Journal of the National Cancer Institute,* Volume 21, October 1958, 713–720.

[18] Haahtela, T., Marttila, O., Vilkka, V., Jappinen, P., and Jaakkola, J. J. K., "The South Karelia Air Pollution Study: Acute Health Effects of Malodorous Sulphur Air Pollutants Released by a Pulp Mill," *American Journal of Public Health,* Volume 82, April 1992, 603–605.

[19] Nischan, P., Ebeling, K., and Schindler, C., "Smoking and Invasive Cervical Cancer Risk: Results from a Case-Control Study," *American Journal of Epidemiology,* Volume 128, July 1988, 74–77.

[20] Coste, J., Job-Spira, N., Fernandez, H., Papiernik, E., and Spira, A., "Risk Factors for Ectopic Pregnancy: A Case-Control Study in France, with Special Focus on Infectious Factors," *American Journal of Epidemiology,* Volume 133, May 1, 1991, 839–849.

[21] Armstrong, B. G., McDonald, A. D., and Sloan, M., "Cigarette, Alcohol, and Coffee Consumption and Spontaneous Abortion," *American Journal of Public Health,* Volume 82, January 1992, 85–87.

[22] Smith, P. F., Mikl, J., Truman, B. I., Lessner, L., Lehman, J. S., Stevens, R. W., Lord, E. A., Broaddus, R. K., and Morse, D. L., "HIV Infection Among Women Entering the New York State Correctional System," *American Journal of Public Health Supplement,* Volume 81, May 1991, 35–40.

[23] Chassin, M. R., Kosecoff, J., Park, R. E., Winslow, C. M., Kahn, K. L., Merrick, N. J., Keesey, J., Fink, A., Solomon, D. H., Brook, R. H., "Does Inappropriate Use Explain Geographic Variations in the Use of Health Care Services?," *Journal of the American Medical Association,* Volume 258, November 13, 1987, 2533–2537.

[24] King, A., "Enhancing the Self-Report of Alcohol Consumption in the Community: Two Questionnaire Formats," *American Journal of Public Health*, Volume 84, February 1994, 294–296.

16

Multiple 2 × 2 Contingency Tables

When the relationship between a pair of dichotomous random variables is being investigated, it is sometimes examined in two or more populations. As a result, the data to be analyzed consist of a number of 2 × 2 contingency tables rather than just one. In some cases, these tables originate from different studies; more often, they are the results of a single investigation that have been subclassified, or *stratified*, by some factor that is believed to influence the outcome. In either event, it is possible to make inference about the relationship between the two variables by examining the association in each table separately. In many cases, however, it is more useful to be able to combine the information across tables to make a single, unifying statement.

16.1 Simpson's Paradox

Consider the following data, taken from a study investigating the relationship between smoking and aortic stenosis, a narrowing or stricture of the aorta that impedes the flow of blood to the body. Since gender is associated with both of these variables, we suspect that it might influence the observed relationship between them. Therefore, we begin our analysis by examining the effects among males and females separately. The stratified data are presented below [1].

Males

Aortic Stenosis	Smoker		Total
	Yes	No	
Yes	37	25	62
No	24	20	44
Total	61	45	106

Females

Aortic Stenosis	Smoker		Total
	Yes	No	
Yes	14	29	43
No	19	47	66
Total	33	76	109

For males, the odds of developing aortic stenosis among smokers relative to nonsmokers are estimated by

$$\widehat{OR}_M = \frac{(37)(20)}{(25)(24)}$$
$$= 1.23.$$

For females, the odds of aortic stenosis among smokers relative to nonsmokers are estimated by

$$\widehat{OR}_F = \frac{(14)(47)}{(29)(19)}$$
$$= 1.19.$$

We observe the same trend in each subgroup of the population: For both males and females, the odds of developing aortic stenosis are higher among smokers than they are among nonsmokers. It is possible that these two quantities are actually estimating the same population value. Consequently, we might attempt to combine the information in the tables to make a single statement summarizing the relationship between smoking and aortic stenosis. If the two individual tables are summed, the results are as follows:

Aortic Stenosis	Smoker		Total
	Yes	No	
Yes	51	54	105
No	43	67	110
Total	94	121	215

For all individuals in the study, regardless of gender, the odds of developing aortic stenosis among smokers relative to nonsmokers are

$$\widehat{OR} = \frac{(51)(67)}{(54)(43)}$$
$$= 1.47.$$

If the effect of gender is ignored, the strength of the association between smoking and aortic stenosis appears greater than it is for either males or females alone. This phenomenon is an example of *Simpson's paradox*. Simpson's paradox occurs when either the magnitude or the direction of the relationship between two variables is influenced by the presence of a third factor. In this case, gender is a confounder in the relationship between exposure and disease; failure to control for its effect has caused the true magnitude of the association to appear greater than it actually is.

16.2 The Mantel-Haenszel Method

The following data come from a study investigating the relationship between the consumption of caffeinated coffee and nonfatal myocardial infarction among adult males under the age of 55. The study provides exposure and disease information for two samples of men: a group of 1559 smokers and a group of 937 nonsmokers [2].

Smokers

Myocardial Infarction	Coffee		Total
	Yes	No	
Yes	1011	81	1092
No	390	77	467
Total	1401	158	1559

Nonsmokers

Myocardial Infarction	Coffee		Total
	Yes	No	
Yes	383	66	449
No	365	123	488
Total	748	189	937

Among the smokers, the relative odds of suffering a nonfatal myocardial infarction for males drinking caffeinated coffee versus males drinking no coffee at all are estimated by

$$\widehat{OR}_S = \frac{(1011)(77)}{(390)(81)}$$

$$= 2.46.$$

Among nonsmokers, the relative odds of experiencing a nonfatal myocardial infarction among coffee drinkers versus nondrinkers are estimated by

$$\widehat{\text{OR}}_{\text{NS}} = \frac{(383)(123)}{(365)(66)}$$
$$= 1.96.$$

In both subgroups of the population, we observe that the odds of suffering a myocardial infarction are greater among coffee drinkers than among non–coffee drinkers. It is possible that the two odds ratios are actually estimating the same population value and differ only because of sampling variability. If this is the case, we would like to be able to combine the information in the two tables to make a single overall statement about the relationship between myocardial infarction and caffeinated coffee.

We have already seen that if smoking is a confounder in the relationship between coffee drinking and myocardial infarction, we cannot simply sum the observations in the two contingency tables. If we were to attempt this, the following table would result.

Myocardial Infarction	Coffee		Total
	Yes	**No**	**Total**
Yes	1394	147	1541
No	755	200	955
Total	2149	347	2496

Based on these unstratified data, the estimate of the odds of nonfatal myocardial infarction among coffee drinkers relative to those who do not drink coffee is

$$\widehat{\text{OR}} = \frac{(1394)(200)}{(755)(147)}$$
$$= 2.51.$$

This odds ratio is larger than the relative odds in either of the individual strata, suggesting that smoking status is indeed a confounder.

Rather than simply summing the observations in different subgroups, we can use another technique—known as the *Mantel-Haenszel method*—to combine the information in a number of 2×2 tables. We begin by determining whether the strength of association between the exposure and the disease is uniform across the tables. Then, if it is appropriate to do so, the method provides a means of calculating both a point estimate and a confidence interval for the overall population relative odds. In addition, it allows us to test the null hypothesis of no association between exposure and disease.

16.2.1 Test of Homogeneity

Before combining the information in two or more contingency tables, we must first verify that the population odds ratios are in fact constant across the different strata. If they are not, it is not beneficial to compute a single summary value for the overall relative odds. Instead, it would be better to treat the data in the various contingency tables as if they had been drawn from distinct populations and report a different odds ratio for each subgroup.

We determine whether the strength of association between exposure and disease is uniform across a series of g 2 × 2 tables—where g is an integer greater than or equal to 2—by conducting what is called a test of homogeneity. The *test of homogeneity* evaluates the null hypothesis

H_0: The population odds ratios for the g tables are identical,

or, equivalently,

$$H_0: OR_1 = OR_2 = \cdots = OR_i = \cdots = OR_g.$$

The alternative hypothesis is that not all odds ratios are the same. To perform the test, we calculate the statistic

$$X^2 = \sum_{i=1}^{g} w_i (y_i - Y)^2.$$

In this expression, y_i is the logarithm of the estimated odds ratio for the ith table, Y is a weighted average of the g separate log odds ratios, and w_i is the weighting factor for the ith table.

Suppose that the ith 2 × 2 table has the following format:

Disease	Exposure		Total
	Yes	**No**	
Yes	a_i	b_i	N_{1i}
No	c_i	d_i	N_{2i}
Total	M_{1i}	M_{2i}	T_i

The estimate of the odds ratio for this table is

$$\widehat{OR}_i = \frac{a_i d_i}{b_i c_i};$$

the logarithm of the estimated odds ratio is

$$y_i = \ln(\widehat{OR}_i)$$
$$= \ln\left(\frac{a_i d_i}{b_i c_i}\right).$$

The weighted average Y is calculated using the formula

$$Y = \frac{\sum_{i=1}^{g} w_i y_i}{\sum_{i=1}^{g} w_i}$$

for $i = 1, 2, \ldots g$. The weights w_i are computed as

$$w_i = \left[\frac{1}{a_i} + \frac{1}{b_i} + \frac{1}{c_i} + \frac{1}{d_i} \right]^{-1}$$

$$= \frac{1}{[(1/a_i) + (1/b_i) + (1/c_i) + (1/d_i)]}.$$

This quantity is actually the reciprocal of the estimated variance of the log odds ratio for the ith table. If any one of the cell entries is equal to 0, w_i is undefined. In this case, we can use the same modification to the variance that we applied in Section 15.3. Adding 0.5 to each of the values a_i, b_i, c_i, and d_i, the weighting factor becomes

$$w_i = \left[\frac{1}{(a_i + 0.5)} + \frac{1}{(b_i + 0.5)} + \frac{1}{(c_i + 0.5)} + \frac{1}{(d_i + 0.5)} \right]^{-1}.$$

Under the null hypothesis that the odds ratio is constant across tables, the sum

$$X^2 = \sum_{i=1}^{g} w_i (y_i - Y)^2$$

has a distribution that is approximately chi-square with $g - 1$ degrees of freedom. If the p-value associated with this statistic is less than the level of significance of the test, we reject the null hypothesis and report the separate estimates for the odds ratios. If p is greater than α, we cannot reject H_0; therefore, we conclude that it is acceptable to combine the information in the g 2×2 tables using the Mantel-Haenszel method.

Recall that the data used to investigate the relationship between coffee consumption and nonfatal myocardial infarction among men under the age of 55 are divided into two distinct strata: a group of smokers and a group of nonsmokers. Since we have only two contingency tables, g is equal to 2. Before attempting to combine the information in the tables, we test the null hypothesis that the population odds ratios for the two groups are identical, or

$$H_0: OR_1 = OR_2,$$

against the alternative

$$H_A: OR_1 \neq OR_2.$$

We conduct a two-sided test of homogeneity at the $\alpha = 0.05$ level of significance.

Recall that the estimate of the relative odds for smokers is

$$\widehat{OR}_1 = \frac{a_1 d_1}{b_1 c_1}$$

$$= \frac{(1011)(77)}{(390)(81)}$$

$$= 2.46.$$

Therefore,

$$y_1 = \ln (\widehat{OR}_1)$$
$$= \ln (2.46)$$
$$= 0.900.$$

The weighting factor for this table is

$$w_1 = \frac{1}{[(1/1011) + (1/390) + (1/81) + (1/77)]}$$
$$= 34.62.$$

Similarly, the estimate of the relative odds for nonsmokers is

$$\widehat{OR}_2 = \frac{a_2 d_2}{b_2 c_2}$$
$$= \frac{(383)(123)}{(365)(66)}$$
$$= 1.96,$$

and

$$y_2 = \ln (\widehat{OR}_2)$$
$$= \ln (1.96)$$
$$= 0.673.$$

For this table, the weighting factor is

$$w_2 = \frac{1}{[(1/383) + (1/365) + (1/66) + (1/123)]}$$
$$= 34.93.$$

Using these values, the weighted average of y_1 and y_2 is found to be

$$Y = \frac{\sum_{i=1}^{g} w_i y_i}{\sum_{i=1}^{g} w_i}$$
$$= \frac{(w_1 y_1 + w_2 y_2)}{(w_1 + w_2)}$$
$$= \frac{(34.62)(0.900) + (34.93)(0.673)}{(34.62 + 34.93)}$$
$$= 0.786.$$

Finally, the test statistic is

$$X^2 = \sum_{i=1}^{g} w_i (y_i - Y)^2$$
$$= w_1 (y_1 - Y)^2 + w_2 (y_2 - Y)^2$$
$$= (34.62)(0.900 - 0.786)^2 + (34.93)(0.673 - 0.786)^2$$
$$= 0.896.$$

Consulting Table A.8, we observe that for a chi-square distribution with 1 degree of freedom, $p > 0.10$. We cannot reject the null hypothesis; the data do not indicate that the population odds ratios relating coffee consumption and nonfatal myocardial infarction differ for smokers and nonsmokers. Consequently, we assume that the odds ratios for the two strata are in fact estimating the same quantity, and proceed with the Mantel-Haenszel method of combining this information.

16.2.2 Summary Odds Ratio

If the individual odds ratios are found to be uniform across tables, the next step in the Mantel-Haenszel method is to compute an estimate of the overall strength of association. This estimate is actually a weighted average of the odds ratios for the g separate strata; it is calculated using the formula

$$\widehat{OR} = \frac{\sum_{i=1}^{g} (a_i d_i / T_i)}{\sum_{i=1}^{g} (b_i c_i / T_i)},$$

where T_i is the total number of observations in the ith table.

For the data relating coffee consumption to nonfatal myocardial infarction, the estimated summary odds ratio is

$$\widehat{OR} = \frac{\sum_{i=1}^{g} (a_i d_i / T_i)}{\sum_{i=1}^{g} (b_i c_i / T_i)}$$
$$= \frac{(a_1 d_1 / T_1) + (a_2 d_2 / T_2)}{(b_1 c_1 / T_1) + (b_2 c_2 / T_2)}$$
$$= \frac{(1011)(77)/1559 + (383)(123)/937}{(390)(81)/1559 + (365)(66)/937}$$
$$= 2.18.$$

Once differences in smoking status have been taken into account, males under the age of 55 who drink caffeinated coffee have odds of experiencing nonfatal myocardial infarction that are 2.18 times greater than the odds for males who do not drink coffee.

In addition to computing a point estimate for the summary odds ratio, we might also want to calculate a confidence interval that represents a range of reasonable values for this quantity. When constructing a confidence interval for the odds ratio using data drawn from a single population, we noted that the sampling distribution of the relative odds is not symmetric; instead, it is skewed to the right. The same is true for the Mantel-Haenszel estimator of the common odds ratio. Since the distribution of the natural logarithm of the odds ratio is more symmetric and is approximately normal, we begin by calculating a confidence interval for ln(OR). In addition, to ensure that our strata sample sizes are large enough to make the technique used valid, we recommend the following restrictions on the expected values of the observations across the g tables:

$$\sum_{i=1}^{g} \frac{M_{1i}N_{1i}}{T_i} \geq 5,$$

$$\sum_{i=1}^{g} \frac{M_{1i}N_{2i}}{T_i} \geq 5,$$

$$\sum_{i=1}^{g} \frac{M_{2i}N_{1i}}{T_i} \geq 5,$$

and

$$\sum_{i=1}^{g} \frac{M_{2i}N_{2i}}{T_i} \geq 5.$$

The quantity Y that we calculated when conducting the test of homogeneity,

$$Y = \frac{\sum_{i=1}^{g} w_i y_i}{\sum_{i=1}^{g} w_i},$$

is the weighted average of g separate log odds ratios and represents an estimator of ln (OR). The estimated standard error of Y is

$$\widehat{se}(Y) = \frac{1}{\sqrt{\sum_{i=1}^{g} w_i}}.$$

Therefore, a 95% confidence interval for ln (OR) takes the form

$$(Y - 1.96\,\widehat{se}(Y),\ Y + 1.96\,\widehat{se}(Y)).$$

If we take the antilogarithm of each limit, a 95% confidence interval for the summary odds ratio itself is

$$(e^{Y - 1.96\,\widehat{se}(Y)},\ e^{Y + 1.96\,\widehat{se}(Y)}).$$

Before computing a confidence interval for the common odds ratio measuring the strength of association between coffee consumption and myocardial infarction, we first verify the restrictions placed on the expected values of the observations to ensure that the sample sizes are large enough; note that

$$\sum_{i=1}^{2} \frac{M_{1i}N_{1i}}{T_i} = \frac{M_{11}N_{11}}{T_1} + \frac{M_{12}N_{12}}{T_2}$$

$$= \frac{(1401)(1092)}{1559} + \frac{(748)(449)}{937}$$

$$= 1339.8,$$

$$\sum_{i=1}^{2} \frac{M_{1i}N_{2i}}{T_i} = \frac{M_{11}N_{21}}{T_1} + \frac{M_{12}N_{22}}{T_2}$$

$$= \frac{(1401)(467)}{1559} + \frac{(748)(488)}{937}$$

$$= 809.2,$$

$$\sum_{i=1}^{2} \frac{M_{2i}N_{1i}}{T_i} = \frac{M_{21}N_{11}}{T_1} + \frac{M_{22}N_{12}}{T_2}$$

$$= \frac{(158)(1092)}{1559} + \frac{(189)(449)}{937}$$

$$= 201.2,$$

and

$$\sum_{i=1}^{2} \frac{M_{2i}N_{2i}}{T_i} = \frac{M_{21}N_{21}}{T_1} + \frac{M_{22}N_{22}}{T_2}$$

$$= \frac{(158)(467)}{1559} + \frac{(189)(488)}{937}$$

$$= 145.8.$$

Since each of these sums is greater than 5, we proceed with the construction of a confidence interval. We previously found that

$$Y = 0.786.$$

Since $w_1 = 34.62$ and $w_2 = 34.93$,

$$\widehat{se}\,(Y) = \frac{1}{\sqrt{\sum_{i=1}^{2} w_i}}$$

$$= \frac{1}{\sqrt{w_1 + w_2}}$$

$$= \frac{1}{\sqrt{34.62 + 34.93}}$$

$$= 0.120.$$

Therefore, a 95% confidence interval for ln (OR) is

$$(0.786 - 1.96(0.120), 0.786 + 1.96(0.120)),$$

or

$$(0.551, 1.021).$$

The 95% confidence interval for the summary odds ratio itself is

$$(e^{0.551}, e^{1.021}),$$

or

$$(1.73, 2.78).$$

After adjusting for the effects of smoking, we are 95% confident that men who drink caffeinated coffee have odds of experiencing nonfatal myocardial infarction that are between 1.73 and 2.78 times greater than the odds for men who do not drink coffee.

16.2.3   Test of Association

The final step in the Mantel-Haenszel method of combining information from two or more 2 × 2 contingency tables is to test whether the summary odds ratio is equal to 1; relative odds of 1 indicate that there is no association between exposure and disease. One way to perform this test is to simply refer to the 95% confidence interval for the summary odds ratio. For the data examining the relationship between coffee and myocardial infarction, the interval does not contain the value 1; an odds ratio of 1 would imply that men who drink coffee and those who do not have the same odds of experiencing myocardial infarction. Instead, this sample would lead us to reject the null hypothesis of no association between exposure and disease at the 0.05 level of significance and conclude that men who drink caffeinated coffee have higher odds of myocardial infarction.

Recall, however, that the confidence interval for the summary odds ratio was constructed under the assumption that the sampling distribution of the log of the odds ratio is approximately normal. To evaluate the null hypothesis

$$H_0: \mathrm{OR} = 1$$

more directly, we could use an alternative method and calculate the test statistic

$$X^2 = \frac{\left[\sum_{i=1}^{g} a_i - \sum_{i=1}^{g} m_i\right]^2}{\sum_{i=1}^{g} \sigma_i^2}.$$

In this expression, a_i is the observed number of exposed individuals who develop disease; it is the upper-left-hand entry in the ith 2×2 table. The term m_i—the expected value of a_i—is computed as

$$m_i = \frac{M_{1i}N_{1i}}{T_i}.$$

Finally, σ_i is the standard deviation of a_i, where

$$\sigma_i^2 = \frac{M_{1i}M_{2i}N_{1i}N_{2i}}{T_i^2(T_i - 1)}.$$

Like the chi-square test statistic for a single 2×2 table, the quantity

$$X^2 = \frac{\left[\sum_{i=1}^{g} a_i - \sum_{i=1}^{g} m_i\right]^2}{\sum_{i=1}^{g} \sigma_i^2}$$

compares the observed frequencies for each table to the corresponding expected frequencies. It has an approximate chi-square distribution with 1 degree of freedom. If the p-value associated with this statistic is less than the level of significance α—suggesting that the deviations between the observed and the expected values are too large to be attributed to chance alone—we reject the null hypothesis that the summary odds ratio is equal to 1. If p is greater than α, we do not reject H_0.

For the data relating coffee consumption to nonfatal myocardial infarction, we would like to evaluate the null hypothesis

$$H_0: \text{OR} = 1$$

against the alternative

$$H_A: \text{OR} \neq 1$$

using a two-sided test and setting the significance level at $\alpha = 0.05$. Note that

$$a_1 = 1011,$$

$$m_1 = \frac{M_{11}N_{11}}{T_1}$$

$$= \frac{(1401)(1092)}{1559}$$

$$= 981.3,$$

and

$$\sigma_1^2 = \frac{M_{11}M_{21}N_{11}N_{21}}{T_1^2(T_1 - 1)}$$

$$= \frac{(1401)(158)(1092)(467)}{(1559)^2(1559 - 1)}$$

$$= 29.81.$$

Similarly,

$$a_2 = 383,$$

$$m_2 = \frac{M_{12}N_{12}}{T_2}$$

$$= \frac{(748)(449)}{937}$$

$$= 358.4,$$

and

$$\sigma_2^2 = \frac{M_{12}M_{22}N_{12}N_{22}}{T_2^2(T_2 - 1)}$$

$$= \frac{(748)(189)(449)(488)}{(937)^2(937 - 1)}$$

$$= 37.69.$$

Therefore, the test statistic is

$$X^2 = \frac{\left[\sum_{i=1}^2 a_i - \sum_{i=1}^2 m_i\right]^2}{\sum_{i=1}^2 \sigma_i^2}$$

$$= \frac{[(a_1 + a_2) - (m_1 + m_2)]^2}{\sigma_1^2 + \sigma_2^2}$$

$$= \frac{[(1011 + 383) - (981.3 + 358.4)]^2}{29.81 + 37.69}$$

$$= 43.68.$$

Consulting Table A.8, we observe that the corresponding p-value is less than 0.001. Accordingly, we reject the null hypothesis of no association between exposure and disease and conclude that the summary odds ratio is not equal to 1. After adjusting for differences in smoking status, we again find that adult males under the age of 55 who drink caffeinated coffee face a significantly higher risk of experiencing nonfatal myocardial infarction than males of the same age who do not drink coffee.

These data represent the results of a single study examining the effects of coffee consumption on human health; other studies have reported conflicting results. At the present time it appears that, in moderate amounts, coffee is a fairly safe beverage for otherwise healthy individuals [3,4].

16.3   Further Applications

Suppose we are presented with a pair of 2×2 contingency tables, both of which provide information about the same dichotomous random variables representing exposure and disease, but that originate from two distinct studies. For example, the following data come from two studies, both conducted in San Francisco, that investigate risk factors for epithelial ovarian cancer [5].

Study 1

Disease Status	Term Pregnancies		Total
	None	One or More	
Cancer	31	80	111
No Cancer	93	379	472
Total	124	459	583

Study 2

Disease Status	Term Pregnancies		Total
	None	One or More	
Cancer	39	149	188
No Cancer	74	465	539
Total	113	614	727

In the first study, the relative odds of developing ovarian cancer for women who have never had a term pregnancy versus women who have had one or more are estimated by

$$\widehat{OR}_1 = \frac{(31)(379)}{(80)(93)}$$
$$= 1.58.$$

In the second study, the same odds ratio is estimated by

$$\widehat{OR}_2 = \frac{(39)(465)}{(149)(74)}$$
$$= 1.64.$$

In both studies, the odds of developing ovarian cancer are greater for women who have never experienced a term pregnancy than for those who have. The two odds ratios are very similar; it is quite possible that they are both estimating the same population value. Can we combine the evidence collected in the two different studies to make a single overall statement about the association between the number of term pregnancies a woman has had and the occurrence of epithelial ovarian cancer?

As we have seen, it is not advisable to simply sum the observations in the two contingency tables to calculate a single odds ratio; instead, we apply the Mantel-Haenszel method for combining the information. This technique consists of four steps: (1) the test of homogeneity, (2) the calculation of a point estimate for the summary odds ratio, (3) the construction of a confidence interval for the summary odds ratio, and (4) a test of hypothesis to determine whether an association exists between exposure and disease.

The test of homogeneity is conducted to verify that the population odds ratio is constant across groups or strata; it evaluates the null hypothesis

H_0: The relative odds of developing ovarian cancer among women who have never had a term pregnancy versus women who have had one or more are identical for the populations represented by the two different studies,

or

$$H_0: \mathrm{OR}_1 = \mathrm{OR}_2.$$

The alternative hypothesis is that the population odds ratios are not the same. To perform the test, we calculate the statistic

$$X^2 = \sum_{i=1}^{2} w_i (y_i - Y)^2.$$

Recall that the estimate of the relative odds for the first study is

$$\widehat{\mathrm{OR}}_1 = \frac{(31)(379)}{(80)(93)}$$
$$= 1.58;$$

therefore,

$$y_1 = \ln(\widehat{\mathrm{OR}}_1)$$
$$= \ln(1.58)$$
$$= 0.457.$$

The weighting factor for this table is

$$w_1 = \frac{1}{[(1/31) + (1/80) + (1/93) + (1/379)]}$$
$$= 17.20.$$

Similarly, the estimate of the odds ratio for the second study is

$$\widehat{OR}_2 = \frac{(39)(465)}{(149)(74)}$$

$$= 1.64;$$

thus,

$$y_2 = \ln(\widehat{OR}_2)$$

$$= \ln(1.64)$$

$$= 0.495,$$

and

$$w_2 = \frac{1}{[(1/39) + (1/149) + (1/74) + (1/465)]}$$

$$= 20.83.$$

The weighted average of y_1 and y_2, or Y, is found to be

$$Y = \frac{\sum_{i=1}^{2} w_i y_i}{\sum_{i=1}^{2} w_i}$$

$$= \frac{(w_1 y_1 + w_2 y_2)}{(w_1 + w_2)}$$

$$= \frac{(17.20)(0.457) + (20.83)(0.495)}{(17.20 + 20.83)}$$

$$= 0.478.$$

Substituting all these values into the formula for the test statistic,

$$X^2 = \sum_{i=1}^{2} w_i (y_i - Y)^2$$

$$= w_1 (y_1 - Y)^2 + w_2 (y_2 - Y)^2$$

$$= (17.20)(0.457 - 0.478)^2 + (20.83)(0.495 - 0.478)^2$$

$$= 0.014.$$

Consulting Table A.8, we note that for a chi-square distribution with 1 degree of freedom, $p > 0.10$. Since p is greater than the significance level $\alpha = 0.05$, we cannot reject the null hypothesis. The data do not suggest that the population odds ratios for the two studies differ; therefore, we can proceed with the Mantel-Haenszel method for combining the information.

The estimate of the summary odds ratio, which is actually a weighted average of the two stratum-specific estimates, is

$$\widehat{OR} = \frac{\sum_{i=1}^{2}(a_i d_i / T_i)}{\sum_{i=1}^{2}(b_i c_i / T_i)}$$

$$= \frac{(a_1 d_1 / T_1) + (a_2 d_2 / T_2)}{(b_1 c_1 / T_1) + (b_2 c_2 / T_2)}$$

$$= \frac{(31)(379)/583 + (39)(465)/727}{(80)(93)/583 + (149)(74)/727}$$

$$= 1.61.$$

Once we have taken study differences into account, we see that women who have never had a term pregnancy have odds of developing ovarian cancer that are 1.61 times greater than the odds for women who have had one or more term pregnancies.

In addition to computing a point estimate for the summary odds ratio, we can also calculate a confidence interval. Before doing so, however, we should first verify that the sample sizes in the two studies are large enough to ensure that the method used is valid. Note that

$$\sum_{i=1}^{2} \frac{M_{1i} N_{1i}}{T_i} = \frac{(124)(111)}{583} + \frac{(113)(188)}{727}$$

$$= 52.8,$$

$$\sum_{i=1}^{2} \frac{M_{1i} N_{2i}}{T_i} = \frac{(124)(472)}{583} + \frac{(113)(539)}{727}$$

$$= 184.2,$$

$$\sum_{i=1}^{2} \frac{M_{2i} N_{1i}}{T_i} = \frac{(459)(111)}{583} + \frac{(614)(188)}{727}$$

$$= 246.2,$$

and

$$\sum_{i=1}^{2} \frac{M_{2i} N_{2i}}{T_i} = \frac{(459)(472)}{583} + \frac{(614)(539)}{727}$$

$$= 826.8.$$

Since each of these sums is greater than 5, we can proceed with the construction of the confidence interval.

Recall that

$$Y = 0.478.$$

The values of the weights are $w_1 = 17.20$ and $w_2 = 20.83$, and the estimated standard error of Y is

$$\widehat{se}\,(Y) = \frac{1}{\sqrt{(w_1 + w_2)}}$$

$$= \frac{1}{\sqrt{(17.20 + 20.83)}}$$

$$= 0.162.$$

Therefore, a 95% confidence interval for ln (OR) is

$$(0.478 - 1.96(0.162),\ 0.478 + 1.96(0.162)),$$

or

$$(0.160, 0.796),$$

and a 95% confidence interval for the summary odds ratio itself is

$$(e^{0.160},\ e^{0.796}),$$

or

$$(1.17, 2.22).$$

After adjusting for the effects of the different studies, we are 95% confident that women who have never had a term pregnancy have odds of developing ovarian cancer that are between 1.17 and 2.22 times greater than the odds for women who have had one or more term pregnancies.

The final step in the Mantel-Haenszel method is to test whether the summary odds ratio is equal to 1, or

$$H_0 \colon OR = 1,$$

which would imply that there is no association between exposure and disease. The alternative hypothesis is that the odds ratio is not equal to 1. By noting that the 95% confidence interval for the summary odds ratio does not contain the value 1, we can conclude that the given sample would lead us to reject H_0 at the 0.05 level of significance. To conduct the test more directly, we calculate the statistic

$$X^2 = \frac{\left[\sum_{i=1}^{2} a_i - \sum_{i=1}^{2} m_i\right]^2}{\sum_{i=1}^{2} \sigma_i^2}.$$

Note that

$$a_1 = 31,$$

$$m_1 = \frac{M_{11}N_{11}}{T_1}$$

$$= \frac{(124)(111)}{583}$$

$$= 23.6,$$

and

$$\sigma_1^2 = \frac{M_{11}M_{21}N_{11}N_{21}}{T_1^2(T_1 - 1)}$$

$$= \frac{(124)(459)(111)(472)}{(583)^2(583 - 1)}$$

$$= 15.07.$$

Similarly,

$$a_2 = 39,$$

$$m_2 = \frac{M_{12}N_{12}}{T_2}$$

$$= \frac{(113)(188)}{727}$$

$$= 29.2,$$

and

$$\sigma_2^2 = \frac{M_{12}M_{22}N_{12}N_{22}}{T_2^2(T_2 - 1)}$$

$$= \frac{(113)(614)(188)(539)}{(727)^2(727 - 1)}$$

$$= 18.32.$$

Therefore, the test statistic is

$$X^2 = \frac{\left[\sum_{i=1}^{2} a_i - \sum_{i=1}^{2} m_i\right]^2}{\sum_{i=1}^{2} \sigma_i^2}$$

$$= \frac{[(a_1 + a_2) - (m_1 + m_2)]^2}{\sigma_1^2 + \sigma_2^2}$$

$$= \frac{[(31 + 39) - (23.6 + 29.2)]^2}{(15.07 + 18.32)}$$

$$= 8.86.$$

16.1

ɔutput displaying the Mantel-Haenszel method for calculating the summary odds

dy	OR	[95% Conf. Interval]		M-H Weight
1	1.5792	0.9871	2.5273	12.7616
2	1.6447	1.0727	2.5224	15.1664
ɪ-H combined	1.6148	1.1751	2.2189	

Test for heterogeneity (M-H)   chi2(1)=0.016   Pr>chi2=0.9006

Test that combined OR=1:
  Mantel-Haenszel chi2(1)=8.83
  Pr>chi2=0.0030

Consulting Table A.8, we observe that $0.001 < p < 0.01$. Therefore, we reject the null hypothesis at the 0.05 level of significance and conclude that the summary odds ratio is not equal to 1. After adjusting for study differences, we find that women who have never had a term pregnancy have significantly higher odds of developing epithelial ovarian cancer than women who have had one or more term pregnancies.

Some computer packages estimate the summary odds ratio for a number of 2×2 tables and also test whether the true odds ratio is equal to 1. The relevant output from Stata is contained in Table 16.1. The top portion of the output shows the estimated odds ratio for each of the individual strata, along with a 95% confidence interval and the corresponding weighting factor. The "M-H combined" estimate is the summary odds ratio; it has a 95% confidence interval as well. The Mantel-Haenszel test statistic for evaluating the null hypothesis that the odds ratio is equal to 1—which differs from the statistic we calculated by hand only because of rounding error—and its corresponding p-value are included in the bottom portion of the output.

16.4 Review Exercises

1. Explain Simpson's paradox.

2. When is it appropriate to combine the information in two or more 2×2 contingency tables to make a single statement about the relationship between an exposure and a disease?

3. What is a summary odds ratio? How is it computed?

4. When constructing a confidence interval for the summary odds ratio, why must you first calculate a confidence interval for the natural logarithm of this quantity?

5. In the 2 × 2 contingency tables below, the data from a Germa... the relationship between smoking status and invasive cervica... stratified by the number of sexual partners that a woman has ha... investigating ...e been

Zero or One Partner

Cancer	Smoker		Total
	Yes	No	
Yes	12	25	37
No	21	118	139
Total	33	143	176

Two or More Partners

Cancer	Smoker		Total
	Yes	No	
Yes	96	92	188
No	142	150	292
Total	238	242	480

(a) Estimate the odds of cervical cancer for smokers relative to nonsmokers for women who have had at most one sexual partner.
(b) Estimate the odds ratio for women who have had two or more sexual partners.
(c) Within each stratum, are the odds of being diagnosed with cervical cancer higher for women who smoke or for those who do not smoke?
(d) If possible, you would like to combine the information in these two strata to make a single overall statement about the relationship between smoking and cervical cancer. What might be the problem if you were to simply sum the entries in the two tables and calculate an odds ratio?
(e) Conduct a test of homogeneity. Based on the results of the test, do you think it is appropriate to use the Mantel-Haenszel method to combine the information in these two tables?
(f) Compute the Mantel-Haenszel estimate of the summary odds ratio.
(g) Construct a 99% confidence interval for the summary odds ratio. Does this confidence interval contain the value 1? What does this mean?
(h) At the 0.01 level of significance, test the null hypothesis that there is no association between smoking status and the presence of invasive cervical cancer. What do you conclude?

6. A group of children five years of age and younger who were free of respiratory problems were enrolled in a cohort study examining the relationship between parental smoking and the subsequent development of asthma [7]. The association between maternal cigarette smoking status and a diagnosis of asthma before the age of twelve was examined separately for boys and for girls.

Boys

Smoking Status	Asthma Diagnosis		Total
	Yes	No	
≥ 1/2 pack/day	17	63	80
< 1/2 pack/day	41	274	315
Total	58	337	395

Girls

Smoking Status	Asthma Diagnosis		Total
	Yes	No	
≥ 1/2 pack/day	8	55	63
< 1/2 pack/day	20	261	281
Total	28	316	344

(a) Estimate the relative odds of developing asthma for boys whose mothers smoke at least one-half pack of cigarettes per day versus those whose mothers smoke less than this.

(b) Estimate the corresponding odds ratio for girls.

(c) Conduct a test of homogeneity to determine whether it is appropriate to combine the information in the two 2×2 tables using the Mantel-Haenszel method. What do you conclude?

(d) If it makes sense to do so, find a point estimate for the summary odds ratio and construct a 95% confidence interval.

(e) What would you do if the results of the test of homogeneity led you to reject the null hypothesis that the odds ratio is identical for boys and girls?

7. In a study investigating various risk factors for heart disease, the relationship between hypertension and coronary artery disease (CAD) was examined for individuals in two different age groups [8].

35–49 Years of Age

Hypertension	CAD		Total
	Yes	No	
Yes	552	212	764
No	941	495	1436
Total	1493	707	2200

More than 65 Years of Age

	CAD		
Hypertension	**Yes**	**No**	**Total**
Yes	1102	87	1189
No	1018	106	1124
Total	2120	193	2313

(a) Within each category of age, are the odds of suffering from coronary artery disease greater or smaller for individuals with hypertension?

(b) Do you feel that it is appropriate to combine the information in these two tables to make a single unifying statement about the relationship between hypertension and coronary artery disease? Why or why not?

(c) Compute the Mantel-Haenszel estimate of the summary odds ratio.

(d) Construct a 95% confidence interval for the summary odds ratio.

(e) At the $\alpha = 0.05$ level of significance, test the null hypothesis that there is no association between hypertension and coronary artery disease. What do you conclude?

8. The data presented in Section 16.1 investigating the association between smoking and the development of aortic stenosis are saved on your disk in the data set `stenosis` [1] (Appendix B, Table B.20). Smoking status is saved under the variable name `smoke`, the presence of aortic stenosis under the name `disease`, and gender under the name `sex`.

(a) Construct a 2 × 2 table of smoking status versus aortic stenosis for males only. Estimate the odds of developing disease for smokers relative to nonsmokers.

(b) Construct the same 2 × 2 table for females. Again estimate the odds ratio.

(c) Within each gender group, are the odds of being diagnosed with aortic stenosis higher for smokers or for nonsmokers?

(d) Conduct a test of homogeneity to determine whether it is appropriate to combine the information for males and females using the Mantel-Haenszel method.

(e) Compute a point estimate of the summary odds ratio.

(f) At the 0.05 level of significance, test the null hypothesis that there is no association between smoking status and aortic stenosis. What do you conclude?

Bibliography

[1] Hoagland, P. M., Cook, E. F., Flatley, M., Walker, C., and Goldman, L., "Case-Control Analysis of Risk Factors for the Presence of Aortic Stenosis in Adults (Age 50 Years or Older)," *American Journal of Cardiology,* Volume 55, March 1, 1985, 744–747.

[2] Rosenberg, L., Palmer, J. R., Kelly, J. P., Kaufman, D. W., and Shapiro, S., "Coffee Drinking and Nonfatal Myocardial Infarction in Men Under 55 Years of Age," *American Journal of Epidemiology*, Volume 128, September 1988, 570–578.

[3] Rosemarin, P. C., "Coffee and Coronary Heart Disease: A Review," *Progress in Cardiovascular Diseases,* Volume 32, November–December 1989, 239–245.

[4] Grobbee, D. E., Rimm, E. B., Giovannucci, E., Colditz, G., Stampfer, M., and Willett, W., "Coffee, Caffeine, and Cardiovascular Disease in Men," *The New England Journal of Medicine,* Volume 323, October 11, 1990, 1026–1032.

[5] Wu, M. L., Whittemore, A. S., Paffenbarger, R. S., Sarles, D. L., Kampert, J. B., Grosser, S., Jung, D. L., Ballon, S., Hendrickson, M., and Mohle-Boetani, J., "Personal and Environmental Characteristics Related to Epithelial Ovarian Cancer, Reproductive and Menstrual Events and Oral Contraceptive Use," *American Journal of Epidemiology,* Volume 128, December 1988, 1216–1227.

[6] Nischan, P., Ebeling, K., and Schindler, C., "Smoking and Invasive Cervical Cancer Risk: Results from a Case-Control Study," *American Journal of Epidemiology,* Volume 128, July 1988, 74–77.

[7] Martinez, F. D., Cline, M., and Burrows, B., "Increased Incidence of Asthma in Children of Smoking Mothers," *Pediatrics,* Volume 89, January 1992, 21–26.

[8] Applegate, W. B., Hughes, J. P., and Zwaag, R. V., "Case-Control Study of Coronary Heart Disease Risk Factors in the Elderly," *Journal of Clinical Epidemiology,* Volume 44, Number 4/5, 1991, 409–415.

17

Correlation

In the preceding chapters, we discussed measures of the strength of association between two dichotomous random variables. We now begin to investigate the relationships that can exist among continuous variables. One statistical technique often employed to measure such an association is known as *correlation analysis. Correlation* is defined as the quantification of the degree to which two random variables are related, provided that the relationship is linear.

17.1   The Two-Way Scatter Plot

Suppose that we are interested in a pair of continuous random variables, each of which is measured on the same set of persons, countries, or other units of study. For example, we might wish to investigate the relationship between the percentage of children who have been immunized against the infectious diseases diphtheria, pertussis, and tetanus (DPT) in a given country and the corresponding mortality rate for children under five years of age in that country. The United Nations Children's Fund considers the under-five mortality rate to be one of the most important indicators of the level of well-being for a population of children.

The data for a random sample of 20 countries are shown in Table 17.1 [1]. If X represents the percentage of children immunized by age one year, and Y represents the under-five mortality rate, we have a pair of outcomes (x_i, y_i) for each nation in the sample. The first country on the list, Bolivia, has a DPT immunization percentage of 77% and an under-five mortality rate of 118 per 1000 live births; therefore, this nation is represented by the data point (77, 118).

Before we conduct any type of analysis, we should always create a two-way scatter plot of the data. If we place the outcomes of the X variable along the horizontal axis and the outcomes of the Y variable along the vertical axis, each point on the graph represents a combination of values (x_i, y_i). We can often determine whether a relationship exists between x and y—the outcomes of the random variables X and Y—simply by examining the graph. As an example, the data from Table 17.1 are plotted in Figure 17.1.

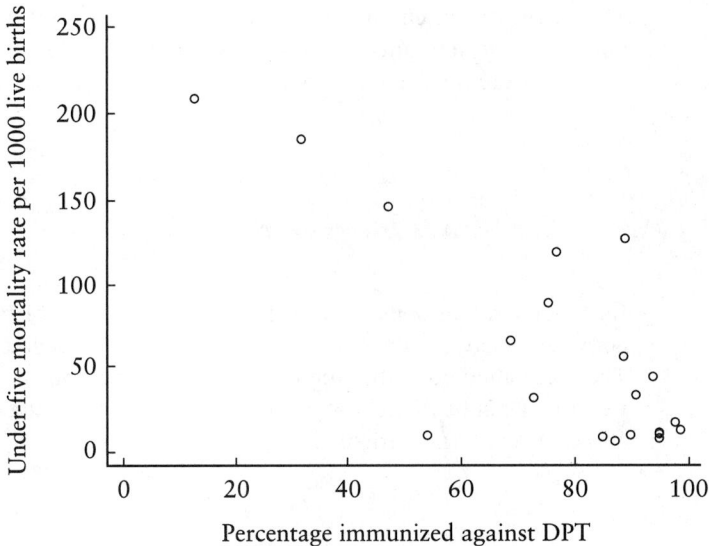

FIGURE 17.1
Under-five mortality rate versus percentage of children immunized against DPT for 20 countries, 1992

TABLE 17.1
Percentage of children immunized against DPT and under-five mortality rate for 20 countries, 1992

Nation	Percentage Immunized	Mortality Rate per 1000 Live Births
Bolivia	77	118
Brazil	69	65
Cambodia	32	184
Canada	85	8
China	94	43
Czech Republic	99	12
Egypt	89	55
Ethiopia	13	208
Finland	95	7
France	95	9
Greece	54	9
India	89	124
Italy	95	10
Japan	87	6
Mexico	91	33
Poland	98	16
Russian Federation	73	32
Senegal	47	145
Turkey	76	87
United Kingdom	90	9

The percentage of children immunized against DPT appears on the horizontal axis and the under-five mortality rate on the vertical axis. Not surprisingly, the mortality rate tends to decrease as the percentage of children immunized increases.

17.2 Pearson's Correlation Coefficient

In the underlying population from which the sample of points (x_i, y_i) is selected, the correlation between the random variables X and Y is denoted by the Greek letter ρ (rho). The correlation quantifies the strength of the linear relationship between the outcomes x and y. It can be thought of as the average of the product of the standard normal deviates of X and Y; in particular,

$$\rho = \text{average}\left[\frac{(X - \mu_x)}{\sigma_x} \frac{(Y - \mu_y)}{\sigma_y}\right].$$

The estimator of the population correlation is known as *Pearson's coefficient of correlation,* or simply the *correlation coefficient.* The correlation coefficient is denoted by r; it is calculated using the formula

$$r = \frac{1}{(n-1)} \sum_{i=1}^{n} \left(\frac{x_i - \bar{x}}{s_x}\right)\left(\frac{y_i - \bar{y}}{s_y}\right)$$

$$= \frac{1}{(n-1)} \frac{\sum_{i=1}^{n} (x_i - \bar{x})(y_i - \bar{y})}{s_x s_y}.$$

Recall that s_x and s_y are the sample standard deviations of the x and y values, respectively. Therefore, an equivalent formula for r is

$$r = \frac{\sum_{i=1}^{n} (x_i - \bar{x})(y_i - \bar{y})}{\sqrt{\left[\sum_{i=1}^{n} (x_i - \bar{x})^2\right]\left[\sum_{i=1}^{n} (y_i - \bar{y})^2\right]}}.$$

The correlation coefficient is a dimensionless number; it has no units of measurement. The maximum value that r can achieve is 1, and its minimum value is -1. Therefore, for any given set of observations, $-1 \leq r \leq 1$. The values $r = 1$ and $r = -1$ occur when there is an exact linear relationship between x and y; if we were to plot all pairs of outcomes (x, y), the points would lie on a straight line. Examples of perfect correlation are illustrated in Figures 17.2 (a) and (b). As the relationship between x and y deviates from perfect linearity, r moves away from 1 or -1 and closer to 0. If y tends to increase in magnitude as x increases, r is greater than 0 and x and y are said to be *positively correlated*; if y decreases as x increases, r is less than 0 and the two variables are *negatively correlated.* If $r = 0$, as it does for the samples of points pictured in Figures 17.2 (c) and (d), there is no linear relationship between x and y and the variables are uncorrelated. However, a nonlinear relationship may exist.

For the data in Table 17.1, the mean percentage of children immunized against DPT is

$$\bar{x} = \frac{1}{n} \sum_{i=1}^{n} x_i$$

$$= \frac{1}{20} \sum_{i=1}^{20} x_i$$

$$= 77.4\%,$$

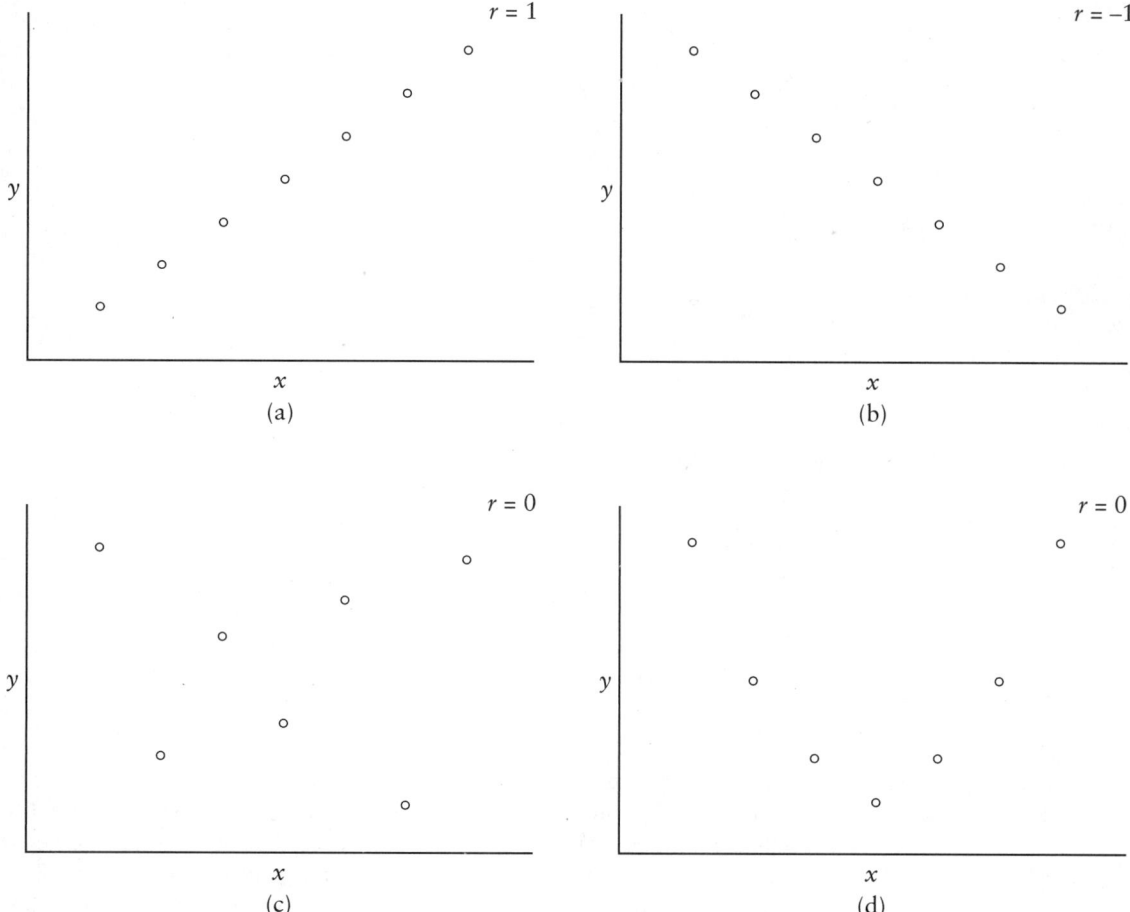

FIGURE 17.2

Scatter plots showing possible relationships between X and Y

and the mean value of the under-five mortality rate is

$$\bar{y} = \frac{1}{n} \sum_{i=1}^{n} y_i$$

$$= \frac{1}{20} \sum_{i=1}^{20} y_i$$

$$= 59.0 \text{ per 1000 live births}.$$

In addition,

$$\sum_{i=1}^{n} (x_i - \bar{x})(y_i - \bar{y}) = \sum_{i=1}^{20} (x_i - 77.4)(y_i - 59.0)$$

$$= -22706,$$

$$\sum_{i=1}^{n} (x_i - \bar{x})^2 = \sum_{i=1}^{20} (x_i - 77.4)^2$$

$$= 10630.8,$$

and

$$\sum_{i=1}^{n} (y_i - \bar{y})^2 = \sum_{i=1}^{20} (y_i - 59.0)^2$$

$$= 77498.$$

As a result, the coefficient of correlation is

$$r = \frac{\sum_{i=1}^{n} (x_i - \bar{x})(y_i - \bar{y})}{\sqrt{\left[\sum_{i=1}^{n} (x_i - \bar{x})^2\right]\left[\sum_{i=1}^{n} (y_i - \bar{y})^2\right]}}$$

$$= \frac{-22706}{\sqrt{(10630.8)(77498)}}$$

$$= -0.79.$$

Based on this sample, there appears to be a strong linear relationship between the percentage of children immunized against DPT in a specified country and its under-five mortality rate; the correlation coefficient is fairly close to its minimum value of -1. Since r is negative, mortality rate decreases in magnitude as percentage of immunization increases. Care must be taken when we interpret this relationship, however. An effective immunization program might be the primary reason for the decrease in mortality, or it might be a ramification of a successful comprehensive health care system that is itself the cause of the decrease. The correlation coefficient merely tells us that a linear relationship exists between two variables; it does not specify whether the relationship is cause-and-effect.

Just as we made inference about a population mean μ based on the sample mean $\bar{x}$, we would also like to be able to draw conclusions about the unknown population correlation ρ using the sample correlation coefficient r. Most frequently, we are interested in determining whether any correlation exists between the random variables X and Y. This can be done by testing the null hypothesis that there is no such correlation in the underlying population, or

$$H_0: \rho = 0.$$

The procedure is similar to other tests of hypotheses that we have encountered; it involves finding the probability of obtaining a sample correlation coefficient as extreme as, or more extreme than, the observed r given that the null hypothesis is true. Since the estimated standard error of r may be expressed as

$$\widehat{se}\,(r) = \sqrt{\frac{1 - r^2}{n - 2}},$$

the test is performed by calculating the statistic

$$t = \frac{r - 0}{\sqrt{(1 - r^2)/(n - 2)}}$$
$$= r\sqrt{\frac{n - 2}{1 - r^2}}.$$

If we assume that the pairs of observations (x_i, y_i) were obtained randomly and that both X and Y are normally distributed, this quantity has a t distribution with $n - 2$ degrees of freedom only when the null hypothesis is true.

Suppose we want to know whether a linear relationship exists between percentage of immunization against DPT and the under-five mortality rate in the population of countries around the world. We conduct a two-sided test of the null hypothesis of no association at the $\alpha = 0.05$ level of significance. Recall that we previously found r to be equal to -0.79. Therefore,

$$t = r\sqrt{\frac{n - 2}{1 - r^2}}$$
$$= -0.79\sqrt{\frac{20 - 2}{1 - (-0.79)^2}}$$
$$= -5.47.$$

Consulting Table A.4, we observe that for a t distribution with 18 degrees of freedom, $p < 2(0.0005) = 0.001$. Given that $\rho = 0$, the probability of observing a sample correlation coefficient as far from 0 as $r = -0.79$ is very small. Therefore, we reject the null hypothesis at the 0.05 level. Based on this sample, there is evidence that the true population correlation is different from 0. Under-five mortality rate decreases as the percentage of children immunized increases; therefore, the correlation is negative. (Note, however, that neither percentage of children immunized nor under-five mortality rate is normally distributed; the percentage is skewed to the left, and the mortality rate skewed to the right. Therefore, the hypothesis testing procedure performed above cannot be assumed to be accurate for these data.)

The testing procedure described is valid only for the special case in which the hypothesized value of the population correlation is equal to 0. If ρ is equal to some other value, represented by ρ_0, the sampling distribution of r is skewed, and the test statistic no longer follows a t distribution. However, methods for testing the more general hypothesis

$$H_0: \rho = \rho_0$$

are available [2].

The coefficient of correlation r has several limitations. First, it quantifies only the strength of the **linear** relationship between two variables; if X and Y have a nonlinear relationship—as they do in Figure 17.2 (d)—it will not provide a valid measure of this association. Second, care must be taken when the data contain any outliers, or pairs of observations that lie considerably outside the range of the other data points. The sample correlation coefficient is highly sensitive to extreme values and, if one or more are present, often gives misleading results. Third, the estimated correlation should never be extrapolated beyond the observed ranges of the variables; the relationship between X and Y may change outside of this region. Finally, it must be kept in mind that a high correlation between two variables does not in itself imply a cause-and-effect relationship.

17.3 Spearman's Rank Correlation Coefficient

Like other parametric techniques, Pearson's correlation coefficient is very sensitive to outlying values. We may be interested in calculating a measure of association that is more robust. One approach is to rank the two sets of outcomes x and y separately and calculate a coefficient of rank correlation. This procedure—which results in a quantity known as *Spearman's rank correlation coefficient*—may be classified among the nonparametric methods presented in Chapter 13.

Spearman's rank correlation coefficient, denoted r_s, is simply Pearson's r calculated for the ranked values of x and y. Therefore,

$$r_s = \frac{\sum_{i=1}^{n} (x_{ri} - \bar{x}_r)(y_{ri} - \bar{y}_r)}{\sqrt{\left[\sum_{i=1}^{n} (x_{ri} - \bar{x}_r)^2\right]\left[\sum_{i=1}^{n} (y_{ri} - \bar{y}_r)^2\right]}},$$

where x_{ri} and y_{ri} are the ranks associated with the ith subject rather than the actual observations. An equivalent method for computing r_s is provided by the formula

$$r_s = 1 - \frac{6\sum_{i=1}^{n} d_i^2}{n(n^2 - 1)};$$

in this case, n is the number of data points in the sample and d_i is the difference between the rank of x_i and the rank of y_i. Like the Pearson correlation coefficient, the Spearman rank correlation coefficient ranges in value from -1 to 1. Values of r_s close to the extremes indicate a high degree of correlation between x and y; values near 0 imply a lack of linear association between the two variables.

Suppose that we were to rank the percentages of children immunized and under-five mortality rates presented in Table 17.1 from the smallest to the largest, separately for each variable, assigning average ranks to tied observations. The results are shown in Table 17.2, along with the difference in ranks for each country and the squares of these differences. Using the second formula for r_s, the Spearman rank correlation coefficient is

$$r_s = 1 - \frac{6 \sum_{i=1}^{n} d_i^2}{n(n^2 - 1)}$$

$$= 1 - \frac{6(2045.5)}{20(399)}$$

$$= -0.54.$$

This value is somewhat smaller in magnitude than the Pearson correlation coefficient—perhaps r is inflated due to the non-normality of the data—but still suggests a moderate relationship between the percentage of children immunized against DPT and the under-five mortality rate. Again the correlation is negative.

Spearman's rank correlation coefficient may also be thought of as a measure of the concordance of the ranks for the outcomes x and y. If the 20 measurements of percent immunization and under-five mortality rate in Table 17.2 happened to be ranked in the same order for each variable—meaning that the country with the ith largest percentage immunization also had the ith largest mortality rate for all values of i—each difference d_i would be equal to 0, and

$$r_s = 1 - \frac{6(0)}{n(n^2 - 1)}$$

$$= 1.$$

If the ranking of the first variable is the inverse of the ranking of the second—so that the country with the largest percentage of children immunized against DPT has the smallest under-five mortality rate, and so forth—it can be shown that

$$r_s = -1.$$

When there is no linear correspondence between the two sets of ranks,

$$r_s = 0.$$

TABLE 17.2

Ranked percentages of children immunized against DPT and under-five mortality rates for 20 countries, 1992

Nation	Percentage Immunized	Rank	Mortality Rate	Rank	d_i	d_i^2
Ethiopia	13	1	208	20	-19	361
Cambodia	32	2	184	19	-17	289
Senegal	47	3	145	18	-15	225
Greece	54	4	9	5	-1	1
Brazil	69	5	65	14	-9	81
Russian Federation	73	6	32	10	-4	16
Turkey	76	7	87	15	-8	64
Bolivia	77	8	118	16	-8	64
Canada	85	9	8	3	6	36
Japan	87	10	6	1	9	81
India	89	11.5	124	17	-5.5	30.25
Egypt	89	11.5	55	13	-1.5	2.25
United Kingdom	90	13	9	5	8	64
Mexico	91	14	33	11	3	9
China	94	15	43	12	3	9
France	95	17	9	5	12	144
Finland	95	17	7	2	15	225
Italy	95	17	10	7	10	100
Poland	98	19	16	9	10	100
Czech Republic	99	20	12	8	12	144
						2045.5

If the sample size n is not too small—in particular, if it is greater than or equal to 10—and if we can assume that pairs of ranks (x_{ri}, y_{ri}) are chosen randomly, we can test the null hypothesis that the unknown population correlation is equal to 0,

$$H_0: \rho = 0,$$

using the same procedure that we used for Pearson's r. If n is less than 10, tables of critical values must be used to evaluate H_0 [2]. For the ranked data of Table 17.2, the test statistic is

$$t_s = r_s \sqrt{\frac{n-2}{1-r_s^2}}$$

$$= -0.54 \sqrt{\frac{20-2}{1-(-0.54)^2}}$$

$$= -2.72.$$

For a t distribution with 18 degrees of freedom, $0.01 < p < 0.02$. Therefore, we reject H_0 at the 0.05 level and conclude that the true population correlation is less than 0. This testing procedure does not require that X and Y be normally distributed.

Like other nonparametric techniques, Spearman's rank correlation coefficient has advantages and disadvantages. It is much less sensitive to outlying values than Pearson's correlation coefficient. In addition, it can be used when one or both of the relevant variables are ordinal. Because it relies on ranks rather than on actual observations, however, the nonparametric method does not use everything that is known about a distribution.

17.4   Further Applications

Suppose that we wish to examine the relationship between the percentage of births attended by trained health care personnel—including physicians, nurses, midwives, and other health care workers—and the maternal mortality rate per 100,000 live births. The values for a random sample of 20 countries appear in Table 17.3 [1]. We would like to determine whether there is any evidence of a linear relationship between these two variables.

We begin our analysis by constructing a two-way scatter plot of the data, placing percentage of births attended on the horizontal axis and maternal mortality rate on the vertical axis. The graph is displayed in Figure 17.3. Note that higher values of percentage of births attended appear to be associated with lower values of maternal mortality rate.

TABLE 17.3

Percentage of births attended by trained health care personnel and maternal mortality rate for 20 countries

Nation	Percentage Attended	Maternal Mortality Rate per 100,000 Live Births
Bangladesh	5	600
Belgium	100	3
Chile	98	67
Ecuador	84	170
Hong Kong	100	6
Hungary	99	15
Iran	70	120
Kenya	50	170
Morocco	26	300
Nepal	6	830
Netherlands	100	10
Nigeria	37	800
Pakistan	35	500
Panama	96	60
Philippines	55	100
Portugal	90	10
Spain	96	5
Switzerland	99	5
United States	99	8
Vietnam	95	120

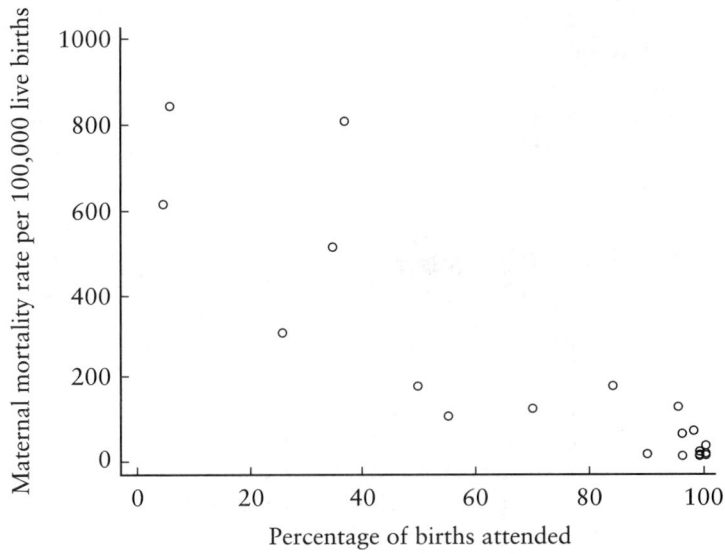

FIGURE 17.3
Maternal mortality rate versus percentage of births attended by trained
health care personnel for 20 countries

Let x represent the observed outcomes of percentage of births attended and y the observed maternal mortality rates. The mean percentage of births attended is

$$\bar{x} = \frac{1}{20} \sum_{i=1}^{20} x_i$$

$$= 72.0\%,$$

and the mean mortality rate is

$$\bar{y} = \frac{1}{20} \sum_{i=1}^{20} y_i$$

$$= 196.3 \text{ per } 100,000 \text{ live births}.$$

Also,

$$\sum_{i=1}^{n} (x_i - \bar{x})(y_i - \bar{y}) = \sum_{i=1}^{20} (x_i - 72.0)(y_i - 196.3)$$

$$= -150,634,$$

$$\sum_{i=1}^{n} (x_i - \bar{x})^2 = \sum_{i=1}^{20} (x_i - 72.0)^2$$

$$= 21,696,$$

and

$$\sum_{i=1}^{n} (y_i - \bar{y})^2 = \sum_{i=1}^{20} (y_i - 196.3)^2$$
$$= 1{,}364{,}390.2.$$

The Pearson coefficient of correlation, which quantifies the degree to which these outcomes are linearly related, is

$$r = \frac{\sum_{i=1}^{n} (x_i - \bar{x})(y_i - \bar{y})}{\sqrt{\left[\sum_{i=1}^{n} (x_i - \bar{x})^2\right]\left[\sum_{i=1}^{n} (y_i - \bar{y})^2\right]}}$$
$$= \frac{-150{,}634}{\sqrt{(21{,}696)(1{,}364{,}390.2)}}$$
$$= -0.88.$$

A value of -1 or 1 would imply that an exact linear relationship exists; if r were equal to 0, that would mean there is no linear relationship at all. Based on the sample of 20 countries, there appears to be a strong relationship between percentage of births attended by trained health care personnel and maternal mortality rate. Since r is negative, mortality rate decreases as percentage of births attended increases.

We might also wish to use the information in the sample to test the null hypothesis that there is no linear relationship between percentage of births attended and maternal mortality rate in the underlying population of countries around the world, or

$$H_0: \rho = 0,$$

using a two-sided test conducted at the 0.05 level of significance. Since $r = -0.88$,

$$t = r\sqrt{\frac{n-2}{1-r^2}}$$
$$= -0.88\sqrt{\frac{20-2}{1-(-0.88)^2}}$$
$$= -7.86.$$

For a t distribution with 18 degrees of freedom, $p < 0.001$. We reject H_0 and conclude that the true population correlation is not equal to 0; since r is negative, maternal mortality rate decreases in magnitude as percentage of births attended increases. (Note that this hypothesis testing procedure assumes that both X and Y are normally distributed, which is not the case for these data.)

If we are interested in calculating a more robust measure of association between two variables, we can order the sets of outcomes x and y from smallest to largest and compute the rank correlation coefficient instead. Spearman's rank correlation is simply Pearson's r calculated using ranks rather than actual observations. The ranked percentages of births attended and maternal mortality rates are displayed in Table 17.4, along

TABLE 17.4

Ranked percentage of births attended by trained health care personnel and maternal mortality rate for 20 countries

Nation	Percentage Attended	Rank	Mortality Rate	Rank	d_i	d_i^2
Bangladesh	5	1	600	18	−17	289
Nepal	6	2	830	20	−18	324
Morocco	26	3	300	16	−13	169
Pakistan	35	4	500	17	−13	169
Nigeria	37	5	800	19	−14	196
Kenya	50	6	170	14.5	−8.5	72.25
Philippines	55	7	100	11	−4	16
Iran	70	8	120	12.5	−4.5	20.25
Ecuador	84	9	170	14.5	−5.5	30.25
Portugal	90	10	10	6.5	3.5	12.25
Vietnam	95	11	120	12.5	−1.5	2.25
Spain	96	12.5	5	2.5	10	100
Panama	96	12.5	60	9	3.5	12.25
Chile	98	14	67	10	4	16
Switzerland	99	16	5	2.5	13.5	182.25
United States	99	16	8	5	11	121
Hungary	99	16	15	8	8	64
Netherlands	100	19	10	6.5	12.5	156.25
Hong Kong	100	19	6	4	15	225
Belgium	100	19	3	1	18	324
						2501

with the differences in ranks for each country and the squares of these differences. The Spearman rank correlation for the data is

$$r_s = 1 - \frac{6 \sum_{i=1}^{n} d_i^2}{n(n^2 - 1)}$$

$$= 1 - \frac{6(2501)}{20(399)}$$

$$= -0.88.$$

This value is identical in magnitude to the Pearson correlation coefficient; it also signifies a strong negative relationship between percentage of births attended and maternal mortality rate.

To test the null hypothesis that the unknown population correlation is equal to 0, or

$$H_0: \rho = 0,$$

using the rank correlation, we calculate that

$$t_s = r_s \sqrt{\frac{n-2}{1-r_s^2}}$$

$$= -0.88 \sqrt{\frac{20-2}{1-(-0.88)^2}}$$

$$= -7.86.$$

For a t distribution with 18 degrees of freedom, $p < 0.001$. We again reject the null hypothesis that the population correlation is equal to 0; as the percentage of births attended by trained health care personnel increases, the maternal mortality rate decreases. This testing procedure does not require that X and Y be normally distributed.

   Finding either the correlation coefficient or the rank correlation coefficient for two random variables involves quite a bit of computation; however, most computer packages will calculate both the Pearson and Spearman correlations for us. Table 17.5 shows the SAS output for the Pearson correlation coefficient. The top portion of the output displays numerical summary measures for each of the two variables involved; the bottom shows the value of r along with the p-value for the test of the null hypothesis that the population correlation is equal to 0. Note that the correlation of any variable with itself is simply 1. Table 17.6 displays the Stata output for the Spearman rank correlation coefficient. In addition to r_s, it also shows the relevant p-value.

TABLE 17.5

SAS output displaying the Pearson correlation coefficient

Correlation Analysis
2 Variables: ATTEND MORTRATE

Simple Statistics

Variable	N	Mean	Std Dev	Sum	Minimum	Maximum
ATTEND	20	72.0000	33.7919	1440	5.0000	100.0000
MORTRATE	20	194.9500	268.9221	3899	3.0000	830.0000

Pearson Correlation Coefficients/Prob > |R| under Ho: Rho=0/N=20

	ATTEND	MORTRATE
ATTEND	1.00000	−0.87681
	0.0	0.0001
MORTRATE	−0.87681	1.00000
	0.0001	0.0

TABLE 17.6

Stata output displaying the Spearman rank correlation coefficient

```
 Number of obs = 20
Spearman's rho = -0.8897

Test of Ho: attend and mortrate independent

      Pr > |t| = 0.0000
```

17.5 Review Exercises

1. When you are investigating the relationship between two continuous random variables, why is it important to create a scatter plot of the data?

2. What are the strengths and limitations of Pearson's correlation coefficient?

3. How does Spearman's rank correlation differ from the Pearson correlation?

4. If a test of hypothesis indicates that the correlation between two random variables is not significantly different from 0, does this necessarily imply that the variables are independent? Explain.

5. In a study conducted in Italy, 10 patients with hypertriglyceridemia were placed on a low-fat, high-carbohydrate diet. Before the start of the diet, cholesterol and triglyceride measurements were recorded for each subject [3].

Patient	Cholesterol Level (mmol/l)	Triglyceride Level (mmol/l)
1	5.12	2.30
2	6.18	2.54
3	6.77	2.95
4	6.65	3.77
5	6.36	4.18
6	5.90	5.31
7	5.48	5.53
8	6.02	8.83
9	10.34	9.48
10	8.51	14.20

(a) Construct a two-way scatter plot for these data.
(b) Does there appear to be any evidence of a linear relationship between cholesterol and triglyceride levels prior to the diet?
(c) Compute r, the Pearson correlation coefficient.
(d) At the 0.05 level of significance, test the null hypothesis that the population correlation ρ is equal to 0. What do you conclude?
(e) Calculate r_s, the Spearman rank correlation coefficient.
(f) How does the value of r_s compare to that of r?
(g) Using r_s, again test the null hypothesis that the population correlation is equal to 0. What do you conclude?

6. Thirty-five patients with ischemic heart disease, a suppression of blood flow to the heart, took part in a series of tests designed to evaluate the perception of pain. In one part of the study, the patients exercised until they experienced angina, or chest pain; time until the onset of angina and the duration of the attack were recorded. The data are saved on your disk in the file ischemic [4] (Appendix B, Table B.6).

Time to angina in seconds is saved under the variable name `time`; the duration of angina, also in seconds, is saved under the name `duration`.

(a) Create a two-way scatter plot for these data.

(b) In the population of patients with ischemic heart disease, does there appear to be any evidence of a linear relationship between time to angina and the duration of the attack?

(c) Calculate Pearson's correlation coefficient.

(d) Does the duration of angina tend to increase or decrease as time to angina increases?

(e) Test the null hypothesis

$$H_0: \rho = 0.$$

What do you conclude?

(f) Compute Spearman's rank correlation.

(g) Using r_s, again test the null hypothesis that the population correlation is equal to 0. What do you conclude?

7. The data set `lowbwt` contains information collected for a sample of 100 low birth weight infants born in two teaching hospitals in Boston, Massachusetts [5] (Appendix B, Table B.7). Measurements of systolic blood pressure are saved under the variable name `sbp`, and values of the Apgar score recorded five minutes after birth—an index of neonatal asphyxia or oxygen deprivation—are saved under the name `apgar5`. The Apgar score is an ordinal random variable that takes values between 0 and 10.

(a) Estimate the correlation of the random variables systolic blood pressure and five-minute Apgar score for this population of low birth weight infants.

(b) Does Apgar score tend to increase or decrease as systolic blood pressure increases?

(c) Test the null hypothesis

$$H_0: \rho = 0.$$

What do you conclude?

8. Suppose that you are interested in determining whether a relationship exists between the fluoride content in a public water supply and the dental caries experience of children using this water. Data from a study examining 7257 children in 21 cities are saved on your disk in the file `water` [6] (Appendix B, Table B.21). The fluoride content of the public water supply in each city, measured in parts per million, is saved under the variable name `fluoride`; the number of dental caries per 100 children examined is saved under the name `caries`. The total dental caries experience is obtained by summing the numbers of filled teeth, teeth with untreated dental caries, teeth requiring extraction, and missing teeth.

(a) Construct a two-way scatter plot for these data.

(b) What is the correlation between the number of dental caries per 100 children and the fluoride content of the water?

(c) Is this correlation significantly different from 0?

(d) For the 21 cities in the study, the highest fluoride content in a given water supply is 2.6 ppm. If you were to increase the fluoride content of the water to more than 4 ppm, do you believe that the number of dental caries per 100 children would decrease?

9. One of the functions of the Federation of State Medical Boards is to collect data summarizing disciplinary actions taken against nonfederal physicians by medical licensing boards. Serious actions include license revocations, suspensions, and probations. For each of the years 1991 through 1995, the number of serious actions per 1000 doctors was ranked by state from highest to lowest. The ranks are contained in a data set called `actions` [7] (Appendix B, Table B.22); the ranks for 1991 are saved under the variable name `rank91`, those for 1992 under `rank92`, and so on.

(a) Which states have the highest rates of serious actions in each of the five years 1991 through 1995? Which states have the lowest rates?

(b) Construct a two-way scatter plot for the ranks of disciplinary actions in 1992 versus the ranks in 1991.

(c) Does there appear to be a relationship between these two quantities?

(d) Calculate the correlation of the two sets of ranks.

(e) Is this correlation significantly different from 0? What do you conclude?

(f) Calculate the correlations of the ranks in 1991 and those in 1993; those in 1991 and 1994; and those in 1991 and 1995. What happens to the magnitude of the correlation as the years being compared get further apart?

(g) Is each of these three correlations significantly different from 0?

(h) Do you believe that all states are equally strict in taking disciplinary action against physicians?

Bibliography

[1] United Nations Children's Fund, *The State of the World's Children 1994,* New York: Oxford University Press.

[2] Snedecor, G. W., and Cochran, W. G., *Statistical Methods,* Ames, Iowa: The Iowa State University Press, 1980.

[3] Cominacini, L., Zocca, I., Garbin, U., Davoli, A., Compri, R., Brunetti, L., and Bosello, O., "Long-Term Effect of a Low-Fat, High-Carbohydrate Diet on Plasma Lipids of Patients Affected by Familial Endogenous Hypertriglyceridemia," *American Journal of Clinical Nutrition,* Volume 48, July 1988, 57–65.

[4] Miller, P. F., Sheps, D. S., Bragdon, E. E., Herbst, M. C., Dalton, J. L., Hinderliter, A. L., Koch, G. G., Maixner, W., and Ekelund, L. G., "Aging and Pain Perception in Ischemic Heart Disease," *American Heart Journal,* Volume 120, July 1990, 22–30.

[5] Leviton, A., Fenton, T., Kuban, K. C. K., and Pagano, M., "Labor and Delivery Characteristics and the Risk of Germinal Matrix Hemorrhage in Low Birth Weight Infants," *Journal of Child Neurology,* Volume 6, October 1991, 35–40.

[6] Dean, H. T., Arnold, F. A., and Elvove, E., "Domestic Water and Dental Caries," *Public Health Reports,* Volume 57, August 7, 1942, 1155–1179.

[7] Public Citizen Health Research Group, "Ranking of Doctor Disciplinary Actions by State Medical Licensing Boards—1992," *Health Letter,* Volume 9, August 1993, 4–5.

Simple Linear Regression

Like correlation analysis, *simple linear regression* is a technique that is used to explore the nature of the relationship between two continuous random variables. The primary difference between these two analytical methods is that regression enables us to investigate the change in one variable, called the *response,* which corresponds to a given change in the other, known as the *explanatory variable.* Correlation analysis makes no such distinction; the two variables involved are treated symmetrically. The ultimate objective of regression analysis is to predict or estimate the value of the response that is associated with a fixed value of the explanatory variable.

An example of a situation in which regression analysis might be preferred to correlation is illustrated by the pediatric growth charts in Figures 18.1 and 18.2. Among children of both sexes, head circumference appears to increase linearly between the ages of 2 and 18 years. Rather than quantifying the strength of this association, we might be interested in predicting the change in head circumference that corresponds to a given change in age. In this case, head circumference is the response, and age is the explanatory variable. An understanding of their relationship helps parents and pediatricians to monitor growth and detect possible cases of macrocephaly and microcephaly.

18.1 Regression Concepts

Suppose that we are interested in the probability distribution of a continuous random variable Y. The outcomes of Y, denoted y, are the head circumference measurements in centimeters for the population of low birth weight infants—defined as those weighing less than 1500 grams—born in two teaching hospitals in Boston, Massachusetts [1]. We are told that the mean head circumference for the infants in this population is

$$\mu_y = 27 \text{ cm}$$

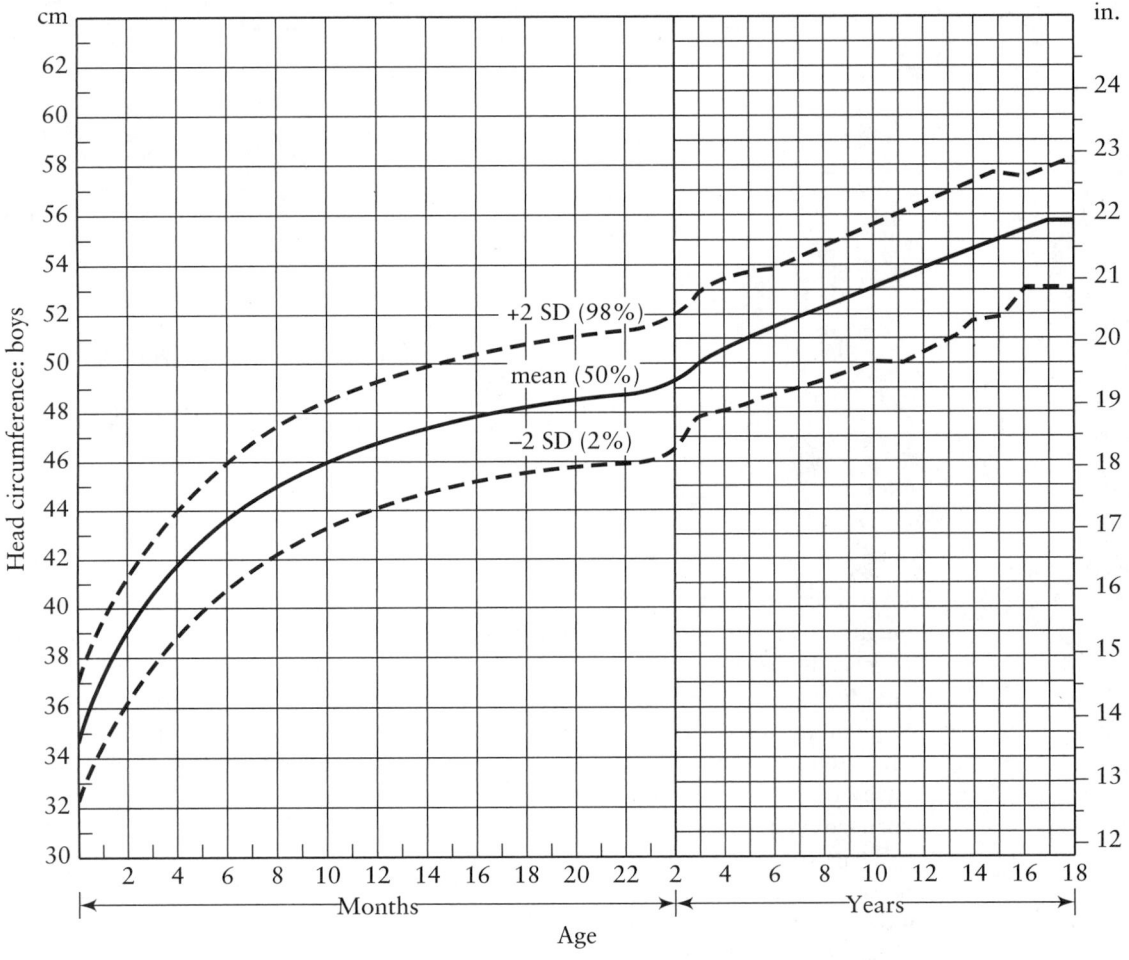

FIGURE 18.1

Head circumference versus age for boys (*Source*: From G. Nellhaus, *Pediatrics*, 1968, 41:106. Copyright 1968 American Academy of Pediatrics)

and that the standard deviation is

$$\sigma_y = 2.5 \text{ cm.}$$

Since the distribution of measurements is roughly normal, we are able to say that approximately 95% of the infants have head circumferences that measure between

$$\mu_y - 1.96\sigma_y = 27 - (1.96)(2.5)$$
$$= 22.1 \text{ cm}$$

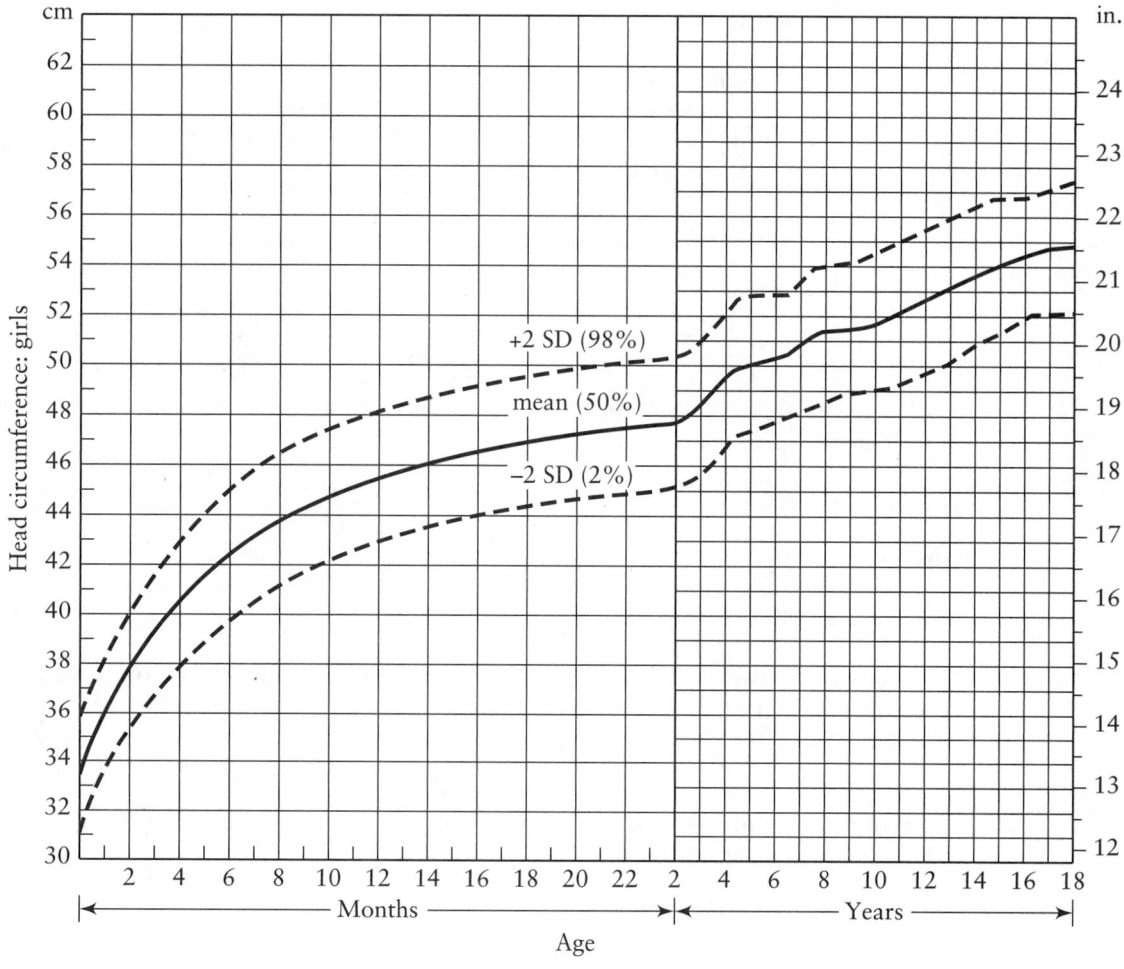

FIGURE 18.2
Head circumference versus age for girls (*Source*: From G. Nellhaus, *Pediatrics*, 1968, 41:106. Copyright 1968 American Academy of Pediatrics)

and

$$\mu_y + 1.96\sigma_y = 27 + (1.96)(2.5)$$

$$= 31.9 \text{ cm}.$$

Suppose we also know that the head circumferences of newborn infants increase with gestational age and that for each specified age x the distribution of measurements is approximately normal. For example, the head circumferences of infants whose gestational age is 26 weeks are normally distributed with mean

$$\mu_{y|26} = 24 \text{ cm}$$

and standard deviation

$$\sigma_{y|26} = 1.6 \text{ cm}.$$

Similarly, the head circumferences of infants whose gestational age is 29 weeks are approximately normal with mean

$$\mu_{y|29} = 26.5 \text{ cm}$$

and standard deviation

$$\sigma_{y|29} = 1.6 \text{ cm},$$

whereas the measurements for infants whose gestational age is 32 weeks are normal with mean

$$\mu_{y|32} = 29 \text{ cm}$$

and standard deviation

$$\sigma_{y|32} = 1.6 \text{ cm}.$$

For each value of gestational age x, the standard deviation $\sigma_{y|x}$ is constant and is less than σ_y. In fact, it can be shown that

$$\sigma_{y|x}^2 = (1 - \rho^2)\sigma_y^2,$$

where ρ is the correlation between X and Y in the underlying population [2]. If X and Y have no linear relationship, $\rho = 0$ and

$$\sigma_{y|x}^2 = (1 - 0)\sigma_y^2$$
$$= \sigma_y^2.$$

For the random variables head circumference and gestational age, $\sigma_y = 2.5$ cm and $\sigma_{y|x} = 1.6$ cm. Therefore,

$$(1.6)^2 = (1 - \rho^2)(2.5)^2,$$

and

$$\rho = \sqrt{1 - \frac{(1.6)^2}{(2.5)^2}}$$
$$= \sqrt{0.5904}$$
$$= \pm 0.77.$$

There is a fairly strong correlation between head circumference and gestational age in the underlying population of low birth weight infants; using this method of calculation, however, we cannot determine whether the correlation is positive or negative.

Because the standard deviation of the distribution of head circumference measurements for infants of a specified gestational age ($\sigma_{y|x} = 1.6$ cm) is smaller than the standard deviation for infants of all ages combined ($\sigma_y = 2.5$ cm), working with a single value of gestational age allows us to be more precise in our descriptions. For example, we can say that approximately 95% of the values of head circumference for the population of infants whose gestational age is 26 weeks lie between

$$\mu_{y|26} - 1.96\sigma_{y|26} = 24 - (1.96)(1.6)$$
$$= 20.9 \text{ cm}$$

and

$$\mu_{y|26} + 1.96\sigma_{y|26} = 24 + (1.96)(1.6)$$
$$= 27.1 \text{ cm.}$$

Similarly, roughly 95% of the infants whose gestational age is 29 weeks have head circumferences between

$$\mu_{y|29} - 1.96\sigma_{y|29} = 26.5 - (1.96)(1.6)$$
$$= 23.4 \text{ cm}$$

and

$$\mu_{y|29} + 1.96\sigma_{y|29} = 26.5 + (1.96)(1.6)$$
$$= 29.6 \text{ cm,}$$

whereas 95% of the infants whose gestational age is 32 weeks have measurements between

$$\mu_{y|32} - 1.96\sigma_{y|32} = 29 - (1.96)(1.6)$$
$$= 25.9 \text{ cm}$$

and

$$\mu_{y|32} + 1.96\sigma_{y|32} = 29 + (1.96)(1.6)$$
$$= 32.1 \text{ cm.}$$

In summary, the respective intervals are as follows:

Gestational Age (weeks)	Interval Containing 95% of the Observations
26	(20.9, 27.1)
29	(23.4, 29.6)
32	(25.9, 32.1)

Each of these intervals is constructed to enclose 95% of the population head circumference values for infants of a particular gestational age. None is as wide as (22.1, 31.9), the interval computed for the entire population of low birth weight infants. In addition, the intervals shift to the right as gestational age increases.

18.2   The Model

18.2.1   The Population Regression Line

As noted in the preceding section, mean head circumference tends to become larger as gestational age increases. Based on the means plotted in Figure 18.3, the relationship is linear. One way to quantify this relationship is to fit a model of the form

$$\mu_{y|x} = \alpha + \beta x,$$

where $\mu_{y|x}$ is the mean head circumference of low birth weight infants whose gestational age is x weeks. This model—known as the *population regression line*—is the equation of a straight line. The parameters α and β are constants called the *coefficients* of the equation; α is the *y*-intercept of the line and β is its slope. The *y-intercept* is the mean value of the response y when x is equal to 0, or $\mu_{y|0}$. The *slope* is the change in the mean value of y that corresponds to a one-unit increase in x. If β is positive, $\mu_{y|x}$ increases in magnitude as x increases; if β is negative, $\mu_{y|x}$ decreases as x increases.

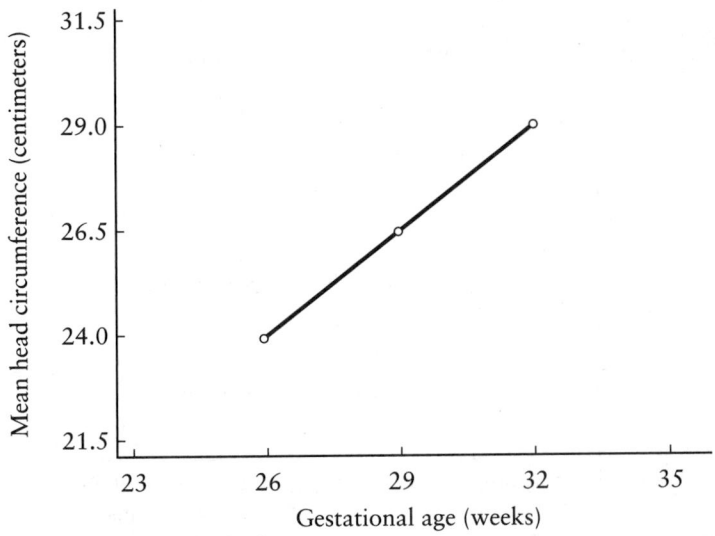

FIGURE 18.3

Population regression line of mean head circumference versus gestational age for low birth weight infants, $\mu_{y|x} = 2.3 + 0.83x$

Even if the relationship between mean head circumference and gestational age is a perfect straight line, the relationship between individual values of head circumference and age is not. As previously noted, the distribution of head circumference measurements for all low birth weight infants of a particular gestational age x is approximately normal with mean $\mu_{y|x}$ and standard deviation $\sigma_{y|x}$. The scatter around the mean is a result of the natural variation among children; we would not expect all low birth weight infants whose gestational age is 29 weeks to have exactly the same head circumference. To accommodate this scatter, we actually fit a model of the form

$$y = \alpha + \beta x + \varepsilon,$$

where ε, known as the *error,* is the distance a particular outcome y lies from the population regression line

$$\mu_{y|x} = \alpha + \beta x.$$

If ε is positive, y is greater than $\mu_{y|x}$. If ε is negative, y is less than $\mu_{y|x}$.

In simple linear regression, the coefficients of the population regression line are estimated using a random sample of observations (x_i, y_i). Before we attempt to fit such a line, however, we must make a few assumptions:

1. For a specified value of x, which is considered to have been measured without error, the distribution of the y values is normal with mean $\mu_{y|x}$ and standard deviation $\sigma_{y|x}$. This concept is illustrated in Figure 18.4.

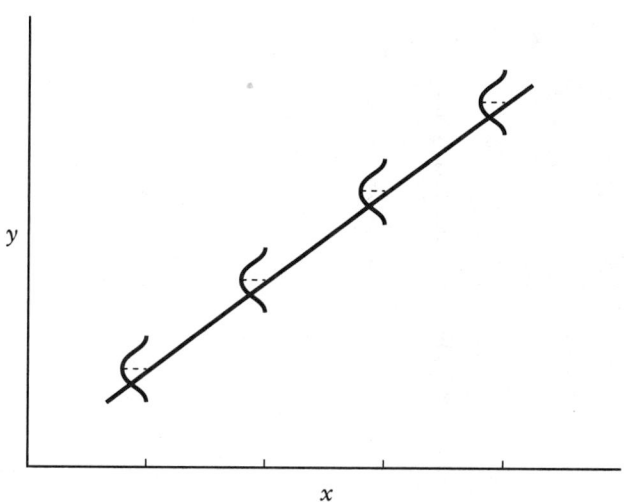

FIGURE 18.4
Normality of the outcomes y for a given value of x

2. The relationship between $\mu_{y|x}$ and x is described by the straight line

$$\mu_{y|x} = \alpha + \beta x.$$

3. For any specified value of x, $\sigma_{y|x}$—the standard deviation of the outcomes y—does not change. This assumption of constant variability across all values of x is known as *homoscedasticity*; it is analogous to the assumption of equal variances in the two-sample t-test or the one-way analysis of variance.

4. The outcomes y are independent.

18.2.2 The Method of Least Squares

Consider Figure 18.5, the two-way scatter plot of head circumference versus gestational age for a sample of 100 low birth weight infants born in Boston, Massachusetts. The explanatory variable is shown on the horizontal axis and the response appears on the vertical axis. The data points themselves vary widely, but the overall pattern suggests that head circumference tends to increase in magnitude as gestational age increases.

In previous chapters, we attempted to estimate a population parameter—such as a mean or an odds ratio—based on the observations in a randomly chosen sample; similarly, we estimate the coefficients of a population regression line using a single sample of measurements. Suppose that we were to draw an arbitrary line through the scatter of points in Figure 18.5. One such line is shown in Figure 18.6. Lines sketched by two different individuals are unlikely to be identical, even though both persons might be attempting to depict the same trend. The question then arises as to which line best de-

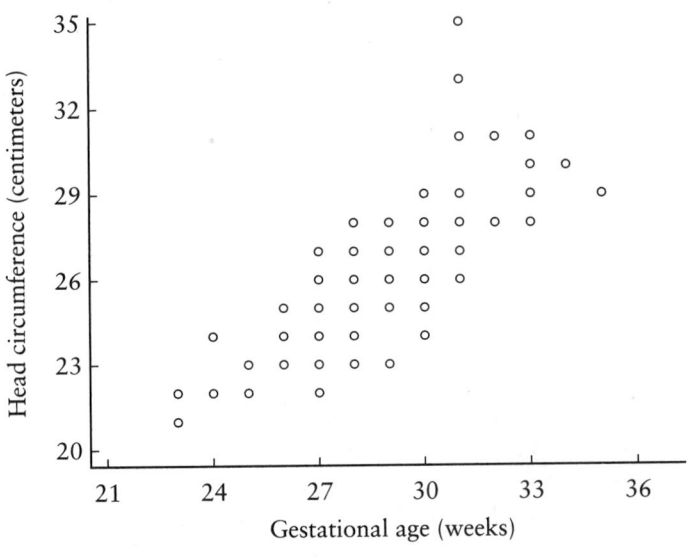

FIGURE 18.5

Head circumference versus gestational age for a sample of 100 low birth
weight infants

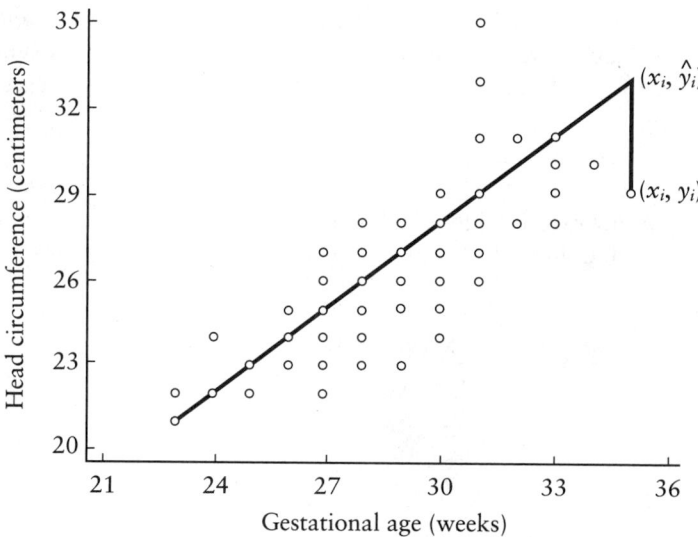

FIGURE 18.6
Arbitrary line depicting a relationship between head circumference and gestational age

scribes the relationship between mean head circumference and gestational age. What is needed is a more objective procedure for estimating the line. (A word of explanation regarding Figure 18.5: Although the graph contains information for 100 infants, there appear to be far fewer data points. Since the outcomes of both random variables are rounded to the nearest integer, many infants end up with identical values of head circumference and gestational age; consequently, some data points are plotted on top of others.)

One mathematical technique for fitting a straight line to a set of points (x_i, y_i) is known as the *method of least squares*. Observe that each of the 100 data points representing measurements of head circumference and gestational age lies some vertical distance from the arbitrary line drawn in Figure 18.6; we label this distance e_i. If y_i is the observed outcome of Y for a particular value x_i, and $\hat{y}_i$—or y_i-hat—is the corresponding point on the fitted line, then

$$e_i = y_i - \hat{y}_i.$$

The distance e_i is known as the *residual*. If all the residuals are equal to 0, this implies that each point (x_i, y_i) lies directly on the fitted line; the points are as close to the line as they can be. Since this is not the case, however, we choose a criterion for fitting a line that makes the residuals as small as possible. The sum of the squares of the residuals,

$$\sum_{i=1}^{n} e_i^2 = \sum_{i=1}^{n} (y_i - \hat{y}_i)^2,$$

is often called the *error sum of squares,* or the *residual sum of squares.* The least-squares regression line is constructed so that the error sum of squares is minimized.

The process of fitting the least-squares line represented by

$$\hat{y} = \hat{\alpha} + \hat{\beta} x$$

involves finding $\hat{\alpha}$ and $\hat{\beta}$, the estimates of the population regression coefficients α and β. Using calculus to minimize the error sum of squares

$$\sum_{i=1}^{n} e_i^2 = \sum_{i=1}^{n} (y_i - \hat{y}_i)^2$$

$$= \sum_{i=1}^{n} (y_i - \hat{\alpha} - \hat{\beta} x_i)^2,$$

we find that

$$\hat{\beta} = \frac{\sum_{i=1}^{n} (x_i - \bar{x})(y_i - \bar{y})}{\sum_{i=1}^{n} (x_i - \bar{x})^2}$$

and

$$\hat{\alpha} = \bar{y} - \hat{\beta} \bar{x}.$$

These equations yield the slope and the *y*-intercept for the fitted least-squares line. In the expression for $\hat{\beta}$, the numerator is the sum of the cross-products of deviations around the mean for x and y; the denominator is the sum of squared deviations around the mean for x alone. The equation for $\hat{\alpha}$ is expressed in terms of the estimated slope $\hat{\beta}$. Once we know $\hat{\alpha}$ and $\hat{\beta}$, we are able to substitute various values of x into the equation for the line, solve for the corresponding values of $\hat{y}$, and plot these points to draw the least-squares regression line.

The least-squares line fitted to the 100 measurements of head circumference and gestational age is

$$\hat{y} = 3.9143 + 0.7801 x.$$

This line—which is plotted in Figure 18.7—has an error sum of squares that is smaller than the sum for any other line that could be drawn through the scatter of points. The *y*-intercept of the fitted line is 3.9143. Theoretically, this is the mean value of head circumference that corresponds to a gestational age of 0 weeks. In this example, however, an age of 0 weeks does not make sense. The slope of the line is 0.7801, implying that for each one-week increase in gestational age, an infant's head circumference increases by 0.7801 centimeters on average.

were confirmed to be accurate, 783 either lacked information or contained inaccuracies but did not require recoding of the underlying cause of death, and 864 were incorrect and required recoding [14].

Date of Study	Certificate Status			Total
	Confirmed Accurate	Inaccurate No Change	Incorrect Recoding	
1955–1965	2040	367	327	2734
1970	149	60	48	257
1970–1971	288	25	70	383
1975–1977	703	197	252	1152
1977–1978	425	62	88	575
1980	121	72	79	272
Total	3726	783	864	5373

(a) Do you believe that the results are homogeneous or consistent across studies?

(b) It should be noted that autopsies are not performed at random; in fact, many are done because the cause of death listed on the certificate is uncertain. What problems may arise if you attempt to use the results of these studies to make inference about the population as a whole?

12. In a study of intraobserver variability in the assessment of cervical smears, 3325 slides were screened for the presence or absence of abnormal squamous cells. Each slide was screened by a particular observer and then rescreened six months later by the same observer. The results of this study are shown below [15].

First Screening	Second Screening		Total
	Present	Absent	
Present	1763	489	2252
Absent	403	670	1073
Total	2166	1159	3325

(a) Do these data support the null hypothesis that there is no association between time of screening and diagnosis?

The data could also be displayed in the following manner:

Abnormal Cells	Screening		Total
	First	Second	
Present	2252	2166	4418
Absent	1073	1159	2232
Total	3325	3325	6650

(e) Does the 95% confidence interval contain the value 0? Would you have expected that it would?

9. A study was conducted to evaluate the relative efficacy of supplementation with calcium versus calcitriol in the treatment of postmenopausal osteoporosis [12]. Calcitriol is an agent that has the ability to increase gastrointestinal absorption of calcium. A number of patients withdrew from this study prematurely due to the adverse effects of treatment, which include thirst, skin problems, and neurologic symptoms. The relevant data appear below.

| Treatment | Withdrawal | | Total |
	Yes	No	
Calcitriol	27	287	314
Calcium	20	288	308
Total	47	575	622

(a) Compute the sample proportion of subjects who withdrew from the study in each treatment group.
(b) Test the null hypothesis that there is no association between treatment group and withdrawal from the study at the 0.05 level of significance. What do you conclude?

10. In a survey conducted in Italy, physicians with different specialties were questioned regarding the surgical treatment of early breast cancer. In particular, they were asked whether they would recommend radical surgery regardless of a patient's age (R), conservative surgery only for younger patients (CR), or conservative surgery regardless of age (C). The results of this survey are presented below [13].

| Physician Specialty | Surgery | | | Total |
	R	CR	C	
Internal	6	22	42	70
Surgery	23	61	127	211
Radiotherapy	2	3	54	59
Oncology	1	12	43	56
Gynecology	1	12	31	44
Total	33	110	297	440

(a) At the 0.05 level of significance, test the null hypothesis that there is no association between physician specialty and recommended treatment.
(b) What do you conclude?

11. The following table compiles data from six studies designed to investigate the accuracy of death certificates. The results of 5373 autopsies were compared to the causes of death listed on the certificates. Of those considered, 3726 certificates

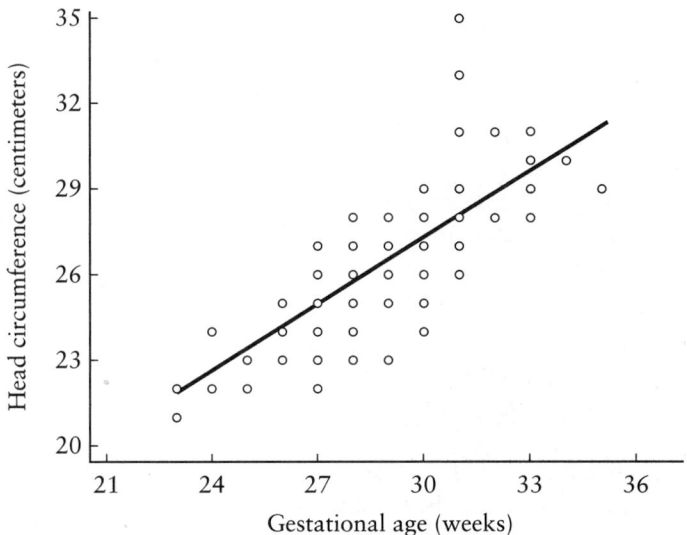

FIGURE 18.7
Least-squares regression of head circumference on gestational age,
$\hat{y} = 3.9143 + 0.7801x$

18.2.3 Inference for Regression Coefficients

We would like to be able to use the least-squares regression line

$$\hat{y} = \hat{\alpha} + \hat{\beta}x$$

to make inference about the population regression line

$$\mu_{y|x} = \alpha + \beta x.$$

We can begin by saying that $\hat{\alpha}$ is a point estimate of the population intercept α and $\hat{\beta}$ is a point estimate of the slope β. If we were to select repeated samples of size n from the underlying population of paired outcomes (x, y) and calculate a least-squares line for each set of observations, the estimated values of α and β would vary from sample to sample. We need the standard errors of these estimators—just as we needed $\sigma/\sqrt{n}$, the standard error of the sample mean $\overline{X}$—to be able to construct confidence intervals and conduct tests of hypotheses.

It can be shown that

$$\text{se}(\hat{\beta}) = \frac{\sigma_{y|x}}{\sqrt{\sum_{i=1}^{n} (x_i - \overline{x})^2}}$$

and

$$se(\hat{\alpha}) = \sigma_{y|x} \sqrt{\frac{1}{n} + \frac{\bar{x}^2}{\sum_{i=1}^{n} (x_i - \bar{x})^2}} \, .$$

The standard errors of the estimated coefficients $\hat{\alpha}$ and $\hat{\beta}$ both depend on $\sigma_{y|x}$, the standard deviation of the y values for a given x. In practice, this value is usually unknown. As a result, we must estimate $\sigma_{y|x}$ by the sample standard deviation $s_{y|x}$, where

$$s_{y|x} = \sqrt{\frac{\sum_{i=1}^{n} (y_i - \hat{y}_i)^2}{n - 2}} \, .$$

Note that this formula involves the sum of the squared deviations of the actual observations y_i from the fitted values $\hat{y}_i$; the sum of squared residuals is the quantity that was minimized when we fit the least-squares line. The estimate $s_{y|x}$ is often called the *standard deviation from regression*. For the least-squares regression of head circumference on gestational age,

$$s_{y|x} = 1.5904.$$

This estimate can be used to compute

$$\widehat{se}(\hat{\beta}) = \frac{s_{y|x}}{\sqrt{\sum_{i=1}^{n} (x_i - \bar{x})^2}}$$
$$= 0.0631$$

and

$$\widehat{se}(\hat{\alpha}) = s_{y|x} \sqrt{\frac{1}{n} + \frac{\bar{x}^2}{\sum_{i=1}^{n} (x_i - \bar{x})^2}}$$
$$= 1.8291.$$

The slope is usually the more important coefficient in the linear regression equation; it quantifies the average change in y that corresponds to each one-unit change in x. We can test the null hypothesis that the population slope is equal to β_0, or

$$H_0: \beta = \beta_0,$$

against the alternative

$$H_A: \beta \neq \beta_0$$

by finding p, the probability of observing an estimated slope as extreme as or more extreme than $\hat{\beta}$, meaning further away from β_0, given that β_0 is the true population value. (By *more extreme*, we mean further away from β_0.) The test is carried out by calculating the statistic

$$t = \frac{\hat{\beta} - \beta_0}{\widehat{se}(\hat{\beta})}.$$

If the null hypothesis is true, this ratio has a t distribution with $n - 2$ degrees of freedom. Using Table A.4, we find the probability p. We then compare p to α—the significance level of the test—to determine whether we should reject or not reject H_0.

Most frequently, we are interested in the case in which $\beta_0 = 0$. If the population slope is equal to 0, then

$$\mu_{y|x} = \alpha + (0)x$$
$$= \alpha.$$

There is no linear relationship between x and y; the mean value of y is the same regardless of the value of x. For the head circumference and gestational age data, this would imply that the mean value of head circumference is the same for infants of all gestational ages. It can be shown that a test of the null hypothesis

$$H_0: \beta = 0$$

is mathematically equivalent to the test of

$$H_0: \rho = 0,$$

where ρ is the correlation between head circumference and gestational age in the underlying population of low birth weight infants. In fact,

$$\hat{\beta} = r\left(\frac{s_y}{s_x}\right),$$

where s_x and s_y are the standard deviations of the x and y values respectively [2]. Both null hypotheses claim that y does not change as x increases.

To conduct a two-sided test of the null hypothesis that the true slope relating head circumference to gestational age is equal to 0 at the 0.05 level of significance, we calculate

$$t = \frac{\hat{\beta} - \beta_0}{\widehat{se}(\hat{\beta})}$$
$$= \frac{0.7801 - 0}{0.0631}$$
$$= 12.36.$$

For a t distribution with $100 - 2 = 98$ degrees of freedom, $p < 0.001$. Therefore, we reject the null hypothesis that the slope β is equal to 0. In the underlying population of low birth weight infants, there is a statistically significant linear relationship between head circumference and gestational age: Head circumference increases in magnitude as gestational age increases.

In addition to conducting a test of hypothesis, we can also calculate a confidence interval for the true population slope. For a t distribution with 98 degrees of freedom, approximately 95% of the observations fall between -1.98 and 1.98. Therefore,

$$(\hat{\beta} - 1.98\,\widehat{se}\,(\hat{\beta}), \hat{\beta} + 1.98\,\widehat{se}\,(\hat{\beta}))$$

is a 95% confidence interval for β. Since we previously found that

$$\widehat{se}\,(\hat{\beta}) = 0.0631,$$

the interval is

$$(0.7801 - 1.98(0.0631), 0.7801 + 1.98(0.0631)),$$

or

$$(0.6564, 0.9038).$$

While 0.7801 is a point estimate for β, we are 95% confident that the above limits cover the true population slope.

If we are interested in testing whether the population intercept is equal to a specified value α_0, we use calculations that are analogous to those for the slope. We compute the statistic

$$t = \frac{\hat{\alpha} - \alpha_0}{\widehat{se}\,(\hat{\alpha})}$$

and compare this value to the t distribution with $n - 2$ degrees of freedom. We can also construct a confidence interval for the true population intercept α just as we calculated an interval for the slope β. However, if the observed data points tend to be far from the intercept—as they are for the head circumference and gestational age data, where the smallest value of gestational age is $x = 23$ weeks—there is very little practical value in making inference about the intercept. As we have already noted, a gestational age of 0 weeks does not make any sense. In fact, it is dangerous to extrapolate the fitted line beyond the range of the observed values x; the relationship between X and Y might be quite different outside this range.

18.2.4 Inference for Predicted Values

In addition to making inference about the population slope and intercept, we might also be interested in using the least-squares regression line to estimate the mean value of y corresponding to a particular value of x, and to construct a 95% confidence interval for the mean. If we have a sample of 100 observations, for instance, the confidence interval will take the form

$$(\hat{y} - 1.98\,\widehat{se}\,(\hat{y}),\ \hat{y} + 1.98\,\widehat{se}\,(\hat{y})),$$

where $\hat{y}$ is the predicted mean of the normally distributed outcomes, and the standard error of $\hat{y}$ is estimated by

$$\widehat{se}\,(\hat{y}) = s_{y|x}\sqrt{\left[\frac{1}{n} + \frac{(x - \bar{x})^2}{\sum_{i=1}^{n}(x_i - \bar{x})^2}\right]}.$$

Note the term $(x - \bar{x})^2$ in the expression for the standard error. This quantity assumes the value 0 when x is equal to $\bar{x}$ and grows larger as x moves farther and farther away. As a result, if x is near $\bar{x}$, the confidence interval is relatively narrow. It grows wider as x moves away from $\bar{x}$. In a sense, we are more confident about the mean value of the response when we are closer to the mean value of the explanatory variable.

Return once again to the head circumference and gestational age data. When $x = 29$ weeks,

$$\begin{aligned}\hat{y} &= \hat{\alpha} + \hat{\beta}x \\ &= 3.9143 + (0.7801)(29) \\ &= 26.54 \text{ cm}.\end{aligned}$$

The value 26.54 cm is a point estimate for the mean value of y when x is equal to 29. The estimated standard error of $\hat{y}$ is

$$\widehat{se}\,(\hat{y}) = 0.159 \text{ cm}.$$

Therefore, a 95% confidence interval for the mean value of y is

$$(26.54 - 1.98(0.159),\ 26.54 + 1.98(0.159)),$$

or

$$(26.23, 26.85).$$

The curved lines in Figure 18.8 represent the 95% confidence limits on the mean value of y for each observed value of x, from 23 weeks to 35 weeks. As we move away from $x = 29$, which is very close to $\bar{x}$, the confidence limits gradually become wider.

Sometimes, instead of predicting the mean value of y for a given value of x, we prefer to predict an individual value of y for a new member of the population. The predicted individual value is denoted by $\tilde{y}$, or y-tilde, and is identical to the predicted mean $\hat{y}$; in particular,

$$\begin{aligned}\tilde{y} &= \hat{\alpha} + \hat{\beta}x \\ &= \hat{y}.\end{aligned}$$

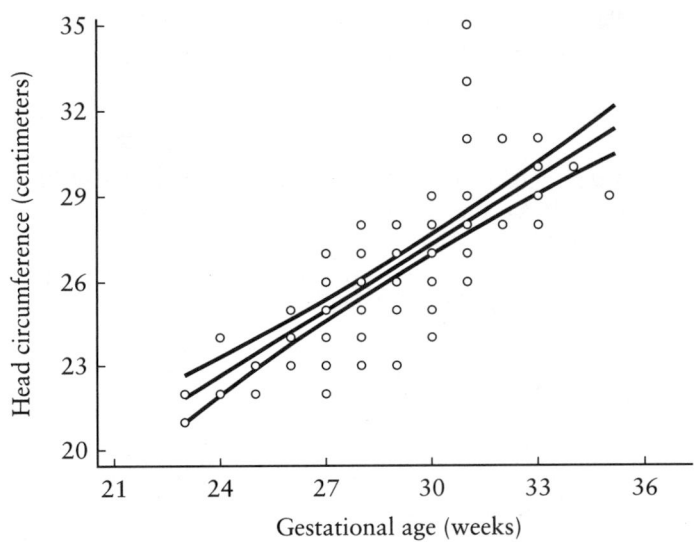

FIGURE 18.8
The 95% confidence limits on the predicted mean of y for a given value of x

The standard error of $\tilde{y}$, however, is not the same as the standard error of $\hat{y}$. When we were computing $\widehat{se}\,(\hat{y})$, we were interested only in the variability of the estimated mean of the y values. When considering an individual y, we have an extra source of variability to account for: the dispersion of the y values themselves around that mean. Recall that for a given value of x, the outcomes y are normally distributed with standard deviation $\sigma_{y|x}$. Therefore, we would expect the expression for the standard error of $\tilde{y}$ to incorporate an extra term involving $\sigma_{y|x}$—or its estimator $s_{y|x}$—which is not included in the expression for the standard error of $\hat{y}$. In fact,

$$\widehat{se}\,(\tilde{y}) = \sqrt{s_{y|x}^2 + \widehat{se}\,(\hat{y})^2}$$

$$= s_{y|x}\sqrt{\left[1 + \frac{1}{n} + \frac{(x - \bar{x})^2}{\sum_{i=1}^{n}(x_i - \bar{x})^2}\right]}.$$

Once again, the term $(x - \bar{x})^2$ implies that the standard error is smallest when x is equal to $\bar{x}$, and grows larger as x moves away from $\bar{x}$. If we have a sample of 100 observations, a 95% prediction interval for an individual outcome y takes the form

$$(\tilde{y} - 1.98\,\widehat{se}\,(\tilde{y}),\ \tilde{y} + 1.98\,\widehat{se}\,(\tilde{y})).$$

Because of the extra source of variability, the limits on a predicted **individual** value of y are wider than the limits on the predicted **mean** y for the same value of x.

Suppose that a new child is selected from the underlying population of low birth weight infants. If this newborn has a gestational age of 29 weeks, then

$$\tilde{y} = \hat{\alpha} + \hat{\beta}x$$
$$= 3.9143 + (0.7801)(29)$$
$$= 26.54 \text{ cm.}$$

The standard error of $\tilde{y}$ is estimated as

$$\widehat{se}\,(\tilde{y}) = \sqrt{s_{y|x}^2 + \widehat{se}\,(\hat{y})^2}$$
$$= \sqrt{(1.5904)^2 + (0.159)^2}$$
$$= 1.598 \text{ cm.}$$

Therefore, a 95% prediction interval for an individual new value of head circumference is

$$(26.54 - 1.98(1.598), \ 26.54 + 1.98(1.598)),$$

or

$$(23.38, 29.70).$$

The curved lines in Figure 18.9 are the 95% limits on an individual value of y for each observed value of x from 23 to 35 weeks. Note that these bands are considerably farther from the least-squares regression line than the 95% confidence limits around the mean value of y.

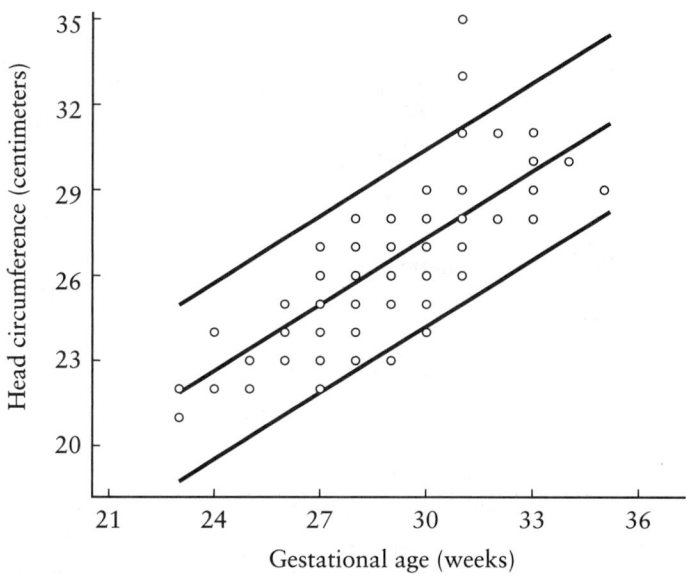

FIGURE 18.9
The 95% confidence limits on an individual predicted y for a given value of x

18.3   Evaluation of the Model

18.3.1   The Coefficient of Determination

After generating a least-squares regression line represented by

$$\hat{y} = \hat{\alpha} + \hat{\beta}x,$$

we might wonder how well this model actually fits the observed data. One way to get a sense of the fit is to compute the coefficient of determination. The *coefficient of determination* is represented by R^2 and is the square of the Pearson correlation coefficient r; consequently,

$$r^2 = R^2.$$

Since r can assume any value in the range -1 to 1, R^2 must lie between 0 and 1. If $R^2 = 1$, all the data points in the sample fall directly on the least-squares line. If $R^2 = 0$, there is no linear relationship between x and y.

The coefficient of determination can also be interpreted as the proportion of the variability among the observed values of y that is explained by the linear regression of y on x. This interpretation derives from the relationship between σ_y, the standard deviation of the outcomes of the response variable Y, and $\sigma_{y|x}$, the standard deviation of y for a specified value of the explanatory variable X, that was presented in Section 18.1:

$$\sigma_{y|x}^2 = (1 - \rho^2)\sigma_y^2.$$

Recall that ρ is the correlation between X and Y in the underlying population. If we replace σ_y and $\sigma_{y|x}$ by their estimators—the sample standard deviations s_y and $s_{y|x}$—and ρ by the Pearson correlation coefficient r, we have

$$\begin{aligned}
s_{y|x}^2 &= (1 - r^2)s_y^2 \\
&= (1 - R^2)s_y^2.
\end{aligned}$$

Solving this equation for R^2,

$$\begin{aligned}
R^2 &= 1 - \frac{s_{y|x}^2}{s_y^2} \\
&= \frac{s_y^2 - s_{y|x}^2}{s_y^2}.
\end{aligned}$$

Since $s_{y|x}^2$ is the variation in the y values that still remains after accounting for the relationship between y and x, $s_y^2 - s_{y|x}^2$ must be the variation in y that is explained by their linear relationship. Thus, R^2 is the proportion of the total observed variability among the y values that is explained by the linear regression of y on x.

For the regression of head circumference on gestational age, the coefficient of determination can be shown to be

$$R^2 = 0.6095.$$

This value implies a moderately strong linear relationship between gestational age and head circumference; in particular, 60.95% of the variability among the observed values of head circumference is explained by the linear relationship between head circumference and gestational age. The remaining

$$100 - 60.95 = 39.05\%$$

of the variation is not explained by this relationship.

18.3.2 Residual Plots

Another strategy for evaluating how well the least-squares regression line fits the observed data in the sample used to construct it is to generate a two-way scatter plot of the residuals against the fitted or predicted values of the response variable. For example, one particular child in the sample of 100 low birth weight infants has a gestational age of 29 weeks and a head circumference of 27 centimeters. The child's predicted head circumference, given that $x_i = 29$ weeks, is

$$\hat{y}_i = \hat{\alpha} + \hat{\beta}x_i$$
$$= 3.9143 + (0.7801)(29)$$
$$= 26.54 \text{ cm.}$$

The residual associated with this observation is

$$e_i = y_i - \hat{y}_i$$
$$= 27 - 26.54$$
$$= 0.46;$$

therefore, the point (26.54, 0.46) would be included on the graph. Figure 18.10 is a scatter plot of the points $(\hat{y}_i, e_i)$ for all 100 observations in the sample of low birth weight infants.

A plot of the residuals serves three purposes. First, it can help us to detect outlying observations in the sample. In Figure 18.10, one residual in particular is somewhat larger than the others; this point is associated with a child whose gestational age is 31 weeks and whose head circumference is 35 centimeters. We would predict the infant's head circumference to be only

$$\hat{y} = 3.914 + 0.7801(31)$$
$$= 28.10 \text{ cm.}$$

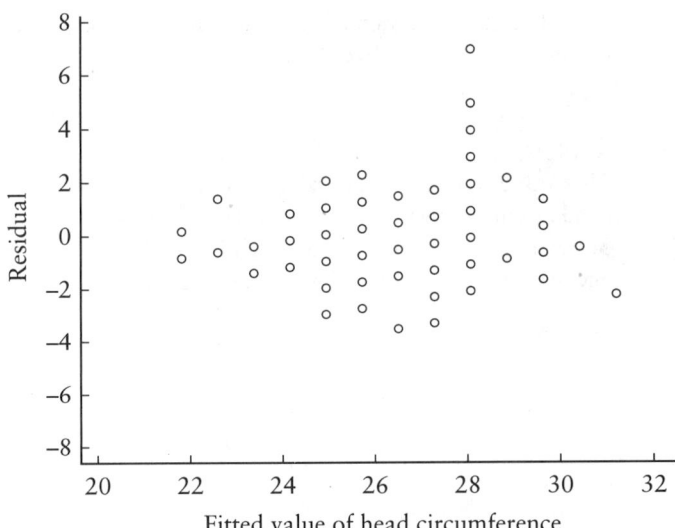

FIGURE 18.10
Residuals versus fitted values of head circumference

The method of least squares—like the sample mean or the Pearson correlation coefficient—can be very sensitive to such outliers in the data, especially if they correspond to relatively large or relatively small values of x. When it is believed that an outlier is the result of an error in measuring or recording a particular observation, removal of this point improves the fit of the regression line. However, care must be taken not to throw away unusual data points that are in fact valid; these observations might be the most interesting ones in the data set.

A plot of the residuals can also suggest a failure in the assumption of homoscedasticity. Recall that homoscedasticity means that the standard deviation of the outcomes y, or $\sigma_{y|x}$, is constant across all values of x. If the range of the magnitudes of the residuals either increases or decreases as $\hat{y}$ becomes larger—producing a fan-shaped scatter such as the one in Figure 18.11—this implies that $\sigma_{y|x}$ does not take the same value for all values of x. In this case, simple linear regression is not the appropriate technique for modeling the relationship between x and y. No such pattern is evident in Figure 18.10, the residual plot for the sample of head circumference and gestational age measurements for 100 low birth weight infants. Thus, the assumption of homoscedasticity does not appear to have been violated. (We should note that it can be difficult to evaluate this and other assumptions based on a residual plot if the number of data points is small.)

Finally, if the residuals do not exhibit a random scatter but instead follow a distinct trend—e_i increases as $\hat{y}_i$ increases, for example—this would suggest that the true relationship between x and y might not be linear. In this situation, a *transformation* of x or y or both might be appropriate. When transforming a variable, we simply measure it on a different scale. In many ways, it is analogous to measuring a variable in different units; height can be measured in either inches or centimeters, for example. Often, a curvilinear relationship between two variables can be transformed into a more straight-

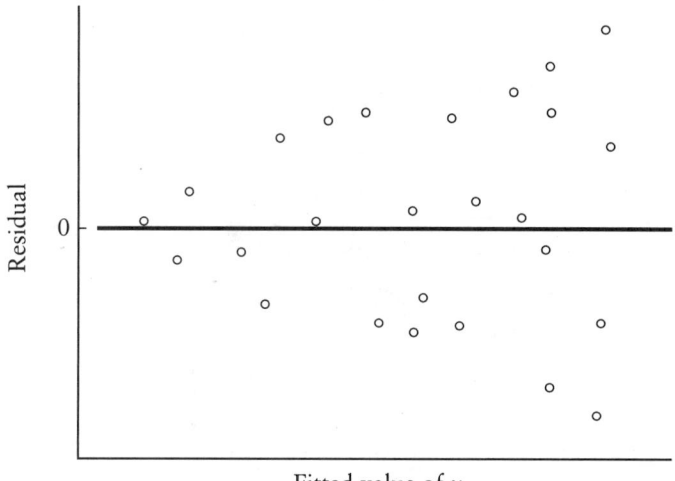

FIGURE 18.11
Violation of the assumption of homoscedasticity

forward linear one. If this is possible, we can use simple linear regression to fit a model to the transformed data.

18.3.3   Transformations

Consider Figure 18.12. This graph is a two-way scatter plot of crude birth rate per 1000 population versus gross national product per capita for 143 countries around the world [3]. The gross national product (GNP) is expressed in United States dollars. Note that birth rate tends to decrease as gross national product increases. The relationship, however, is not a linear one. Instead, birth rate drops off rapidly at first; when the GNP per capita reaches approximately $5000, it begins to level off. Consequently, if we wish to describe the relationship between birth rate and gross national product, we cannot use simple linear regression without applying some type of transformation first.

When the relationship between x and y is not linear, we begin by looking at transformations of the form x^p or y^p, where

$$p = \ldots -3, -2, -1, -\frac{1}{2}, \ln, \frac{1}{2}, 1, 2, 3, \ldots.$$

Note that "ln" refers to the natural logarithm of x or y rather than an exponent. Thus, possible transformations might be $\ln(y)$, $x^{1/2} = \sqrt{x}$, or x^2.

The *circle of powers*—or the *ladder of powers,* as it is sometimes called—provides a general guideline for choosing a transformation. The strategy is illustrated in Figure 18.13. If the plotted data resemble the pattern in Quadrant I, for instance, an appropriate transformation would be either "up" on x or "up" on y. In other words, either x or y would be raised to a power greater than $p = 1$; the more curvature in the data, the

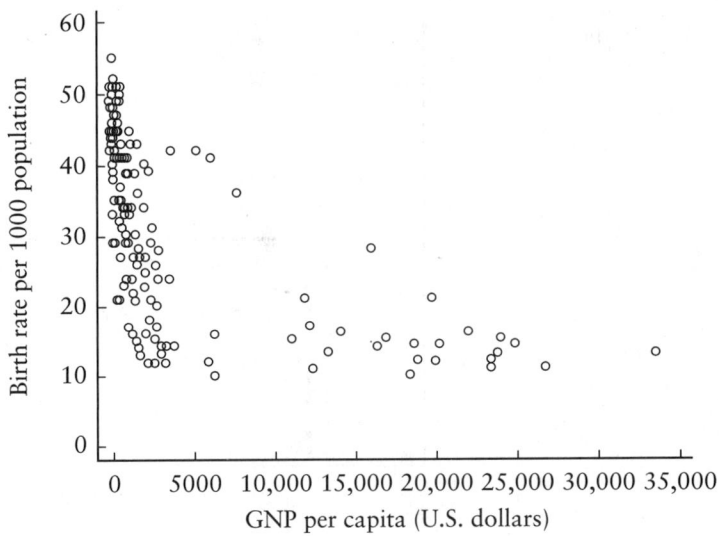

FIGURE 18.12

Birth rate per 1000 population versus gross national product per capita for 143 countries, 1992

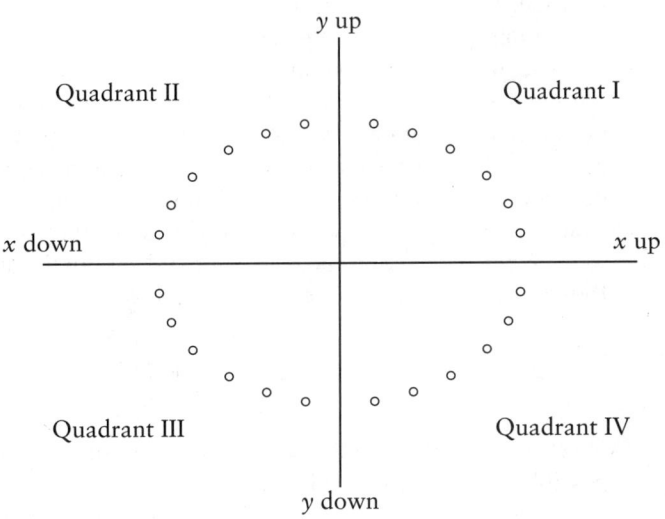

FIGURE 18.13

The circle of powers

higher the value of p needed to achieve linearity. We might try replacing x with x^2, for example. If a two-way scatter plot suggests that the relationship between y and x^2 is linear, we would fit a model of the form

$$\hat{y} = \hat{\alpha} + \hat{\beta}x^2$$

instead of the usual

$$\hat{y} = \hat{\alpha} + \hat{\beta}x.$$

If the data follow the trend in Quadrant II, we would want to transform "up" on y or "down" on x, meaning that we would either raise x to a power less than 1 or raise y to a power greater than 1. Therefore, we might try replacing x by $\sqrt{x}$ or $\ln(x)$. Whichever transformation is chosen, we must always verify that the assumption of homoscedasticity is valid.

The data in Figure 18.12 most closely resemble the pattern in Quadrant III; therefore, we would want to raise either x or y to a power that is less than 1. We might try replacing gross national product with its natural logarithm, for instance. The effect of this transformation is illustrated in Figure 18.14; note that the relationship between birth rate and the logarithm of GNP appears much more linear than the relationship between birth rate and GNP itself. Therefore, we would fit a simple linear regression model of the form

$$\hat{y} = \hat{\alpha} + \hat{\beta}\ln(x).$$

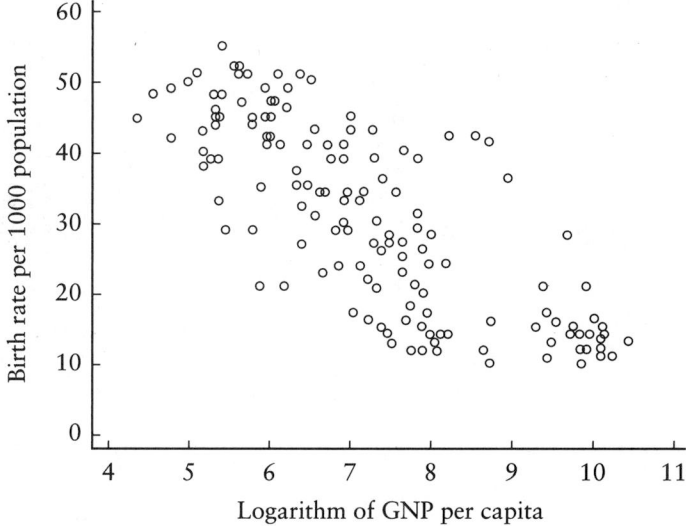

FIGURE 18.14
Birth rate per 1000 population versus the natural logarithm of gross national product per capita

Although the units are unfamiliar—gross national product is measured in ln (U.S. dollars) instead of dollars—this transformation allows us to apply a method that would otherwise be inappropriate.

18.4 Further Applications

Suppose that we are now interested in the relationship between length and gestational age for the population of low birth weight infants, again defined as those weighing less than 1500 grams. We begin our analysis by constructing a two-way scatter plot of length versus gestational age for the sample of 100 low birth weight infants born in Boston, Massachusetts. The plot is displayed in Figure 18.15. The points on the graph exhibit a great deal of scatter; however, it is clear that length increases as gestational age increases. The relationship appears to be a linear one.

To estimate the true population regression line

$$\mu_{y|x} = \alpha + \beta x,$$

where $\mu_{y|x}$ is the mean length for low birth weight infants of the specified gestational age x, α is the y-intercept of the line, and β is its slope, we apply the method of least squares to fit the model

$$\hat{y} = \hat{\alpha} + \hat{\beta} x.$$

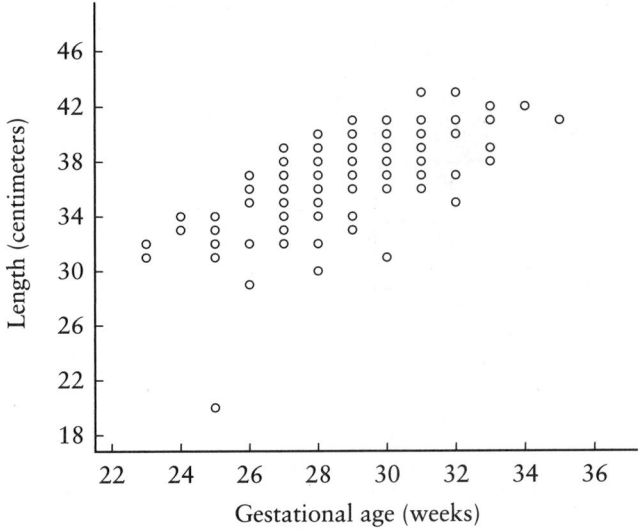

FIGURE 18.15

Length versus gestational age for a sample of 100 low birth weight infants

Rather than use calculus to minimize the sum of the squared residuals

$$\sum_{i=1}^{n} e_i^2 = \sum_{i=1}^{n} (y_i - \hat{y}_i)^2$$

$$= \sum_{i=1}^{n} (y_i - \hat{\alpha} - \hat{\beta}x_i)^2,$$

we rely on the computer to do the calculations for us. Table 18.1 shows the relevant output from Stata.

The top portion of Table 18.1 displays an analysis of variance table on the left and additional information about the model on the right; we will return to some of this information later on. The bottom portion contains the estimated coefficients for the least-squares regression line. The column on the far left lists the names of the response (length) and explanatory (gestage) variables; the label "_cons" refers to the y-intercept, or the constant term in the equation. The point estimates for the regression coefficients, $\hat{\alpha}$ and $\hat{\beta}$, appear in the second column. Rounding these values to four decimal places, the fitted least-squares regression line for this sample of 100 low birth weight infants is

$$\hat{y} = 9.3282 + 0.9516x.$$

The y-intercept of 9.3282 is the estimated mean value of length that corresponds to an x value of 0; in this example, however, a gestational age of 0 weeks does not make sense. The slope of the line indicates that for each one-week increase in gestational age, an infant's length increases by 0.9516 centimeters on average.

The third column in the bottom portion of Table 18.1 displays the estimated standard errors of $\hat{\alpha}$ and $\hat{\beta}$. Suppose that we wish to test the null hypothesis that the population slope is equal to 0, or

$$H_0: \beta = 0.$$

TABLE 18.1

Stata output displaying the simple linear regression of length on gestational age

Source	SS	df	MS			
				Number of obs	=	100
				F(1,98)	=	82.13
Model	575.73916	1	575.73916	Prob > F	=	0.0000
Residual	687.02084	98	7.01041674	R-square	=	0.4559
				Adj R-square	=	0.4504
Total	1262.76	99	12.7551515	Root MSE	=	2.6477

Variable	Coefficient	Std. Error	t	P > |t|	[95% Conf. Interval]	
length						
gestage	0.9516035	0.1050062	9.062	0.000	.7432221	1.159985
_cons	9.3281740	3.0451630	3.063	0.003	3.285149	15.3712

The appropriate test statistic is

$$t = \frac{\hat{\beta} - \beta_0}{\widehat{se}(\hat{\beta})}$$
$$= \frac{0.9516 - 0}{0.1050}$$
$$= 9.062;$$

this statistic is provided in the fourth column of the table. The adjacent column contains the p-value that corresponds to a two-sided test. Since p is less than 0.001, we reject the null hypothesis that β is equal to 0. Based on this sample of low birth weight infants, length increases as gestational age increases.

Since the data set contains measurements for 100 infants, and we know that for a t distribution with $100 - 2 = 98$ degrees of freedom 95% of the observations fall between -1.98 and 1.98, a 95% confidence interval for the population slope takes the form

$$(\hat{\beta} - 1.98\,\widehat{se}(\hat{\beta}), \hat{\beta} + 1.98\,\widehat{se}(\hat{\beta})).$$

If we substitute the values of $\hat{\beta}$ and $\widehat{se}(\hat{\beta})$ from the table, the 95% confidence interval is

$$(0.9516 - 1.98(0.1050), 0.9516 + 1.98(0.1050)),$$

or

$$(0.7432, 1.1600).$$

These confidence limits are provided in the sixth and seventh columns of Table 18.1.

We might also be interested in using the least-squares regression line to estimate the mean value of length corresponding to a specified value of gestational age, and to construct a 95% confidence interval for this mean. If we have a sample of 100 observations, the confidence interval takes the form

$$(\hat{y} - 1.98\,\widehat{se}(\hat{y}), \hat{y} + 1.98\,\widehat{se}(\hat{y})).$$

For each of the 100 low birth weight infants in the sample—each with a particular gestational age x—we can use the computer to obtain the corresponding predicted length $\hat{y}$, as well as its estimated standard error $\widehat{se}(\hat{y})$. The data for the first 10 infants in the sample are shown in Table 18.2. When x is equal to 29 weeks, we observe that

$$\hat{y} = 9.3282 + 0.9516(29)$$
$$= 36.93 \text{ cm.}$$

The estimated standard error of $\hat{y}$ is 0.265 cm. Therefore, a 95% confidence interval for the mean value of length is

$$(36.93 - 1.98(0.265), 36.93 + 1.98(0.265)),$$

TABLE 18.2
Predicted values of length and estimated standard errors for the
first 10 infants in the sample

Gestational Age	Predicted Length $\hat{y}$	Standard Error of $\hat{y}$
29	36.92467	0.2650237
31	38.82788	0.3452454
33	40.73109	0.5063217
31	38.82788	0.3452454
30	37.87628	0.2892917
25	33.11826	0.4867806
27	35.02147	0.3308946
29	36.92467	0.2650237
28	35.97307	0.2807812
29	36.92467	0.2650237

or

(36.41, 37.45).

Analogous confidence intervals can be calculated for each observed value of x from 23 weeks to 35 weeks. The confidence intervals become wider as we move further away from $\bar{x}$, the mean of the x values.

The predicted individual value of y for a new member of the population is identical to the predicted mean of y; for an infant whose gestational age is 29 weeks,

$$\tilde{y} = 9.3282 + 0.9516(29)$$
$$= 36.93 \text{ cm.}$$

Its standard error, however, is not the same. In addition to the variability of the estimated mean value of y, it incorporates the variation of the y values around that mean, the standard deviation from regression $s_{y|x}$. The standard deviation from regression is shown in the top portion of the output in Table 18.1, beside the label Root MSE. Note that

$$s_{y|x} = 2.6477,$$

and the estimated standard error of $\tilde{y}$ is

$$\widehat{se}(\tilde{y}) = \sqrt{s_{y|x}^2 + \widehat{se}(\hat{y})^2}$$
$$= \sqrt{(2.6477)^2 + (0.265)^2}$$
$$= 2.661.$$

Therefore, a 95% prediction interval for the individual new value of length is

$$(36.93 - 1.98(2.661), 36.93 + 1.98(2.661)),$$

or

(31.66, 42.20).

Because of the extra source of variability, this interval is quite a bit wider than the 95% confidence interval for the predicted mean value of y.

After generating the least-squares regression line, we might wish to have some idea about how well this model fits the observed data. One way to evaluate the fit is to examine the coefficient of determination. In Table 18.1, the value of R^2 is displayed in the top portion of the output on the right-hand side. For the simple linear regression of length on gestational age, the coefficient of determination is

$$R^2 = 0.4559;$$

this means that approximately 45.59% of the variability among the observed values of length is explained by the linear relationship between length and gestational age. The remaining 54.41% is not explained by this relationship. The line of output directly below R^2, labeled Adj R-square (adjusted R^2), will be discussed in the following chapter.

A second technique for evaluating the fit of the least-squares regression line to the sample data involves looking at a two-way scatter plot of the residuals versus the predicted values of length. The residuals are obtained by subtracting the fitted values $\hat{y}_i$ from the actual observations y_i; the calculations may be performed using a computer package. Table 18.3 shows the observed and predicted values of length, along with the differences between them, for the first 10 infants in the sample. Figure 18.16 is a scatter plot of the points $(\hat{y}_i, e_i)$ for all 100 low birth weight infants.

Looking at the residual plot, we see that there is one point with a particularly low residual that appears to be an outlier. We might try removing this point, fitting a new line, and then comparing the two models to see how much of an effect the point has on the estimated regression coefficients. However, there is no evidence that the assumption of homoscedasticity has been violated, or that a transformation of either the response or the explanatory variable is necessary.

TABLE 18.3

Residuals for the first 10 infants in the sample

Length	Predicted Length $\hat{y}$	Residual
41	36.92467	4.075324
40	38.82788	1.172117
38	40.73109	−2.731091
38	38.82788	−0.827883
38	37.87628	0.123720
32	33.11826	−1.118262
33	35.02147	−2.021469
38	36.92467	1.075324
30	35.97307	−5.973073
34	36.92467	−2.924676

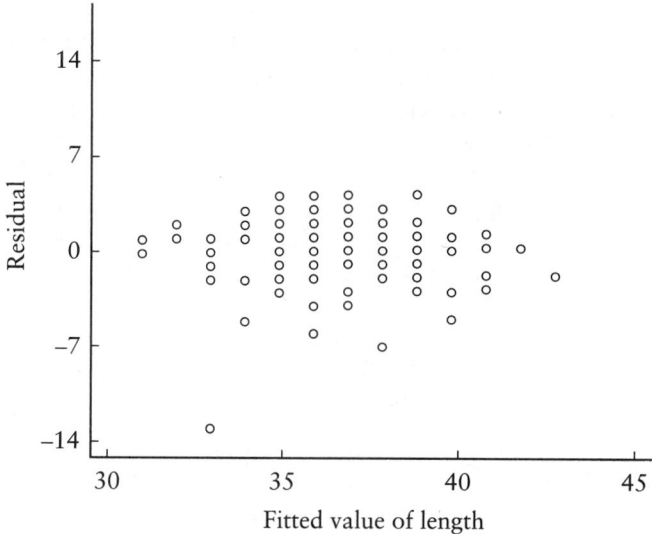

FIGURE 18.16
Residuals versus fitted values of length for a sample of 100 low birth
weight infants

18.5   Review Exercises

1. What is the main distinction between correlation analysis and simple linear regression?

2. What assumptions do you make when using the method of least squares to estimate a population regression line?

3. Explain the least-squares criterion for obtaining estimates of the regression coefficients.

4. Why is it dangerous to extrapolate an estimated linear regression line outside the range of the observed data values?

5. Given a specified value of the explanatory variable, how does a confidence interval constructed for the mean of the response differ from a prediction interval constructed for a new, individual value? Explain.

6. Why might you need to consider transforming either the response or the explanatory variable when fitting a simple linear regression model? How is the circle of powers used in this situation?

7. For a given sample of data, how can a two-way scatter plot of the residuals versus the fitted values of the response be used to evaluate the fit of a least-squares regression line?

8. Figure 18.17 displays a two-way scatter plot of CBVR—the response of cerebral blood volume in the brain to changes in carbon dioxide tension in the arteries—versus gestational age for a sample of 17 newborn infants [4]. The graph also shows the fitted least-squares regression line for these data. The investigators who constructed the model determined that the slope of the line β is significantly larger than 0.

   (a) Suppose that you are interested in only those infants who are born prematurely. If you were to eliminate the four data points corresponding to newborns whose gestational age is 38 weeks or greater, would you still believe that there is a significant increase in CBVR as gestational age increases?

   (b) In an earlier study, the same investigators found no obvious relationship between CBVR and gestational age in newborn infants; gestational age was not useful in predicting CBVR. Would this information cause you to modify your answer above?

9. The data set `lowbwt` contains information for the sample of 100 low birth weight infants born in Boston, Massachusetts [1] (Appendix B, Table B.7). Measurements of systolic blood pressure are saved under the variable name `sbp`, and values of gestational age under the name `gestage`.

   (a) Construct a two-way scatter plot of systolic blood pressure versus gestational age. Does the graph suggest anything about the nature of the relationship between these variables?

   (b) Using systolic blood pressure as the response and gestational age as the explanatory variable, compute the least-squares regression line. Interpret the estimated slope and y-intercept of the line; what do they mean in words?

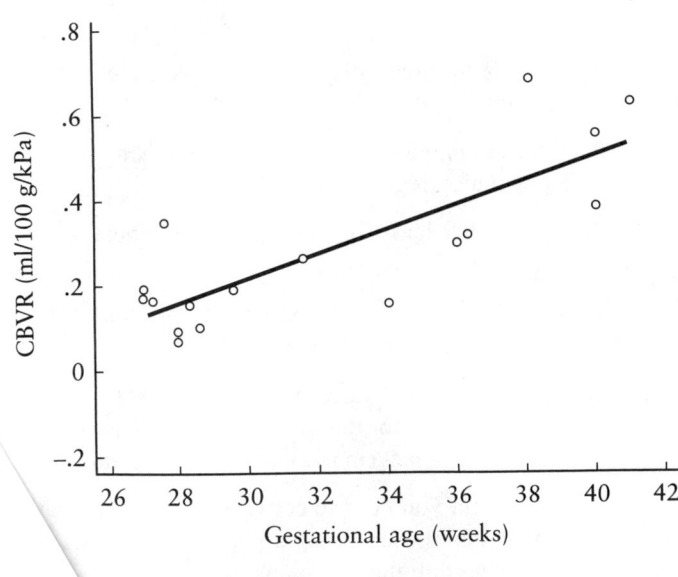

FIGURE 18.17

cerebral blood volume versus gestational age for a sample of 17 newborns

(c) At the 0.05 level of significance, test the null hypothesis that the true population slope β is equal to 0. What do you conclude?

(d) What is the estimated mean systolic blood pressure for the population of low birth weight infants whose gestational age is 31 weeks?

(e) Construct a 95% confidence interval for the true mean value of systolic blood pressure when $x = 31$ weeks.

(f) Suppose that you randomly select a new child from the population of low birth weight infants with gestational age 31 weeks. What is the predicted systolic blood pressure for this child?

(g) Construct a 95% prediction interval for this new value of systolic blood pressure.

(h) Does the least-squares regression model seem to fit the observed data? Comment on the coefficient of determination and a plot of the residuals versus the fitted values of systolic blood pressure.

10. Measurements of length and weight for a sample of 20 low birth weight infants are contained in the data set `twenty` [1] (Appendix B, Table B.23). The length measurements are saved under the variable name `length`, and the corresponding birth weights under `weight`.

(a) Construct a two-way scatter plot of birth weight versus length for the 20 infants in the sample. Without doing any calculations, sketch your best guess for the least-squares regression line directly on the scatter plot.

(b) Now compute the true least-squares regression line. Draw this line on the scatter plot. Does the actual least-squares line concur with your guess?

Based on the two-way scatter plot, it is clear that one point lies outside the range of the remainder of the data. This point corresponds to the ninth infant in the sample of size 20. To illustrate the effect that the outlier has on the model, remove this point from the data set.

(c) Compute the new least-squares regression line based on the sample of size 19, and sketch this line on the original scatter plot. How does the least-squares line change? In particular, comment on the values of the slope and the intercept.

(d) Compare the coefficients of determination (R^2) and the standard deviations from regression ($s_{y|x}$) for the two least-squares regression lines. Explain how these values changed when you removed the outlier from the original data set. Why did they change?

11. The relationship between total fertility rate and the prevalence of contraceptive practice was investigated for a large number of countries around the world [5]. Measuring fertility rate in births per woman 15 to 49 years of age and the proportion of currently married women using any form of contraception as a percentage, the least-squares regression line relating these two quantities is

$$\hat{y} = 6.83 - 0.062x.$$

For the subset of 17 countries in the sub-Saharan region of Africa, fertility rate and the prevalence of contraceptive practice are saved in a data set called `africa` (Appendix B, Table B.24). Measures of fertility rate are saved under the variable name `fertrate`, and values of percentage contraception under `contra`.

(a) Interpret the slope and the intercept of the least-squares regression line generated for the countries around the world. What do they imply in words?

(b) Using the data for the 17 sub-Saharan African countries, construct a two-way scatter plot of total fertility rate versus prevalence of contraceptive practice. Does there appear to be a relationship between these two quantities?

(c) On the scatter plot, sketch the least-squares line estimated based on countries throughout the world.

(e) How do the actual fertility rates for the African nations compare with those that would be predicted based on the fitted regression line?

12. In the 11 years before the passage of the Federal Coal Mine Health and Safety Act of 1969, the fatality rates for underground miners varied little. After the implementation of that act, however, fatality rates decreased steadily until 1979. The fatality rates for the years 1970 through 1981 are provided below [6]; for computational purposes, calendar years have been converted to a scale beginning at 1. This information is contained in the data set `miner` (Appendix B, Table B.25). Values of the response, fatality rate, are saved under the name `rate`, and values of the explanatory variable, calendar year, under the name `year`.

Calendar Year	Year	Fatality Rate per 1000 Employees
1970	1	2.419
1971	2	1.732
1972	3	1.361
1973	4	1.108
1974	5	0.996
1975	6	0.952
1976	7	0.904
1977	8	0.792
1978	9	0.701
1979	10	0.890
1980	11	0.799
1981	12	1.084

(a) Construct a two-way scatter plot of fatality rate versus year. What does this plot suggest about the relationship between these two variables?

(b) To model the trend in fatality rates, fit the least-squares regression line

$$\hat{y} = \hat{\alpha} + \hat{\beta}x,$$

where x represents year. Using both the coefficient of determination R^2 and a plot of the residuals versus the fitted values of fatality rate, comment on the fit of the model to the observed data.

(c) Now transform the explanatory variable x to $\ln(x)$. Create a scatter plot of fatality rate versus the natural logarithm of year.

(d) Fit the least-squares model

$$\hat{y} = \hat{\alpha} + \hat{\beta}\ln(x).$$

Use the coefficient of determination and a plot of the residuals versus the fitted values of fatality rate to compare the fit of this model to the model constructed in (b).

(e) Transform x to $1/x$. Construct a two-way scatter plot of fatality rate versus the reciprocal of year.

(f) Fit the least-squares model

$$\hat{y} = \hat{\alpha} + \hat{\beta}\left(\frac{1}{x}\right).$$

Using the coefficient of determination and a plot of the residuals, comment on the fit of this model and compare it to the previous ones.

(g) Which of the three models appears to fit the data best? Defend your selection.

13. Statistics that summarize personal health care expenditures by state for the years 1966 through 1982 have been examined in an attempt to understand issues related to rising health care costs. Suppose that you are interested in focusing on the relationship between expense per admission into a community hospital and average length of stay in the facility. The data set `hospital` contains information for each state in the United States (including the District of Columbia) for the year 1982 [7] (Appendix B, Table B.26). The measures of mean expense per admission are saved under the variable name `expadm`; the corresponding average lengths of stay are saved under `los`.

(a) Generate numerical summary statistics for the variables expense per admission and length of stay in the hospital. What are the means and medians of each variable? What are their minimum and maximum values?

(b) Construct a two-way scatter plot of expense per admission versus length of stay. What does the scatter plot suggest about the nature of the relationship between these variables?

(c) Using expense per admission as the response and length of stay as the explanatory variable, compute the least-squares regression line. Interpret the estimated slope and y-intercept of this line in words.

(d) Construct a 95% confidence interval for β, the true slope of the population regression line. What does this interval tell you about the linear relationship between expense per admission and length of stay in the hospital?

(e) What is the coefficient of determination for the least-squares line? How is R^2 related to the Pearson correlation coefficient r?

(f) Construct a plot of the residuals versus the fitted values of expense per admission. In what three ways does the residual plot help you to evaluate the fit of the model to the observed data?

Bibliography

[1] Leviton, A., Fenton, T., Kuban, K. C. K., and Pagano, M., "Labor and Delivery Characteristics and the Risk of Germinal Matrix Hemorrhage in Low Birth Weight Infants," *Journal of Child Neurology,* Volume 6, October 1991, 35–40.

[2] Kleinbaum, D. G., Kupper, L. L., and Muller, K. E., *Applied Regression Analysis and Other Multivariable Methods,* Boston: PWS-Kent, 1988.

[3] United Nations Children's Fund, *The State of the World's Children 1994,* New York: Oxford University Press.

[4] Wyatt, J. S., Edwards, A. D., Cope, M., Delpy, D. T., McCormick, D. C., Potter, A., and Reynolds, E. O. R., "Response of Cerebral Blood Volume to Changes in Arterial Carbon Dioxide Tension in Preterm and Term Infants," *Pediatric Research,* Volume 29, June 1991, 553–557.

[5] Frank, O., and Bongaarts, J., "Behavioural and Biological Determinants of Fertility Transition in Sub-Saharan Africa," *Statistics in Medicine,* Volume 10, February 1991, 161–175.

[6] Weeks, J. L., and Fox, M., "Fatality Rates and Regulatory Policies in Bituminous Coal Mining, United States, 1959–1981," *American Journal of Public Health,* Volume 73, November 1983, 1278–1280.

[7] Levit, K. R., "Personal Health Care Expenditures, by State: 1966–1982," *Health Care Financing Review*, Volume 6, Summer 1985, 1–25.

19

Multiple Regression

In the preceding chapter, we saw how simple linear regression can be used to explore the nature of the relationship between two continuous random variables. In particular, it allows us to predict or estimate the value of a response that corresponds to a given value of an explanatory variable. If knowing the value of a single explanatory variable improves our ability to predict the response, we might suspect that additional explanatory variables could be used to our advantage. To investigate the more complicated relationship among a number of different variables, we use a natural extension of simple linear regression analysis known as *multiple regression*.

19.1 The Model

Using multiple regression, we attempt to estimate the population equation

$$\mu_{y|x_1, x_2, \ldots, x_q} = \alpha + \beta_1 x_1 + \beta_2 x_2 + \cdots + \beta_q x_q$$

where $x_1, x_2, \ldots,$ and x_q are the outcomes of q distinct explanatory variables, and $\mu_{y|x_1, x_2, \ldots, x_q}$ is the mean value of y when the explanatory variables assume these values. The parameters $\alpha, \beta_1, \beta_2, \ldots,$ and β_q are constants that are again called the coefficients of the equation. The intercept α is the mean value of the response y when all explanatory variables take the value 0, or $\mu_{y|0, 0, \ldots, 0}$; the slope β_i is the change in the mean value of y that corresponds to a one-unit increase in x_i, given that all other explanatory variables remain constant.

To accommodate the natural variation in measures of the response, we actually fit a model of the form

$$y = \alpha + \beta_1 x_1 + \beta_2 x_2 + \cdots + \beta_q x_q + \varepsilon,$$

where ε is the error term. The coefficients of the population regression equation are estimated using a random sample of observations represented as $(x_{1i}, x_{2i}, \ldots, x_{qi}, y_i)$.

However, just as we had to make a number of assumptions for the model involving a single explanatory variable, we make a set of analogous assumptions for the more complex multiple regression model. These assumptions are as follows:

1. For specified values of x_1, x_2,..., and x_q, all of which are considered to be measured without error, the distribution of the y values is normal with mean $\mu_{y|x_1, x_2,...,x_q}$ and standard deviation $\sigma_{y|x_1, x_2,...,x_q}$.

2. The relationship between $\mu_{y|x_1,x_2,...,x_q}$ and x_1, x_2,..., and x_q is represented by the equation

$$\mu_{y|x_1,x_2,...,x_q} = \alpha + \beta_1 x_1 + \beta_2 x_2 + \cdots + \beta_q x_q.$$

3. For any set of values x_1, x_2,..., and x_q, $\sigma_{y|x_1,x_2,...,x_q}$ is constant. As in simple linear regression, this is referred to as homoscedasticity.

4. The outcomes y are independent.

19.1.1 The Least-Squares Regression Equation

To estimate the population regression equation

$$\mu_{y|x_1,x_2,...,x_q} = \alpha + \beta_1 x_1 + \beta_2 x_2 + \cdots + \beta_q x_q,$$

we use the method of least squares to fit the model

$$\hat{y} = \hat{\alpha} + \hat{\beta}_1 x_1 + \hat{\beta}_2 x_2 + \cdots + \hat{\beta}_q x_q.$$

This technique requires that we minimize the sum of the squares of the residuals, or, in this case,

$$\sum_{i=1}^{n} e_i^2 = \sum_{i=1}^{n} (y_i - \hat{y}_i)^2$$

$$= \sum_{i=1}^{n} (y_i - \hat{\alpha} - \hat{\beta}_1 x_{1i} - \hat{\beta}_2 x_{2i} - \cdots - \hat{\beta}_q x_{qi})^2.$$

Recall that y_i is the observed outcome of the response Y for particular values x_{1i}, x_{2i},..., and x_{qi}, while $\hat{y}_i$ is the corresponding value from the fitted equation. When a single explanatory variable was involved, the fitted model was simply a straight line. With two explanatory variables, the model represents a plane in three-dimensional space; with three or more variables, it becomes a hyperplane in higher-dimensional space. Although the calculations are more complicated than they were for models with a single explanatory variable, they do not present a problem as long as a computer is available.

In Chapter 18, we found a significant linear relationship between head circumference and gestational age for the population of low birth weight infants; the fitted least-squares regression line was

$$\hat{y} = 3.9143 + 0.7801x.$$

We might wonder whether head circumference also depends on the birth weight of an infant. Figure 19.1 is a two-way scatter plot of head circumference versus birth weight for a sample of 100 low birth weight infants born in Boston, Massachusetts [1]. The graph suggests that head circumference increases as weight increases. Given that we have already accounted for gestational age, does birth weight further improve our ability to predict the head circumference of a child?

Suppose that we let x_1 represent gestational age and x_2 designate birth weight. The fitted least-squares regression equation is

$$\hat{y} = 8.3080 + 0.4487x_1 + 0.0047x_2.$$

The intercept of 8.3080 is, in theory, the mean value of head circumference for low birth weight infants with gestational age 0 weeks and birth weight 0 grams. In this example, neither an age of 0 nor a weight of 0 makes sense. The estimated coefficient of gestational age is not what it was when age was the only explanatory variable in the model; its value has decreased from 0.7801 to 0.4487. This implies that, given that a child's birth weight remains constant, each one-week increase in gestational age corresponds to a 0.4487-centimeter increase in head circumference, on average. Equivalently, given two infants with the same birth weight but such that the gestational age of the first child is one week greater than the gestational age of the second, the first child would have a head circumference approximately 0.4487 centimeters larger. Similarly, the coefficient of birth weight indicates that if a child's gestational age does not change, each one-gram increase in birth weight results in a 0.0047-centimeter increase in head circumference, on average.

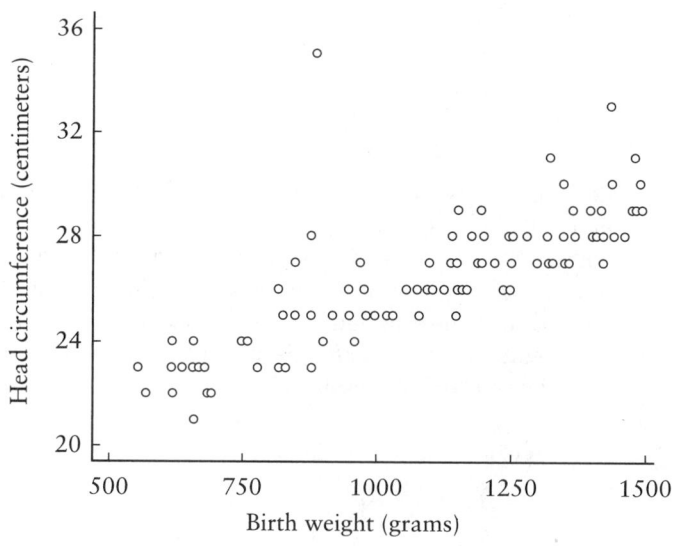

FIGURE 19.1
Head circumference versus birth weight for a sample of 100 low birth weight infants

19.1.2 Inference for Regression Coefficients

Just as when applying simple linear regression analysis, we would like to be able to use the least-squares regression model

$$\hat{y} = \hat{\alpha} + \hat{\beta}_1 x_1 + \hat{\beta}_2 x_2 + \cdots + \hat{\beta}_q x_q$$

to make inference about the population regression equation

$$\mu_{y|x_1, x_2, \ldots, x_q} = \alpha + \beta_1 x_1 + \beta_2 x_2 + \cdots + \beta_q x_q.$$

The regression coefficients $\hat{\alpha}$ and $\hat{\beta}_1$ through $\hat{\beta}_q$ are estimated using a sample of data drawn from the underlying population; their values would change if a different sample were selected. Therefore, we need the standard errors of these estimators to be able to make inference about the true population parameters.

Tests of hypotheses for the population intercept and slopes can be carried out just as they were for the model containing a single explanatory variable, with two differences. First, when testing the null hypothesis

$$H_0: \beta_i = \beta_{i0}$$

against the alternative

$$H_A: \beta_i \neq \beta_{i0},$$

we assume that the values of all other explanatory variables $x_j \neq x_i$ remain constant. Second, if the null hypothesis is true, the test statistic

$$t = \frac{\hat{\beta}_i - \beta_{i0}}{\widehat{se}(\hat{\beta}_i)}$$

does not follow a t distribution with $n - 2$ degrees of freedom. Instead, it has a t distribution with $n - q - 1$ degrees of freedom, where q is the number of explanatory variables in the model. For the model containing gestational age and birth weight, q is equal to 2 and the appropriate t distribution has $100 - 2 - 1 = 97$ degrees of freedom. This t distribution is used to find p, the probability of observing an estimated slope as extreme as or more extreme than $\hat{\beta}_i$, given that the true population slope is β_{i0}.

For the 100 low birth weight infants born in Boston, it can be shown that

$$\widehat{se}(\hat{\alpha}) = 1.5789,$$
$$\widehat{se}(\hat{\beta}_1) = 0.0672,$$

and

$$\widehat{se}(\hat{\beta}_2) = 0.00063.$$

To conduct a two-sided test of the null hypothesis that β_1—the true slope relating head circumference to gestational age, assuming that the value of birth weight remains constant—is equal to 0, we calculate the statistic

$$t = \frac{\hat{\beta}_1 - \beta_{10}}{\widehat{se}(\hat{\beta}_1)}$$
$$= \frac{0.4487 - 0}{0.0672}$$
$$= 6.68.$$

For a t distribution with 97 degrees of freedom, $p < 0.001$; therefore, we reject the null hypothesis at the 0.05 level of significance and conclude that β_1 is greater than 0. Similarly, to test the null hypothesis

$$H_0: \beta_2 = 0$$

against the alternative

$$H_A: \beta_2 \neq 0,$$

assuming that gestational age remains constant, we calculate

$$t = \frac{\hat{\beta}_2 - \beta_{20}}{\widehat{se}(\hat{\beta}_2)}$$
$$= \frac{0.0047 - 0}{0.00063}$$
$$= 7.47.$$

Once again, $p < 0.001$, and we conclude that β_2 is significantly greater than 0. Therefore, head circumference increases as either gestational age or birth weight increases. We must bear in mind, however, that multiple tests of hypothesis based on the same set of data are not independent; if each individual test is conducted at the α level of significance, the overall probability of making a type I error—or rejecting a null hypothesis that is true—is in fact larger than α.

In addition to conducting tests of hypotheses, we also can calculate confidence intervals for the population regression coefficients. Furthermore, we can construct a confidence interval for the predicted mean value of Y and a prediction interval for the predicted individual y corresponding to a given set of values for the explanatory variables. In all cases, the procedures are analogous to those used when a single explanatory variable was involved.

19.1.3 Evaluation of the Model

Using techniques such as the coefficient of determination and a plot of the residuals, we are able to assess how well a particular least-squares model actually fits the observed data. For example, it can be shown that the model containing gestational age and weight explains 75.20% of the variation in the observed head circumference measurements; the model containing gestational age alone explained 60.95%. This increase in R^2 suggests that adding the explanatory variable weight to the model improves our ability to predict head circumference for the population of low birth weight infants.

We must be careful when comparing coefficients of determination from two different models. The inclusion of an additional variable in a model can never cause R^2 to decrease; knowledge of both gestational age and birth weight, for example, can never explain less of the observed variability in head circumference than knowledge of gestational age alone. To get around this problem, we can use a second measure, called the *adjusted R^2*, that compensates for the added complexity of a model. The adjusted R^2 increases when the inclusion of a variable improves our ability to predict the response and decreases when it does not. Consequently, the adjusted R^2 allows us to make a more valid comparison between models that contain different numbers of explanatory variables. Like the coefficient of determination, the adjusted R^2 is an estimator of the population correlation ρ; unlike R^2, however, it cannot be directly interpreted as the proportion of the variability among the observed values of y that is explained by the linear regression model.

Figure 19.2 displays a scatter plot of the residuals from the model containing both gestational age and birth weight versus the fitted values of head circumference from the same model for the sample of 100 low birth weight infants. There is one residual with a particularly large value that could be considered an outlier; this point corresponds to a child with gestational age 31 weeks, birth weight 900 grams, and head circumference 35 centimeters. We would predict the infant's head circumference to be only

$$\hat{y} = 8.3080 + 0.4487(31) + 0.0047(900)$$
$$= 26.5 \text{ cm.}$$

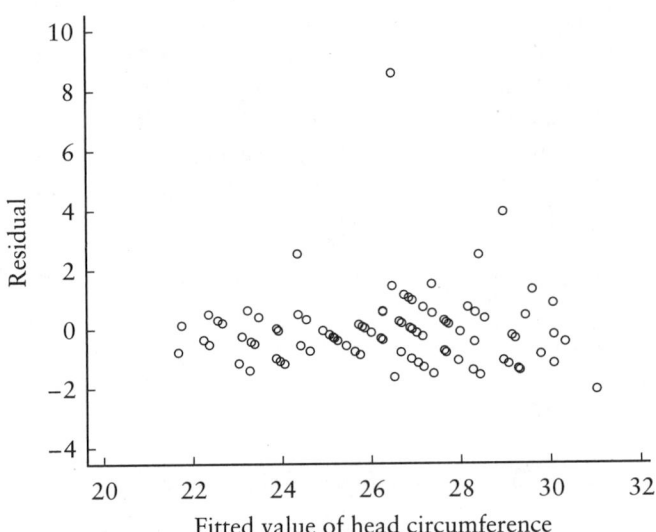

FIGURE 19.2

Residuals versus fitted values of head circumference

Note that this outlier was also evident in Figure 19.1, the scatter plot of head circumference versus birth weight. We might try removing the point, refitting the least-squares model, and determining how much of an effect this outlier has on the estimated coefficients. (If we were to do this, we would find that the point has only a small effect on the values of $\hat{\alpha}$, $\hat{\beta}_1$, and $\hat{\beta}_2$; the relationships between head circumference and both gestational age and birth weight remain statistically significant with $p < 0.001$ in each case.) There is no evidence that the assumption of homoscedasticity has been violated—note the absence of a fan-shaped scatter—or that a transformation of either the response or one of the explanatory variables is necessary.

19.1.4   *Indicator Variables*

All the explanatory variables we have considered up to this point have been measured on a continuous scale. However, regression analysis can be generalized to incorporate discrete or nominal explanatory variables as well. For example, we might wonder whether an expectant mother's diagnosis of toxemia during pregnancy—a condition characterized by high blood pressure and other potentially serious complications—affects the head circumference of her child. The diagnosis of toxemia is a dichotomous random variable; a woman either had it or she did not. We would like to be able to quantify the effect of toxemia on head circumference by comparing infants whose mothers suffered from this condition to infants whose mothers did not.

Since the explanatory variables in a regression analysis must assume numerical values, we designate the presence of toxemia during pregnancy by 1 and its absence by 0. These numbers do not represent any actual measurements; they simply identify the categories of the nominal random variable. Because its values do not have any quantitative meaning, an explanatory variable of this sort is called an *indicator variable* or a *dummy variable*.

Suppose that we add the indicator variable toxemia to the regression equation that already contains gestational age. For the sake of simplicity, we ignore birth weight for the moment. The fitted least-squares regression model is

$$\hat{y} = 1.4956 + 0.8740x_1 - 1.4123x_3,$$

where x_1 represents gestational age and x_3 represents toxemia. The coefficient of toxemia is negative, indicating that mean head circumference decreases as the value of toxemia increases from 0 to 1. A test of the null hypothesis

$$H_0: \beta_3 = 0$$

against the alternative

$$H_A: \beta_3 \neq 0,$$

assuming that gestational age does not change, results in a test statistic of $t = -3.48$ and $p = 0.001$. Therefore, we reject the null hypothesis at the 0.05 level of significance and conclude that β_3 is less than 0; given two infants with identical gestational ages, head circumference would be smaller on average for the child whose mother experienced toxemia during pregnancy than for the child whose mother did not.

In order to better understand a regression model containing one continuous explanatory variable and one dichotomous explanatory variable, we can think about the least-squares regression equation fitted to the sample of 100 low birth weight infants as two different models, corresponding to the two possible values of the dichotomous random variable toxemia. When $x_3 = 1$, for instance, indicating that a woman was diagnosed with toxemia during pregnancy,

$$\hat{y} = 1.4956 + 0.8740x_1 - 1.4123(1)$$
$$= 0.0833 + 0.8740x_1.$$

When $x_3 = 0$,

$$\hat{y} = 1.4956 + 0.8740x_1 - 1.4123(0)$$
$$= 1.4956 + 0.8740x_1.$$

The two lines are plotted in Figure 19.3. Note that the equations for the infants whose mothers were diagnosed with toxemia and those whose mothers were not have identical slopes; in either group, a one-week increase in gestational age is associated with a 0.8740-centimeter increase in head circumference, on average. This is the consequence of fitting a single regression model to the two different groups of infants. Since one line lies entirely above the other—as determined by the different y-intercepts—the equations also suggest that across all values of gestational age, children whose mothers were not

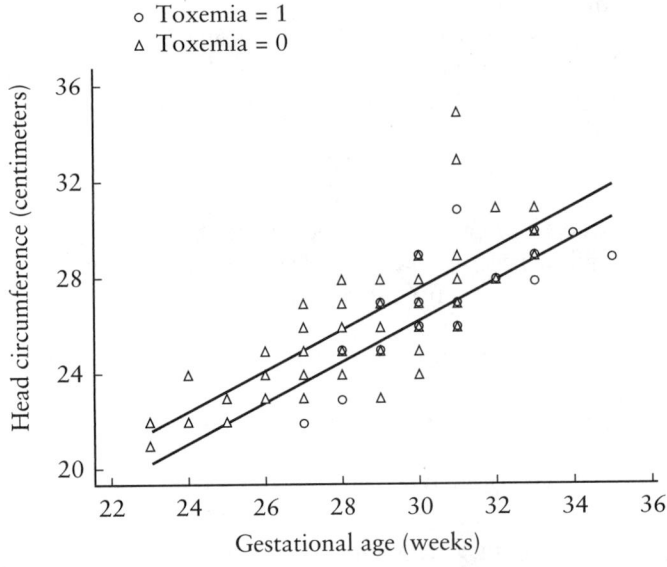

FIGURE 19.3

Fitted least-squares regression lines for different levels of toxemia diagnosed with toxemia tend to have larger head circumference measurements than children whose mothers were diagnosed with toxemia.

19.1.5 Interaction Terms

In some situations, one explanatory variable has a different effect on the predicted response y depending on the value of a second explanatory variable. As an example, a one-week increase in gestational age might have a different effect on a child's head circumference depending on whether the infant's mother had experienced toxemia during pregnancy or not. To model a relationship of this kind, we create what is known as an *interaction term*. An interaction term is generated by multiplying together the outcomes of two variables x_i and x_j to create a third variable $x_i x_j$.

Suppose we wish to add an interaction between gestational age and toxemia to the regression model that already contains these two variables individually. We would multiply the outcomes for gestational age, x_1, by the outcomes for toxemia, x_3, to create a new variable $x_1 x_3$. (Because x_3 can assume only two possible values, $x_1 x_3$ would be equal to 0 when $x_3 = 0$ and equal to x_1 when $x_3 = 1$.) In the population of low birth weight infants, this new explanatory variable would have a corresponding slope β_{13}.

Based on the sample of 100 infants, the fitted least-squares model is

$$\hat{y} = 1.7629 + 0.8646x_1 - 2.8150x_3 + 0.0462x_1 x_3.$$

Testing the null hypothesis

$$H_0: \beta_{13} = 0$$

versus the alternative

$$H_A: \beta_{13} \neq 0,$$

we are unable to reject H_0 at the 0.05 level of significance. We conclude that this sample does not provide evidence that gestational age has a different effect on head circumference depending on whether a mother experienced toxemia during pregnancy or not.

Because the interaction term is not statistically significant, we would not want to retain it in the regression model. If it had achieved significance, however, we might again wish to evaluate the separate models corresponding to the two possible values of the dichotomous random variable toxemia. When $x_3 = 1$, the least-squares equation would be

$$\hat{y} = 1.7629 + 0.8646x_1 - 2.8150(1) + 0.0462x_1(1)$$
$$= -1.0521 + 0.9108x_1.$$

When $x_3 = 0$,

$$\hat{y} = 1.7629 + 0.8646x_1 - 2.8150(0) + 0.0462x_1(0)$$
$$= 1.7629 + 0.8646x_1.$$

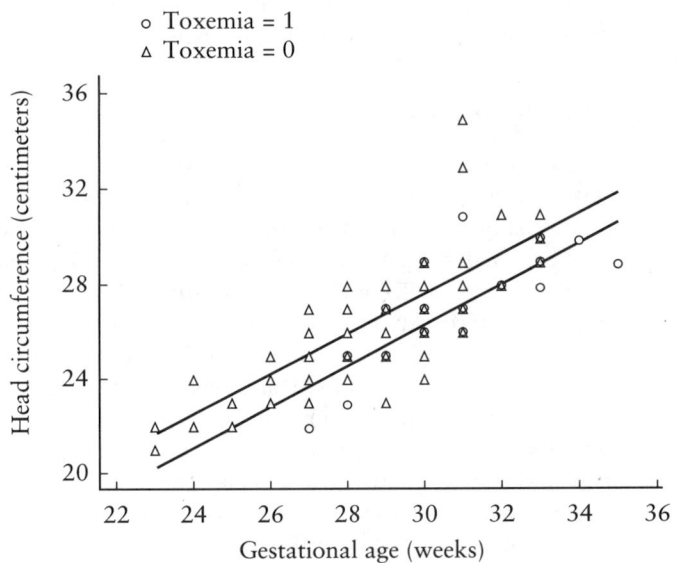

FIGURE 19.4
Fitted least-squares regression lines for different levels of toxemia,
interaction term included

These two lines are plotted in Figure 19.4; note that they have different intercepts **and** different slopes. In the range of interest, however, one line still lies completely above the other. This implies that across all relevant values of gestational age, infants whose mothers did not experience toxemia average larger head circumference measurements than infants whose mothers were diagnosed with this condition.

19.2 Model Selection

As a general rule, we usually prefer to include in a regression model only those explanatory variables that help us to predict or to explain the observed variability in the response y, the coefficients of which can be accurately estimated. Consequently, if we are presented with a number of potential explanatory variables, how do we decide which ones to retain in the model and which to leave out? This decision is usually made based on a combination of statistical and nonstatistical considerations. Initially, we should have some prior knowledge as to which variables might be important. To study the full effect of each of these explanatory variables, however, it would be necessary to perform a separate regression analysis for each possible combination of the variables. The resulting models could then be evaluated according to some statistical criteria. This strategy for finding the "best" regression equation is known as the *all possible models* approach. While it is the most thorough method, it is also extremely time-consuming. If we have a large number of potential explanatory variables, the procedure may not be feasible. As a result, we frequently resort to one of several alternative approaches for choosing a regression model. The two most commonly used procedures are known as forward selection and backward elimination.

Forward selection proceeds by introducing variables into the model one at a time. The model is evaluated at each step, and the process continues until some specified statistical criterion is achieved. For example, we might begin by including the single explanatory variable that yields the largest coefficient of determination, and thus explains the greatest proportion of the observed variability in y. We next put into the equation the variable that increases R^2 the most, assuming that the first variable remains in the model and that the increase in R^2 is statistically significant. The procedure continues until we reach a point where none of the remaining variables explains a significant amount of the additional variability in y.

Backward elimination begins by including all explanatory variables in the model. Variables are dropped one at a time, beginning with the one that reduces R^2 by the least amount and thus explains the smallest proportion of the observed variability in y, given the other variables in the model. If the decrease in R^2 is not statistically significant, the variable is left out of the model permanently. The equation is evaluated at each step, and the procedure is repeated until each of the variables remaining in the model explains a significant portion of the observed variation in the response.

When features of both the forward selection and backward elimination techniques are used together, the method is called *stepwise selection*. We begin as if we were using the forward selection procedure, introducing variables into the model one at a time; as each new explanatory variable is entered into the equation, however, all previously entered variables are checked to ensure that they maintain their statistical significance. Consequently, a variable entered into the model in one step might be dropped out again at a later step. Note that it is possible that we could end up with different final models, depending on which procedure is applied.

Regardless of the strategy that we choose to fit a particular model, we should always check for the presence of collinearity. *Collinearity* occurs when two or more of the explanatory variables are correlated to the extent that they convey essentially the same information about the observed variation in y. One symptom of collinearity is the instability of the estimated coefficients and their standard errors. In particular, the standard errors often become very large; this implies that there is a great deal of sampling variability in the estimated coefficients.

In the regression model that contains gestational age, toxemia, and the interaction between the two, toxemia and the gestational age–toxemia interaction are highly correlated; in fact, the Pearson correlation coefficient quantifying the linear relationship between these two variables is equal to 0.997. This model and the model that did not include the interaction term are contrasted below.

	Interaction Term Not Included	Interaction Term Included
Coefficient	−1.412	−2.815
Standard Error	0.406	4.985
Test Statistic	−3.477	−0.565
p-value	0.001	0.574
R^2	0.653	0.653
Adjusted R^2	0.646	0.642

When the interaction term is included in the equation, the estimated coefficient of toxemia doubles in magnitude. In addition, its standard error increases by a factor of 12. In the model without the interaction term, the coefficient of toxemia is significantly different from 0 at the 0.05 level; when the interaction term is present, it no longer achieves statistical significance. The coefficient of determination does not change when the interaction is included; it remains 65.3%. Furthermore, the adjusted R^2 decreases slightly. These facts, taken together, indicate that the inclusion of the gestational age–toxemia interaction term in the regression model does not explain any additional variability in the observed values of head circumference, beyond that which is explained by gestational age and toxemia alone. The information supplied by this term is redundant; in this population, toxemia occurs relatively late in the pregnancy.

19.3 Further Applications

In Chapter 18, we used gestational age to help us predict length for a sample of 100 low birth weight infants—defined as those weighing less than 1500 grams—born in Boston, Massachusetts. We found that a significant linear relationship exists between these two variables; in particular, length increases as gestational age increases.

We now wish to determine whether the length of an infant also depends on the age of its mother. To begin the analysis, we create a two-way scatter plot of length versus mother's age for the 100 infants in the sample. The plot is displayed in Figure 19.5. Based on the graph, and disregarding the one outlying value, length does not appear to either increase or decrease as mother's age increases. Given that we have already ac-

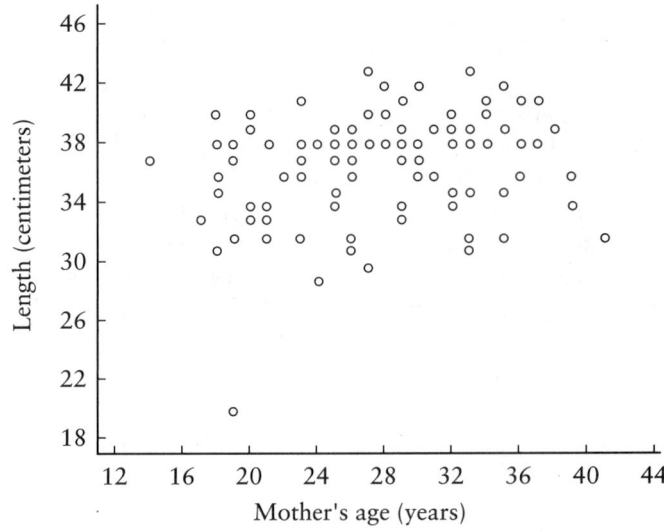

FIGURE 19.5
Length versus mother's age for a sample of 100 low birth weight infants

counted for gestational age, does the inclusion of mother's age in the regression model further improve our ability to predict a child's length?

To estimate the true population regression of length on gestational age and mother's age,

$$\mu_{y|x_1,x_2} = \alpha + \beta_1 x_1 + \beta_2 x_2,$$

we fit a least-squares model of the form

$$\hat{y} = \hat{\alpha} + \hat{\beta}_1 x_1 + \hat{\beta}_2 x_2.$$

Table 19.1 shows the relevant output from Minitab.

The top portion of the output contains information related to the estimated coefficients $\hat{\alpha}$, $\hat{\beta}_1$, and $\hat{\beta}_2$. As noted, the fitted least-squares regression equation is

$$\hat{y} = 9.09 + 0.936x_1 + 0.0247x_2.$$

The coefficient of gestational age is 0.936, implying that if mother's age remains constant, each one-week increase in gestational age corresponds to a 0.936-centimeter increase in length, on average. Similarly, the coefficient of mother's age suggests that if gestational age remains constant, a one-year increase in mother's age would result in an approximate 0.0247-centimeter increase in length. A test of the null hypothesis

$$H_0: \beta_1 = 0$$

is rejected at the 0.05 level of significance, while

$$H_0: \beta_2 = 0$$

is not; consequently, based on this sample of low birth weight infants, we conclude that length increases as gestational age increases, but that length is not affected by mother's age. The relationships between the response length and the two explanatory variables are each adjusted for the other.

The bottom piece of the Minitab output displays information for several observations that are considered to be "unusual." They are marked this way either because they have large residuals, and thus are potential outliers, or because the value of one of their explanatory variables is particularly high or particularly low relative to the other observations, meaning that these points could have a large effect on the estimated regression coefficients. The column on the far right contains standardized residuals, or residuals that have been divided by their standard errors.

Recall from Chapter 18 that gestational age alone explains 45.6% of the variability in the observed values of length; gestational age and mother's age together explain 45.8%. The adjusted coefficient of determination—which appears beside R^2 in Table 19.1—has actually decreased slightly, from 45.0% to 44.6%. This lack of change in R^2, combined with our failure to reject the null hypothesis that the coefficient of mother's age is equal to 0, demonstrates that adding this explanatory variable to the model does not improve our ability to predict length for the low birth weight infants in this population.

TABLE 19.1

Minitab output displaying the regression of length on gestational age and mother's age

```
                The regression equation is
                LENGTH = 9.09 + 0.936 GESTAGE + 0.0247 MOMAGE

                Predictor        Coef       Stdev     t-ratio          p
                Constant        9.091       3.088        2.94      0.004
                GESTAGE        0.9361      0.1093        8.56      0.000
                MOMAGE        0.02472     0.04631        0.53      0.595

                s = 2.657      R-sq = 45.8%      R-sq (adj) = 44.6%

        Analysis of Variance

                SOURCE          DF          SS          MS        F          p
                Regression       2      577.75      288.88    40.91      0.000
                Error           97      685.01        7.06
                Total           99     1262.76

                SOURCE          DF      SEQ SS
                GESTAGE          1      575.74
                MOMAGE           1        2.01


    Unusual Observations

    Obs.     GESTAGE     LENGTH        Fit   Stdev.Fit      Residual      St.Resid
       9        28.0       30.0     35.969       0.282        -5.969        -2.26R
      32        30.0       31.0     37.989       0.359        -6.989        -2.65R
      57        25.0       20.0     32.963       0.569       -12.963        -4.99R
      58        23.0       32.0     31.635       1.036         0.365         0.15X
      92        25.0       34.0     33.457       0.801         0.543         0.21X

R denotes an obs. with a large st. resid.
X denotes an obs. whose X value gives it large influence
```

We now wish to investigate whether an expectant mother's diagnosis of toxemia during pregnancy affects the length of her child. To do this, we add an indicator variable representing toxemia status—where a diagnosis of toxemia is represented by 1 and no such diagnosis by 0—to the model that already contains gestational age. As shown in the SAS output in Table 19.2, the least-squares regression equation for this model is

$$\hat{y} = 6.284 + 1.070x_1 - 1.777x_3.$$

The coefficient of toxemia, where toxemia status is represented by x_3, is negative and is significantly different from 0 at the 0.05 level ($p = 0.012$); given two infants with identical gestational ages, the child whose mother had experienced toxemia would tend to be 1.78 centimeters shorter, on average, than the child whose mother had not. Also note that the coefficient of determination has increased from 45.6% for gestational age alone to 49.0% for the model containing both gestational age and toxemia; the adjusted R^2 has also increased from 45.0% to 48.0%.

To better understand this least-squares model, we can actually obtain two different models corresponding to the two possible values of the dichotomous random vari-

TABLE 19.2

SAS output displaying the regression of length on gestational age and toxemia

Model: MODEL1
Dependent Variable: LENGTH

Analysis of Variance

Source	DF	Sum of Squares	Mean Square	F Value	Prob>F
Model	2	619.25362	309.62681	46.672	0.0001
Error	97	643.50638	6.63409		
C Total	99	1262.76000			

Root MSE	2.57567	R-square	0.4904	
Dep Mean	36.82000	Adj R-sq	0.4799	
C.V.	6.99531			

Parameter Estimates

Variable	DF	Parameter Estimate	Standard Error	T for H0: Parameter=0	Prob>|T|
INTERCEP	1	6.284326	3.19182416	1.969	0.0518
GESTAGE	1	1.069883	0.11210390	9.544	0.0001
TOX	1	-1.777381	0.69399183	-2.561	0.0120

able toxemia. When $x_3 = 1$, indicating that a mother did experience toxemia during pregnancy,

$$\hat{y} = 6.284 + 1.070x_1 - 1.777(1)$$
$$= 4.507 + 1.070x_1.$$

When $x_3 = 0$,

$$\hat{y} = 6.284 + 1.070x_1 - 1.777(0)$$
$$= 6.284 + 1.070x_1.$$

The two lines are plotted in Figure 19.6. Note that the lines have identical slopes; for either group, a one-week increase in gestational age corresponds to a 1.07-centimeter increase in length, on average.

To determine whether an increase in gestational age has a different effect on length for infants whose mothers were diagnosed with toxemia versus infants whose mothers were not, we could add to the model an additional variable, which is the interaction between gestational age and toxemia. The interaction term is obtained by multiplying together the outcomes of the two random variables representing gestational age and toxemia status. The Stata output corresponding to this model is presented in Table 19.3. Based on the sample of 100 low birth weight infants, the fitted least-squares model is

$$\hat{y} = 6.608 + 1.058x_1 - 3.477x_3 + 0.0559x_1x_3.$$

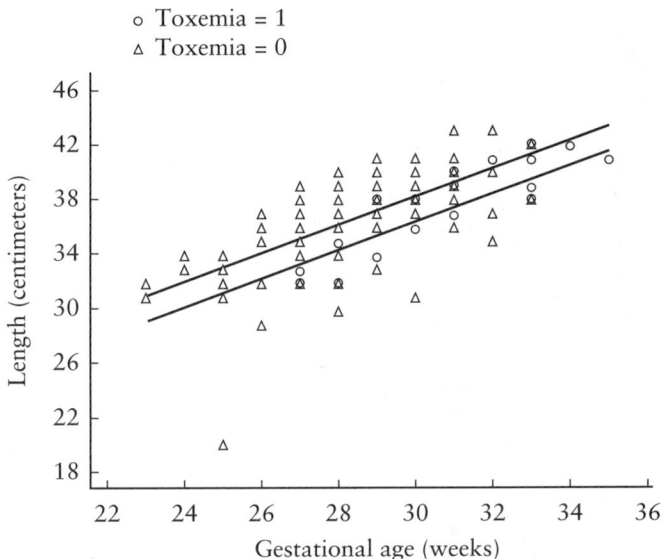

FIGURE 19.6

Fitted least-squares regression lines for different levels of toxemia

TABLE 19.3

Stata output displaying the linear regression of length on gestational age, toxemia, and their interaction

Source	SS	df	MS		
Model	619.522097	3	206.507366	Number of obs = 100
Residual	643.237903	96	6.70039483	F(3,96) = 30.82
				Prob > F = 0.0000
				R-square = 0.4906
Total	1262.76	99	12.7551515	Adj R-square = 0.4747
				Root MSE = 2.5885

length	Coef.	Std. Err.	t	P>|t|	[95% Conf. Interval]	
gestage	1.058458	.1262952	8.381	0.000	.8077647	1.309152
tox	-3.477085	8.5198381	-0.408	0.684	-20.38883	13.43466
gesttox	.0559409	.2794651	0.200	0.842	-4.4987929	.6106747
_cons	6.608269	3.592847	1.893	0.069	-.5234754	13.74001

We are unable to reject the null hypothesis that β_{13}, the coefficient of the interaction term, is equal to 0 ($p = 0.84$); the adjusted R^2 has decreased from 48.0% to 47.5%. Furthermore, the high correlation between toxemia and the gestational age–toxemia interaction—the Pearson correlation coefficient is equal to 0.997—has introduced collinearity into the model. Note that the standard error of the estimated coefficient of toxemia is approximately 12 times larger than it was in the model that did not contain

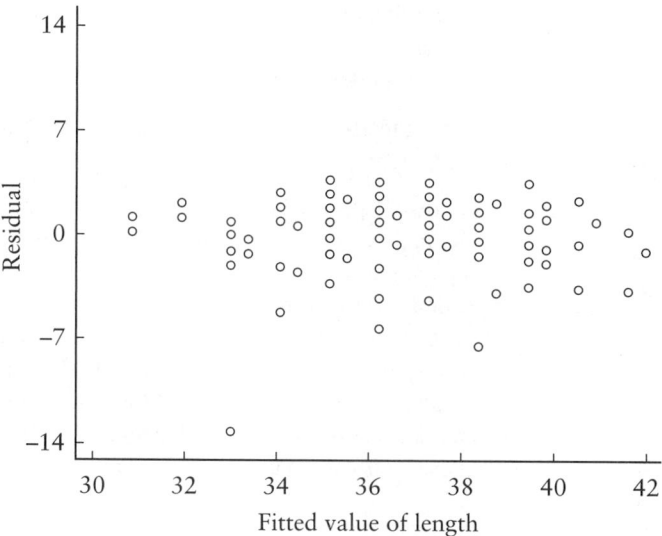

FIGURE 19.7
Residuals versus fitted values of length

the interaction term. Therefore, we conclude that there is no evidence that gestational age has a different effect on length depending on whether a mother experienced toxemia during pregnancy or not.

Returning to the model that contains gestational age and toxemia status but not their interaction term, a plot of the residuals is displayed in Figure 19.7. There appears to be one outlier in the data set. We might consider dropping this observation, refitting the least-squares equation, and comparing the two models to determine how much of an effect the point has on the estimated coefficients. However, the assumption of homoscedasticity has not been violated, and a transformation of variables does not appear to be necessary.

19.4   Review Exercises

1. What assumptions do you make when using the method of least squares to estimate a population regression equation containing two or more explanatory variables?

2. Given a multiple regression model with a total of q distinct explanatory variables, how would you make inference about a single coefficient β_j?

3. Explain how the coefficient of determination and the adjusted R^2 can be used to help evaluate the fit of a multiple regression model to the observed data.

4. What is the function of an interaction term in a regression model? How is an interaction term created?

5. If you are performing an analysis with a single response and several potential explanatory variables, how would you decide which variables to include in a multiple regression model and which to leave out?

6. How can collinearity between two explanatory variables affect the estimated coefficients in a regression model?

7. In a study designed to examine the effects of adding oats to the typical American diet, individuals were randomly divided into two different groups. Twice a day, the first group substituted oats for other foods containing carbohydrates; the members of the second group did not make any changes to their diet. One outcome of interest is the serum cholesterol level of each individual eight weeks after the start of the study. Explanatory variables that might affect this response include diet group, serum cholesterol level at the start of the study, body mass index, and gender. The estimated coefficients and standard errors from the multiple regression model containing these four explanatory variables are displayed below [2].

Variable	Coefficient	Standard Error
Diet Group	−11.25	4.33
Baseline Cholesterol	0.85	0.07
Body Mass Index	0.23	0.65
Gender	−3.02	4.42

(a) Conduct tests of the null hypotheses that each of the four coefficients in the population regression equation is equal to 0. At the 0.05 level of significance, which of the explanatory variables have an effect on serum cholesterol level eight weeks after the start of the study?

(b) If an individual's body mass index were to increase by 1 kg/m^2 while the values of all other explanatory variables remained constant, what would happen to his or her serum cholesterol level?

(c) If an individual's body mass index were to increase by 10 kg/m^2 while the values of all other explanatory variables remained constant, what would happen to his or her serum cholesterol level?

(d) The indicator variable gender is coded so that 1 represents a male and 0 a female. Who is more likely to have a higher serum cholesterol level eight weeks after the start of the study, a man or a woman? How much higher would it be, on average?

8. For the population of low birth weight infants, a significant linear relationship was found to exist between systolic blood pressure and gestational age. Recall that the relevant data are in the file lowbwt [1] (Appendix B, Table B.7). The measurements of systolic blood pressure are saved under the variable name sbp, and the corresponding gestational ages under gestage. Also contained in the data set is apgar5, the five-minute apgar score for each infant. (The apgar score is an indicator of a child's general state of health five minutes after it is born; although it is actually an ordinal measurement, it is often treated as if it were continuous.)

(a) Construct a two-way scatter plot of systolic blood pressure versus five-minute apgar score. Does there appear to be a linear relationship between these two variables?

(b) Using systolic blood pressure as the response and gestational age and apgar score as the explanatory variables, fit the least-squares model

$$\hat{y} = a + \hat{\beta}_1 x_1 + \hat{\beta}_2 x_2.$$

Interpret $\hat{\beta}_1$, the estimated coefficient of gestational age. What does it mean in words? Similarly, interpret $\hat{\beta}_2$, the estimated coefficient of five-minute apgar score.

(c) What is the estimated mean systolic blood pressure for the population of low birth weight infants whose gestational age is 31 weeks and whose five-minute apgar score is 7?

(d) Construct a 95% confidence interval for the true mean value of systolic blood pressure when $x_1 = 31$ weeks and $x_2 = 7$.

(e) Test the null hypothesis

$$H_0: \beta_2 = 0$$

at the 0.05 level of significance. What do you conclude?

(f) Comment on the magnitude of R^2. Does the inclusion of five-minute apgar score in the model already containing gestational age improve your ability to predict systolic blood pressure?

(g) Construct a plot of the residuals versus the fitted values of systolic blood pressure. What does this plot tell you about the fit of the model to the observed data?

9. The data set lowbwt also contains sex, a dichotomous random variable designating the gender of each infant.

(a) Add the indicator variable sex—where 1 represents a male and 0 a female—to the model that contains gestational age. Given two infants with identical gestational ages, one male and the other female, which would tend to have the higher systolic blood pressure? By how much, on average?

(b) Construct a two-way scatter plot of systolic blood pressure versus gestational age. On the graph, draw the two separate least-squares regression lines corresponding to males and to females. Is the gender difference in systolic blood pressure at each value of gestational age significantly different from 0?

(c) Add to the model a third explanatory variable that is the interaction between gestational age and sex. Does gestational age have a different effect on systolic blood pressure depending on the gender of the infant?

(d) Would you choose to include sex and the gestational age–sex interaction term in the regression model simultaneously? Why or why not?

10. The Bayley Scales of Infant Development produce two scores—the Psychomotor Development Index (PDI) and the Mental Development Index (MDI)—which can be used to assess a child's level of functioning. As part of a study examining the development and neurologic status of children who underwent reparative heart surgery during the first three months of life, the Bayley Scales were administered

to a sample of one-year-old infants born with congenital heart disease. Prior to heart surgery, the children had been randomized to one of two different treatment groups, called "circulatory arrest" and "low-flow bypass," which differed in the specific way in which the operation was performed. The data for this study are saved in the data set heart [3] (Appendix B, Table B.12). PDI scores are saved under the variable name pdi, MDI scores under mdi, and indicators of treatment group under trtment. For the treatment group variable, 0 represents circulatory arrest and 1 is low-flow bypass.

(a) In Chapter 11, the two-sample *t*-test was used to compare mean PDI and MDI scores for infants assigned to the circulatory arrest and low-flow bypass treatment groups. These analyses could also be performed using linear regression. Fit two simple linear regression models—one with PDI score as the response and the other with MDI score—that both have the indicator of treatment group as the explanatory variable.

(b) Who is more likely to have a higher PDI score, a child assigned to the circulatory arrest treatment group or one assigned to the low-flow bypass group? How much higher would the score be, on average?

(c) Who is more likely to have a higher MDI score? How much higher, on average?

(d) Is the treatment group difference in either PDI or MDI scores statistically significant at the 0.05 level? What do you conclude?

11. In Chapter 18, the relationship between expense per admission into a community hospital and average length of stay in the facility was examined for each state in the United States in 1982. The relevant data are in the file hospital [4] (Appendix B, Table B.26); mean expense per admission is saved under the variable name expadm, and average length of stay under the name los. Also included in the data set is salary, the average salary per employee in 1982.

(a) Summarize the average salary per employee both graphically and numerically. What are the mean and median average salaries? What are the minimum and maximum values?

(b) Construct a two-way scatter plot of mean expense per admission versus average salary. What does the graph suggest about the relationship between these two variables?

(c) Fit the least-squares model where mean expense per admission is the response and average length of stay and average salary are the explanatory variables. Interpret the estimated regression coefficients.

(d) What happens to the estimated coefficient of length of stay when average salary is added to the model?

(e) Does the inclusion of salary in addition to average length of stay improve your ability to predict mean expense per admission? Explain.

(f) Examine a plot of the residuals versus the fitted values of expense per admission. What does this plot tell you about the fit of the model to the observed data?

12. A study was conducted to examine the roles of firearms and various other factors in the rate of homicides in the city of Detroit. Information for the years 1961 to 1973 is provided in the data set detroit [5] (Appendix B, Table B.27); the number of homicides per 100,000 population is saved under the variable name homi-

cide. Other variables in the data set include `police`, the number of full-time police officers per 100,000 population; `unemp`, the percentage of adults who are unemployed; `register`, the number of handgun registrations per 100,000 population; and `weekly`, the average weekly earnings for city residents.

(a) For each of the four explanatory variables listed—`police`, `unemp`, `register`, and `weekly`—construct a two-way scatter plot of homicide rate versus that variable. Are any linear relationships apparent from the graphs?

(b) Fit four simple linear regression models using homicide rate as the response and each of the other variables as the single explanatory variable. Individually, which of the variables have an effect on homicide rate that is significant at the 0.05 level?

(c) List the coefficient of determination for each regression model. Which variable explains the greatest proportion of the observed variation among the values of homicide rate?

(d) Using the method of forward selection, find the "best" multiple regression equation. Each variable contained in the final model should explain a significant amount of the observed variability in homicide rate. What do you conclude?

Bibliography

[1] Leviton, A., Fenton, T., Kuban, K. C. K., and Pagano, M., "Labor and Delivery Characteristics and the Risk of Germinal Matrix Hemorrhage in Low Birth Weight Infants," *Journal of Child Neurology,* Volume 6, October 1991, 35–40.

[2] Van Horn, L., Moag-Stahlberg, A., Liu, K., Ballew, C., Ruth, K., Hughes, R., and Stamler, J., "Effects on Serum Lipids of Adding Instant Oats to Usual American Diets," *American Journal of Public Health,* Volume 81, February 1991, 183–188.

[3] Bellinger, D. C., Jonas, R. A., Rappaport, L. A., Wypij, D., Wernovsky, G., Kuban, K. C. K., Barnes, P. D., Holmes, G. L., Hickey, P. R., Strand, R. D., Walsh, A. Z., Helmers, S. L., Constantinou, J. E., Carrazana, E. J., Mayer, J. E., Hanley, F. L., Castaneda, A. R., Ware, J. H., and Newburger, J. W., "Development and Neurologic Status of Children After Heart Surgery with Hypothermic Circulatory Arrest or Low-Flow Cardiopulmonary Bypass," *The New England Journal of Medicine,* Volume 332, March 2, 1995, 549–555.

[4] Levit, K. R., "Personal Health Care Expenditures, by State: 1966–1982," *Health Care Financing Review*, Volume 6, Summer 1985, 1–25.

[5] Fisher, J. C., "Homicide in Detroit: The Role of Firearms," *Criminology,* Volume 14, 1976, 387–400.

20

Logistic Regression

When studying linear regression, we attempted to estimate a population regression equation

$$\mu_{y|x_1, x_2, \ldots, x_q} = \alpha + \beta_1 x_1 + \beta_2 x_2 + \cdots + \beta_q x_q$$

by fitting a model of the form

$$y = \alpha + \beta_1 x_1 + \beta_2 x_2 + \cdots + \beta_q x_q + \varepsilon.$$

The response Y was continuous, and was assumed to follow a normal distribution. We were concerned with predicting or estimating the mean value of the response corresponding to a given set of values for the explanatory variables.

There are many situations, however, in which the response of interest is dichotomous rather than continuous. Examples of variables that assume only two possible values are disease status (the disease is either present or absent) and survival following surgery (a patient is either alive or dead). In general, the value 1 is used to represent a "success," or the outcome we are most interested in, and 0 represents a "failure." The mean of the dichotomous random variable Y, designated p, is the proportion of times that it takes the value 1. Equivalently,

$$p = P(Y = 1)$$
$$= P(\text{"success"}).$$

Just as we estimated the mean value of the response when Y was continuous, we would like to be able to estimate the probability p associated with a dichotomous response (which, of course, is also its mean) for various values of an explanatory variable. To do this, we use a technique known as *logistic regression*.

20.1 The Model

Consider the population of low birth weight infants—in this case, defined as those weighing less than 1750 grams—who satisfy the following criteria: They are confined to a neonatal intensive care unit, they require intubation during the first 12 hours of life, and they survive for at least 28 days. In a sample of 223 such infants drawn from the underlying population, 76 were diagnosed with bronchopulmonary dysplasia (BPD), a chronic type of lung disease [1]. The remaining 147 were not. Let Y be a dichotomous random variable for which the value 1 represents the presence of BPD in a child and 0 its absence. We would estimate the probability that an infant in this population develops bronchopulmonary dysplasia by the sample proportion

$$\tilde{p} = \frac{76}{223}$$
$$= 0.341.$$

Overall, 34.1% of these low birth weight children are diagnosed with BPD.

We might suspect that there are certain factors—both maternal and neonatal—that affect the likelihood that a particular infant will develop BPD. If we can classify a child according to these characteristics, it is possible we could estimate his or her probability of developing the lung disease with greater precision than that afforded by the single value $\tilde{p}$, and subsequently take measures to decrease this probability.

One factor of interest might be the birth weight of an infant. If the response Y were continuous, we would begin an analysis by constructing a two-way scatter plot of the response versus the continuous explanatory variable. A graph of bronchopulmonary dysplasia versus birth weight is displayed in Figure 20.1. Note that in this case, all points lie on one of two parallel lines, depending on whether Y takes the value 0 or 1. There does appear to be a tendency for infants who develop BPD to have somewhat lower birth weights, on average; however, the nature of this relationship is not clear.

Since the two-way scatter plot is not particularly helpful, we might instead begin to explore whether an association exists between a diagnosis of bronchopulmonary dysplasia and birth weight by subdividing the population of infants into three categories: those weighing 950 grams or less, those weighing between 951 and 1350 grams, and those weighing 1351 grams or more. We could then estimate the probability that a child will develop BPD in each of these subgroups individually.

Birth Weight (grams)	Sample Size	Number with BPD	$\tilde{p}$
0–950	68	49	0.721
951–1350	80	18	0.225
1351–1750	75	9	0.120
	223	76	0.341

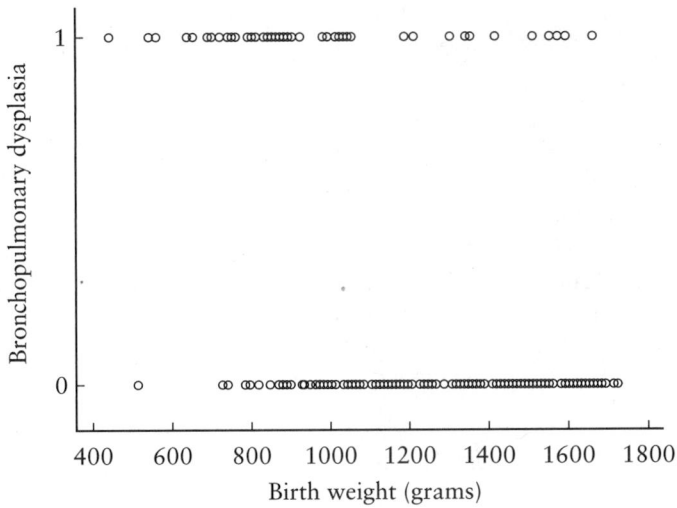

FIGURE 20.1
Diagnosis of bronchopulmonary dysplasia versus birth weight for a sample
of 223 low birth weight infants

The estimated probability of being diagnosed with bronchopulmonary dysplasia decreases as the birth weight of an infant increases, from a high of 0.721 for children weighing 950 grams or less to a low of 0.120 for those weighing 1351 grams or more. Since there does appear to be a relationship between these two variables, we would like to be able to use a child's birth weight to help us predict the likelihood that he or she will develop BPD.

20.1.1 The Logistic Function

Our first strategy might be to fit a model of the form

$$p = \alpha + \beta x,$$

where x represents birth weight. This is simply the standard linear regression model in which y—the outcome of a continuous, normally distributed random variable—has been replaced by p. As before, α is the intercept of the line and β is its slope. On inspection, however, this model is not feasible. Since p is a probability, it is restricted to taking values between 0 and 1. The term $\alpha + \beta x$, in contrast, could easily yield a value that lies outside this range.

We might try to solve this problem by instead fitting the model

$$p = e^{\alpha + \beta x}.$$

This equation guarantees that the estimate of p is positive. We would soon realize, however, that this model is also unsuitable. Although the term $e^{\alpha + \beta x}$ cannot produce a negative estimate of p, it can result in a value that is greater than 1.

To accommodate this final constraint, we fit a model of the form

$$p = \frac{e^{\alpha + \beta x}}{1 + e^{\alpha + \beta x}}.$$

The expression on the right, called a *logistic function,* cannot yield a value that is either negative or greater than 1; consequently, it restricts the estimated value of p to the required range.

Recall that if an event occurs with probability p, the odds in favor of the event are $p/(1 - p)$ to 1. Thus, if a success occurs with probability

$$p = \frac{e^{\alpha + \beta x}}{1 + e^{\alpha + \beta x}},$$

the odds in favor of success are

$$\frac{p}{1 - p} = \frac{e^{\alpha + \beta x}/(1 + e^{\alpha + \beta x})}{1/(1 + e^{\alpha + \beta x})}$$

$$= e^{\alpha + \beta x}.$$

Taking the natural logarithm of each side of this equation,

$$\ln\left[\frac{p}{1 - p}\right] = \ln\left[e^{\alpha + \beta x}\right]$$

$$= \alpha + \beta x.$$

Thus, modeling the probability p with a logistic function is equivalent to fitting a linear regression model in which the continuous response y has been replaced by the logarithm of the odds of success for a dichotomous random variable. Instead of assuming that the relationship between p and x is linear, we assume that the relationship between $\ln[p/(1 - p)]$ and x is linear. The technique of fitting a model of this form is known as logistic regression.

20.1.2 *The Fitted Equation*

In order to use a child's birth weight to help us predict the likelihood that he or she will develop the chronic lung disease bronchopulmonary dysplasia, we fit the model

$$\ln\left[\frac{\hat{p}}{1 - \hat{p}}\right] = \hat{\alpha} + \hat{\beta}x.$$

Although we categorized birth weight into three different intervals when exploring its relationship with BPD, we use the original continuous random variable for the logistic regression analysis. As with a linear regression model, $\hat{\alpha}$ and $\hat{\beta}$ are estimates of the population coefficients. However, we cannot apply the method of least squares, which assumes that the response is continuous and normally distributed, to fit a logistic model; instead, we use maximum likelihood estimation [2]. Recall that this technique uses the information in a sample to find the parameter estimates that are most likely to have produced the observed data.

For the sample of 223 low birth weight infants, the estimated logistic regression equation is

$$\ln\left[\frac{\hat{p}}{1-\hat{p}}\right] = 4.0343 - 0.0042x.$$

The coefficient of weight implies that for each one-gram increase in birth weight, the log odds that the infant develops BPD decrease by 0.0042, on average. When the log odds decrease, the probability p decreases as well. In order to test

$$H_0: \beta = 0,$$

the null hypothesis that there is no relationship between p and x, against the alternative

$$H_A: \beta \neq 0,$$

we need to know the standard error of the estimator $\hat{\beta}$. Then if H_0 is true, the test statistic

$$z = \frac{\hat{\beta}}{\widehat{se}(\hat{\beta})}$$

follows a standard normal distribution. It turns out that the coefficient of birth weight is in fact significantly different from 0 at the 0.05 level; thus we conclude that in the underlying population of low birth weight infants meeting the necessary medical criteria, the probability of developing BPD decreases as birth weight increases.

In order to estimate the probability that an infant with a particular weight develops chronic lung disease, we simply substitute the appropriate value of x into the preceding equation. To estimate the probability that a child weighing 750 grams at birth develops BPD, for example, we substitute the value $x = 750$ grams to find

$$\ln\left[\frac{\hat{p}}{1-\hat{p}}\right] = 4.0343 - 0.0042(750)$$

$$= 0.8843.$$

Taking the antilogarithm of each side of the equation,

$$\frac{\hat{p}}{1-\hat{p}} = e^{0.8843}$$

$$= 2.4213.$$

Finally, solving for $\hat{p}$,

$$\hat{p} = \frac{2.4213}{1 + 2.4213}$$
$$= \frac{2.4213}{3.4213}$$
$$= 0.708.$$

The estimated probability that a child weighing 750 grams at birth develops BPD is 0.708.

Using similar calculations, we find that the estimated probability of developing BPD for an infant weighing 1150 grams at birth is

$$\hat{p} = 0.311,$$

whereas the probability for a child weighing 1550 grams is

$$\hat{p} = 0.078.$$

If we calculated the estimated probability $\hat{p}$ for each observed value of birth weight and plotted $\hat{p}$ versus weight, the result would be the curve in Figure 20.2. According to the logistic regression model, the estimated value of p decreases as birth weight increases. As previously noted, however, the relationship between p and x is not linear.

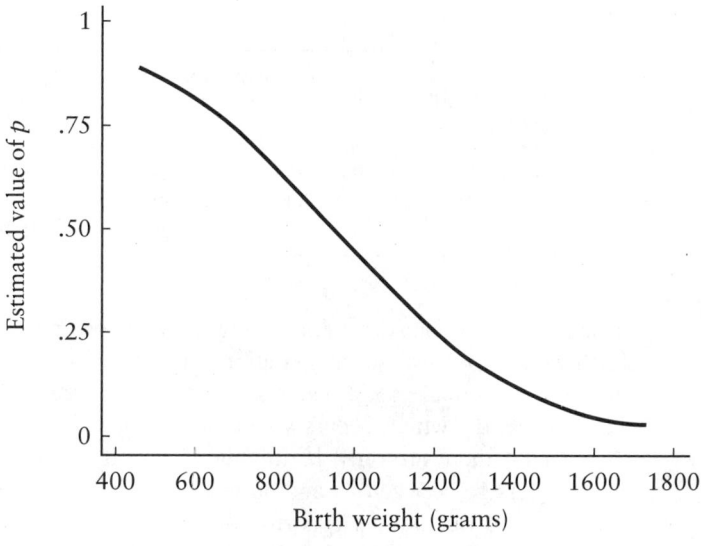

FIGURE 20.2
Logistic regression of bronchopulmonary dysplasia on birth weight,
$\ln[\hat{p}/(1 - \hat{p})] = 4.0343 - 0.0042x$

20.2   Multiple Logistic Regression

Now that we have determined that birth weight influences the probability that an infant will develop BPD, we might wonder whether knowing a child's gestational age would further improve our ability to predict p. To begin to explore this possibility, we again subdivide the population of low birth weight infants into three categories: those whose gestational age is less than or equal to 28 weeks, those whose gestational age is between 29 and 30 weeks, and those whose gestational age is 31 weeks or more.

Gestational Age (weeks)	Sample Size	Number with BPD	$\tilde{p}$
≤ 28	58	40	0.690
29–30	73	26	0.356
≥ 31	92	10	0.109
	223	76	0.341

The estimated probability of developing bronchopulmonary dysplasia decreases as gestational age increases.

We next cross-classify the sample of 223 infants according to the three categories of birth weight and three categories of gestational age and estimate the probability of developing BPD within each combination of categories. The numbers in parentheses are the sample sizes in each subgroup.

Birth Weight (grams)	Gestational Age (weeks)		
	≤ 28	29–30	≥ 31
0–950	0.805 (41)	0.714 (21)	0.167 (6)
951–1350	0.412 (17)	0.194 (36)	0.148 (27)
1351–1750	— (0)	0.250 (16)	0.085 (59)

There appears to be something of a trend in the estimates of p. For a given category of birth weight, $\tilde{p}$ decreases as gestational age increases. In addition, for a specified category of gestational age, $\tilde{p}$ decreases as weight increases. There is a slight discrepancy in these trends, which occurs when the number of infants in a particular subgroup is very small; there are only 16 infants who weigh more than 1350 grams and have a gestational age between 29 and 30 weeks. In an instance such as this, $\tilde{p}$ is subject to a great deal of fluctuation. Ignoring this aberration, however, it appears that we might be able to estimate the probability that a child develops bronchopulmonary dysplasia with greater precision when we know the values of both birth weight and gestational age.

To model the probability p as a function of the two explanatory variables, we fit a model of the form

$$\ln\left[\frac{\hat{p}}{1-\hat{p}}\right] = \hat{\alpha} + \hat{\beta}_1 x_1 + \hat{\beta}_2 x_2,$$

where x_1 designates weight and x_2 represents age. Like birth weight, gestational age is now treated as a continuous random variable. The estimated logistic regression equation is

$$\ln\left[\frac{\hat{p}}{1-\hat{p}}\right] = 13.8273 - 0.0024x_1 - 0.3983x_2.$$

The coefficient of birth weight has decreased somewhat now that another explanatory variable has been added to the model; however, it is still significantly different from 0 at the 0.05 level. The coefficient of gestational age implies that if birth weight remains constant while gestational age increases by one week, the log odds of developing BPD decrease by 0.3983 on average. A test of the null hypothesis

$$H_0: \beta_2 = 0$$

indicates that this coefficient is also different from 0 at the 0.05 level of significance. Therefore, given that we already know the birth weight of an infant, knowledge of his or her gestational age will further improve our ability to predict the probability that he or she develops BPD. This probability decreases as either birth weight or gestational age increases.

To estimate the probability that an infant weighing 750 grams at birth with gestational age 27 weeks will develop BPD, we substitute the values $x_1 = 750$ grams and $x_2 = 27$ weeks into the estimated equation to find

$$\ln\left[\frac{\hat{p}}{1-\hat{p}}\right] = 13.8273 - 0.0024(750) - 0.3983(27)$$

$$= 1.2732.$$

Taking the antilogarithm of each side results in

$$\frac{\hat{p}}{1-\hat{p}} = e^{1.2732}$$

$$= 3.5723,$$

and solving for $\hat{p}$, we have

$$\hat{p} = \frac{3.5723}{1 + 3.5723}$$

$$= \frac{3.5723}{4.5723}$$

$$= 0.781.$$

Thus, the estimated probability is 0.781. For an infant who weighs 1150 grams and has gestational age 32 weeks,

$$\ln\left[\frac{\hat{p}}{1-\hat{p}}\right] = 13.8273 - 0.0024(1150) - 0.3983(32)$$
$$= -1.6783.$$

Therefore,

$$\frac{\hat{p}}{1-\hat{p}} = e^{-1.6783}$$
$$= 0.1867,$$

and

$$\hat{p} = \frac{0.1867}{1+0.1867}$$
$$= \frac{0.1867}{1.1867}$$
$$= 0.157.$$

As previously noted, the estimated probability that an infant will develop bronchopulmonary dysplasia decreases as either birth weight or gestational age increases.

20.3   Indicator Variables

Like the linear regression model, the logistic regression model can be generalized to include discrete or nominal explanatory variables in addition to continuous ones. Suppose that we fit the model

$$\ln\left[\frac{\hat{p}}{1-\hat{p}}\right] = \hat{\alpha} + \hat{\beta}_3 x_3,$$

where x_3 is the outcome of a dichotomous random variable indicating whether a mother was diagnosed with toxemia during her pregnancy. If the presence of toxemia is represented by 1 and its absence by 0, the equation estimated from the sample of 223 low birth weight infants is

$$\ln\left[\frac{\hat{p}}{1-\hat{p}}\right] = -0.5718 - 0.7719 x_3.$$

Note that the coefficient of toxemia is negative, implying that the log odds of developing BPD—and thus the probability p itself—is lower for children whose mothers experienced toxemia than for those whose mothers did not.

It can be shown that if an explanatory variable x_i is dichotomous, its estimated coefficient $\hat{\beta}_i$ has a special interpretation. In this case, the antilogarithm of $\hat{\beta}_i$, or $e^{\hat{\beta}_i}$, is the estimated odds ratio of the response for the two possible levels of x_i. For example, the relative odds of developing bronchopulmonary dysplasia for children whose mothers suffered from toxemia versus children whose mothers did not is

$$\widehat{OR} = e^{\hat{\beta}_3}$$
$$= e^{-0.7719}$$
$$= 0.46.$$

The odds of developing chronic lung disease are actually lower for children whose mothers experienced toxemia during pregnancy.

The same results could have been obtained by arranging the sample data as a 2×2 contingency table. Of the 76 infants who developed BPD, 6 had mothers who were diagnosed with toxemia during pregnancy; of the 147 children who did not, 23 had mothers who suffered from toxemia.

	Toxemia		
BPD	**Yes**	**No**	**Total**
Yes	6	70	76
No	23	124	147
Total	29	194	223

Note that the odds ratio estimated by computing the cross-product of the entries in the contingency table,

$$\widehat{OR} = \frac{(6)(124)}{(70)(23)}$$
$$= 0.46,$$

is identical to that obtained from the logistic regression model.

A confidence interval for the odds ratio can be calculated from the model by computing a confidence interval for the coefficient β_3 and taking the antilogarithm of its upper and lower limits. If $\widehat{se}(\hat{\beta}_3) = 0.4822$, a 95% confidence interval for β_3 is

$$(-0.7719 - 1.96(0.4822), \; -0.7719 + 1.96(0.4822)),$$

or

$$(-1.7170, 0.1732).$$

Furthermore, a 95% confidence interval for e^{β_3}, the relative odds of developing bronchopulmonary dysplasia, is

$$(e^{-1.7170}, e^{0.1732}),$$

or

$$(0.18, 1.19).$$

We are 95% confident that these limits cover the true population odds ratio for infants whose mothers experienced toxemia during pregnancy versus infants whose mothers did not. Note that the interval contains the value 1; therefore, the sample of low birth weight infants does not provide evidence that the probability of developing BPD is different depending on the toxemia status of a child's mother.

Suppose that we add a second dichotomous explanatory variable indicating whether a mother was administered steroids during her pregnancy to the model that already contains toxemia. If the variable x_4 takes the value 1 if a woman was prescribed steroids and 0 if she was not, the equation estimated from the sample of 223 low birth weight infants is

$$\ln\left[\frac{\hat{p}}{1 - \hat{p}}\right] = -0.7172 - 0.7883x_3 + 0.3000x_4.$$

The coefficient of toxemia changes very little with the inclusion of the second explanatory variable. The estimated coefficient of steroid use is positive, implying that the probability of developing BPD is higher for infants whose mothers were administered the drugs; like the coefficient of toxemia, however, it is not significantly different from 0 at the 0.05 level.

Including a second dichotomous random variable in the logistic regression model is analogous to performing a stratified analysis using the Mantel-Haenszel method; the exponentiated coefficient of toxemia provides an estimate of the relative odds of developing BPD for children whose mothers were diagnosed with toxemia versus children whose mothers were not that has been adjusted for the effect of steroid use. (Similarly, the coefficient of steroid use allows us to estimate the relative odds of developing lung disease for infants whose mothers used steroids versus those whose mothers did not that has been adjusted for toxemia status.) In this case, the adjusted odds ratio for toxemia,

$$\widehat{OR} = e^{-0.7883}$$
$$= 0.45,$$

is nearly identical to the unadjusted estimate of 0.46. If we want to determine whether toxemia status has a different effect on the probability of developing BPD depending on whether or not a mother was administered steroids during her pregnancy, it would be necessary to include in the logistic regression model an interaction term that is the product of the two dichotomous explanatory variables.

20.4   **Further Applications**

Suppose that we are interested in identifying factors that influence the probability that a low birth weight infant will experience a germinal matrix hemorrhage—a hemorrhage in the brain. Germinal matrix hemorrhage is a dichotomous random variable that takes the value 1 if this event occurs and 0 if it does not. We use the sample of 100 low birth weight infants born in Boston, Massachusetts, to estimate the probability of a hemorrhage [3].

To begin, we would like to determine whether the head circumference of an infant affects the probability that he or she will suffer a brain hemorrhage. Because the response is dichotomous, we fit a logistic regression model of the form

$$\ln\left[\frac{\hat{p}}{1-\hat{p}}\right] = \hat{\alpha} + \hat{\beta}_1 x_1,$$

where x_1 represents head circumference. Table 20.1 shows the relevant output from Minitab.

At the top of the table, we see that among the 100 low birth weight infants, 15 experienced a germinal matrix hemorrhage and 85 did not. Beneath this is information relating to the estimated regression coefficients $\hat{\alpha}$, the intercept term or constant, and $\hat{\beta}_1$, the coefficient of head circumference. The fitted equation is

$$\ln\left[\frac{\hat{p}}{1-\hat{p}}\right] = 1.193 - 0.1117 x_1.$$

TABLE 20.1

Minitab output displaying the logistic regression of germinal matrix hemorrhage on head circumference

```
Link Function: Logit

Response Information

Variable    Value    Count
grmhem          1       15      (Event)
                0       85
            Total      100
```

Logistic Regression Table

Predictor	Coef	StDev	z	P	Odds Ratio	95% CI Lower	95% CI Upper
Constant	1.193	3.006	0.40	0.692			
headcirc	-0.1117	0.1153	-0.97	0.332	0.89	0.71	1.12

```
Log-Likelihood = -41.788
Test that all slopes are zero: G=0.966, DF=1, P-Value=0.326
```

The coefficient $\hat{\beta}_1$ implies that for each one-centimeter increase in head circumference, the log odds of experiencing a hemorrhage decrease by 0.1117 on average. Since the log odds decrease, the probability p decreases as well. However, the null hypothesis

$$H_0: \beta_1 = 0$$

fails to be rejected at the 0.05 level of significance ($p = 0.33$); therefore, this sample does not provide evidence that the probability of a brain hemorrhage differs depending on a child's head circumference.

Note that the Minitab output also displays an odds ratio of 0.89 for head circumference. For a continuous explanatory variable, this is the estimated relative odds of experiencing a hemorrhage associated with a one-unit increase in the variable.

If we now wish to estimate the probability that an infant with a particular head circumference will suffer a germinal matrix hemorrhage—keeping in mind that any differences in probabilities for various values of head circumference are not statistically significant at the 0.05 level—we would substitute the appropriate value of x_1 into the estimated equation and solve for $\hat{p}$. Given a child whose head circumference is 28 centimeters, for instance,

$$\ln\left[\frac{\hat{p}}{1-\hat{p}}\right] = 1.193 - 0.1117(28)$$
$$= -1.9346.$$

Therefore,

$$\frac{\hat{p}}{1-\hat{p}} = e^{-1.9346}$$
$$= 0.1445,$$

and

$$\hat{p} = \frac{0.1445}{1 + 0.1445}$$
$$= \frac{0.1445}{1.1445}$$
$$= 0.126.$$

The predicted probabilities of experiencing a hemorrhage for the first 10 infants in the sample are listed in Table 20.2; note that the calculated probabilities decrease slightly as head circumference increases.

We now attempt to determine whether the gender of an infant influences the probability that he or she experiences a germinal matrix hemorrhage. To do this, we fit a logistic regression model of the form

$$\ln\left[\frac{\hat{p}}{1-\hat{p}}\right] = \hat{\alpha} + \hat{\beta}x_2,$$

where x_2 takes the value 1 for a male and 0 for a female. Table 20.3 shows two different versions of output from Stata. From the first version, the fitted equation is

$$\ln\left[\frac{\hat{p}}{1-\hat{p}}\right] = -2.303 + 0.8938x_2.$$

Because the estimated coefficient of gender is positive, both the log odds of experiencing a hemorrhage and the probability p itself are higher for males than for females.

TABLE 20.2

Predicted probabilities of experiencing a germinal matrix hemorrhage for the first 10 infants in the sample

Head Circumference (cm)	Hemorrhage	Predicted $\hat{p}$
29	0	0.114
23	1	0.202
28	0	0.126
27	0	0.139
26	0	0.153
26	1	0.153
27	0	0.139
28	0	0.126
28	0	0.126
26	0	0.153

TABLE 20.3

Stata output displaying the logistic regression of germinal matrix hemorrhage on gender

```
Iteration 0: Log Likelihood = -42.270909
Iteration 1: Log Likelihood = -41.174001
Iteration 2: Log Likelihood = -41.147057
Iteration 3: Log Likelihood = -41.147023

Logit Estimates                              Number of obs =      100
                                             chi2(1)       =     2.25
                                             Prob > chi2   = 0.1338
Log Likelihood = -41.147023                  Pseudo R2     = 0.0266
```

grmhem	Coef.	Std. Err.	z	P > |z|	[95% Conf. Interval]	
sex	.8938179	.6230005	1.435	0.151	-.3272407	2.114876
_cons	-2.302585	.5244028	-4.391	0.000	-3.330396	-1.274774

```
Logit Estimates                              Number of obs =      100
                                             chi2(1)       =     2.25
                                             Prob > chi2   = 0.1338
Log Likelihood = -41.147023                  Pseudo R2     = 0.0266
```

grmhem	Odds Ratio	Std. Err.	z	P > |z|	[95% Conf. Interval]	
sex	2.444444	1.52289	1.435	0.151	.7209102	8.288561

From the second version of the output, the relative odds of experiencing a hemorrhage for boys versus girls is estimated by

$$\widehat{OR} = e^{0.8938}$$
$$= 2.44.$$

Males appear to be more likely to suffer a hemorrhage than females; however, a test of the null hypothesis

$$H_0: \beta_2 = 0$$

results in a p-value of 0.15. At the 0.05 level of significance, the probability of a hemorrhage does not vary depending on gender.

20.5   Review Exercises

1. When the response variable of interest is dichotomous rather than continuous, why is it not advisable to fit a standard linear regression model using the probability of success as the outcome?

2. What is a logistic function?

3. How does logistic regression differ from linear regression?

4. How can a logistic regression model be used to estimate the relative odds of success of one dichotomous random variable for the two possible levels of another?

5. In a study investigating maternal risk factors for congenital syphilis, syphilis is treated as a dichotomous response variable, where 1 represents the presence of disease in a newborn and 0 its absence. The estimated coefficients from a logistic regression model containing the explanatory variables cocaine or crack use, marital status, number of prenatal visits to a doctor, alcohol use, and level of education are listed below [4]. The estimated intercept $\hat{\alpha}$ is not given.

Variable	Coefficient
Cocaine/Crack Use	1.354
Marital Status	0.779
Number of Prenatal Visits	−0.098
Alcohol Use	0.723
Level of Education	0.298

(a) As an expectant mother's number of prenatal visits to the doctor increases, what happens to the probability that her child will be born with congenital syphilis?

(b) Marital status is a dichotomous random variable, where the value 1 indicates that a woman is unmarried and 0 that she is married. What are the relative odds that a newborn will suffer from syphilis for unmarried versus married mothers?

(c) Cocaine or crack use is also a dichotomous random variable; the value 1 indicates that a woman used drugs during her pregnancy and 0 that she did not. What is the estimated odds ratio that a child will be born with congenital syphilis for women who used cocaine or crack versus those who did not?

(d) The estimated coefficient of cocaine or crack use has standard error 0.162. Construct a 95% confidence interval for the population odds ratio. What do you conclude?

6. Suppose that you are interested in studying intravenous drug use among high school students in the United States. Drug use is characterized as a dichotomous random variable, where 1 indicates that an individual has injected drugs within the past year and 0 that he or she has not. Factors that might be related to drug use are instruction about the human immunodeficiency virus (HIV) in school, age of the student, gender, and general knowledge about HIV, including the various modes of transmission and ways to reduce risk. The estimated coefficients and standard errors from a logistic regression model containing each of these explanatory variables as well as the interaction between instruction and gender are displayed below [5].

Variable	Coefficient	Standard Error
Intercept	−1.183	0.859
HIV Instruction	0.039	0.421
Age	−0.164	0.092
Gender	1.212	0.423
HIV Knowledge	−0.187	0.048
HIV Instruction/Gender	−0.663	0.512

(a) Conduct tests of the null hypotheses that each of the coefficients in the population regression equation is equal to 0. At the 0.05 level of significance, which of the explanatory variables influence the probability of intravenous drug use in the past year?

(b) As a student becomes older, does the probability that he or she has used intravenous drugs in the past year increase or decrease?

(c) The dichotomous random variable gender is coded so that 1 represents a male and 0 a female. What are the estimated odds of injecting drugs for males relative to females?

(d) Does HIV instruction have a different relationship with the probability of intravenous drug use for males as opposed to females?

7. The data set lowbwt contains information for the sample of 100 low birth weight infants born in Boston, Massachusetts [3] (Appendix B, Table B.7). The variable grmhem is a dichotomous random variable indicating whether an infant experienced a germinal matrix hemorrhage. The value 1 indicates that a hemorrhage occurred and 0 that it did not. The infants' five-minute apgar scores are saved under the name apgar5, and indicators of toxemia—where 1 represents a diagnosis of toxemia during pregnancy for the child's mother and 0 no such diagnosis—under the variable name tox.

(a) Using germinal matrix hemorrhage as the response, fit a logistic regression model of the form

$$\ln\left[\frac{\hat{p}}{1-\hat{p}}\right] = \hat{\alpha} + \hat{\beta}_1 x_1,$$

where x_1 is five-minute apgar score. Interpret $\hat{\beta}_1$, the estimated coefficient of apgar score.

(b) If a particular child has a five-minute apgar score of 3, what is the predicted probability that this child will experience a brain hemorrhage? What is the probability if the child's score is 7?

(c) At the 0.05 level of significance, test the null hypothesis that the population parameter β_1 is equal to 0. What do you conclude?

(d) Now fit the regression model

$$\ln\left[\frac{\hat{p}}{1-\hat{p}}\right] = \hat{\alpha} + \hat{\beta}_2 x_2,$$

where x_2 represents toxemia status. Interpret $\hat{\beta}_2$, the estimated coefficient of toxemia.

(e) For a child whose mother was diagnosed with toxemia during pregnancy, what is the predicted probability of experiencing a germinal matrix hemorrhage? What is the probability for a child whose mother was not diagnosed with toxemia?

(f) What are the estimated odds of suffering a germinal matrix hemorrhage for children whose mothers were diagnosed with toxemia relative to children whose mothers were not?

(g) Construct a 95% confidence interval for the population odds ratio. Does this interval contain the value 1? What does this tell you?

8. In Chapter 16, the Mantel-Haenszel method was used to investigate the association between smoking and the development of aortic stenosis for males and females. The data are contained in the data set stenosis [6] (Appendix B, Table B.20). Smoking status is saved under the variable name smoke, the presence of aortic stenosis under the name disease, and gender under the name sex.

(a) Using the presence of aortic stenosis as the response, fit a logistic regression model with smoking status as the single explanatory variable. Interpret the estimated coefficient of smoking status.

(b) What are the estimated odds of suffering from aortic stenosis for individuals who smoke relative to those who do not?

(c) Construct a 95% confidence interval for the population odds ratio. Does this interval contain the value 1? What does this tell you?

(d) Add the explanatory variable gender to the model that already contains smoking status. What is the estimated relative odds of aortic stenosis for smokers versus nonsmokers, adjusting for gender?

(e) Construct a 95% confidence interval for the population odds ratio that adjusts for gender. What do you conclude?

(e) Do you believe that the relationship between the presence of aortic stenosis and smoking status differs for males and females? Explain.

9. In a group of patients undergoing dialysis for chronic renal failure for a period of at least two years, it was determined which of the individuals had experienced at least one episode of peritonitis, an inflammation of the membrane lining the abdominal cavity, and which had not. The results are contained in a data set called `dialysis` [7] (Appendix B, Table B.28). The variable `perito` is a dichotomous random variable taking the value 1 if an individual experienced an infection and 0 otherwise. Potential explanatory variables are age, gender, and racial background. The variable `age` is continuous; `sex` and `race` are dichotomous and take the value 1 for female and non-white patients, respectively. Male and white individuals are represented by 0.

(a) Fit three separate logistic regression models investigating the effects of age, gender, and racial group on the probability that an individual experiences peritonitis. Interpret the estimated coefficients of each explanatory variable.

(b) What is the predicted probability that a white patient undergoing dialysis for chronic renal failure will experience peritonitis? What is the probability for a non-white patient?

(c) What are the estimated odds of developing peritonitis for females versus males?

(d) At the 0.05 level of significance, which of the explanatory variables helps to predict peritonitis in patients undergoing dialysis?

(e) Do you see any problems with the way in which the response is categorized?

Bibliography

[1] Van Marter, L. J., Leviton, A., Kuban, K. C. K., Pagano, M., and Allred, E. N., "Maternal Glucocorticoid Therapy and Reduced Risk of Bronchopulmonary Dysplasia," *Pediatrics,* Volume 86, September 1990, 331–336.

[2] Hosmer, D. W., and Lemeshow, S., *Applied Logistic Regression*, New York: Wiley, 1989.

[3] Leviton, A., Fenton, T., Kuban, K. C. K., and Pagano, M., "Labor and Delivery Characteristics and the Risk of Germinal Matrix Hemorrhage in Low Birth Weight Infants," *Journal of Child Neurology,* Volume 6, October 1991, 35–40.

[4] Zweig, M. S., Singh, T., Htoo, M., and Schultz, S., "The Association Between Congenital Syphilis and Cocaine/Crack Use in New York City: A Case-Control Study," *American Journal of Public Health,* Volume 81, October 1991, 1316–1318.

[5] Holtzman, D., Anderson, J. E., Kann, L., Arday, S. L., Truman, B. I., and Kohbe, L. J., "HIV Instruction, HIV Knowledge, and Drug Injection Among High School Students in the United States," *American Journal of Public Health,* Volume 81, December 1991, 1596–1601.

[6] Hoagland, P. M., Cook, E. F., Flatley, M., Walker, C., and Goldman, L., "Case-Control Analysis of Risk Factors for the Presence of Aortic Stenosis in Adults (Age 50 Years or Older)," *American Journal of Cardiology,* Volume 55, March 1, 1985, 744–747.

[7] Vonesh, E. F., "Modelling Peritonitis Rates and Associated Risk Factors for Individuals on Continuous Ambulatory Peritoneal Dialysis," *Statistics in Medicine,* Volume 9, March 1990, 263–271.

21

Survival Analysis

In some studies, the response variable of interest is the amount of time from an initial observation until the occurrence of a subsequent event. Examples include the time from birth until death, the time from transplant surgery until the new organ fails, and the time from the start of maintenance therapy for a patient whose cancer has gone into remission until the relapse of disease. This time interval between a starting point and a subsequent event, often called a failure, is known as the *survival time*. Although the measurements of survival times are continuous, their distributions are rarely normal; instead, they tend to be skewed to the right. The analysis of this type of data generally focuses on estimating the probability that an individual will survive for a given length of time.

One common circumstance in working with survival data is that not all the individuals in a sample are observed until their respective times of failure. If the time interval between the starting point and the subsequent failure can be quite long, the data may be analyzed before this second event of interest has occurred in all patients. Other patients who either move away before the study is complete or refuse to participate any longer are said to be *lost to follow-up*. The incomplete observation of a time to failure is known as *censoring*; the presence of censored observations distinguishes the analysis of survival data from other types of analyses.

A distribution of survival times can be characterized by a *survival function*, represented by $S(t)$. $S(t)$ is defined as the probability that an individual survives beyond time t. Equivalently, for a given t, $S(t)$ specifies the proportion of individuals who have not yet failed. If T is a continuous random variable representing survival time, then

$$S(t) = P(T > t).$$

The graph of $S(t)$ versus t is called a *survival curve*.

Survival curves have been used for many years. For example, a study published in 1938 investigated the effects of tobacco on human longevity among white males over the age of 30 [1]. Three categories of individuals were considered: nonusers of tobacco, moderate smokers, and heavy smokers. The results of the study are presented in Figure 21.1. As is evident from the graph, the smoking of tobacco is associated with a

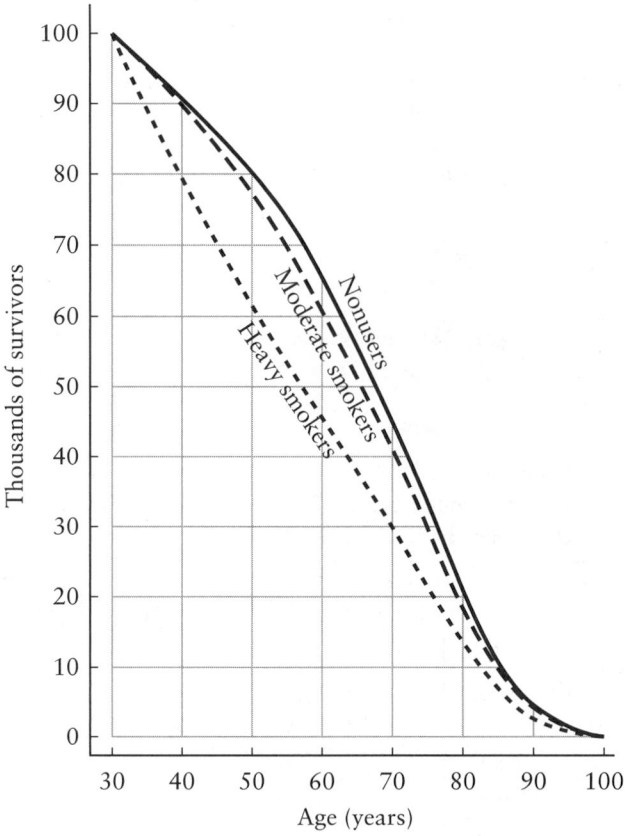

FIGURE 21.1
Survival curves for three categories of white males: nonusers of tobacco, moderate smokers, and heavy smokers, 1938

shorter duration of life; in addition, longevity is affected by the amount of tobacco used. It seems that these results were ignored not only at the time of publication, but in subsequent years as well.

21.1   The Life Table Method

The classical method of estimating a survival curve is known as the *life table method.* This technique parallels the study of life tables in Chapter 5. Whereas we previously examined descriptive statistics for an entire population, however, in this chapter we explore methods of statistical inference based on a sample of observations drawn from the population.

To begin, survival times are grouped within intervals of fixed length. The first three columns of the table enumerate the time interval t to $t + n$, the proportion of individuals alive at time t who fail prior to $t + n$, and the number of individuals still alive at time t. If l_0 is the number of individuals alive at time 0 and l_t is the number of these still alive at time t, the proportion of individuals who have not yet failed at time t can be calculated as

$$S(t) = \frac{l_t}{l_0}.$$

Table 21.1 reproduces a portion of the first three columns of the complete United States life table for 1979–1981; this table was provided at the end of Chapter 5 and rep-

TABLE 21.1
Complete United States life table for individuals less than 30 years of age, 1979–1981

t to $t + n$	$_nq_t$	l_t	$S(t)$
0–1	0.01260	100,000	1.0000
1–2	0.00093	98,740	0.9874
2–3	0.00065	98,648	0.9865
3–4	0.00050	98,584	0.9858
4–5	0.00040	98,535	0.9854
5–6	0.00037	98,495	0.9850
6–7	0.00033	98,459	0.9846
7–8	0.00030	98,426	0.9843
8–9	0.00027	98,396	0.9840
9–10	0.00023	98,370	0.9837
10–11	0.00020	98,347	0.9835
11–12	0.00019	98,328	0.9833
12–13	0.00025	98,309	0.9831
13–14	0.00037	98,285	0.9829
14–15	0.00053	98,248	0.9825
15–16	0.00069	98,196	0.9820
16–17	0.00083	98,129	0.9813
17–18	0.00095	98,047	0.9805
18–19	0.00105	97,953	0.9795
19–20	0.00112	97,851	0.9785
20–21	0.00120	97,741	0.9774
21–22	0.00127	97,623	0.9762
22–23	0.00132	97,499	0.9750
23–24	0.00134	97,370	0.9737
24–25	0.00133	97,240	0.9724
25–26	0.00132	97,110	0.9711
26–27	0.00131	96,982	0.9698
27–28	0.00130	96,856	0.9686
28–29	0.00130	96,730	0.9673
29–30	0.00131	96,604	0.9660

resents the entire United States population. The fourth column is the survival function at time t, where

$$S(t) = \frac{l_t}{100,000}.$$

The corresponding survival curve is plotted in Figure 21.2. Table 21.1 is an example of a *current,* or *cross-sectional, life table.* It is constructed from data gathered over a relatively short period of time; the persons represented in the different intervals are not the same group of individuals throughout. For Table 21.1, it is assumed that the entire population is represented and hence that there is no sampling variability in $S(t)$. However, the life table method can also be applied to a sample of individuals drawn from a population.

Ideally, we would prefer to work with a *longitudinal life table,* which tracks an actual cohort of individuals over their entire lifetimes. This method is not practical for large population studies; it would involve following a sizable group of individuals for over 100 years. However, it is often used in smaller clinical studies in which patients are enrolled sequentially and followed for shorter periods of time.

Consider the data presented in Table 21.2. A total of 12 hemophiliacs, all 40 years of age or younger at HIV seroconversion, were followed from the time of primary AIDS diagnosis between 1984 and 1989 until death [2]. In all cases, transmission of HIV had occurred through infected blood products. We would actually prefer that our starting point be the time at which an individual contracted AIDS rather than the time of diagnosis, but this information was not known. For most of the patients, treatment was not available. What are we able to infer about the survival of the population of hemophiliacs diagnosed in the mid- to late 1980s on the basis of this sample of 12 individuals?

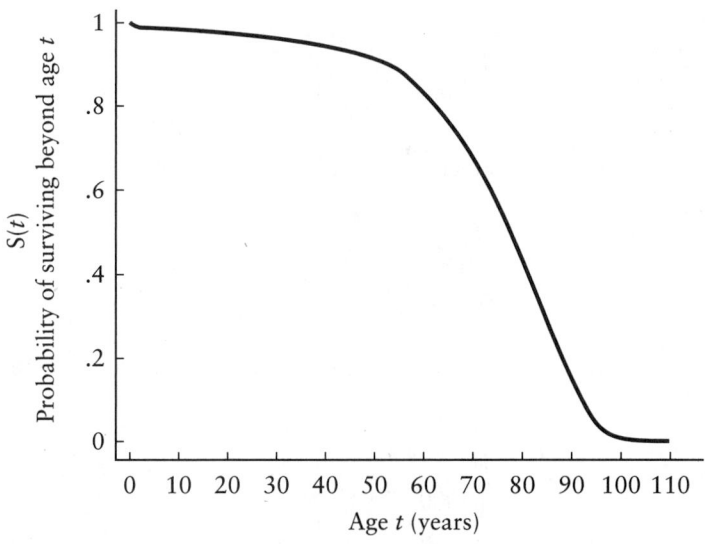

FIGURE 21.2
Survival curve for the United States population, 1979–1981

TABLE 21.2
Interval from primary AIDS diagnosis until death for a sample of 12 hemophiliac patients at most 40 years of age at HIV seroconversion

Patient Number	Survival (months)
1	2
2	3
3	6
4	6
5	7
6	10
7	15
8	15
9	16
10	27
11	30
12	32

Using the life table procedure described above, we could summarize the data for the 12 hemophiliacs as in Table 21.3. A survival time of t months means that an individual survived until time t and then died. Since 1 out of the 12 individuals in the initial cohort died at 2 months, the proportion of patients dying in the interval 2–3 months is estimated as

$$_1q_2 = \frac{1}{12}$$
$$= 0.0833.$$

One of the remaining 11 individuals died at 3 months; consequently,

$$_1q_3 = \frac{1}{11}$$
$$= 0.0909.$$

Similarly, 2 of the remaining 10 patients died at 6 months, and

$$_1q_6 = \frac{2}{10}$$
$$= 0.2000.$$

Recall that $_nq_t$ is also called the hazard function. In time intervals not containing a death, such as 0–1 and 1–2, the estimated hazard function is equal to 0.

TABLE 21.3

Life table method of estimating S(t) for hemophiliac patients at most 40 years of age at HIV seroconversion

t to $t + n$	$_n q_t$	l_t	$1 - _n q_t$	$\hat{S}(t)$
0–1	0.0000	12	1.0000	1.0000
1–2	0.0000	12	1.0000	1.0000
2–3	0.0833	12	0.9167	0.9167
3–4	0.0909	11	0.9091	0.8333
4–5	0.0000	10	1.0000	0.8333
5–6	0.0000	10	1.0000	0.8333
6–7	0.2000	10	0.8000	0.6667
7–8	0.1250	8	0.8750	0.5833
8–9	0.0000	7	1.0000	0.5833
9–10	0.0000	7	1.0000	0.5833
10–11	0.1429	7	0.8571	0.5000
11–12	0.0000	6	1.0000	0.5000
12–13	0.0000	6	1.0000	0.5000
13–14	0.0000	6	1.0000	0.5000
14–15	0.0000	6	1.0000	0.5000
15–16	0.3333	6	0.6667	0.3333
16–17	0.2500	4	0.7500	0.2500
17–18	0.0000	3	1.0000	0.2500
18–19	0.0000	3	1.0000	0.2500
19–20	0.0000	3	1.0000	0.2500
20–21	0.0000	3	1.0000	0.2500
21–22	0.0000	3	1.0000	0.2500
22–23	0.0000	3	1.0000	0.2500
23–24	0.0000	3	1.0000	0.2500
24–25	0.0000	3	1.0000	0.2500
25–26	0.0000	3	1.0000	0.2500
26–27	0.0000	3	1.0000	0.2500
27–28	0.3333	3	0.6667	0.1667
28–29	0.0000	2	1.0000	0.1667
29–30	0.0000	2	1.0000	0.1667
30–31	0.5000	2	0.5000	0.0833
31–32	0.0000	1	1.0000	0.0833
32–33	1.0000	1	0.0000	0.0000

The fourth column of Table 21.3 is something new; it contains the proportion of individuals who do not fail during a given interval. In the interval 2–3 months, for instance, the proportion of patients who died is $_1 q_2 = 0.0833$ and the proportion who survived is

$$1 - _1 q_2 = 1 - 0.0833$$
$$= 0.9167.$$

In time intervals not containing a death, the estimated proportion of patients who do not fail is equal to 1.

The proportions of individuals who do not fail in each interval can be used to estimate the survival function. Note that since no one in the sample died at time 0 months, the estimate of $\widehat{S}(0) = P(T > 0)$ is

$$\widehat{S}(0) = 1.$$

Subsequent values of $\widehat{S}(t)$ can be calculated using the multiplicative rule of probability. For example, let A be the event that an individual is alive during the interval 0–1 months, and B the event that he or she survives at time 1. Therefore, the event that the patient survives longer than 1 month can be represented by $A \cap B$. The multiplicative rule of probability states that

$$P(A \cap B) = P(A)\,P(B \mid A).$$

The probability that an individual survives longer than 1 month is $P(T > 1) = S(1)$, and the probability that he or she is alive during the interval 0–1 is $S(0)$. In addition, the probability that a patient survives at time 1 given that he or she was alive up until this point is $1 - {}_1 q_1$. As a result,

$$\begin{aligned} \widehat{S}(1) &= \widehat{S}(0)\,[1 - {}_1 q_1] \\ &= (1.0000)(1.0000) \\ &= 1.0000. \end{aligned}$$

Similarly, the probability that a patient survives longer than 2 months is the probability of being alive during the interval 1–2 months multiplied by the probability of not failing at time 2, given that he or she was alive until this point, or

$$\begin{aligned} \widehat{S}(2) &= \widehat{S}(1)\,[1 - {}_1 q_2] \\ &= (1.0000)(0.9167) \\ &= 0.9167. \end{aligned}$$

The probability of living longer than 3 months is estimated by

$$\begin{aligned} \widehat{S}(3) &= \widehat{S}(2)\,[1 - {}_1 q_3] \\ &= (0.9167)(0.9091) \\ &= 0.8333, \end{aligned}$$

and so on. After 32 months, all the patients in the sample have died; as a result, $\widehat{S}(32) = 0$. At this point, there are no individuals remaining who have not yet failed.

A survival curve can be approximated by plotting the survival function $\widehat{S}(t)$ generated using the life table method versus the point representing the start of each interval, and then connecting the points with straight lines. A survival curve for hemophiliacs at most 40 years of age at HIV seroconversion in the mid- to late 1980s is shown in Figure 21.3. We must keep in mind, however, that this curve was estimated based on a very small sample of 12 patients at the start of the epidemic. In general, current survival is much better.

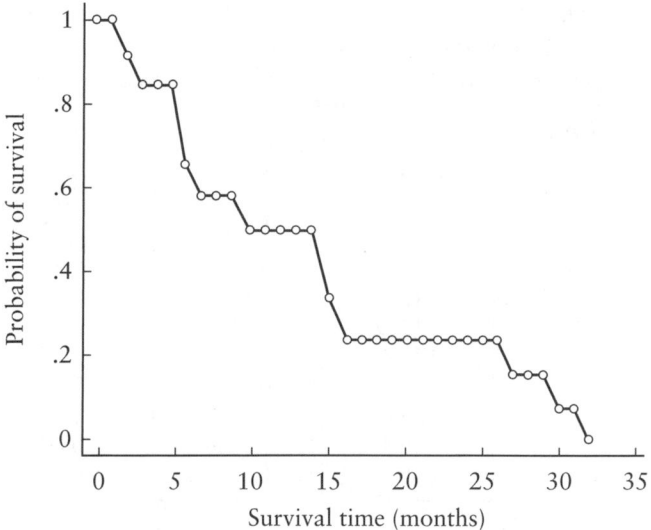

FIGURE 21.3
Survival curve for hemophiliac patients at most 40 years of age at
HIV seroconversion

21.2   The Product-Limit Method

When we use the life table method, the estimated survival function $\widehat{S}(t)$ changes only during the time intervals in which at least one death occurs. For smaller data sets, such as the sample of 12 hemophiliac patients diagnosed with AIDS, there can be many intervals without a single death. In these instances, it might not make sense to present the survival function in this way. The *product-limit method* of estimating a survival function, also called the *Kaplan-Meier method,* is a nonparametric technique that uses the exact survival time for each individual in a sample instead of grouping the times into intervals.

Table 21.4 displays the product-limit estimate of the survival function based on the sample of 12 hemophiliacs under the age of 40 at the time of HIV seroconversion. Instead of time intervals, the first column of the table contains the exact times at which failures occurred; patients died 2 months after diagnosis, 3 months after diagnosis, 6 months after diagnosis, and so on. The patient with the longest survival died 32 months after primary AIDS diagnosis. The second column of the table lists the proportions of patients alive just prior to each time t who fail at that time, and the third column the proportions of individuals who do not fail at t. Using the multiplicative rule of probability, the proportions of individuals who do not fail can be used to estimate the survival function; the technique is the same as it was for the life table method.

The survival curve corresponding to the function in Table 21.4 is plotted in Figure 21.4. When the product-limit method is used, $\widehat{S}(t)$ is assumed to remain the same over the time periods between deaths; it changes precisely when a subject fails.

Keep in mind that $\widehat{S}(t)$ was calculated using the data in a single sample of observations; if we were to select a second sample of 12 hemophiliacs and calculate a second

TABLE 21.4
Product-limit method of estimating S(t) for hemophiliac
patients at most 40 years of age at HIV seroconversion

Time	q_t	$1 - q_t$	$\hat{S}(t)$
0	0.0000	1.0000	1.0000
2	0.0833	0.9167	0.9167
3	0.0909	0.9091	0.8333
6	0.2000	0.8000	0.6667
7	0.1250	0.8750	0.5833
10	0.1429	0.8571	0.5000
15	0.3333	0.6667	0.3333
16	0.2500	0.7500	0.2500
27	0.3333	0.6667	0.1667
30	0.5000	0.5000	0.0833
32	1.0000	0.0000	0.0000

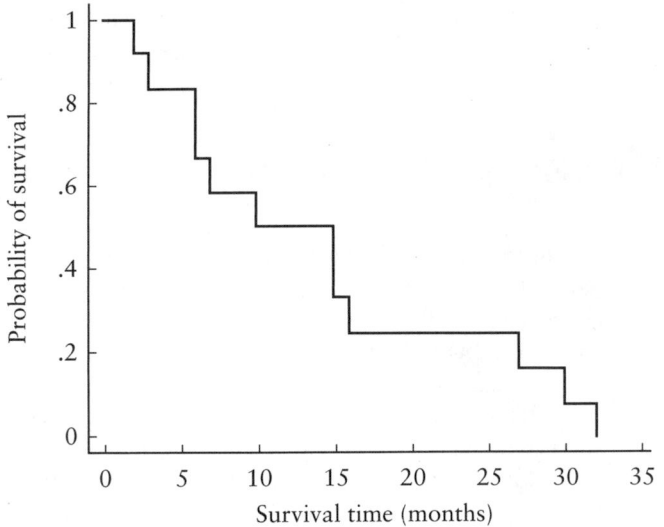

FIGURE 21.4
Survival curve for hemophiliac patients at most 40 years of age at
HIV seroconversion

survival function using the product-limit method, the results would differ from those in Figure 21.4. $\hat{S}(t)$ is merely an estimate of the true population survival function for all hemophiliacs diagnosed with AIDS in the mid- to late 1980s who were at most 40 years of age at HIV seroconversion. To quantify the sampling variability in this estimate, we can calculate the standard error of $\hat{S}(t)$ and construct a pair of confidence bands around the survival curve [3]. Figure 21.5 displays 95% confidence bands for the product-limit estimate.

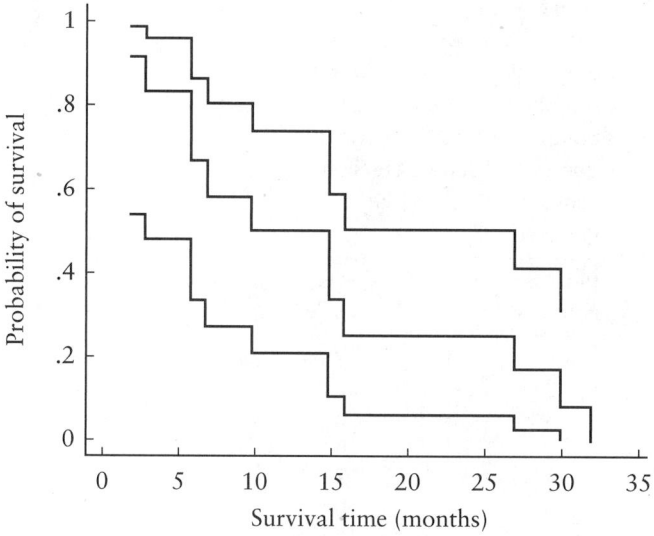

FIGURE 21.5
Survival curve for hemophiliac patients at most 40 years of age at HIV seroconversion, with 95% confidence bands

The product-limit method for estimating a survival curve can be adjusted to account for the partial information about survival times that is available from censored observations. Suppose that when the data for the 12 hemophiliac AIDS patients were analyzed, the individuals with the second and sixth longest survival times had not yet died. Instead, they were still alive after 3 and 10 months of follow-up, respectively. A censored observation is denoted by a plus ($+$) sign, as in Table 21.5. The corresponding product-limit estimate of the survival function is calculated in Table 21.6 and plotted in Figure 21.6; each small x on the graph denotes a censored survival time. Note that $\widehat{S}(t)$ does not change from its previous value if the observation at time t is censored; however, this observation is not used to calculate the probability of failure at any subsequent time point. At time 3, for instance, a patient is censored but no one dies. Therefore,

$$q_3 = \frac{0}{11}$$
$$= 0,$$

and $\widehat{S}(3) = \widehat{S}(2)$. At time 6, 1 individual out of 12 died at 2 months and another was censored at 3 months; only 10 individuals remain at risk, and, since 2 of these die,

$$q_6 = \frac{2}{10}$$
$$= 0.2000.$$

TABLE 21.5

Interval from primary AIDS diagnosis until death for a sample of 12 hemophiliac patients at most 40 years of age at HIV seroconversion, censored observations included

Patient Number	Survival (months)
1	2
2	3+
3	6
4	6
5	7
6	10+
7	15
8	15
9	16
10	27
11	30
12	32

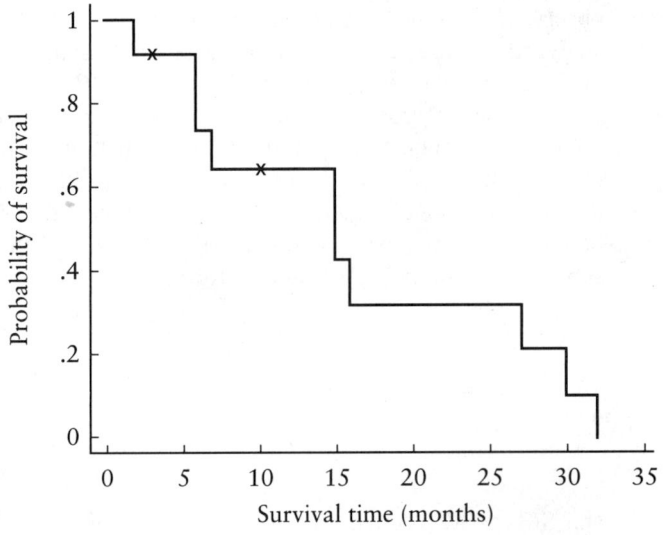

FIGURE 21.6

Survival curve for hemophiliac patients at most 40 years of age at HIV seroconversion, censored observations included

TABLE 21.6

Product-limit method of estimating S(*t*) for hemophiliac patients at most 40 years of age at HIV seroconversion, censored observations included

Time	q_t	$1 - q_t$	$\hat{S}(t)$
0	0.0000	1.0000	1.0000
2	0.0833	0.9167	0.9167
3	0.0000	1.0000	0.9167
6	0.2000	0.8000	0.7333
7	0.1250	0.8750	0.6417
10	0.0000	1.0000	0.6417
15	0.3333	0.6667	0.4278
16	0.2500	0.7500	0.3208
27	0.3333	0.6667	0.2139
30	0.5000	0.5000	0.1069
32	1.0000	0.0000	0.0000

21.3   The Log-Rank Test

Instead of simply characterizing the survival times for a single group of subjects, we often want to compare the distributions of survival times for two different populations. Our goal is to determine whether survival differs systematically between the groups. Recall the data for the 12 hemophiliacs—all 40 years of age or younger at the time of HIV seroconversion—that were presented in Table 21.2. We might wish to compare this distribution of survival times from primary AIDS diagnosis until death to the distribution of times for another group of hemophiliacs who were over 40 at seroconversion. Survival times for both groups are listed in Table 21.7, and the product-limit estimates of the survival curves are plotted in Figure 21.7. Survival for patients undergoing HIV seroconversion at an earlier age is represented by the upper curve in the figure, and survival for patients undergoing seroconversion at a later age is represented by the lower curve. At any point in time following AIDS diagnosis, the estimated probability of survival is higher for individuals who were younger at seroconversion. We would of course expect some sampling variability in these estimates; however, is the difference between the two curves greater than might be observed by chance alone?

If there are no censored observations in either group, the Wilcoxon rank sum test described in Chapter 13 could be used to compare median survival times. If censored data are present, however, other procedures must be used. One of a number of different methods available for testing the null hypothesis that two distributions of survival times are identical is a nonparametric technique known as the *log-rank test*. The idea behind the log-rank test is that we construct a 2 × 2 contingency table displaying group (in this example, age at seroconversion) versus survival status for each time *t* at which a death occurs. When *t* is equal to 1 month, for example, none of the 12 patients who were

TABLE 21.7
Interval from primary AIDS diagnosis until death for a sample of 21 hemophiliac patients, stratified by age at HIV seroconversion

Age ≤ 40 Years		Age > 40 Years	
Patient Number	**Survival (months)**	**Patient Number**	**Survival (months)**
1	2	1	1
2	3	2	1
3	6	3	1
4	6	4	1
5	7	5	2
6	10	6	3
7	15	7	3
8	15	8	9
9	16	9	22
10	27		
11	30		
12	32		

younger at seroconversion die, but 4 of the 9 older patients fail. Therefore, the appropriate 2×2 table for $t = 1$ is as follows:

	Failure		
Group	**Yes**	**No**	**Total**
Age ≤ 40	0	12	12
Age > 40	4	5	9
Total	4	17	21

Similarly, when t is equal to 2 months, 1 of the 12 younger patients and 1 of the 5 remaining older patients die; consequently, the 2×2 table for $t = 2$ is as follows:

	Failure		
Group	**Yes**	**No**	**Total**
Age ≤ 40	1	11	12
Age > 40	1	4	5
Total	2	15	17

Once the entire sequence of 2×2 tables has been generated, the information contained in the tables is accumulated using the Mantel-Haenszel test statistic from Chapter 16. This statistic compares the observed number of failures at each time to the expected number of failures given that the distributions of survival times for the two age groups

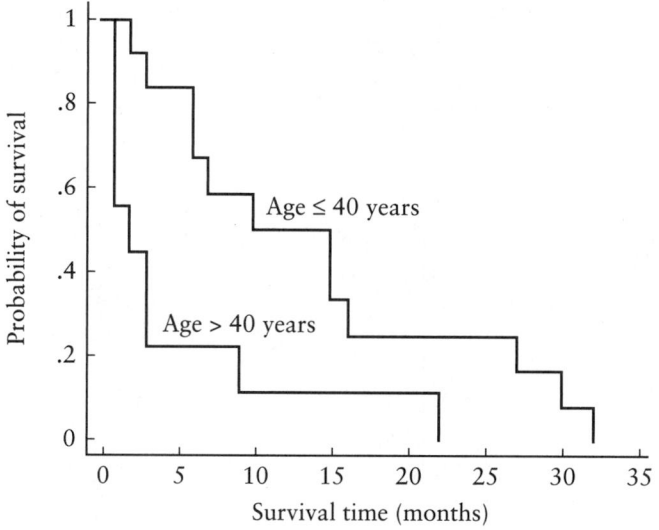

FIGURE 21.7

Survival curves for two groups of hemophiliac patients, stratified by age
at HIV seroconversion

are identical. If the null hypothesis is true, the test statistic has an approximate chi-square distribution with 1 degree of freedom.

For the two groups of hemophiliacs diagnosed with AIDS in the mid- to late 1980s, a test of the null hypothesis

$$H_0: S_{\leq 40}(t) = S_{>40}(t)$$

against the alternative hypothesis that the two survival functions are not equal results in a p-value of 0.025. This is the probability of finding a difference in survival as great as or greater than that observed, given that the null hypothesis is true. Since p is less than 0.05, we reject H_0 and conclude that individuals experiencing seroconversion at an earlier age lived longer after primary AIDS diagnosis than individuals undergoing seroconversion at a later age.

As another example, consider the data from a clinical trial comparing two different treatment regimens for moderate-risk breast cancer [4]. We wish to compare the distributions of survival times after diagnosis for women receiving treatment A versus women receiving treatment B, and determine whether either treatment prolongs survival relative to the other. Plots of the two product-limit survival curves, with time since breast cancer diagnosis on the horizontal axis, are displayed in Figure 21.8. There does not appear to be a difference in survival for patients in the two treatment groups; note the great deal of overlap in the curves. Furthermore, a log-rank test of the null hypothesis

$$H_0: S_A(t) = S_B(t)$$

results in a p-value of 0.88. We are unable to reject H_0 at the 0.05 level of significance; the data do not provide evidence of a difference in survival functions for the two treatments.

However, when the individuals enrolled in the clinical trial are separated into two distinct subpopulations—premenopausal women and postmenopausal women—treatment does appear to have a substantial effect on survival. As shown in Figure 21.9, treatment A improves survival for premenopausal women. The product-limit survival curve

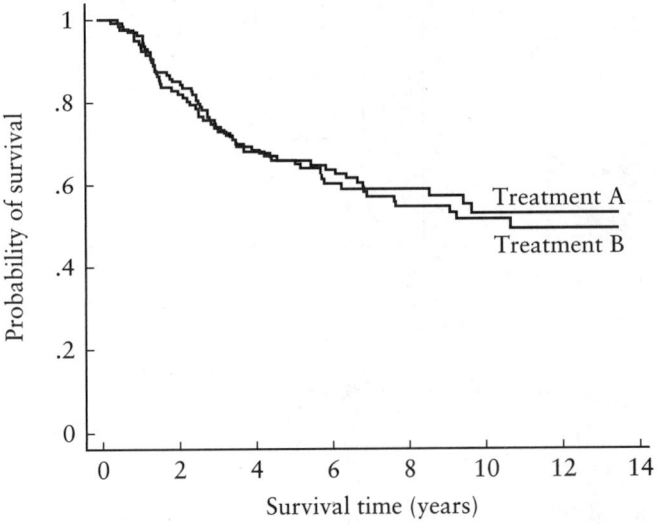

FIGURE 21.8
Survival curves for moderate-risk breast cancer patients in two treatment groups

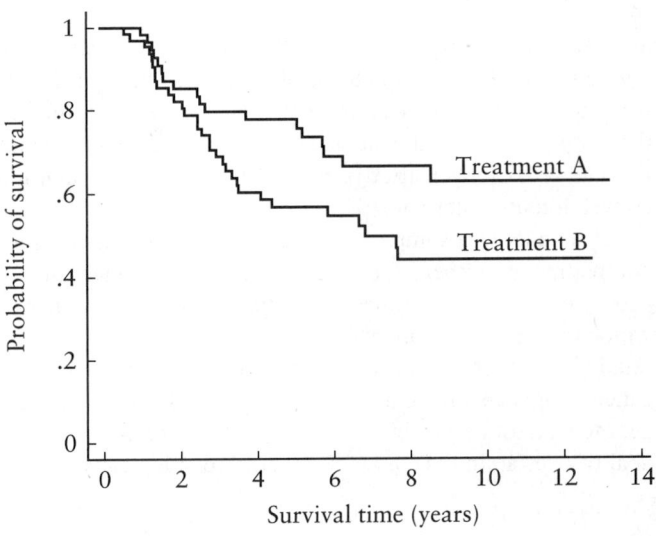

FIGURE 21.9
Survival curves for premenopausal, moderate-risk breast cancer patients in two treatment groups

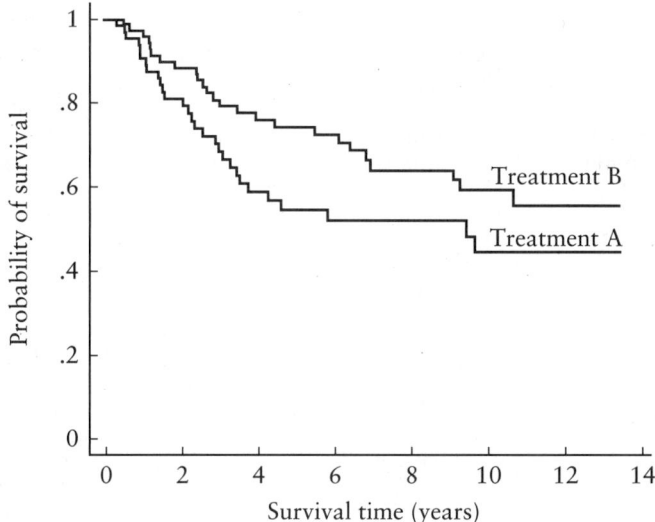

FIGURE 21.10

Survival curves for postmenopausal, moderate-risk breast cancer patients in two treatment groups

for women receiving treatment A lies above the survival curve for those receiving treatment B; at any point in time following diagnosis, the estimated probability of survival is higher for women receiving treatment A. The log-rank test yields a p-value of 0.052. In contrast, Figure 21.10 suggests that treatment B is more effective in prolonging survival for postmenopausal women; in this case, the survival curve for women receiving treatment B lies above the curve for those receiving treatment A. The p-value of this log-rank test is 0.086. Since the treatment effects are going in opposite directions in the two different subpopulations, they cancel each other out when the groups are combined. This serves as a reminder that care must be taken not to ignore important confounding variables.

21.4   *Further Applications*

Suppose that we are interested in studying patients with systemic cancer who subsequently develop a brain metastasis; our ultimate goal is to prolong their lives by controlling the disease. A sample of 23 such patients, all of whom were treated with radiotherapy, were followed from the first day of their treatment until recurrence of the original brain tumor. Recurrence is defined as the reappearance of a metastasis at exactly the same site, or, in the case of patients whose tumor never completely disappeared, enlargement of the original lesion. Times to recurrence for the 23 patients are presented in Table 21.8 [5]. What can we infer about the reappearance of brain metastases based on the information in this sample?

We could begin our analysis by summarizing the recurrence time data using the classic life table method. For intervals of length 2 weeks, we first determine the

TABLE 21.8

Time to recurrence of brain metastasis for a sample of 23 patients treated with radiotherapy

Patient Number	Recurrence (weeks)	Patient Number	Recurrence (weeks)
1	2	13	14
2	2	14	14
3	2	15	18
4	3	16	19
5	4	17	20
6	5	18	22
7	5	19	22
8	6	20	31
9	7	21	33
10	8	22	39
11	9	23	195
12	10		

proportion of patients who experienced a recurrence in each interval. The results are presented in the second column of Table 21.9. Since 4 out of 23 individuals had a recurrence of the original brain metastasis at least 2 weeks but not more than 4 weeks after the start of treatment, the proportion of individuals failing in the interval 2–4 is estimated as

$$_2q_2 = \frac{4}{23}$$
$$= 0.1739.$$

Similarly, 3 of the remaining 19 patients had a recurrence between 4 and 6 weeks after treatment; therefore,

$$_2q_4 = \frac{3}{19}$$
$$= 0.1579.$$

In time intervals not containing a failure, the estimated proportion of individuals experiencing a recurrence is equal to 0.

The fourth column of the life table contains the proportion of patients who did not fail during a given interval, or $1 - {}_nq_t$. These proportions of individuals who did not experience a recurrence can be used to estimate the survival function $S(t)$. Since none of the patients in the sample failed at time 0 months, the estimate of $S(0) = P(T > 0)$ is

$$\widehat{S}(0) = 1.$$

Subsequent values of $\widehat{S}(t)$ are calculated using the multiplicative rule of probability. The probability that an individual had not experienced a recurrence during the interval 0–2 is $S(0)$, and the probability that the patient did not fail in the interval 2–4 given that he

TABLE 21.9

Life table method of estimating S(t) for patients with brain metastasis treated with radiotherapy

t to $t + n$	$_nq_t$	l_t	$1 - {_nq_t}$	$\hat{S}(t)$
0–2	0.0000	23	1.0000	1.0000
2–4	0.1739	23	0.8261	0.8261
4–6	0.1579	19	0.8421	0.6957
6–8	0.1250	16	0.8750	0.6087
8–10	0.1429	14	0.8571	0.5217
10–12	0.0833	12	0.9167	0.4783
12–14	0.0000	11	1.0000	0.4783
14–16	0.1818	11	0.8182	0.3913
16–18	0.0000	9	1.0000	0.3913
18–20	0.2222	9	0.7778	0.3043
20–22	0.1429	7	0.8571	0.2609
22–24	0.3333	6	0.6667	0.1739
24–26	0.0000	4	1.0000	0.1739
26–28	0.0000	4	1.0000	0.1739
28–30	0.0000	4	1.0000	0.1739
30–32	0.2500	4	0.7500	0.1304
32–34	0.3333	3	0.6667	0.0870
34–36	0.0000	2	1.0000	0.0870
36–38	0.0000	2	1.0000	0.0870
38–40	0.5000	2	0.5000	0.0435
40+	1.0000	1	0.0000	0.0000

or she did not fail prior to this is $1 - {_2q_2}$. Therefore, the probability that an individual experiences a recurrence some time after 2 weeks is estimated by

$$\hat{S}(2) = \hat{S}(0)[1 - {_2q_2}]$$
$$= (1.0000)(0.8261)$$
$$= 0.8261.$$

Similarly, the probability that a patient fails after 4 weeks is estimated by

$$\hat{S}(4) = \hat{S}(2)[1 - {_2q_4}]$$
$$= (0.8261)(0.8421)$$
$$= 0.6957.$$

By the last interval in the table—the only one whose length is not 2 weeks—every patient in the study has experienced a recurrence of the original metastasis; consequently,

$$\hat{S}(40) = 0.$$

Since we are dealing with a relatively small group of patients, we might prefer to estimate the survival function using the product-limit method. The product-limit method of estimating S(t) is a nonparametric technique that uses the exact recurrence time for each individual instead of grouping the times into intervals. In this case, therefore, the three patients who experienced a tumor recurrence 2 weeks after their initial treatment would not be grouped with the individual who failed 3 weeks after the start of treatment.

Table 21.10 displays the product-limit estimate of S(t) for the sample of 23 patients treated for brain metastasis. The first column of the table contains the exact times at which the failures occurred, rather than time intervals. The second column lists the proportions of patients who had not failed prior to time *t* who experience a recurrence at that time, and the third column contains the proportions of individuals who do not fail at *t*. Using the multiplicative rule of probability, these proportions are used to estimate the survival function S(t); the technique is the same as it was for the life table. The corresponding survival curve is plotted in Figure 21.11.

The product-limit method for estimating a survival curve can be modified to take into account the partial information about recurrence times that is available from censored observations. In Table 21.11, censored survival times for the sample of 23 patients treated with radiotherapy are denoted by a plus (+) sign. These patients either died before experiencing a recurrence of their original brain metastasis or remained tumor-free at the end of the follow-up period. The product-limit estimate of the survival function is calculated in Table 21.12, and the corresponding survival curve is plotted in Figure 21.12. Note that if the longest survival time in a sample is censored, the curve does not drop down to the horizontal axis to indicate an estimated survival probability equal to 0. In addition, the survival function does not change if the observation at time *t* is

TABLE 21.10

Product-limit method of estimating S(t) for patients with brain metastasis treated with radiotherapy

Time	q_t	$1 - q_t$	$\hat{S}(t)$
0	0.0000	1.0000	1.0000
2	0.1304	0.8696	0.8696
3	0.0500	0.9500	0.8261
4	0.0526	0.9474	0.7826
5	0.1111	0.8889	0.6957
6	0.0625	0.9375	0.6522
7	0.0667	0.9333	0.6087
8	0.0714	0.9286	0.5652
9	0.0769	0.9231	0.5217
10	0.0833	0.9167	0.4783
14	0.1818	0.8182	0.3913
18	0.1111	0.8889	0.3478
19	0.1250	0.8750	0.3043
20	0.1429	0.8571	0.2609
22	0.3333	0.6667	0.1739
31	0.2500	0.7500	0.1304
33	0.3333	0.6667	0.0870
39	0.5000	0.5000	0.0435
195	1.0000	0.0000	0.0000

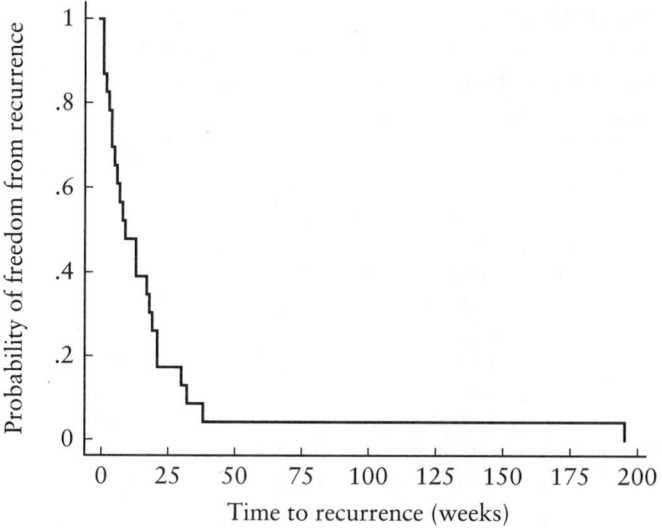

FIGURE 21.11
Survival curve for patients with brain metastasis treated with radiotherapy

TABLE 21.11
Time to recurrence of brain metastasis for a sample of 23 patients
treated with radiotherapy, censored observations included

Patient Number	Recurrence (weeks)	Patient Number	Recurrence (weeks)
1	2+	13	14
2	2+	14	14+
3	2+	15	18+
4	3	16	19+
5	4	17	20
6	5	18	22
7	5+	19	22+
8	6	20	31+
9	7	21	33
10	8	22	39
11	9+	23	195+
12	10		

censored. However, this observation is not used to calculate the probability of failure at any subsequent time point. At time 2 weeks, 3 patients are censored but no one experiences a tumor recurrence. Therefore,

$$q_2 = \frac{0}{23}$$
$$= 0.$$

TABLE 21.12
Product-limit method of estimating $S(t)$ for patients with brain metastasis treated with radiotherapy, censored observations included

Time	q_t	$1 - q_t$	$\hat{S}(t)$
0	0.0000	1.0000	1.0000
2	0.0000	1.0000	1.0000
3	0.0500	0.9500	0.9500
4	0.0526	0.9474	0.9000
5	0.0556	0.9444	0.8500
6	0.0625	0.9375	0.7969
7	0.0667	0.9333	0.7437
8	0.0714	0.9286	0.6906
9	0.0000	1.0000	0.6906
10	0.0833	0.9167	0.6331
14	0.0909	0.9091	0.5755
18	0.0000	1.0000	0.5755
19	0.0000	1.0000	0.5755
20	0.1429	0.8571	0.4933
22	0.1667	0.8333	0.4111
31	0.0000	1.0000	0.4111
33	0.3333	0.6667	0.2741
39	0.5000	0.5000	0.1370
195	0.0000	1.0000	0.1370

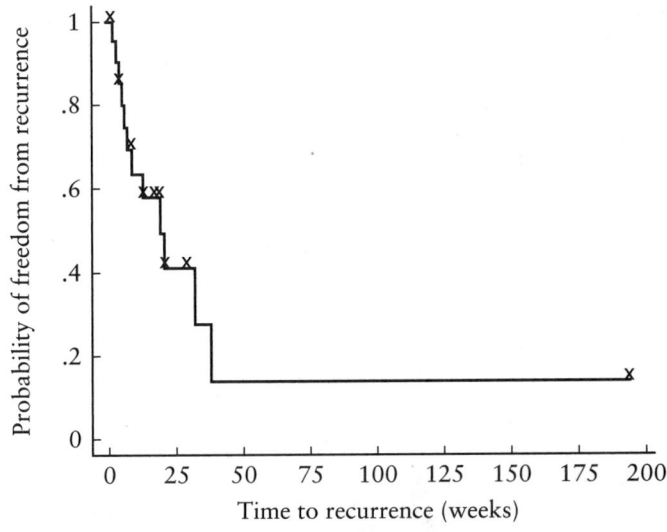

FIGURE 21.12
Survival curve for patients with brain metastasis treated with radiotherapy, censored observations included

At time 3, 1 of the remaining 20 patients dies, and

$$q_3 = \frac{1}{20}$$
$$= 0.0500.$$

Rather than work with survival times drawn from a single population, we often want to compare the distributions of times for two different groups. For example, we might wish to compare the times to recurrence of brain metastasis for patients treated with radiotherapy alone and for those undergoing surgical removal of the tumor and subsequent radiotherapy. Survival times for both groups are presented in Table 21.13; the corresponding product-limit survival curves are plotted in Figure 21.13. Based on the curves, it appears that individuals treated with both surgery and postoperative

TABLE 21.13

Time to recurrence of brain metastasis for a sample of 48 patients, stratified by treatment

Radiotherapy Alone		Surgery/Radiotherapy	
Patient Number	Recurrence (weeks)	Patient Number	Recurrence (weeks)
1	2+	1	2+
2	2+	2	2+
3	2+	3	6+
4	3	4	6+
5	4	5	6+
6	5	6	10+
7	5+	7	14+
8	6	8	21+
9	7	9	23
10	8	10	29+
11	9+	11	32+
12	10	12	34+
13	14	13	34+
14	14+	14	37
15	18+	15	37+
16	19+	16	42+
17	20	17	51
18	22	18	57
19	22+	19	59
20	31+	20	63+
21	33	21	66+
22	39	22	71+
23	195+	23	71+
		24	73+
		25	85+

radiotherapy have fewer recurrences of brain metastases and that the recurrences that do take place happen at a later time.

The log-rank test can be used to evaluate the null hypothesis that the distributions of recurrence times are identical in the two treatment groups. The test statistic calculated compares the observed number of recurrences at each time to the expected number given that H_0 is true. Although the calculations are somewhat complicated, they do not present a problem as long as a computer is available. The appropriate output from Stata is presented in Table 21.14. For each treatment group—where Group 1 contains patients treated with radiotherapy alone and Group 2 contains those treated with both surgery and radiotherapy—the table contains the observed and expected numbers of events, or, in this case, recurrences of brain tumor. It also displays the test statistic and its corresponding *p*-value. Since $p = 0.0001$, we reject the null hypothesis and conclude that surgical removal of the brain metastasis followed by radiotherapy results in a longer time to recurrence of the original tumor than radiotherapy alone.

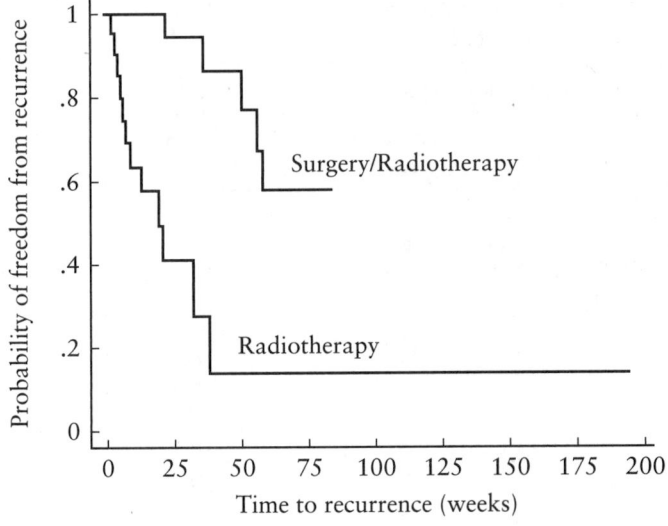

FIGURE 21.13

Survival curves for patients with brain metastasis, stratified by treatment

TABLE 21.14

Stata output displaying the log-rank test

Log-rank test for equality of survivor functions		
	Events	
Group	observed	expected
1	12	4.90
2	5	12.10
Total	17	17.00
chi2(1) =		14.44
Pr>chi2 =		0.0001

21.5 Review Exercises

1. What is a survival function?

2. What are censored observations? How do these observations occur?

3. Describe the difference between a cross-sectional life table and a longitudinal one.

4. How does the life table method of estimating a survival curve differ from the product-limit method?

5. Could the two-sample t-test be used to compare survival times in two different groups of individuals? Why or why not?

6. Suppose that you are interested in examining the survival times of individuals who receive bone marrow transplants for non-neoplastic disease [6]. Presented below are the survival times in months for 8 such patients. Assume that none of the observations is censored.

   | 3.0 | 4.5 | 5.0 | 10.0 | 15.5 | 18.5 | 25.0 | 34.0 |

   (a) What is the median survival time for these patients?
   (b) For fixed intervals of length 2 weeks, use the life table method to estimate the survival function $S(t)$.
   (c) Is the life table cross-sectional or longitudinal?
   (d) Construct a survival curve based on the life table estimate of $S(t)$.
   (e) Use the product-limit method to estimate the survival function.
   (f) Construct a survival curve based on the product-limit estimate.

7. Displayed below are the survival times in months since diagnosis for 10 AIDS patients suffering from concomitant esophageal candidiasis, an infection due to candida yeast, and cytomegalovirus, a herpes infection that can cause serious illness [7]. Censored observations are denoted by a plus (+) sign.

Patient Number	Survival (months)
1	0.5+
2	1
3	1
4	1
5	2
6	5+
7	8+
8	9
9	10+
10	12+

   (a) How many deaths were observed in this sample of patients?
   (b) Use the product-limit method to estimate the survival function $S(t)$.
   (c) What is $\widehat{S}(1)$, the estimated probability of survival at 1 month? What is the estimated probability of survival at 5 months? At 6 months?
   (d) Construct a survival curve based on the product-limit estimate.

8. In the 1980s, a study was conducted to examine the effects of the drug ganciclovir on AIDS patients suffering from disseminated cytomegalovirus infection. Two groups of patients were followed; 18 were treated with the drug, and 11 were not. The results of this study are contained in the data set `cyto` (Appendix B, Table B.29) [8]. Survival times in months after diagnosis are saved under the variable name `time`, and indicators of censoring status—where 0 designates that an observation was censored and 1 that a death occurred—under the name `censor`. Values of treatment group, where 1 indicates that a patient took the drug and 0 that he or she did not, are saved under the name `group`.

   (a) How many deaths occurred in each treatment group?

   (b) Use the product-limit method to estimate the survival function for each treatment group.

   (c) Construct survival curves for the two treatment groups based on the product-limit estimate of S(t).

   (d) Does it appear that the individuals in one group survive longer than those in the other group?

   (e) Use the log-rank test to evaluate the null hypothesis that the distributions of survival times are identical in the two groups. What do you conclude?

9. In a study of bladder cancer, tumors were removed from the bladders of 86 patients. Subsequently, the individuals were assigned to be treated either with a placebo or with the drug thiopeta. Time to the first recurrence of tumor in months is saved under the variable name `time` in the data set `bladder` (Appendix B, Table B.30) [9]. Treatment status is saved under the name `group`; the value 1 represents placebo. Indicators of censoring status—where 1 designates that a tumor did recur and 0 that it did not and that the observation was censored—are saved under the name `censor`.

   (a) Use the product-limit method to estimate the survival function in each treatment group.

   (b) Construct survival curves based on the product-limit estimates.

   (c) Does it appear that the individuals in one group have a longer time to first recurrence of tumor than those in the other group?

   (d) Test the null hypothesis that the distributions of recurrence times are identical in the two treatment groups. What do you conclude?

   (e) The variable `number` is an indicator of the number of tumors initially removed from the bladder; 1 indicates that a patient had a single tumor, and 2 that the individual had two or more tumors. For patients treated with the placebo, test the null hypothesis that the distributions of recurrence times are identical for individuals who had one tumor and for those who had two or more tumors. What do you conclude?

Bibliography

[1] Pearl, R., "Tobacco Smoking and Longevity," *Science,* Volume 87, March 4, 1938, 216–217.

[2] Ragni, M. V., and Kingsley, L. A., "Cumulative Risk for AIDS and Other HIV Outcomes in a Cohort of Hemophiliacs in Western Pennsylvania," *Journal of Acquired Immune Deficiency Syndromes,* Volume 3, July 1990, 708–713.

[3] Brown, B. W., "Estimation in Survival Analysis: Parametric Models, Product-Limit and Life Table Methods," *Statistics in Medical Research,* New York: Wiley, 1982.

[4] Shapiro, C. L., Henderson, I. C., Gelman, R. S., Harris, J. R., Canellos, G. P., and Frei, E., "A Randomized Trial of Cyclophosphamide, Methotrexate, and Fluorouracil Versus Methotrexate, Fluorouracil Adjuvant Chemotherapy in Moderate Risk Breast Cancer Patients," *Proceedings of the American Association for Cancer Research,* Volume 31, March 1990, 185.

[5] Patchell, R. A., Tibbs, P. A., Walsh, J. W., Dempsey, R. J., Maruyama, Y., Kryscio, R. J., Markesbery, W. R., MacDonald, J. S., and Young, B., "A Randomized Trial of Surgery in the Treatment of Single Metastases to the Brain," *The New England Journal of Medicine,* Volume 322, February 22, 1990, 494–500.

[6] Ash, R. C., Casper, J. T., Chitambar, C. R., Hansen, R., Bunin, N., Truitt, R. L., Lawton, C., Murray, K., Hunter, J., Baxter-Lowe, L. A., Gottschall, J. L., Oldham, K., Anderson, T., Camitta, B., and Menitove, J., "Successful Allogeneic Transplantation of T-Cell Depleted Bone Marrow from Closely HLA-Matched Unrelated Donors," *The New England Journal of Medicine,* Volume 322, February 22, 1990, 485–494.

[7] Laine, L., Bonacini, M., Sattler, F., Young, T., and Sherrod, A., "Cytomegalovirus and *Candida* Esophagitis in Patients with AIDS," *Journal of Acquired Immune Deficiency Syndromes,* Volume 5, June 1992, 605–609.

[8] Kotler, D. P., "Cytomegalovirus Colitis and Wasting," *Journal of Acquired Immune Deficiency Syndromes,* Volume 4, Supplement 1, 1991, S36–S41.

[9] Wei, L. J., Lin, D. Y., and Weissfeld, L., "Regression Analysis of Multivariate Incomplete Failure Time Data by Modeling Marginal Distributions," *Journal of the American Statistical Association,* Volume 84, December 1989, 1065–1073.

22

Sampling Theory

When studying inference, we learned that one of the fundamental goals of statistics is to describe some characteristic of a population using the information contained in a sample of observations. In previous chapters in which we were attempting to estimate a mean, the underlying population—such as the serum cholesterol levels of all adult males in the United States—was assumed to be infinite with mean μ and standard deviation σ. From this population, a random sample of size n was selected. The central limit theorem told us that the distribution of the mean of the sample values was approximately normal with mean μ and standard deviation $\sigma/\sqrt{n}$. It was critical that the sample was representative of the population so that the conclusions drawn were valid. This chapter provides further detail on some of the important issues regarding sampling theory.

22.1 Sampling Schemes

Suppose that, instead of being infinite, our underlying population is finite and consists of a total of N subjects or elements. If N is large, it may still not be feasible to evaluate all elements of the population. Therefore, we would again like to make inference about a specified population characteristic using the information contained in a sample of the subjects.

The individual elements in the population of interest are called *study units,* or *sampling units*; a study unit may be a person, a family, a city, an object, or anything else that is the unit of analysis in a population. For example, suppose that we wish to determine the average amount of alcohol consumed each week by 15- to 17-year-olds living in the state of Massachusetts. In this case, the study units would be teenagers between the ages of 15 and 17 residing in Massachusetts at a particular time.

The ideal population we would like to describe is known as the *target population.* In the preceding example, the target population consists of all 15- to 17-year-olds living in Massachusetts. In many situations, the entire target population is not accessible. If we are using school records to select our sample of teenagers, for instance, individuals who do not attend high school would have no chance of being included. After we account for practical constraints, the group from which we can actually sample is known as the *study population.* A list of the elements in the study population is called a *sampling frame.* Note that a random sample, although representative of the study population from which it is selected, may not be representative of the target population. If the two groups differ in some important way—perhaps the study population is younger on average than the target population—the selected sample is said to be biased. *Selection bias* is a systematic tendency to exclude certain members of the target population.

22.1.1 *Simple Random Sampling*

The most elementary type of sample that can be drawn from the study population is a *simple random sample.* In simple random sampling, units are independently selected one at a time until the desired sample size is achieved. Since a given unit may be chosen only once, this strategy is an example of *sampling without replacement.* Each study unit in the finite population has an equal chance of being included in the sample; the probability that a particular unit is chosen is n/N, where n is the size of the sample and N is the size of the underlying population. The quantity n/N is the *sampling fraction* of the population.

When the underlying population of size N has mean μ and standard deviation σ, a finite version of the central limit theorem states that the distribution of the sample mean $\bar{x}$ has mean μ and standard deviation $\sqrt{1 - (n/N)}(\sigma/\sqrt{n})$. Note that the standard deviation of the sample mean for a finite population differs from that for an infinite population by a factor of $\sqrt{1 - (n/N)}$. The square of this quantity, or $1 - (n/N)$, is called the *finite population correction factor.* If n has some fixed value and N is very large, n/N is close to 0. In this case, the finite population correction factor is approximately 1, and we return to the familiar situation in which the standard deviation of the sample mean is $\sigma/\sqrt{n}$. If the whole population is included in the sample, then n/N is equal to 1, and the standard deviation is 0. When the entire population is evaluated, there is no sampling variability in the mean.

One way to choose a simple random sample is to list and number each study unit, mix them up thoroughly, and then select units from this sampling frame until the required sample size is achieved. Another way is to use a computer or a table of random numbers to identify the units to be included in the sample. In either case, each unit should have an equal chance of being chosen. In this way, the possibility of bias is greatly reduced. For example, to determine the average amount of alcohol consumed each week by 15- to 17-year-olds in Massachusetts, it would not be feasible to interview every such teenager. Instead, a numbered list of these individuals could be compiled and one of the preceding methods used to select a sample of size n. The required information would then be obtained from the members of this group only, rather than from the entire population.

22.1.2 *Systematic Sampling*

If a complete list of the N elements in a population is available, *systematic sampling* can be used. Systematic sampling is similar to simple random sampling, but is easier to apply in practice. If a sample of size n is desired, the sampling fraction for the population is n/N. This is equivalent to a sampling fraction of $1/(N/n)$, or 1 in N/n. Therefore, the initial sample unit is randomly selected from the first k units on the list, where k is equal to N/n. If N/n is not an integer, its value is rounded to the nearest whole number. Next, as we move down the list, every kth consecutive unit is chosen. For example, to select a sample of n 15- to 17-year-olds living in Massachusetts, we would first randomly select a number between 1 and $k = N/n$. Suppose that we choose 4. We would then obtain information about the 4th person on the list, as well as persons $4 + k$, $4 + 2k$, $4 + 3k$, and so on.

Ideally, a sampling frame should be a complete list of all members of the target population. In reality, however, this is rarely the case. In some situations, it is impossible to devise a sampling frame for the population we wish to study. Suppose that we are interested in the individuals who will use a particular health care service over the next year. Even when a sampling frame is not available, systematic random sampling may often be applied. We still wish to sample a fraction of 1 in N/n. If the population size N is unknown, it must be estimated. In this case, the initial study unit is randomly selected from the first k units that become available. After this, each kth consecutive unit is chosen. Therefore, the sampling frame is compiled as the study progresses.

Unlike simple random sampling, systematic sampling requires the selection of only a single random number. Also, it distributes the sample evenly over the entire population list. Bias may arise if there is some type of periodic or cyclic sequence in the list; such patterns are rare, however. If we can assume that the list is randomly ordered, we can treat the resulting sample as a simple random sample.

22.1.3 Stratified Sampling

Simple random sampling does not take into consideration any information that is known about the elements of a population and that might affect the characteristic of interest. Under this sampling scheme, it is also possible that a particular subgroup of the population would not be represented in a given sample purely by chance. When selecting teenagers in Massachusetts, for instance, we might not choose any 17-year-old males. This could occur merely as the result of sampling variation. If we feel that it is important to include 17-year-old males, we can avoid this problem by selecting a *stratified random sample*. Our strata might consist of various combinations of gender and age: 15-year-old males, 15-year-old females, 16-year-old males, 16-year-old females, 17-year-old males, and 17-year-old females. Using this method, we divide the population into H distinct subgroups, or strata, such that the hth stratum has size N_h. A separate simple random sample of size n_h is then selected from each stratum, resulting in a sampling fraction of n_h/N_h for the hth subgroup. This method ensures that each stratum is represented in the overall sample. It does not require that all study units have an equal chance of being selected, however. Certain small subgroups of a population might be oversampled to provide sufficient numbers for more in-depth analysis.

When stratified sampling is used, the population mean is estimated as a weighted average of the stratum-specific sample means. In addition, the variance is calculated as a weighted average of the within-strata variances. The variances between study units in different subgroups do not make a contribution. Therefore, we can make the overall variance—or standard deviation—smaller by choosing subgroups so that the study units within a particular stratum are as homogeneous as possible, decreasing the within-strata variances, whereas the units in distinct strata are as different as possible, increasing the between-strata variances. If the stratum-specific sample sizes are chosen properly, the estimated mean of a stratified sample has a smaller variance and is consequently more precise than the mean of a simple random sample.

22.1.4 Cluster Sampling

If study units form natural groups or if an adequate list of the entire population is diffi-cult to compile, *cluster sampling* may be considered. This involves selecting a random sample of groups or clusters and then looking at all study units within the chosen groups. In *two-stage sampling,* a random sample of clusters is selected and then, within each group, a random sample of study units. If our clusters were schools in Massachu-setts, for instance, we could begin by selecting a sample of schools followed by a sam-ple of 15- to 17-year-olds within each of the schools. Cluster sampling is usually more economical than other types of sampling; it saves both time and money. For samples of the same size, however, it does produce a larger variance than simple random sampling.

22.1.5 Nonprobability Sampling

All the sampling strategies described result in *probability samples.* Because the proba-bility of being included in the sample is known for each subject in a population, valid and reliable inferences can be made. This cannot be said of *nonprobability samples,* in which the probability that an individual subject is included is not known. Examples of nonprobability samples are convenience samples and samples made up of volunteers, such as individuals autopsied and blood donors. These types of samples are prone to bias and cannot be assumed to be representative of any target population.

In general, the choice of a sampling strategy depends on a number of factors, in-cluding the objectives of the study and the available resources [1]. The costs and bene-fits of the various methods should be weighed carefully. In practice, an investigator often combines two or more different strategies.

22.2 Sources of Bias

No matter what the sampling scheme, when we are choosing a sample, selection bias is not the only potential source of error. A second source of bias is *nonresponse.* In situa-tions where the units of study are persons, there are typically individuals who cannot be reached, or who cannot or will not provide the information requested. Bias is present if these nonrespondents differ systematically from the individuals who do respond.

Consider the results of the following study, in which a sample of 5574 psychia-trists practicing in the United States were surveyed. A total of 1442, or only 26%, re-turned the questionnaire [2]. Of the 1057 males responding to the question, 7.1% admitted to having sexual contact with one or more patients. Of the 257 females, 3.1% confessed to this practice. How might nonresponse affect these estimates?

Given that the Hippocratic oath expressly forbids sexual contact between physicians and their patients, it seems unlikely that there were claims of sexual contact that did not actually occur. It also seems probable that there were psychiatrists who did in fact have sexual contact with a patient and subsequently refused to return the survey. Therefore, it is likely that the percentages calculated from the survey data underestimate the true population proportions; by how much, we do not know.

A further potential source of bias in a sample survey results from the fact that a respondent may choose to lie rather than reveal something that is sensitive or incriminating; this might be true of the psychiatrists mentioned above, for example. Another situation in which an individual might not tell the truth is in a study investigating substance abuse patterns during pregnancy. In some states, a woman who confesses to using cocaine during her pregnancy runs the risk of having her child taken away from her; this risk might provide sufficient incentive to lie.

In other circumstances, a person may lie even if the consequences are not as dire. For decades, public opinion polls have consistently reported that 40% of Americans attend a worship service at a church or synagogue at least once a week. This percentage is far higher than in most other Western nations. A study conducted in 1993 checked the attendance figures at religious services in a selected county in Ohio as well as in a number of other churches around the country and found that true attendance was closer to 20% than to 40% [3]. Follow-up studies have suggested that it is often the most committed members of a religious group who exaggerate their involvement; even if they did not attend a service during the week in question, they feel they can answer in the affirmative because they usually do attend.

One way to minimize the problem of lying in sample surveys is to apply the technique of *randomized response*. By introducing an extra degree of uncertainty into the data, we can mask the responses of specific individuals while still making inference about the population as a whole. If it works, randomized response reduces the motivation to lie.

For example, suppose that the quantity we wish to estimate is the population prevalence of some characteristic, represented by π. An example might be the proportion of psychiatrists having sexual contact with one or more patients. A random sample of individuals from the population are questioned as to whether they possess this characteristic or not. Rather than being told to answer the question in a straightforward manner, a certain anonymous proportion of the respondents—represented by a, where $0 < a < 1$—is instructed to answer "yes" under all circumstances. The remaining individuals are asked to tell the truth. Therefore, in a sample of size n, approximately na persons will always give an affirmative answer; the other $n(1 - a)$ will reply truthfully. Of these $n(1 - a)$, $n(1 - a)\pi$ will say "yes," and $n(1 - a)(1 - \pi)$ will say "no." If n^* is the total number of respondents answering "yes," then, on average,

$$n^* = na + n(1 - a)\pi.$$

If we subtract na from each side of this equation and divide by $n(1 - a)$, the population prevalence π may be estimated as

$$\hat{\pi} = \frac{(n^*/n) - a}{1 - a}.$$

As an example, a study conducted in New York City compared telephone responses obtained by direct questioning to those obtained through the use of randomized response. The study investigated the illicit use of four different drugs: cocaine, heroin, PCP, and LSD. Each individual questioned was asked to have three coins available; he or she was to toss the coins before being asked a question and to respond according to the outcome of the toss. The rules were a little more complicated than those described above. If all three coins were heads, the respondent was instructed to answer in the affirmative; if all three were tails, he or she had to answer "no." If the coins were a mixture of both heads and tails, the respondent was asked to tell the truth. Therefore, the proportion of individuals always replying "yes" was

$$a_1 = \frac{1}{2} \times \frac{1}{2} \times \frac{1}{2}$$
$$= \frac{1}{8}.$$

Similarly, the proportion always providing a negative response was

$$a_2 = \frac{1}{2} \times \frac{1}{2} \times \frac{1}{2}$$
$$= \frac{1}{8}.$$

The remaining

$$1 - \frac{1}{8} - \frac{1}{8} = \frac{6}{8}$$
$$= \frac{3}{4}$$

were instructed to tell the truth. In a sample of size n, approximately $(3/4)n\pi$ of these would reply "yes," and $(3/4)n(1 - \pi)$ would reply "no." If n^* is the total number of individuals answering "yes," then

$$n^* = \frac{1}{8}n + \frac{3}{4}n\pi.$$

Consequently,

$$\hat{\pi} = \frac{8n^* - n}{6n}$$

would be the estimate of the proportion using a particular drug. For three of the four drugs in question, the proportions of individuals acknowledging use were higher when the answers were obtained by means of randomized response than they were when direct questioning was used; for cocaine the percentage increased from 11% to 21%, and for heroin it went from 3% to 10% [4]. This suggests that some individuals were not being completely truthful when they were questioned directly.

The primary advantage of randomized response is that it reduces the proportion of individuals providing untruthful answers. Although it is impossible to identify individual responses, aggregate information can still be obtained. However, since this technique introduces an extra source of uncertainty into the analysis, the estimator $\hat{\pi}$ has a larger variance than it would in the situation in which no masking device is used and everyone is assumed to answer the question honestly. What we lose in precision, however, we gain in accuracy.

22.3 *Further Applications*

In previous chapters, we studied a sample of low birth weight infants born in two teaching hospitals in Boston, Massachusetts. To illustrate some of the practical issues of sampling, we now treat these children as if they constituted a finite population of size $N = 100$. We might wish to describe a particular characteristic of this population—their average gestational age, perhaps. The 100 measures of gestational age for the infants are displayed in Table 22.1 [5]; the true mean for the population is $\mu = 28.9$ weeks. Suppose that we do not know this and we also do not have the resources to obtain the necessary information for each child. Instead, we must estimate the mean by obtaining information about a representative fraction of the newborns. How would we proceed?

To select a simple random sample of size n, we choose study units independently from a list of the population elements—known as the sampling frame—until we achieve the desired sample size. Suppose that we wish to draw a sample of size $n = 10$. One way to go about this would be to write the integers from 1 to 100 on slips of paper. After mixing them up thoroughly, we would select 10 different numbers. If we were working with a very large population, this method would be impractical; instead, we could use a computer to generate the random numbers. In either case, each study unit has an equal chance of being selected. The probability that a particular unit is chosen is

$$\frac{n}{N} = \frac{10}{100}$$
$$= 0.10.$$

The ratio n/N is the sampling fraction of the population.

Suppose that we follow this procedure for drawing a simple random sample and select the following set of numbers:

93 11 28 6 90 51 10 22 36 48

Returning to the population of low birth weight infants, we determine the gestational age of each of the newborns chosen; the appropriate values are marked in Table 22.1 and are listed below.

32 26 28 25 24 23 29 30 27 28

TABLE 22.1
Measures of gestational age for a population of 100 low birth weight infants

ID	Age	ID	Age	ID	Age	ID	Age
1	29	26	28	[51]	23	76	31
2	31	27	29	52	27	77	30
3	33	[28]	28	53	28	78	27
4	31	29	29	54	27	79	25
5	30	30	30	55	27	80	25
[6]	25	31	31	56	26	81	26
7	27	32	30	57	25	82	29
8	29	33	31	58	23	83	29
9	28	34	29	59	26	84	34
[10]	29	35	27	60	24	85	30
[11]	26	[36]	27	61	29	86	29
12	30	37	27	62	29	87	33
13	29	38	32	63	27	88	30
14	29	39	31	64	30	89	29
15	29	40	28	65	30	[90]	24
16	29	41	30	66	32	91	33
17	29	42	29	67	33	92	25
18	33	43	28	68	27	[93]	32
19	33	44	31	69	31	94	31
20	29	45	27	70	26	95	31
21	28	46	25	71	27	96	31
[22]	30	47	30	72	27	97	29
23	27	[48]	28	73	35	98	32
24	33	49	28	74	28	99	33
25	32	50	25	75	30	100	28

Note that the observations selected in a simple random sample need not be distributed evenly over the entire sampling frame. Using this random sample, we would estimate the population mean as

$$\bar{x} = \frac{32 + 26 + 28 + 25 + 24 + 23 + 29 + 30 + 27 + 28}{10}$$

$$= 27.2 \text{ weeks.}$$

This value is a little smaller than the true population mean of 28.9 weeks.

As an alternative to the procedure described above, we might prefer to apply the technique of systematic sampling. When a complete list of the N elements in a population is available, systematic sampling is easier to carry out; it requires the selection of only a single random number. As noted above, the desired sampling fraction for the population of low birth weight infants is 0.10, or 1 in 10. Therefore, we would begin by randomly selecting the initial study unit from the first 10 units on the list.

Suppose that we write the integers from 1 to 10 on slips of paper and randomly choose the number 5. In addition to identifying the gestational age of the 5th infant on the list, we would determine the gestational age of every 10th consecutive child—the 15th, the 25th, and so on. The appropriate ages are displayed below.

ID	Age
5	30
15	29
25	32
35	27
45	27
55	27
65	30
75	30
85	30
95	31

These observations are evenly distributed over the sampling frame. As long as the population list is randomly ordered—and we have no reason to believe that it is not—a systematic sample can be treated as a simple random sample. In this case, therefore, we would estimate the population mean as

$$\bar{x} = \frac{30 + 29 + 32 + 27 + 27 + 27 + 30 + 30 + 30 + 31}{10}$$

$$= 29.3 \text{ weeks.}$$

This time, our estimate is slightly larger than the true population mean.

If we feel it is important to include representative numbers of male and female infants in our sample—we might think that gender is associated with gestational age—we could select a stratified random sample. To do this, we must first divide the population of low birth weight infants into two distinct subgroups of 44 boys and 56 girls. The 100 values of gestational age for the population, sorted by gender, are displayed in Table 22.2. Even though we are working with two separate subpopulations, we would still like to have an overall sampling fraction of 1/10. Therefore, we should select a simple random sample of size

$$44 \times \frac{1}{10} = 4.4$$

$$\approx 4$$

from the group of males and a sample of size

$$56 \times \frac{1}{10} = 5.6$$

$$\approx 6$$

from the group of females.

TABLE 22.2

Measures of gestational age for a population of 100 low birth weight infants, stratified by gender

Males				Females			
ID	**Age**	**ID**	**Age**	**ID**	**Age**	**ID**	**Age**
1	29	72	27	3	33	49	28
2	31	75	30	4	31	50	25
6	25	76	31	5	30	51	23
7	27	77	30	8	29	55	27
15	29	85	30	9	28	57	25
16	29	86	29	10	29	58	23
21	28	87	33	11	26	59	26
23	27	88	30	12	30	60	24
24	33	89	29	13	29	62	29
26	28	90	24	14	29	65	30
28	28	91	33	17	29	66	32
31	31	92	25	18	33	67	33
34	29	95	31	19	33	68	27
37	27	96	31	20	29	69	31
39	31	97	29	22	30	70	26
41	30	98	32	25	32	73	35
42	29			27	29	74	28
43	28			29	29	78	27
47	30			30	30	79	25
48	28			32	30	80	25
52	27			33	31	81	26
53	28			35	27	82	29
54	27			36	27	83	29
56	26			38	32	84	34
61	29			40	28	93	32
63	27			44	31	96	31
64	30			45	27	99	33
71	27			46	25	100	28

Using simple random sampling, we choose observations 2, 85, 61, and 54 for the males. For the females, we select 51, 14, 33, 25, 62, and 74. These observations are marked in Table 22.2. Thus, we see that the stratum-specific sample means are

$$\bar{x}_{\text{males}} = \frac{31 + 30 + 29 + 27}{4}$$

$$= 29.3 \text{ weeks}$$

and

$$\bar{x}_{\text{females}} = \frac{23 + 29 + 31 + 32 + 29 + 28}{6}$$

$$= 28.7 \text{ weeks.}$$

The true population mean is estimated as a weighted average of these quantities; therefore,

$$\bar{x} = \frac{4(29.3) + 6(28.6)}{10}$$

$$= 28.9 \text{ weeks.}$$

By chance, this value is identical to the true population mean μ.

22.4   Review Exercises

1. When you are conducting a survey, how is the study population related to the target population? What is the sampling frame?

2. How does the finite version of the central limit theorem differ from the more commonly used version, in which the underlying population is assumed to be infinite?

3. When might you prefer to use systematic sampling rather than simple random sampling? When would you prefer stratified sampling?

4. How can nonresponse result in a biased sample? What could you do to attempt to minimize nonresponse?

5. A study was conducted to examine the effects of maternal marijuana and cocaine use on fetal growth. Drug exposure was assessed in two different ways: the mothers were questioned directly during an interview, and urinalysis was performed [6].
   (a) Suppose that it is necessary to rely entirely on the information provided by the mothers. How might nonresponse affect the results of the study?
   (b) An alternative strategy might be to interview only those women who agree to be questioned. Do you feel that this method would provide a representative sample of the underlying population of expectant mothers? Why or why not?

6. Each year, the United States Department of Agriculture uses the revenue collected from excise taxes to estimate the number of cigarettes consumed in this country. Over the 11-year period 1974 to 1985, however, repeated surveys of smoking practices can account for only about 72% of the total consumption [7].
   (a) How would you explain this discrepancy between the estimates of cigarette consumption?
   (b) Which source are you more likely to believe, the excise tax revenue or the surveys of smoking practices?

7. The data set lowbwt contains information describing 100 low birth weight infants born in Boston, Massachusetts [5] (Appendix B, Table B.7). Assume that these infants constitute a finite population. Their measures of systolic blood pressure are saved under the variable name sbp; the mean systolic blood pressure is $\mu = 47.1$ mm Hg. Suppose that we do not know the true population mean and wish to estimate it using a sample of 20 newborns.

(a) What is the sampling fraction of the population?

(b) Select a simple random sample and use it to estimate the true mean systolic blood pressure for this population of low birth weight infants.

(c) Draw a systematic sample from the same population and again estimate the mean systolic blood pressure.

(d) Suppose you believe that a diagnosis of toxemia in an expectant mother might affect the systolic blood pressure of her child. Divide the population of low birth weight infants into two groups: those whose mothers were diagnosed with toxemia, and those whose mothers were not. Select a stratified random sample of size 20. Use these blood pressures to estimate the true population mean.

(e) What are the sampling fractions in each of the two strata?

(f) Could cluster sampling be applied in this problem? If so, how?

8. Suppose that you are interested in conducting your own survey to estimate the proportion of psychiatrists who have had sexual contact with one or more patients. How would you carry out this study? Justify your method of data collection. Include a discussion of how you would attempt to minimize bias.

9. Return to the first exercise in Chapter 1, which asked you to design a study aimed at investigating an issue that you believe might influence the health of the world. Reread your original proposal and critique your study design. What would you do differently now?

Bibliography

[1] Mendenhall, W., Ott, L., and Scheaffer, R. L., *Elementary Survey Sampling,* Belmont, Calif.: Wadsworth, 1971.

[2] Gatrell, N., Herman, J., Olarte, S., Feldstein, M., and Localio, R., "Psychiatrist-Patient Sexual Contact: Results of a National Survey, Prevalence," *American Journal of Psychiatry,* Volume 143, September 1986, 1126–1131.

[3] Owen, Karen, "Honesty No Longer Assumed Where Religion Is Concerned," *The Owensboro Messenger-Inquirer,* January 16, 1999.

[4] Weissman, A. N., Steer, R. A., and Lipton, D. S., "Estimating Illicit Drug Use Through Telephone Interviews and the Randomized Response Technique," *Drug and Alcohol Dependence,* Volume 18, 1986, 225–233.

[5] Leviton, A., Fenton, T., Kuban, K. C. K., and Pagano, M., "Labor and Delivery Characteristics and the Risk of Germinal Matrix Hemorrhage in Low Birth Weight Infants," *Journal of Child Neurology,* Volume 6, October 1991, 35–40.

[6] Zuckerman, B., Frank, D. A., Hingson, R., Amaro, H., Levenson, S. M., Kayne, H., Parker, S., Vinci, R., Aboagye, K., Fried, L. E., Cabral, H., Timperi, R., and Bauchner, H., "Effects of Maternal Marijuana and Cocaine Use on Fetal Growth," *The New England Journal of Medicine,* Volume 320, June 1989, 762–768.

[7] Hatziandreu, E. J., Pierce, J. P., Fiore, M. C., Grise, V., Novotny, T. E., and Davis, R. M., "The Reliability of Self-Reported Cigarette Consumption in the United States," *American Journal of Public Health,* Volume 79, August 1989, 1020–1023.

Appendix A

Tables

TABLE A.1
Binomial probabilities

n	k	.05	.10	.15	.20	.25	.30	.35	.40	.45	.50
2	0	.9025	.8100	.7225	.6400	.5625	.4900	.4225	.3600	.3025	.2500
	1	.0950	.1800	.2550	.3200	.3750	.4200	.4550	.4800	.4950	.5000
	2	.0025	.0100	.0225	.0400	.0625	.0900	.1225	.1600	.2025	.2500
3	0	.8574	.7290	.6141	.5120	.4219	.3430	.2746	.2160	.1664	.1250
	1	.1354	.2430	.3251	.3840	.4219	.4410	.4436	.4320	.4084	.3750
	2	.0071	.0270	.0574	.0960	.1406	.1890	.2389	.2880	.3341	.3750
	3	.0001	.0010	.0034	.0080	.0156	.0270	.0429	.0640	.0911	.1250
4	0	.8145	.6561	.5220	.4096	.3164	.2401	.1785	.1296	.0915	.0625
	1	.1715	.2916	.3685	.4096	.4219	.4116	.3845	.3456	.2995	.2500
	2	.0135	.0486	.0975	.1536	.2109	.2646	.3105	.3456	.3675	.3750
	3	.0005	.0036	.0115	.0256	.0469	.0756	.1115	.1536	.2005	.2500
	4	.0000	.0001	.0005	.0016	.0039	.0081	.0150	.0256	.0410	.0625
5	0	.7738	.5905	.4437	.3277	.2373	.1681	.1160	.0778	.0503	.0313
	1	.2036	.3280	.3915	.4096	.3955	.3602	.3124	.2592	.2059	.1563
	2	.0214	.0729	.1382	.2048	.2637	.3087	.3364	.3456	.3369	.3125
	3	.0011	.0081	.0244	.0512	.0879	.1323	.1811	.2304	.2757	.3125
	4	.0000	.0004	.0022	.0064	.0146	.0283	.0488	.0768	.1128	.1563
	5	.0000	.0000	.0001	.0003	.0010	.0024	.0053	.0102	.0185	.0313
6	0	.7351	.5314	.3771	.2621	.1780	.1176	.0754	.0467	.0277	.0156
	1	.2321	.3543	.3993	.3932	.3560	.3025	.2437	.1866	.1359	.0938
	2	.0305	.0984	.1762	.2458	.2966	.3241	.3280	.3110	.2780	.2344
	3	.0021	.0146	.0415	.0819	.1318	.1852	.2355	.2765	.3032	.3125
	4	.0001	.0012	.0055	.0154	.0330	.0595	.0951	.1382	.1861	.2344
	5	.0000	.0001	.0004	.0015	.0044	.0102	.0205	.0369	.0609	.0938
	6	.0000	.0000	.0000	.0001	.0002	.0007	.0018	.0041	.0083	.0156
7	0	.6983	.4783	.3206	.2097	.1335	.0824	.0490	.0280	.0152	.0078
	1	.2573	.3720	.3960	.3670	.3115	.2471	.1848	.1306	.0872	.0547
	2	.0406	.1240	.2097	.2753	.3115	.3177	.2985	.2613	.2140	.1641
	3	.0036	.0230	.0617	.1147	.1730	.2269	.2679	.2903	.2918	.2734
	4	.0002	.0026	.0109	.0287	.0577	.0972	.1442	.1935	.2388	.2734
	5	.0000	.0002	.0012	.0043	.0115	.0250	.0466	.0774	.1172	.1641
	6	.0000	.0000	.0001	.0004	.0013	.0036	.0084	.0172	.0320	.0547
	7	.0000	.0000	.0000	.0000	.0001	.0002	.0006	.0016	.0037	.0078
8	0	.6634	.4305	.2725	.1678	.1001	.0576	.0319	.0168	.0084	.0039
	1	.2793	.3826	.3847	.3355	.2670	.1977	.1373	.0896	.0548	.0313

(continued)

n	k	.05	.10	.15	.20	.25	.30	.35	.40	.45	.50
	2	.0515	.1488	.2376	.2936	.3115	.2965	.2587	.2090	.1569	.1094
	3	.0054	.0331	.0839	.1468	.2076	.2541	.2786	.2787	.2568	.2188
	4	.0004	.0046	.0185	.0459	.0865	.1361	.1875	.2322	.2627	.2734
	5	.0000	.0004	.0026	.0092	.0231	.0467	.0808	.1239	.1719	.2188
	6	.0000	.0000	.0002	.0011	.0038	.0100	.0217	.0413	.0703	.1094
	7	.0000	.0000	.0000	.0001	.0004	.0012	.0033	.0079	.0164	.0313
	8	.0000	.0000	.0000	.0000	.0000	.0001	.0002	.0007	.0017	.0039
9	0	.6302	.3874	.2316	.1342	.0751	.0404	.0207	.0101	.0046	.0020
	1	.2985	.3874	.3679	.3020	.2253	.1556	.1004	.0605	.0339	.0176
	2	.0629	.1722	.2597	.3020	.3003	.2668	.2162	.1612	.1110	.0703
	3	.0077	.0446	.1069	.1762	.2336	.2668	.2716	.2508	.2119	.1641
	4	.0006	.0074	.0283	.0661	.1168	.1715	.2194	.2508	.2600	.2461
	5	.0000	.0008	.0050	.0165	.0389	.0735	.1181	.1672	.2128	.2461
	6	.0000	.0001	.0006	.0028	.0087	.0210	.0424	.0743	.1160	.1641
	7	.0000	.0000	.0000	.0003	.0012	.0039	.0098	.0212	.0407	.0703
	8	.0000	.0000	.0000	.0000	.0001	.0004	.0013	.0035	.0083	.0176
	9	.0000	.0000	.0000	.0000	.0000	.0000	.0001	.0003	.0008	.0020
10	0	.5987	.3487	.1969	.1074	.0563	.0282	.0135	.0060	.0025	.0010
	1	.3151	.3874	.3474	.2684	.1877	.1211	.0725	.0403	.0207	.0098
	2	.0746	.1937	.2759	.3020	.2816	.2335	.1757	.1209	.0763	.0439
	3	.0105	.0574	.1298	.2013	.2503	.2668	.2522	.2150	.1665	.1172
	4	.0010	.0112	.0401	.0881	.1460	.2001	.2377	.2508	.2384	.2051
	5	.0001	.0015	.0085	.0264	.0584	.1029	.1536	.2007	.2340	.2461
	6	.0000	.0001	.0012	.0055	.0162	.0368	.0689	.1115	.1596	.2051
	7	.0000	.0000	.0001	.0008	.0031	.0090	.0212	.0425	.0746	.1172
	8	.0000	.0000	.0000	.0001	.0004	.0014	.0043	.0106	.0229	.0439
	9	.0000	.0000	.0000	.0000	.0000	.0001	.0005	.0016	.0042	.0098
	10	.0000	.0000	.0000	.0000	.0000	.0000	.0000	.0001	.0003	.0010
11	0	.5688	.3138	.1673	.0859	.0422	.0198	.0088	.0036	.0014	.0005
	1	.3293	.3835	.3248	.2362	.1549	.0932	.0518	.0266	.0125	.0054
	2	.0867	.2131	.2866	.2953	.2581	.1998	.1395	.0887	.0513	.0269
	3	.0137	.0710	.1517	.2215	.2581	.2568	.2254	.1774	.1259	.0806
	4	.0014	.0158	.0536	.1107	.1721	.2201	.2428	.2365	.2060	.1611
	5	.0001	.0025	.0132	.0388	.0803	.1321	.1830	.2207	.2360	.2256
	6	.0000	.0003	.0023	.0097	.0268	.0566	.0985	.1471	.1931	.2256
	7	.0000	.0000	.0003	.0017	.0064	.0173	.0379	.0701	.1128	.1611
	8	.0000	.0000	.0000	.0002	.0011	.0037	.0102	.0234	.0462	.0806
	9	.0000	.0000	.0000	.0000	.0001	.0005	.0018	.0052	.0126	.0269
	10	.0000	.0000	.0000	.0000	.0000	.0000	.0002	.0007	.0021	.0054
	11	.0000	.0000	.0000	.0000	.0000	.0000	.0000	.0000	.0002	.0005
12	0	.5404	.2824	.1422	.0687	.0317	.0138	.0057	.0022	.0008	.0002
	1	.3413	.3766	.3012	.2062	.1267	.0712	.0368	.0174	.0075	.0029
	2	.0988	.2301	.2924	.2835	.2323	.1678	.1088	.0639	.0339	.0161
	3	.0173	.0852	.1720	.2362	.2581	.2397	.1954	.1419	.0923	.0537
	4	.0021	.0213	.0683	.1329	.1936	.2311	.2367	.2128	.1700	.1208
	5	.0002	.0038	.0193	.0532	.1032	.1585	.2039	.2270	.2225	.1934
	6	.0000	.0005	.0040	.0155	.0401	.0792	.1281	.1766	.2124	.2256
	7	.0000	.0000	.0006	.0033	.0115	.0291	.0591	.1009	.1489	.1934

n	k	.05	.10	.15	.20	.25	.30	.35	.40	.45	.50
	8	.0000	.0000	.0001	.0005	.0024	.0078	.0199	.0420	.0762	.1208
	9	.0000	.0000	.0000	.0001	.0004	.0015	.0048	.0125	.0277	.0537
	10	.0000	.0000	.0000	.0000	.0000	.0002	.0008	.0025	.0068	.0161
	11	.0000	.0000	.0000	.0000	.0000	.0000	.0001	.0003	.0010	.0029
	12	.0000	.0000	.0000	.0000	.0000	.0000	.0000	.0000	.0001	.0002
13	0	.5133	.2542	.1209	.0550	.0238	.0097	.0037	.0013	.0004	.0001
	1	.3512	.3672	.2774	.1787	.1029	.0540	.0259	.0113	.0045	.0016
	2	.1109	.2448	.2937	.2680	.2059	.1388	.0836	.0453	.0220	.0095
	3	.0214	.0997	.1900	.2457	.2517	.2181	.1651	.1107	.0660	.0349
	4	.0028	.0277	.0838	.1535	.2097	.2337	.2222	.1845	.1350	.0873
	5	.0003	.0055	.0266	.0691	.1258	.1803	.2154	.2214	.1989	.1571
	6	.0000	.0008	.0063	.0230	.0559	.1030	.1546	.1968	.2169	.2095
	7	.0000	.0001	.0011	.0058	.0186	.0442	.0833	.1312	.1775	.2095
	8	.0000	.0000	.0001	.0011	.0047	.0142	.0336	.0656	.1089	.1571
	9	.0000	.0000	.0000	.0001	.0009	.0034	.0101	.0243	.0495	.0873
	10	.0000	.0000	.0000	.0000	.0001	.0006	.0022	.0065	.0162	.0349
	11	.0000	.0000	.0000	.0000	.0000	.0001	.0003	.0012	.0036	.0095
	12	.0000	.0000	.0000	.0000	.0000	.0000	.0000	.0001	.0005	.0016
	13	.0000	.0000	.0000	.0000	.0000	.0000	.0000	.0000	.0000	.0001
14	0	.4877	.2288	.1028	.0440	.0178	.0068	.0024	.0008	.0002	.0001
	1	.3593	.3559	.2539	.1539	.0832	.0407	.0181	.0073	.0027	.0009
	2	.1229	.2570	.2912	.2501	.1802	.1134	.0634	.0317	.0141	.0056
	3	.0259	.1142	.2056	.2501	.2402	.1943	.1366	.0845	.0462	.0222
	4	.0037	.0349	.0998	.1720	.2202	.2290	.2022	.1549	.1040	.0611
	5	.0004	.0078	.0352	.0860	.1468	.1963	.2178	.2066	.1701	.1222
	6	.0000	.0013	.0093	.0322	.0734	.1262	.1759	.2066	.2088	.1833
	7	.0000	.0002	.0019	.0092	.0280	.0618	.1082	.1574	.1952	.2095
	8	.0000	.0000	.0003	.0020	.0082	.0232	.0510	.0918	.1398	.1833
	9	.0000	.0000	.0000	.0003	.0018	.0066	.0183	.0408	.0762	.1222
	10	.0000	.0000	.0000	.0000	.0003	.0014	.0049	.0136	.0312	.0611
	11	.0000	.0000	.0000	.0000	.0000	.0002	.0010	.0033	.0093	.0222
	12	.0000	.0000	.0000	.0000	.0000	.0000	.0001	.0005	.0019	.0056
	13	.0000	.0000	.0000	.0000	.0000	.0000	.0000	.0001	.0002	.0009
	14	.0000	.0000	.0000	.0000	.0000	.0000	.0000	.0000	.0000	.0001
15	0	.4633	.2059	.0874	.0352	.0134	.0047	.0016	.0005	.0001	.0000
	1	.3658	.3432	.2312	.1319	.0668	.0305	.0126	.0047	.0016	.0005
	2	.1348	.2669	.2856	.2309	.1559	.0916	.0476	.0219	.0090	.0032
	3	.0307	.1285	.2184	.2501	.2252	.1700	.1110	.0634	.0318	.0139
	4	.0049	.0428	.1156	.1876	.2252	.2186	.1792	.1268	.0780	.0417
	5	.0006	.0105	.0449	.1032	.1651	.2061	.2123	.1859	.1404	.0916
	6	.0000	.0019	.0132	.0430	.0917	.1472	.1906	.2066	.1914	.1527
	7	.0000	.0003	.0030	.0138	.0393	.0811	.1319	.1771	.2013	.1964
	8	.0000	.0000	.0005	.0035	.0131	.0348	.0710	.1181	.1647	.1964
	9	.0000	.0000	.0001	.0007	.0034	.0116	.0298	.0612	.1048	.1527
	10	.0000	.0000	.0000	.0001	.0007	.0030	.0096	.0245	.0515	.0916
	11	.0000	.0000	.0000	.0000	.0001	.0006	.0024	.0074	.0191	.0417
	12	.0000	.0000	.0000	.0000	.0000	.0001	.0004	.0016	.0052	.0139
	13	.0000	.0000	.0000	.0000	.0000	.0000	.0001	.0003	.0010	.0032

(continued)

n	k	.05	.10	.15	.20	.25	.30	.35	.40	.45	.50
	14	.0000	.0000	.0000	.0000	.0000	.0000	.0000	.0000	.0001	.0005
	15	.0000	.0000	.0000	.0000	.0000	.0000	.0000	.0000	.0000	.0000
16	0	.4401	.1853	.0743	.0281	.0100	.0033	.0010	.0003	.0001	.0000
	1	.3706	.3294	.2097	.1126	.0535	.0228	.0087	.0030	.0009	.0002
	2	.1463	.2745	.2775	.2111	.1336	.0732	.0353	.0150	.0056	.0018
	3	.0359	.1423	.2285	.2463	.2079	.1465	.0888	.0468	.0215	.0085
	4	.0061	.0514	.1311	.2001	.2252	.2040	.1553	.1014	.0572	.0278
	5	.0008	.0137	.0555	.1201	.1802	.2099	.2008	.1623	.1123	.0667
	6	.0001	.0028	.0180	.0550	.1101	.1649	.1982	.1983	.1684	.1222
	7	.0000	.0004	.0045	.0197	.0524	.1010	.1524	.1889	.1969	.1746
	8	.0000	.0001	.0009	.0055	.0197	.0487	.0923	.1417	.1812	.1964
	9	.0000	.0000	.0001	.0012	.0058	.0185	.0442	.0840	.1318	.1746
	10	.0000	.0000	.0000	.0002	.0014	.0056	.0167	.0392	.0755	.1222
	11	.0000	.0000	.0000	.0000	.0002	.0013	.0049	.0142	.0337	.0667
	12	.0000	.0000	.0000	.0000	.0000	.0002	.0011	.0040	.0115	.0278
	13	.0000	.0000	.0000	.0000	.0000	.0000	.0002	.0008	.0029	.0085
	14	.0000	.0000	.0000	.0000	.0000	.0000	.0000	.0001	.0005	.0018
	15	.0000	.0000	.0000	.0000	.0000	.0000	.0000	.0000	.0001	.0002
	16	.0000	.0000	.0000	.0000	.0000	.0000	.0000	.0000	.0000	.0000
17	0	.4181	.1668	.0631	.0225	.0075	.0023	.0007	.0002	.0000	.0000
	1	.3741	.3150	.1893	.0957	.0426	.0169	.0060	.0019	.0005	.0001
	2	.1575	.2800	.2673	.1914	.1136	.0581	.0260	.0102	.0035	.0010
	3	.0415	.1556	.2359	.2393	.1893	.1245	.0701	.0341	.0144	.0052
	4	.0076	.0605	.1457	.2093	.2209	.1868	.1320	.0796	.0411	.0182
	5	.0010	.0175	.0668	.1361	.1914	.2081	.1849	.1379	.0875	.0472
	6	.0001	.0039	.0236	.0680	.1276	.1784	.1991	.1839	.1432	.0944
	7	.0000	.0007	.0065	.0267	.0668	.1201	.1685	.1927	.1841	.1484
	8	.0000	.0001	.0014	.0084	.0279	.0644	.1134	.1606	.1883	.1855
	9	.0000	.0000	.0003	.0021	.0093	.0276	.0611	.1070	.1540	.1855
	10	.0000	.0000	.0000	.0004	.0025	.0095	.0263	.0571	.1008	.1484
	11	.0000	.0000	.0000	.0001	.0005	.0026	.0090	.0242	.0525	.0944
	12	.0000	.0000	.0000	.0000	.0001	.0006	.0024	.0081	.0215	.0472
	13	.0000	.0000	.0000	.0000	.0000	.0001	.0005	.0021	.0068	.0182
	14	.0000	.0000	.0000	.0000	.0000	.0000	.0001	.0004	.0016	.0052
	15	.0000	.0000	.0000	.0000	.0000	.0000	.0000	.0001	.0003	.0010
	16	.0000	.0000	.0000	.0000	.0000	.0000	.0000	.0000	.0000	.0001
	17	.0000	.0000	.0000	.0000	.0000	.0000	.0000	.0000	.0000	.0000
18	0	.3972	.1501	.0536	.0180	.0056	.0016	.0004	.0001	.0000	.0000
	1	.3763	.3002	.1704	.0811	.0338	.0126	.0042	.0012	.0003	.0001
	2	.1683	.2835	.2556	.1723	.0958	.0458	.0190	.0069	.0022	.0006
	3	.0473	.1680	.2406	.2297	.1704	.1046	.0547	.0246	.0095	.0031
	4	.0093	.0700	.1592	.2153	.2130	.1681	.1104	.0614	.0291	.0117
	5	.0014	.0218	.0787	.1507	.1988	.2017	.1664	.1146	.0666	.0327
	6	.0002	.0052	.0301	.0816	.1436	.1873	.1941	.1655	.1181	.0708
	7	.0000	.0010	.0091	.0350	.0820	.1376	.1792	.1892	.1657	.1214
	8	.0000	.0002	.0022	.0120	.0376	.0811	.1327	.1734	.1864	.1669
	9	.0000	.0000	.0004	.0033	.0139	.0386	.0794	.1284	.1694	.1855
	10	.0000	.0000	.0001	.0008	.0042	.0149	.0385	.0771	.1248	.1669

n	k	.05	.10	.15	.20	.25	.30	.35	.40	.45	.50
	11	.0000	.0000	.0000	.0001	.0010	.0046	.0151	.0374	.0742	.1214
	12	.0000	.0000	.0000	.0000	.0002	.0012	.0047	.0145	.0354	.0708
	13	.0000	.0000	.0000	.0000	.0000	.0002	.0012	.0045	.0134	.0327
	14	.0000	.0000	.0000	.0000	.0000	.0000	.0002	.0011	.0039	.0117
	15	.0000	.0000	.0000	.0000	.0000	.0000	.0000	.0002	.0009	.0031
	16	.0000	.0000	.0000	.0000	.0000	.0000	.0000	.0000	.0001	.0006
	17	.0000	.0000	.0000	.0000	.0000	.0000	.0000	.0000	.0000	.0001
	18	.0000	.0000	.0000	.0000	.0000	.0000	.0000	.0000	.0000	.0000
19	0	.3774	.1351	.0456	.0144	.0042	.0011	.0003	.0001	.0000	.0000
	1	.3774	.2852	.1529	.0685	.0268	.0093	.0029	.0008	.0002	.0000
	2	.1787	.2852	.2428	.1540	.0803	.0358	.0138	.0046	.0013	.0003
	3	.0533	.1796	.2428	.2182	.1517	.0869	.0422	.0175	.0062	.0018
	4	.0112	.0798	.1714	.2182	.2023	.1491	.0909	.0467	.0203	.0074
	5	.0018	.0266	.0907	.1636	.2023	.1916	.1468	.0933	.0497	.0222
	6	.0002	.0069	.0374	.0955	.1574	.1916	.1844	.1451	.0949	.0518
	7	.0000	.0014	.0122	.0443	.0974	.1525	.1844	.1797	.1443	.0961
	8	.0000	.0002	.0032	.0166	.0487	.0981	.1489	.1797	.1771	.1442
	9	.0000	.0000	.0007	.0051	.0198	.0514	.0980	.1464	.1771	.1762
	10	.0000	.0000	.0001	.0013	.0066	.0220	.0528	.0976	.1449	.1762
	11	.0000	.0000	.0000	.0003	.0018	.0077	.0233	.0532	.0970	.1442
	12	.0000	.0000	.0000	.0000	.0004	.0022	.0083	.0237	.0529	.0961
	13	.0000	.0000	.0000	.0000	.0001	.0005	.0024	.0085	.0233	.0518
	14	.0000	.0000	.0000	.0000	.0000	.0001	.0006	.0024	.0082	.0222
	15	.0000	.0000	.0000	.0000	.0000	.0000	.0001	.0005	.0022	.0074
	16	.0000	.0000	.0000	.0000	.0000	.0000	.0000	.0001	.0005	.0018
	17	.0000	.0000	.0000	.0000	.0000	.0000	.0000	.0000	.0001	.0003
	18	.0000	.0000	.0000	.0000	.0000	.0000	.0000	.0000	.0000	.0000
	19	.0000	.0000	.0000	.0000	.0000	.0000	.0000	.0000	.0000	.0000
20	0	.3585	.1216	.0388	.0115	.0032	.0008	.0002	.0000	.0000	.0000
	1	.3774	.2702	.1368	.0576	.0211	.0068	.0020	.0005	.0001	.0000
	2	.1887	.2852	.2293	.1369	.0669	.0278	.0100	.0031	.0008	.0002
	3	.0596	.1901	.2428	.2054	.1339	.0716	.0323	.0123	.0040	.0011
	4	.0133	.0898	.1821	.2182	.1897	.1304	.0738	.0350	.0139	.0046
	5	.0022	.0319	.1028	.1746	.2023	.1789	.1272	.0746	.0365	.0148
	6	.0003	.0089	.0454	.1091	.1686	.1916	.1712	.1244	.0746	.0370
	7	.0000	.0020	.0160	.0546	.1124	.1643	.1844	.1659	.1221	.0739
	8	.0000	.0004	.0046	.0222	.0609	.1144	.1614	.1797	.1623	.1201
	9	.0000	.0001	.0011	.0011	.0271	.0654	.1158	.1597	.1771	.1602
	10	.0000	.0000	.0002	.0074	.0099	.0308	.0686	.1171	.1593	.1762
	11	.0000	.0000	.0000	.0020	.0030	.0120	.0336	.0710	.1185	.1602
	12	.0000	.0000	.0000	.0005	.0008	.0039	.0136	.0355	.0727	.1201
	13	.0000	.0000	.0000	.0001	.0002	.0010	.0045	.0146	.0366	.0739
	14	.0000	.0000	.0000	.0000	.0000	.0002	.0012	.0049	.0150	.0370
	15	.0000	.0000	.0000	.0000	.0000	.0000	.0003	.0013	.0049	.0148
	16	.0000	.0000	.0000	.0000	.0000	.0000	.0000	.0000	.0013	.0046
	17	.0000	.0000	.0000	.0000	.0000	.0000	.0000	.0000	.0002	.0011
	18	.0000	.0000	.0000	.0000	.0000	.0000	.0000	.0000	.0000	.0002
	19	.0000	.0000	.0000	.0000	.0000	.0000	.0000	.0000	.0000	.0000
	20	.0000	.0000	.0000	.0000	.0000	.0000	.0000	.0000	.0000	.0000

TABLE A.2
Poisson probabilities

	μ									
k	0.5	1.0	1.5	2.0	2.5	3.0	3.5	4.0	4.5	5.0
0	0.6065	0.3679	0.2231	0.1353	0.0821	0.0498	0.0302	0.0183	0.0111	0.0067
1	0.3033	0.3679	0.3347	0.2707	0.2052	0.1494	0.1057	0.0733	0.0500	0.0337
2	0.0758	0.1839	0.2510	0.2707	0.2565	0.2240	0.1850	0.1465	0.1125	0.0842
3	0.0126	0.0613	0.1255	0.1804	0.2138	0.2240	0.2158	0.1954	0.1687	0.1404
4	0.0016	0.0153	0.0471	0.0902	0.1336	0.1680	0.1888	0.1954	0.1898	0.1755
5	0.0002	0.0031	0.0141	0.0361	0.0668	0.1008	0.1322	0.1563	0.1708	0.1755
6	0.0000	0.0005	0.0035	0.0120	0.0278	0.0504	0.0771	0.1042	0.1281	0.1462
7	0.0000	0.0001	0.0008	0.0034	0.0099	0.0216	0.0385	0.0595	0.0824	0.1044
8	0.0000	0.0000	0.0001	0.0009	0.0031	0.0081	0.0169	0.0298	0.0463	0.0653
9	0.0000	0.0000	0.0000	0.0002	0.0009	0.0027	0.0066	0.0132	0.0232	0.0363
10	0.0000	0.0000	0.0000	0.0000	0.0002	0.0008	0.0023	0.0053	0.0104	0.0181
11	0.0000	0.0000	0.0000	0.0000	0.0000	0.0002	0.0007	0.0019	0.0043	0.0082
12	0.0000	0.0000	0.0000	0.0000	0.0000	0.0001	0.0002	0.0006	0.0016	0.0034
13	0.0000	0.0000	0.0000	0.0000	0.0000	0.0000	0.0001	0.0002	0.0006	0.0013
14	0.0000	0.0000	0.0000	0.0000	0.0000	0.0000	0.0000	0.0001	0.0002	0.0005
15	0.0000	0.0000	0.0000	0.0000	0.0000	0.0000	0.0000	0.0000	0.0001	0.0002
16	0.0000	0.0000	0.0000	0.0000	0.0000	0.0000	0.0000	0.0000	0.0000	0.0000

	μ									
k	5.5	6.0	6.5	7.0	7.5	8.0	8.5	9.0	9.5	10.0
0	0.0041	0.0025	0.0015	0.0009	0.0006	0.0003	0.0002	0.0001	0.0001	0.0000
1	0.0225	0.0149	0.0098	0.0064	0.0041	0.0027	0.0017	0.0011	0.0007	0.0005
2	0.0618	0.0446	0.0318	0.0223	0.0156	0.0107	0.0074	0.0050	0.0034	0.0023
3	0.1133	0.0892	0.0688	0.0521	0.0389	0.0286	0.0208	0.0150	0.0107	0.0076
4	0.1558	0.1339	0.1118	0.0912	0.0729	0.0573	0.0443	0.0337	0.0254	0.0189
5	0.1714	0.1606	0.1454	0.1277	0.1094	0.0916	0.0752	0.0607	0.0483	0.0378
6	0.1571	0.1606	0.1575	0.1490	0.1367	0.1221	0.1066	0.0911	0.0764	0.0631
7	0.1234	0.1377	0.1462	0.1490	0.1465	0.1396	0.1294	0.1171	0.1037	0.0901
8	0.0849	0.1033	0.1188	0.1304	0.1373	0.1396	0.1375	0.1318	0.1232	0.1126
9	0.0519	0.0688	0.0858	0.1014	0.1144	0.1241	0.1299	0.1318	0.1300	0.1251
10	0.0285	0.0413	0.0558	0.0710	0.0858	0.0993	0.1104	0.1186	0.1235	0.1251
11	0.0143	0.0225	0.0330	0.0452	0.0585	0.0722	0.0853	0.0970	0.1067	0.1137
12	0.0065	0.0113	0.0179	0.0263	0.0366	0.0481	0.0604	0.0728	0.0844	0.0948
13	0.0028	0.0052	0.0089	0.0142	0.0211	0.0296	0.0395	0.0504	0.0617	0.0729
14	0.0011	0.0022	0.0041	0.0071	0.0113	0.0169	0.0240	0.0324	0.0419	0.0521
15	0.0004	0.0009	0.0018	0.0033	0.0057	0.0090	0.0136	0.0194	0.0265	0.0347
16	0.0001	0.0003	0.0007	0.0014	0.0026	0.0045	0.0072	0.0109	0.0157	0.0217
17	0.0000	0.0001	0.0003	0.0006	0.0012	0.0021	0.0036	0.0058	0.0088	0.0128
18	0.0000	0.0000	0.0001	0.0002	0.0005	0.0009	0.0017	0.0029	0.0046	0.0071
19	0.0000	0.0000	0.0000	0.0001	0.0002	0.0004	0.0008	0.0014	0.0023	0.0037
20	0.0000	0.0000	0.0000	0.0000	0.0001	0.0002	0.0003	0.0006	0.0011	0.0019
21	0.0000	0.0000	0.0000	0.0000	0.0000	0.0001	0.0001	0.0003	0.0005	0.0009
22	0.0000	0.0000	0.0000	0.0000	0.0000	0.0000	0.0001	0.0001	0.0002	0.0004
23	0.0000	0.0000	0.0000	0.0000	0.0000	0.0000	0.0000	0.0000	0.0001	0.0002
24	0.0000	0.0000	0.0000	0.0000	0.0000	0.0000	0.0000	0.0000	0.0000	0.0001
25	0.0000	0.0000	0.0000	0.0000	0.0000	0.0000	0.0000	0.0000	0.0000	0.0000

TABLE A.2

(continued)

					μ					
k	10.5	11.0	11.5	12.0	12.5	13.0	13.5	14.0	14.5	15.0
0	0.0000	0.0000	0.0000	0.0000	0.0000	0.0000	0.0000	0.0000	0.0000	0.0000
1	0.0003	0.0002	0.0001	0.0001	0.0000	0.0000	0.0000	0.0000	0.0000	0.0000
2	0.0015	0.0010	0.0007	0.0004	0.0003	0.0002	0.0001	0.0001	0.0001	0.0000
3	0.0053	0.0037	0.0026	0.0018	0.0012	0.0008	0.0006	0.0004	0.0003	0.0002
4	0.0139	0.0102	0.0074	0.0053	0.0038	0.0027	0.0019	0.0013	0.0009	0.0006
5	0.0293	0.0224	0.0170	0.0127	0.0095	0.0070	0.0051	0.0037	0.0027	0.0019
6	0.0513	0.0411	0.0325	0.0255	0.0197	0.0152	0.0115	0.0087	0.0065	0.0048
7	0.0769	0.0646	0.0535	0.0437	0.0353	0.0281	0.0222	0.0174	0.0135	0.0104
8	0.1009	0.0888	0.0769	0.0655	0.0551	0.0457	0.0375	0.0304	0.0244	0.0194
9	0.1177	0.1085	0.0982	0.0874	0.0765	0.0661	0.0563	0.0473	0.0394	0.0324
10	0.1236	0.1194	0.1129	0.1048	0.0956	0.0859	0.0760	0.0663	0.0571	0.0486
11	0.1180	0.1194	0.1181	0.1144	0.1087	0.1015	0.0932	0.0844	0.0753	0.0663
12	0.1032	0.1094	0.1131	0.1144	0.1132	0.1099	0.1049	0.0984	0.0910	0.0829
13	0.0834	0.0926	0.1001	0.1056	0.1089	0.1099	0.1089	0.1060	0.1014	0.0956
14	0.0625	0.0728	0.0822	0.0905	0.0972	0.1021	0.1050	0.1060	0.1051	0.1024
15	0.0438	0.0534	0.0630	0.0724	0.0810	0.0885	0.0945	0.0989	0.1016	0.1024
16	0.0287	0.0367	0.0453	0.0543	0.0633	0.0719	0.0798	0.0866	0.0920	0.0960
17	0.0177	0.0237	0.0306	0.0383	0.0465	0.0550	0.0633	0.0713	0.0785	0.0847
18	0.0104	0.0145	0.0196	0.0255	0.0323	0.0397	0.0475	0.0554	0.0632	0.0706
19	0.0057	0.0084	0.0119	0.0161	0.0213	0.0272	0.0337	0.0409	0.0483	0.0557
20	0.0030	0.0046	0.0068	0.0097	0.0133	0.0177	0.0228	0.0286	0.0350	0.0418
21	0.0015	0.0024	0.0037	0.0055	0.0079	0.0109	0.0146	0.0191	0.0242	0.0299
22	0.0007	0.0012	0.0020	0.0030	0.0045	0.0065	0.0090	0.0121	0.0159	0.0204
23	0.0003	0.0006	0.0010	0.0016	0.0024	0.0037	0.0053	0.0074	0.0100	0.0133
24	0.0001	0.0003	0.0005	0.0008	0.0013	0.0020	0.0030	0.0043	0.0061	0.0083
25	0.0001	0.0001	0.0002	0.0004	0.0006	0.0010	0.0016	0.0024	0.0035	0.0050
26	0.0000	0.0000	0.0001	0.0002	0.0003	0.0005	0.0008	0.0013	0.0020	0.0029
27	0.0000	0.0000	0.0000	0.0001	0.0001	0.0002	0.0004	0.0007	0.0011	0.0016
28	0.0000	0.0000	0.0000	0.0000	0.0001	0.0001	0.0002	0.0003	0.0005	0.0009
29	0.0000	0.0000	0.0000	0.0000	0.0000	0.0001	0.0001	0.0002	0.0003	0.0004
30	0.0000	0.0000	0.0000	0.0000	0.0000	0.0000	0.0000	0.0001	0.0001	0.0002
31	0.0000	0.0000	0.0000	0.0000	0.0000	0.0000	0.0000	0.0000	0.0001	0.0001
32	0.0000	0.0000	0.0000	0.0000	0.0000	0.0000	0.0000	0.0000	0.0000	0.0001
33	0.0000	0.0000	0.0000	0.0000	0.0000	0.0000	0.0000	0.0000	0.0000	0.0000

(continued)

TABLE A.2

(continued)

k	15.5	16.0	16.5	17.0	17.5	18.0	18.5	19.0	19.5	20.0
0	0.0000	0.0000	0.0000	0.0000	0.0000	0.0000	0.0000	0.0000	0.0000	0.0000
1	0.0000	0.0000	0.0000	0.0000	0.0000	0.0000	0.0000	0.0000	0.0000	0.0000
2	0.0000	0.0000	0.0000	0.0000	0.0000	0.0000	0.0000	0.0000	0.0000	0.0000
3	0.0001	0.0001	0.0001	0.0000	0.0000	0.0000	0.0000	0.0000	0.0000	0.0000
4	0.0004	0.0003	0.0002	0.0001	0.0001	0.0001	0.0000	0.0000	0.0000	0.0000
5	0.0014	0.0010	0.0007	0.0005	0.0003	0.0002	0.0002	0.0001	0.0001	0.0001
6	0.0036	0.0026	0.0019	0.0014	0.0010	0.0007	0.0005	0.0004	0.0003	0.0002
7	0.0079	0.0060	0.0045	0.0034	0.0025	0.0019	0.0014	0.0010	0.0007	0.0005
8	0.0153	0.0120	0.0093	0.0072	0.0055	0.0042	0.0031	0.0024	0.0018	0.0013
9	0.0264	0.0213	0.0171	0.0135	0.0107	0.0083	0.0065	0.0050	0.0038	0.0029
10	0.0409	0.0341	0.0281	0.0230	0.0186	0.0150	0.0120	0.0095	0.0074	0.0058
11	0.0577	0.0496	0.0422	0.0355	0.0297	0.0245	0.0201	0.0164	0.0132	0.0106
12	0.0745	0.0661	0.0580	0.0504	0.0432	0.0368	0.0310	0.0259	0.0214	0.0176
13	0.0888	0.0814	0.0736	0.0658	0.0582	0.0509	0.0441	0.0378	0.0322	0.0271
14	0.0983	0.0930	0.0868	0.0800	0.0728	0.0655	0.0583	0.0514	0.0448	0.0387
15	0.1016	0.0992	0.0955	0.0906	0.0849	0.0786	0.0719	0.0650	0.0582	0.0516
16	0.0984	0.0992	0.0985	0.0963	0.0929	0.0884	0.0831	0.0772	0.0710	0.0646
17	0.0897	0.0934	0.0956	0.0963	0.0956	0.0936	0.0904	0.0863	0.0814	0.0760
18	0.0773	0.0830	0.0876	0.0909	0.0929	0.0936	0.0930	0.0911	0.0882	0.0844
19	0.0630	0.0699	0.0761	0.0814	0.0856	0.0887	0.0905	0.0911	0.0905	0.0888
20	0.0489	0.0559	0.0628	0.0692	0.0749	0.0798	0.0837	0.0866	0.0883	0.0888
21	0.0361	0.0426	0.0493	0.0560	0.0624	0.0684	0.0738	0.0783	0.0820	0.0846
22	0.0254	0.0310	0.0370	0.0433	0.0496	0.0560	0.0620	0.0676	0.0727	0.0769
23	0.0171	0.0216	0.0265	0.0320	0.0378	0.0438	0.0499	0.0559	0.0616	0.0669
24	0.0111	0.0144	0.0182	0.0226	0.0275	0.0328	0.0385	0.0442	0.0500	0.0557
25	0.0069	0.0092	0.0120	0.0154	0.0193	0.0237	0.0285	0.0336	0.0390	0.0446
26	0.0041	0.0057	0.0076	0.0101	0.0130	0.0164	0.0202	0.0246	0.0293	0.0343
27	0.0023	0.0034	0.0047	0.0063	0.0084	0.0109	0.0139	0.0173	0.0211	0.0254
28	0.0013	0.0019	0.0028	0.0038	0.0053	0.0070	0.0092	0.0117	0.0147	0.0181
29	0.0007	0.0011	0.0016	0.0023	0.0032	0.0044	0.0058	0.0077	0.0099	0.0125
30	0.0004	0.0006	0.0009	0.0013	0.0019	0.0026	0.0036	0.0049	0.0064	0.0083
31	0.0002	0.0003	0.0005	0.0007	0.0010	0.0015	0.0022	0.0030	0.0040	0.0054
32	0.0001	0.0001	0.0002	0.0004	0.0006	0.0009	0.0012	0.0018	0.0025	0.0034
33	0.0000	0.0001	0.0001	0.0002	0.0003	0.0005	0.0007	0.0010	0.0015	0.0020
34	0.0000	0.0000	0.0001	0.0001	0.0002	0.0002	0.0004	0.0006	0.0008	0.0012
35	0.0000	0.0000	0.0000	0.0000	0.0001	0.0001	0.0002	0.0003	0.0005	0.0007
36	0.0000	0.0000	0.0000	0.0000	0.0000	0.0001	0.0001	0.0002	0.0003	0.0004
37	0.0000	0.0000	0.0000	0.0000	0.0000	0.0000	0.0001	0.0001	0.0001	0.0002
38	0.0000	0.0000	0.0000	0.0000	0.0000	0.0000	0.0000	0.0000	0.0001	0.0001
39	0.0000	0.0000	0.0000	0.0000	0.0000	0.0000	0.0000	0.0000	0.0000	0.0001
40	0.0000	0.0000	0.0000	0.0000	0.0000	0.0000	0.0000	0.0000	0.0000	0.0000

TABLE A.3
Areas in the upper tail of the standard normal distribution

z	0.00	0.01	0.02	0.03	0.04	0.05	0.06	0.07	0.08	0.09
0.0	0.500	0.496	0.492	0.488	0.484	0.480	0.476	0.472	0.468	0.464
0.1	0.460	0.456	0.452	0.448	0.444	0.440	0.436	0.433	0.429	0.425
0.2	0.421	0.417	0.413	0.409	0.405	0.401	0.397	0.394	0.390	0.386
0.3	0.382	0.378	0.374	0.371	0.367	0.363	0.359	0.356	0.352	0.348
0.4	0.345	0.341	0.337	0.334	0.330	0.326	0.323	0.319	0.316	0.312
0.5	0.309	0.305	0.302	0.298	0.295	0.291	0.288	0.284	0.281	0.278
0.6	0.274	0.271	0.268	0.264	0.261	0.258	0.255	0.251	0.248	0.245
0.7	0.242	0.239	0.236	0.233	0.230	0.227	0.224	0.221	0.218	0.215
0.8	0.212	0.209	0.206	0.203	0.200	0.198	0.195	0.192	0.189	0.187
0.9	0.184	0.181	0.179	0.176	0.174	0.171	0.169	0.166	0.164	0.161
1.0	0.159	0.156	0.154	0.152	0.149	0.147	0.145	0.142	0.140	0.138
1.1	0.136	0.133	0.131	0.129	0.127	0.125	0.123	0.121	0.119	0.117
1.2	0.115	0.113	0.111	0.109	0.107	0.106	0.104	0.102	0.100	0.099
1.3	0.097	0.095	0.093	0.092	0.090	0.089	0.087	0.085	0.084	0.082
1.4	0.081	0.079	0.078	0.076	0.075	0.074	0.072	0.071	0.069	0.068
1.5	0.067	0.066	0.064	0.063	0.062	0.061	0.059	0.058	0.057	0.056
1.6	0.055	0.054	0.053	0.052	0.051	0.049	0.048	0.047	0.046	0.046
1.7	0.045	0.044	0.043	0.042	0.041	0.040	0.039	0.038	0.038	0.037
1.8	0.036	0.035	0.034	0.034	0.033	0.032	0.031	0.031	0.030	0.029
1.9	0.029	0.028	0.027	0.027	0.026	0.026	0.025	0.024	0.024	0.023
2.0	0.023	0.022	0.022	0.021	0.021	0.020	0.020	0.019	0.019	0.018
2.1	0.018	0.017	0.017	0.017	0.016	0.016	0.015	0.015	0.015	0.014
2.2	0.014	0.014	0.013	0.013	0.013	0.012	0.012	0.012	0.011	0.011
2.3	0.011	0.010	0.010	0.010	0.010	0.009	0.009	0.009	0.009	0.008
2.4	0.008	0.008	0.008	0.008	0.007	0.007	0.007	0.007	0.007	0.006
2.5	0.006	0.006	0.006	0.006	0.006	0.005	0.005	0.005	0.005	0.005
2.6	0.005	0.005	0.004	0.004	0.004	0.004	0.004	0.004	0.004	0.004
2.7	0.003	0.003	0.003	0.003	0.003	0.003	0.003	0.003	0.003	0.003
2.8	0.003	0.002	0.002	0.002	0.002	0.002	0.002	0.002	0.002	0.002
2.9	0.002	0.002	0.002	0.002	0.002	0.002	0.002	0.001	0.001	0.001
3.0	0.001	0.001	0.001	0.001	0.001	0.001	0.001	0.001	0.001	0.001
3.1	0.001	0.001	0.001	0.001	0.001	0.001	0.001	0.001	0.001	0.001
3.2	0.001	0.001	0.001	0.001	0.001	0.001	0.001	0.001	0.001	0.001
3.3	0.000	0.000	0.000	0.000	0.000	0.000	0.000	0.000	0.000	0.000
3.4	0.000	0.000	0.000	0.000	0.000	0.000	0.000	0.000	0.000	0.000

TABLE A.4
Percentiles of the *t* distribution

	Area in Upper Tail					
df	0.10	0.05	0.025	0.01	0.005	0.0005
1	3.078	6.314	12.706	31.821	63.657	636.619
2	1.886	2.920	4.303	6.965	9.925	31.599
3	1.638	2.353	3.182	4.541	5.841	12.924
4	1.533	2.132	2.776	3.747	4.604	8.610
5	1.476	2.015	2.571	3.365	4.032	6.869
6	1.440	1.943	2.447	3.143	3.707	5.959
7	1.415	1.895	2.365	2.998	3.499	5.408
8	1.397	1.860	2.306	2.896	3.355	5.041
9	1.383	1.833	2.262	2.821	3.250	4.781
10	1.372	1.812	2.228	2.764	3.169	4.587
11	1.363	1.796	2.201	2.718	3.106	4.437
12	1.356	1.782	2.179	2.681	3.055	4.318
13	1.350	1.771	2.160	2.650	3.012	4.221
14	1.345	1.761	2.145	2.624	2.977	4.140
15	1.341	1.753	2.131	2.602	2.947	4.073
16	1.337	1.746	2.120	2.583	2.921	4.015
17	1.333	1.740	2.110	2.567	2.898	3.965
18	1.330	1.734	2.101	2.552	2.878	3.922
19	1.328	1.729	2.093	2.539	2.861	3.883
20	1.325	1.725	2.086	2.528	2.845	3.850
21	1.323	1.721	2.080	2.518	2.831	3.819
22	1.321	1.717	2.074	2.508	2.819	3.792
23	1.319	1.714	2.069	2.500	2.807	3.768
24	1.318	1.711	2.064	2.492	2.797	3.745
25	1.316	1.708	2.060	2.485	2.787	3.725
26	1.315	1.706	2.056	2.479	2.779	3.707
27	1.314	1.703	2.052	2.473	2.771	3.690
28	1.313	1.701	2.048	2.467	2.763	3.674
29	1.311	1.699	2.045	2.462	2.756	3.659
30	1.310	1.697	2.042	2.457	2.750	3.646
40	1.303	1.684	2.021	2.423	2.704	3.551
50	1.299	1.676	2.009	2.403	2.678	3.496
60	1.296	1.671	2.000	2.390	2.660	3.460
70	1.294	1.667	1.994	2.381	2.648	3.435
80	1.292	1.664	1.990	2.374	2.639	3.416
90	1.291	1.662	1.987	2.368	2.632	3.402
100	1.290	1.660	1.984	2.364	2.626	3.390
110	1.289	1.659	1.982	2.361	2.621	3.381
120	1.289	1.658	1.980	2.358	2.617	3.373
∞	1.282	1.645	1.960	2.327	2.576	3.291

TABLE A.5
Percentiles of the *F* distribution

Denominator df	Area in Upper Tail	Numerator Degrees of Freedom (df)										
		1	2	3	4	5	6	7	8	12	24	∞
2	0.100	8.53	9.00	9.16	9.24	9.29	9.33	9.35	9.37	9.41	9.45	9.49
	0.050	18.51	19.00	19.16	19.25	19.30	19.33	19.35	19.37	19.41	19.45	19.50
	0.025	38.51	39.00	39.17	39.25	39.30	39.33	39.36	39.37	39.41	39.46	39.50
	0.010	98.50	99.00	99.17	99.25	99.30	99.33	99.36	99.37	99.42	99.46	99.50
	0.005	198.5	199.0	199.2	199.3	199.3	199.3	199.4	199.4	199.4	199.5	199.5
	0.001	998.5	999.0	999.2	999.3	999.3	999.3	999.4	999.4	999.4	999.5	999.5
3	0.100	5.54	5.46	5.39	5.34	5.31	5.28	5.27	5.25	5.22	5.18	5.13
	0.050	10.13	9.55	9.28	9.12	9.01	8.94	8.89	8.85	8.74	8.64	8.53
	0.025	17.44	16.04	15.44	15.10	14.88	14.73	14.62	14.54	14.34	14.12	13.90
	0.010	34.12	30.82	29.46	28.71	28.24	27.91	27.67	27.49	27.05	26.60	26.13
	0.005	55.55	49.80	47.47	46.19	45.39	44.84	44.43	44.13	43.39	42.62	41.83
	0.001	167.0	148.5	141.1	137.1	134.6	132.9	131.6	130.6	128.3	125.9	123.5
4	0.100	4.54	4.32	4.19	4.11	4.05	4.01	3.98	3.95	3.90	3.83	3.76
	0.050	7.71	6.94	6.59	6.39	6.26	6.16	6.09	6.04	5.91	5.77	5.63
	0.025	12.22	10.65	9.98	9.60	9.36	9.20	9.07	8.98	8.75	8.51	8.26
	0.010	21.20	18.00	16.69	15.98	15.52	15.21	14.98	14.80	14.37	13.93	13.46
	0.005	31.33	26.28	24.26	23.15	22.46	21.97	21.62	21.35	20.70	20.03	19.32
	0.001	74.14	61.25	56.18	53.44	51.71	50.53	49.66	49.00	47.41	45.77	44.05
5	0.100	4.06	3.78	3.62	3.52	3.45	3.40	3.37	3.34	3.27	3.19	3.10
	0.050	6.61	5.79	5.41	5.19	5.05	4.95	4.88	4.82	4.68	4.53	4.36
	0.025	10.01	8.43	7.76	7.39	7.15	6.98	6.85	6.76	6.52	6.28	6.02
	0.010	16.26	13.27	12.06	11.39	10.97	10.67	10.46	10.29	9.89	9.47	9.02
	0.005	22.78	18.31	16.53	15.56	14.94	14.51	14.20	13.96	13.38	12.78	12.14
	0.001	47.18	37.12	33.20	31.09	29.75	28.83	28.16	27.65	26.42	25.13	23.79
6	0.100	3.78	3.46	3.29	3.18	3.11	3.05	3.01	2.98	2.90	2.82	2.72
	0.050	5.99	5.14	4.76	4.53	4.39	4.28	4.21	4.15	4.00	3.84	3.67
	0.025	8.81	7.26	6.60	6.23	5.99	5.82	5.70	5.60	5.37	5.12	4.85
	0.010	13.75	10.92	9.78	9.15	8.75	8.47	8.26	8.10	7.72	7.31	6.88
	0.005	18.63	14.54	12.92	12.03	11.46	11.07	10.79	10.57	10.03	9.47	8.88
	0.001	35.51	27.00	23.70	21.92	20.80	20.03	19.46	19.03	17.99	16.90	15.75

(continued)

Denominator df	Area in Upper Tail	Numerator Degrees of Freedom (df)										
		1	2	3	4	5	6	7	8	12	24	∞
7	0.100	3.59	3.26	3.07	2.96	2.88	2.83	2.78	2.75	2.67	2.58	2.47
	0.050	5.59	4.74	4.35	4.12	3.97	3.87	3.79	3.73	3.57	3.41	3.23
	0.025	8.07	6.54	5.89	5.52	5.29	5.12	4.99	4.90	4.67	4.41	4.14
	0.010	12.25	9.55	8.45	7.85	7.46	7.19	6.99	6.84	6.47	6.07	5.65
	0.005	16.24	12.40	10.88	10.05	9.52	9.16	8.89	8.68	8.18	7.64	7.08
	0.001	29.25	21.69	18.77	17.20	16.21	15.52	15.02	14.63	13.71	12.73	11.70
8	0.100	3.46	3.11	2.92	2.81	2.73	2.67	2.62	2.59	2.50	2.40	2.29
	0.050	5.32	4.46	4.07	3.84	3.69	3.58	3.50	3.44	3.28	3.12	2.93
	0.025	7.57	6.06	5.42	5.05	4.82	4.65	4.53	4.43	4.20	3.95	3.67
	0.010	11.26	8.65	7.59	7.01	6.63	6.37	6.18	6.03	5.67	5.28	4.86
	0.005	14.69	11.04	9.60	8.81	8.30	7.95	7.69	7.50	7.01	6.50	5.95
	0.001	25.41	18.49	15.83	14.39	13.48	12.86	12.40	12.05	11.19	10.3	9.33
9	0.100	3.36	3.01	2.81	2.69	2.61	2.55	2.51	2.47	2.38	2.28	2.16
	0.050	5.12	4.26	3.86	3.63	3.48	3.37	3.29	3.23	3.07	2.90	2.71
	0.025	7.21	5.71	5.08	4.72	4.48	4.32	4.20	4.10	3.87	3.61	3.33
	0.010	10.56	8.02	6.99	6.42	6.06	5.80	5.61	5.47	5.11	4.73	4.31
	0.005	13.61	10.11	8.72	7.96	7.47	7.13	6.88	6.69	6.23	5.73	5.19
	0.001	22.86	16.39	13.90	12.56	11.71	11.13	10.70	10.37	9.57	8.72	7.81
10	0.100	3.29	2.92	2.73	2.61	2.52	2.46	2.41	2.38	2.28	2.18	2.06
	0.050	4.96	4.10	3.71	3.48	3.33	3.22	3.14	3.07	2.91	2.74	2.54
	0.025	6.94	5.46	4.83	4.47	4.24	4.07	3.95	3.85	3.62	3.37	3.08
	0.010	10.04	7.56	6.55	5.99	5.64	5.39	5.20	5.06	4.71	4.33	3.91
	0.005	12.83	9.43	8.08	7.34	6.87	6.54	6.30	6.12	5.66	5.17	4.64
	0.001	21.04	14.91	12.55	11.28	10.48	9.93	9.52	9.20	8.45	7.64	6.76
12	0.100	3.18	2.81	2.61	2.48	2.39	2.33	2.28	2.24	2.15	2.04	1.90
	0.050	4.75	3.89	3.49	3.26	3.11	3.00	2.91	2.85	2.69	2.51	2.30
	0.025	6.55	5.10	4.47	4.12	3.89	3.73	3.61	3.51	3.28	3.02	2.72
	0.010	9.33	6.93	5.95	5.41	5.06	4.82	4.64	4.50	4.16	3.78	3.36
	0.005	11.75	8.51	7.23	6.52	6.07	5.76	5.52	5.35	4.91	4.43	3.90
	0.001	18.64	12.97	10.80	9.63	8.89	8.38	8.00	7.71	7.00	6.25	5.42

TABLE A.5
(continued)

df_2	α											
14	0.100	3.10	2.73	2.52	2.39	2.31	2.24	2.19	2.15	2.05	1.94	1.80
	0.050	4.60	3.74	3.34	3.11	2.96	2.85	2.76	2.70	2.53	2.35	2.13
	0.025	6.30	4.86	4.24	3.89	3.66	3.50	3.38	3.29	3.05	2.79	2.49
	0.010	8.86	6.51	5.56	5.04	4.69	4.46	4.28	4.14	3.80	3.43	3.00
	0.005	11.06	7.92	6.68	6.00	5.56	5.26	5.03	4.86	4.43	3.96	3.44
	0.001	17.14	11.78	9.73	8.62	7.92	7.44	7.08	6.80	6.13	5.41	4.60
16	0.100	3.05	2.67	2.46	2.33	2.24	2.18	2.13	2.09	1.99	1.87	1.72
	0.050	4.49	3.63	3.24	3.01	2.85	2.74	2.66	2.59	2.42	2.24	2.01
	0.025	6.12	4.69	4.08	3.73	3.50	3.34	3.22	3.12	2.89	2.63	2.32
	0.010	8.53	6.23	5.29	4.77	4.44	4.20	4.03	3.89	3.55	3.18	2.75
	0.005	10.58	7.51	6.30	5.64	5.21	4.91	4.69	4.52	4.10	3.64	3.11
	0.001	16.12	10.97	9.01	7.94	7.27	6.80	6.46	6.19	5.55	4.85	4.06
18	0.100	3.01	2.62	2.42	2.29	2.20	2.13	2.08	2.04	1.93	1.81	1.66
	0.050	4.41	3.55	3.16	2.93	2.77	2.66	2.58	2.51	2.34	2.15	1.92
	0.025	5.98	4.56	3.95	3.61	3.38	3.22	3.10	3.01	2.77	2.50	2.19
	0.010	8.29	6.01	5.09	4.58	4.25	4.01	3.84	3.71	3.37	3.00	2.57
	0.005	10.22	7.21	6.03	5.37	4.96	4.66	4.44	4.28	3.86	3.40	2.87
	0.001	15.38	10.39	8.49	7.46	6.81	6.35	6.02	5.76	5.13	4.45	3.67
20	0.100	2.97	2.59	2.38	2.25	2.16	2.09	2.04	2.00	1.89	1.77	1.61
	0.050	4.35	3.49	3.10	2.87	2.71	2.60	2.51	2.45	2.28	2.08	1.84
	0.025	5.87	4.46	3.86	3.51	3.29	3.13	3.01	2.91	2.68	2.41	2.09
	0.010	8.10	5.85	4.94	4.43	4.10	3.87	3.70	3.56	3.23	2.86	2.42
	0.005	9.94	6.99	5.82	5.17	4.76	4.47	4.26	4.09	3.68	3.22	2.69
	0.001	14.82	9.95	8.10	7.10	6.46	6.02	5.69	5.44	4.82	4.15	3.38
30	0.100	2.88	2.49	2.28	2.14	2.05	1.98	1.93	1.88	1.77	1.64	1.46
	0.050	4.17	3.32	2.92	2.69	2.53	2.42	2.33	2.27	2.09	1.89	1.62
	0.025	5.57	4.18	3.59	3.25	3.03	2.87	2.75	2.65	2.41	2.14	1.79
	0.010	7.56	5.39	4.51	4.02	3.70	3.47	3.30	3.17	2.84	2.47	2.01
	0.005	9.18	6.35	5.24	4.62	4.23	3.95	3.74	3.58	3.18	2.73	2.18
	0.001	13.29	8.77	7.05	6.12	5.53	5.12	4.82	4.58	4.00	3.36	2.59
40	0.100	2.84	2.44	2.23	2.09	2.00	1.93	1.87	1.83	1.71	1.57	1.38
	0.050	4.08	3.23	2.84	2.61	2.45	2.34	2.25	2.18	2.00	1.79	1.51
	0.025	5.42	4.05	3.46	3.13	2.90	2.74	2.62	2.53	2.29	2.01	1.64
	0.010	7.31	5.18	4.31	3.83	3.51	3.29	3.12	2.99	2.66	2.29	1.80
	0.005	8.83	6.07	4.98	4.37	3.99	3.71	3.51	3.35	2.95	2.50	1.93
	0.001	12.61	8.25	6.59	5.70	5.13	4.73	4.44	4.21	3.64	3.01	2.23

(continued)

TABLE A.5

(continued)

Denominator df	Area in Upper Tail	Numerator Degrees of Freedom (df)										
		1	2	3	4	5	6	7	8	12	24	∞
60	0.100	2.79	2.39	2.18	2.04	1.95	1.87	1.82	1.77	1.66	1.51	1.29
	0.050	4.00	3.15	2.76	2.53	2.37	2.25	2.17	2.10	1.92	1.70	1.39
	0.025	5.29	3.93	3.34	3.01	2.79	2.63	2.51	2.41	2.17	1.88	1.48
	0.010	7.08	4.98	4.13	3.65	3.34	3.12	2.95	2.82	2.50	2.12	1.60
	0.005	8.49	5.79	4.73	4.14	3.76	3.49	3.29	3.13	2.74	2.29	1.69
	0.001	11.97	7.77	6.17	5.31	4.76	4.37	4.09	3.86	3.32	2.69	1.89
80	0.100	2.77	2.37	2.15	2.02	1.92	1.85	1.79	1.75	1.63	1.48	1.24
	0.050	3.96	3.11	2.72	2.49	2.33	2.21	2.13	2.06	1.88	1.65	1.32
	0.025	5.22	3.86	3.28	2.95	2.73	2.57	2.45	2.35	2.11	1.82	1.40
	0.010	6.96	4.88	4.04	3.56	3.26	3.04	2.87	2.74	2.42	2.03	1.49
	0.005	8.33	5.67	4.61	4.03	3.65	3.39	3.19	3.03	2.64	2.19	1.56
	0.001	11.67	7.54	5.97	5.12	4.58	4.20	3.92	3.70	3.16	2.54	1.72
100	0.100	2.76	2.36	2.14	2.00	1.91	1.83	1.78	1.73	1.61	1.46	1.21
	0.050	3.94	3.09	2.70	2.46	2.31	2.19	2.10	2.03	1.85	1.63	1.28
	0.025	5.18	3.83	3.25	2.92	2.70	2.54	2.42	2.32	2.08	1.78	1.35
	0.010	6.90	4.82	3.98	3.51	3.21	2.99	2.82	2.69	2.37	1.98	1.43
	0.005	8.24	5.59	4.54	3.96	3.59	3.33	3.13	2.97	2.58	2.13	1.49
	0.001	11.50	7.41	5.86	5.02	4.48	4.11	3.83	3.61	3.07	2.46	1.62
120	0.100	2.75	2.35	2.13	1.99	1.90	1.82	1.77	1.72	1.60	1.45	1.19
	0.050	3.92	3.07	2.68	2.45	2.29	2.18	2.09	2.02	1.83	1.61	1.25
	0.025	5.15	3.80	3.23	2.89	2.67	2.52	2.39	2.30	2.05	1.76	1.31
	0.010	6.85	4.79	3.95	3.48	3.17	2.96	2.79	2.66	2.34	1.95	1.38
	0.005	8.18	5.54	4.50	3.92	3.55	3.28	3.09	2.93	2.54	2.09	1.43
	0.001	11.38	7.32	5.78	4.95	4.42	4.04	3.77	3.55	3.02	2.40	1.54
∞	0.100	2.71	2.30	2.08	1.94	1.85	1.77	1.72	1.67	1.55	1.38	1.00
	0.050	3.84	3.00	2.60	2.37	2.21	2.10	2.01	1.94	1.75	1.52	1.00
	0.025	5.02	3.69	3.12	2.79	2.57	2.41	2.29	2.19	1.94	1.64	1.00
	0.010	6.63	4.61	3.78	3.32	3.02	2.80	2.64	2.51	2.18	1.79	1.00
	0.005	7.88	5.30	4.28	3.72	3.35	3.09	2.90	2.74	2.36	1.90	1.00
	0.001	10.83	6.91	5.42	4.62	4.10	3.74	3.47	3.27	2.74	2.13	1.00

TABLE A.6
Distribution function of T, the Wilcoxon signed-rank test

T_0	2	3	4	5	6	7	8	9	10	11	12
									Sample Size		
1	0.5000	0.2500	0.1250	0.0625	0.0313	0.0157	0.0079	0.0040	0.0020	0.0010	0.0005
2		0.3750	0.1875	0.0938	0.0469	0.0235	0.0118	0.0059	0.0030	0.0015	0.0008
3		0.6250	0.3125	0.1563	0.0782	0.0391	0.0196	0.0098	0.0049	0.0025	0.0013
4			0.4375	0.2188	0.1094	0.0547	0.0274	0.0137	0.0069	0.0035	0.0018
5			0.5625	0.3125	0.1563	0.0782	0.0391	0.0196	0.0098	0.0049	0.0025
6				0.4063	0.2188	0.1094	0.0547	0.0274	0.0137	0.0069	0.0035
7				0.5000	0.2813	0.1485	0.0743	0.0372	0.0186	0.0093	0.0047
8					0.3438	0.1875	0.0977	0.0489	0.0245	0.0123	0.0062
9					0.4219	0.2344	0.1250	0.0645	0.0323	0.0162	0.0081
10					0.5000	0.2891	0.1563	0.0821	0.0420	0.0210	0.0105
11						0.3438	0.1915	0.1016	0.0528	0.0269	0.0135
12						0.4063	0.2305	0.1250	0.0655	0.0337	0.0171
13						0.4688	0.2735	0.1504	0.0801	0.0416	0.0213
14						0.5313	0.3204	0.1797	0.0967	0.0508	0.0262
15							0.3711	0.2129	0.1163	0.0616	0.0320
16							0.4219	0.2481	0.1377	0.0738	0.0386
17							0.4727	0.2852	0.1612	0.0875	0.0462
18							0.5274	0.3262	0.1875	0.1031	0.0550
19								0.3672	0.2159	0.1202	0.0647
20								0.4102	0.2461	0.1392	0.0757
21								0.4551	0.2784	0.1602	0.0882
22								0.5000	0.3125	0.1827	0.1019
23									0.3477	0.2066	0.1167
24									0.3848	0.2325	0.1331
25									0.4229	0.2598	0.1507
26									0.4610	0.2886	0.1697
27									0.5000	0.3189	0.1902
28										0.3501	0.2120
29										0.3824	0.2349
30										0.4156	0.2593
31										0.4493	0.2847
32										0.4830	0.3111
33										0.5171	0.3387
34											0.3667
35											0.3956
36											0.4251
37											0.4549
38											0.4849
39											0.5152

TABLE A.7

Distribution functions of W, the Wilcoxon rank sum test

W_0	$n_2 = 3$		
	$n_1 = 1$	2	3
1	0.25		
2	0.50		
3		0.10	
4		0.20	
5		0.40	
6		0.60	0.05
7			0.10
8			0.20
9			0.35
10			0.50

W_0	$n_2 = 4$			
	$n_1 = 1$	2	3	4
1	0.20			
2	0.40			
3	0.60	0.0667		
4		0.1333		
5		0.2667		
6		0.4000	0.0286	
7		0.6000	0.0571	
8			0.1143	
9			0.2000	
10			0.3143	0.0143
11			0.4286	0.0286
12			0.5714	0.0571
13				0.1000
14				0.1714
15				0.2429
16				0.3429
17				0.4429
18				0.5571

TABLE A.7
(continued)

W_0	$n_1 = 1$	2	3	4	5
			$n_2 = 5$		
1	0.1667				
2	0.3333				
3	0.5000	0.0476			
4		0.0952			
5		0.1905			
6		0.2857	0.0179		
7		0.4286	0.0357		
8		0.5714	0.0714		
9			0.1250		
10			0.1964	0.0079	
11			0.2857	0.0159	
12			0.3929	0.0317	
13			0.5000	0.0556	
14				0.0952	
15				0.1429	0.0040
16				0.2063	0.0079
17				0.2778	0.0159
18				0.3651	0.0278
19				0.4524	0.0476
20				0.5476	0.0754
21					0.1111
22					0.1548
23					0.2103
24					0.2738
25					0.3452
26					0.4206
27					0.5000

(continued)

TABLE A.7
(continued)

W_0	$n_1 = 1$	2	3	4	5	6
			$n_2 = 6$			
1	0.1429					
2	0.2857					
3	0.4286	0.0357				
4	0.5714	0.0714				
3		0.1429				
6		0.2143	0.0119			
7		0.3214	0.0238			
8		0.4286	0.0476			
9		0.5714	0.0833			
10			0.1310	0.0048		
11			0.1905	0.0095		
12			0.2738	0.0190		
13			0.3571	0.0333		
14			0.4524	0.0571		
15			0.5476	0.0857	0.0022	
16				0.1286	0.0043	
17				0.1762	0.0087	
18				0.2381	0.0152	
19				0.3048	0.0260	
20				0.3810	0.0411	
21				0.4571	0.0628	0.0011
22				0.5429	0.0887	0.0022
23					0.1234	0.0043
24					0.1645	0.0076
25					0.2143	0.0130
26					0.2684	0.0206
27					0.3312	0.0325
28					0.3961	0.0465
29					0.4654	0.0660
30					0.5346	0.0898
31						0.1201
32						0.1548
33						0.1970
34						0.2424
35						0.2944
36						0.3496
37						0.4091
38						0.4686
39						0.5314

W_0	$n_1 = 1$	2	3	4	5	6	7
				$n_2 = 7$			
1	0.125						
2	0.250						
3	0.375	0.0278					
4	0.500	0.0556					
5		0.1111					
6		0.1667	0.0083				
7		0.2500	0.0167				
8		0.3333	0.0333				
9		0.4444	0.0583				
10		0.5556	0.0917	0.0030			
11			0.1333	0.0061			
12			0.1917	0.0121			
13			0.2583	0.0212			
14			0.3333	0.0364			
15			0.4167	0.0545	0.0013		
16			0.5000	0.0818	0.0025		
17				0.1152	0.0051		
18				0.1576	0.0088		
19				0.2061	0.0152		
20				0.2636	0.0240		
21				0.3242	0.0366	0.0006	
22				0.3939	0.0530	0.0012	
23				0.4636	0.0745	0.0023	
24				0.5364	0.1010	0.0041	
25					0.1338	0.0070	
26					0.1717	0.0111	
27					0.2159	0.0175	
28					0.2652	0.0256	0.0003
29					0.3194	0.0367	0.0006
30					0.3775	0.0507	0.0012
31					0.4381	0.0688	0.0020
32					0.5000	0.0903	0.0035
33						0.1171	0.0055
34						0.1474	0.0087
35						0.1830	0.0131
36						0.2226	0.0189
37						0.2669	0.0265
38						0.3141	0.0364
39						0.3654	0.0487
40						0.4178	0.0641
41						0.4726	0.0825
42						0.5274	0.1043
43							0.1297
44							0.1588
45							0.1914
46							0.2279
47							0.2675
48							0.3100
49							0.3552
50							0.4024
51							0.4508
52							0.5000

(continued)

TABLE A.7

(continued)

					$n_2 = 8$			
W_0	$n_1 = 1$	2	3	4	5	6	7	8
1	0.1111							
2	0.2222							
3	0.3333	0.0222						
4	0.4444	0.0444						
5	0.5556	0.0889						
6		0.1333	0.0061					
7		0.2000	0.0121					
8		0.2667	0.0242					
9		0.3556	0.0424					
10		0.4444	0.0667	0.0020				
11		0.5556	0.0970	0.0040				
12			0.1394	0.0081				
13			0.1879	0.0141				
14			0.2485	0.0242				
15			0.3152	0.0364	0.0008			
16			0.3879	0.0545	0.0016			
17			0.4606	0.0768	0.0031			
18			0.5394	0.1071	0.0054			
19				0.1414	0.0093			
20				0.1838	0.0148			
21				0.2303	0.0225	0.0003		
22				0.2848	0.0326	0.0007		
23				0.3414	0.0466	0.0013		
24				0.4040	0.0637	0.0023		
25				0.4667	0.0855	0.0040		
26				0.5333	0.1111	0.0063		
27					0.1422	0.0100		
28					0.1772	0.0147	0.0002	
29					0.2176	0.0213	0.0003	
30					0.2618	0.0296	0.0006	
31					0.3108	0.0406	0.0011	
32					0.3621	0.0539	0.0019	
33					0.4165	0.0709	0.0030	
34					0.4716	0.0906	0.0047	
35					0.5284	0.1142	0.0070	
36						0.1412	0.0103	0.0001
37						0.1725	0.0145	0.0002
38						0.2068	0.0200	0.0003
39						0.2454	0.0270	0.0005
40						0.2864	0.0361	0.0009
41						0.3310	0.0469	0.0015
42						0.3773	0.0603	0.0023
43						0.4259	0.0760	0.0035
44						0.4749	0.0946	0.0052
45						0.5251	0.1159	0.0074

TABLE A.7
(continued)

W_0	$n_1 = 1$	2	3	4	5	6	7	8
				$n_2 = 8$ (continued)				
46							0.1405	0.0103
47							0.1678	0.0141
48							0.1984	0.0190
49							0.2317	0.0249
50							0.2679	0.0325
51							0.3063	0.0415
52							0.3472	0.0524
53							0.3894	0.0652
54							0.4333	0.0803
55							0.4775	0.0974
56							0.5225	0.1172
57								0.1393
58								0.1641
59								0.1911
60								0.2209
61								0.2527
62								0.2869
63								0.3227
64								0.3605
65								0.3992
66								0.4392
67								0.4796
68								0.5204

(continued)

TABLE A.7

(continued)

W_0	$n_1 = 1$	2	3	4	5	6	7	8	9
					$n_2 = 9$				
1	0.1000								
2	0.2000								
3	0.3000	0.0182							
4	0.4000	0.0364							
5	0.5000	0.0727							
6		0.1091	0.0045						
7		0.1636	0.0091						
8		0.2182	0.0182						
9		0.2909	0.0318						
10		0.3636	0.0500	0.0014					
11		0.4545	0.0727	0.0028					
12		0.5455	0.1045	0.0056					
13			0.1409	0.0098					
14			0.1864	0.0168					
15			0.2409	0.0252	0.0005				
16			0.3000	0.0378	0.0010				
17			0.3636	0.0531	0.0020				
18			0.4318	0.0741	0.0035				
19			0.5000	0.0993	0.0060				
20				0.1301	0.0095				
21				0.1650	0.0145	0.0002			
22				0.2070	0.0210	0.0004			
23				0.2517	0.0300	0.0008			
24				0.3021	0.0415	0.0014			
25				0.3552	0.0559	0.0024			
26				0.4126	0.0734	0.0038			
27				0.4699	0.0949	0.0060			
28				0.5301	0.1199	0.0088	0.0001		
29					0.1489	0.0128	0.0002		
30					0.1818	0.0180	0.0003		
31					0.2188	0.0248	0.0006		
32					0.2592	0.0332	0.0010		
33					0.3032	0.0440	0.0017		
34					0.3497	0.0567	0.0026		
35					0.3986	0.0723	0.0039		
36					0.4491	0.0905	0.0058	0.0000	
37					0.5000	0.1119	0.0082	0.0001	
38						0.1361	0.0115	0.0002	
39						0.1638	0.0156	0.0003	
40						0.1942	0.0209	0.0005	
41						0.2280	0.0274	0.0008	
42						0.2643	0.0356	0.0012	
43						0.3035	0.0454	0.0019	
44						0.3445	0.0571	0.0028	

TABLE A.7

(continued)

						$n_2 = 9$ (continued)			
W_0	$n_1 = 1$	2	3	4	5	6	7	8	9
45						0.3878	0.0708	0.0039	0.0000
46						0.4320	0.0869	0.0056	0.0000
47						0.4773	0.1052	0.0076	0.0001
48						0.5227	0.1261	0.0103	0.0001
49							0.1496	0.0137	0.0002
50							0.1755	0.0180	0.0004
51							0.2039	0.0232	0.0006
52							0.2349	0.0296	0.0009
53							0.2680	0.0372	0.0014
54							0.3032	0.0464	0.0020
55							0.3403	0.0570	0.0028
56							0.3788	0.0694	0.0039
57							0.4185	0.0836	0.0053
58							0.4591	0.0998	0.0071
59							0.5000	0.1179	0.0094
60								0.1383	0.0122
61								0.1606	0.0157
62								0.1852	0.0200
63								0.2117	0.0252
64								0.2404	0.0313
65								0.2707	0.0385
66								0.3029	0.0470
67								0.3365	0.0567
68								0.3715	0.0680
69								0.4074	0.0807
70								0.4442	0.0951
71								0.4813	0.1112
72								0.5187	0.1290
73									0.1487
74									0.1701
75									0.1933
76									0.2181
77									0.2447
78									0.2729
79									0.3024
80									0.3332
81									0.3652
82									0.3981
83									0.4317
84									0.4657
85									0.5000

(continued)

W_0	$n_1 = 1$	2	3	4	5	6	7	8	9	10
					$n_2 = 10$					
1	0.0909									
2	0.1818									
3	0.2727	0.0152								
4	0.3636	0.0303								
5	0.4545	0.0606								
6	0.5455	0.0909	0.0035							
7		0.1364	0.0070							
8		0.1818	0.0140							
9		0.2424	0.0245							
10		0.3030	0.0385	0.0010						
11		0.3788	0.0559	0.0020						
12		0.4545	0.0804	0.0040						
13		0.5455	0.1084	0.0070						
14			0.1434	0.0120						
15			0.1853	0.0180	0.0003					
16			0.2343	0.0270	0.0007					
17			0.2867	0.0380	0.0013					
18			0.3462	0.0529	0.0023					
19			0.4056	0.0709	0.0040					
20			0.4685	0.0939	0.0063					
21			0.5315	0.1199	0.0097	0.0001				
22				0.1518	0.0140	0.0002				
23				0.1868	0.0200	0.0005				
24				0.2268	0.0276	0.0009				
25				0.2697	0.0376	0.0015				
26				0.3177	0.0496	0.0024				
27				0.3666	0.0646	0.0037				
28				0.4196	0.0823	0.0055	0.0001			
29				0.4725	0.1032	0.0080	0.0001			
30				0.5275	0.1272	0.0112	0.0002			
31					0.1548	0.0156	0.0004			
32					0.1855	0.0210	0.0006			
33					0.2198	0.0280	0.0010			
34					0.2567	0.0363	0.0015			
35					0.2970	0.0467	0.0023			
36					0.3393	0.0589	0.0034	0.0000		
37					0.3839	0.0736	0.0048	0.0000		
38					0.4296	0.0903	0.0068	0.0001		
39					0.4765	0.1099	0.0093	0.0002		
40					0.5235	0.1317	0.0125	0.0003		
41						0.1566	0.0165	0.0004		
42						0.1838	0.0215	0.0007		
43						0.2139	0.0277	0.0010		
44						0.2461	0.0351	0.0015		
45						0.2811	0.0439	0.0022	0.0000	
46						0.3177	0.0544	0.0031	0.0000	
47						0.3564	0.0665	0.0043	0.0000	
48						0.3962	0.0806	0.0058	0.0001	
49						0.4374	0.0966	0.0078	0.0001	
50						0.4789	0.1148	0.0103	0.0002	
51						0.5211	0.1349	0.0133	0.0003	
52							0.1574	0.0171	0.0005	

W_0	$n_1 = 1$	2	3	4	5	6	7	8	9	10
					$n_2 = 10$ (continued)					
53							0.1819	0.0217	0.0007	
54							0.2087	0.0273	0.0011	
55							0.2374	0.0338	0.0015	0.0000
56							0.2681	0.0416	0.0021	0.0000
57							0.3004	0.0506	0.0028	0.0000
58							0.3345	0.0610	0.0038	0.0000
59							0.3698	0.0729	0.0051	0.0001
60							0.4063	0.0864	0.0066	0.0001
61							0.4434	0.1015	0.0086	0.0002
62							0.4811	0.1185	0.0110	0.0002
63							0.5189	0.1371	0.0140	0.0004
64								0.1577	0.0175	0.0005
65								0.1800	0.0217	0.0008
66								0.2041	0.0267	0.0010
67								0.2299	0.0326	0.0014
68								0.2574	0.0394	0.0019
69								0.2863	0.0474	0.0026
70								0.3167	0.0564	0.0034
71								0.3482	0.0667	0.0045
72								0.3809	0.0782	0.0057
73								0.4143	0.0912	0.0073
74								0.4484	0.1055	0.0093
75								0.4827	0.1214	0.0116
76								0.5173	0.1388	0.0144
77									0.1577	0.0177
78									0.1781	0.0216
79									0.2001	0.0262
80									0.2235	0.0315
81									0.2483	0.0376
82									0.2745	0.0446
83									0.3019	0.0526
84									0.3304	0.0615
85									0.3598	0.0716
86									0.3901	0.0827
87									0.4211	0.0952
88									0.4524	0.1088
89									0.4841	0.1237
90									0.5159	0.1399
91										0.1575
92										0.1763
93										0.1965
94										0.2179
95										0.2406
96										0.2644
97										0.2894
98										0.3153
99										0.3421
100										0.3697
101										0.3980
102										0.4267
103										0.4559
104										0.4853
105										0.5147

TABLE A.8

Percentiles of the chi-square distribution

	Area in Upper Tail				
df	0.100	0.050	0.025	0.010	0.001
1	2.71	3.84	5.02	6.63	10.83
2	4.61	5.99	7.38	9.21	13.82
3	6.25	7.81	9.35	11.34	16.27
4	7.78	9.49	11.14	13.28	18.47
5	9.24	11.07	12.83	15.09	20.52
6	10.64	12.59	14.45	16.81	22.46
7	12.02	14.07	16.01	18.48	24.32
8	13.36	15.51	17.53	20.09	26.12
9	14.68	16.92	19.02	21.67	27.88
10	15.99	18.31	20.48	23.21	29.59
11	17.28	19.68	21.92	24.72	31.26
12	18.55	21.03	23.34	26.22	32.91
13	19.81	22.36	24.74	27.69	34.53
14	21.06	23.68	26.12	29.14	36.12
15	22.31	25.00	27.49	30.58	37.70
16	23.54	26.30	28.85	32.00	39.25
17	24.77	27.59	30.19	33.41	40.79
18	25.99	28.87	31.53	34.81	42.31
19	27.20	30.14	32.85	36.19	43.82
20	28.41	31.41	34.17	37.57	45.31
21	29.62	32.67	35.48	38.93	46.80
22	30.81	33.92	36.78	40.29	48.27
23	32.01	35.17	38.08	41.64	49.73
24	33.20	36.42	39.36	42.98	51.18
25	34.38	37.65	40.65	44.31	52.62

Appendix B

Data Sets

TABLE B.1

Data set `serzinc`; variable `zinc`

50	70	74	77	81	85	88	93	97	103	109
51	70	74	77	81	85	88	93	97	103	110
53	70	74	77	81	85	88	93	97	103	110
55	70	74	77	81	85	88	93	97	103	111
56	70	74	77	82	85	88	93	97	103	111
58	70	74	77	82	85	88	93	97	103	111
60	70	74	77	82	85	89	93	97	104	111
60	70	74	78	82	85	89	93	98	104	112
60	71	74	78	82	85	89	94	98	104	112
61	71	74	78	82	85	89	94	98	104	112
61	71	74	78	82	85	89	94	98	104	113
61	71	75	78	82	86	89	94	98	104	113
61	71	75	78	82	86	89	94	98	104	113
62	71	75	78	82	86	89	94	98	105	114
62	71	75	78	82	86	89	94	98	105	114
62	71	75	78	82	86	90	94	98	105	114
62	71	75	78	82	86	90	94	99	105	114
63	71	75	78	83	86	90	94	99	105	115
63	71	75	78	83	86	90	94	99	105	115
63	72	75	79	83	86	90	95	99	105	115
64	72	75	79	83	86	90	95	99	105	116
64	72	75	79	83	86	90	95	99	106	116
64	72	75	79	83	86	90	95	100	106	116
64	72	75	79	83	86	91	95	100	106	116
64	72	75	79	83	87	91	95	100	106	117
65	72	75	80	83	87	91	95	100	106	117
65	72	76	80	83	87	91	95	100	106	117
65	72	76	80	83	87	91	95	101	107	118
66	72	76	80	84	87	91	95	101	107	118
66	72	76	80	84	87	91	95	101	107	119
66	72	76	80	84	87	91	96	101	107	119
67	72	76	80	84	87	91	96	101	107	121
67	73	76	80	84	87	91	96	101	107	123
67	73	76	80	84	87	92	96	101	107	124
67	73	76	80	84	88	92	96	102	108	125
67	73	76	80	84	88	92	96	102	108	128
68	73	76	81	84	88	92	96	102	108	131
68	73	76	81	84	88	92	96	102	108	135
68	73	76	81	84	88	92	96	102	108	142
68	73	77	81	84	88	92	96	102	108	147
69	73	77	81	84	88	92	97	102	108	151
70	74	77	81	84	88	93	97	102	109	153

TABLE B.2
Data set `unicef`; variables `nation`, `lowbwt`, `life60`, and `life92`

nation	lowbwt	life60	life92	nation	lowbwt	life60	life92	nation	lowbwt	life60	life92
Afghanistan	20	33	43	Guatemala	14	46	64	Nigeria	16	40	52
Albania	7	62	73	Guinea	21	34	44	Norway	4	73	77
Algeria	9	47	66	Guinea-Bissau	20	34	43	Oman	10	40	69
Angola	19	33	46	Haiti	15	42	56	Pakistan	25	43	59
Argentina	8	65	71	Honduras	9	46	66	Panama	10	61	73
Armenia	.	.	72	Hong Kong	8	66	78	Papua New Guinea	23	41	56
Australia	6	71	77	Hungary	9	68	70	Paraguay	8	64	67
Austria	6	69	76	India	33	44	60	Peru	11	48	64
Azerbaijan	.	.	71	Indonesia	14	41	62	Philippines	15	53	65
Bangladesh	50	40	53	Iran	9	50	67	Poland	.	67	72
Belarus	.	.	71	Iraq	15	48	66	Portugal	5	63	75
Belgium	6	70	76	Ireland	4	70	75	Romania	7	65	70
Benin	.	35	46	Israel	7	69	76	Russian Fed.	.	.	69
Bhutan	.	37	48	Italy	5	69	77	Rwanda	17	42	46
Bolivia	12	43	61	Jamaica	11	63	73	Saudi Arabia	7	44	69
Botswana	8	46	61	Japan	6	68	79	Senegal	11	37	49
Brazil	11	55	66	Jordan	7	47	68	Sierra Leone	17	32	43
Bulgaria	6	68	72	Kazakhstan	.	.	69	Singapore	7	64	74
Burkina Faso	21	36	48	Kenya	16	45	59	Slovakia	.	.	72
Burundi	.	41	48	Korea, Dem.	.	54	71	Somalia	16	36	47
Cambodia	.	42	51	Korea, Rep.	9	54	71	South Africa	.	49	63
Cameroon	13	39	56	Kuwait	7	60	75	Spain	4	69	77
Canada	6	71	77	Kyrgyzstan	.	.	66	Sri Lanka	25	62	71
Central African Rep.	15	39	47	Lao PDR	18	40	51	Sudan	15	39	52
Chad	.	35	47	Latvia	.	70	71	Sweden	5	73	78
Chile	7	57	72	Lebanon	10	60	68	Switzerland	5	71	78
China	9	47	71	Lesotho	11	43	60	Syrian Arab Rep.	11	50	67
Colombia	10	57	69	Liberia	.	41	55	Tanzania	14	41	51
Congo	16	42	52	Libyan Arab Jama.	.	47	63	Thailand	13	52	69
Costa Rica	6	62	76	Lithuania	.	69	73	Togo	20	39	55
Cote d'Ivoire	14	39	52	Madagascar	10	41	55	Trinidad and Tobago	10	63	71
Cuba	8	64	76	Malawi	20	38	44	Tunisia	8	48	68
Czech Rep.	.	.	72	Malaysia	10	54	71	Turkey	8	50	67
Denmark	6	72	76	Mali	17	35	46	Turkmenistan	.	.	66
Dominican Rep.	16	52	67	Mauritania	11	35	48	USA	7	70	76
Ecuador	11	53	66	Mauritius	9	59	70	Uganda	.	43	42
Egypt	10	46	61	Mexico	12	57	70	Ukraine	.	.	70
El Salvador	11	50	66	Moldova	.	.	68	United Arab Emirates	7	53	71
Eritrea	.	.	47	Mongolia	10	47	63	United Kingdom	7	71	76
Estonia	.	69	71	Morocco	9	47	63	Uruguay	8	68	72
Ethiopia	16	36	47	Mozambique	20	37	47	Uzbekistan	.	.	69
Finland	4	68	76	Myanmar	16	44	57	Venezuela	9	60	70
France	5	70	77	Namibia	12	42	59	Viet Nam	17	44	64
Gabon	.	41	53	Nepal	.	38	53	Yemen	19	36	52
Georgia	.	.	73	Netherlands	.	73	77	Yugoslavia (former)	.	63	72
Germany	.	70	76	New Zealand	6	71	76	Zaire	15	41	52
Ghana	17	45	56	Nicaragua	15	47	66	Zambia	13	42	45
Greece	6	69	77	Niger	15	35	46	Zimbabwe	14	45	56

TABLE B.3

Data set `nurshome`; variables `state` and `resident`

Alabama	35.7	Idaho	31.3	Missouri	57.6	Pennsylvania	40.6
Alaska	22.5	Illinois	52.1	Montana	44.2	Rhode Island	60.3
Arizona	21.7	Indiana	59.2	Nebraska	67.4	South Carolina	28.8
Arkansas	50.5	Iowa	70.5	Nevada	18.6	South Dakota	74.9
California	26.7	Kansas	65.4	New Hampshire	52.7	Tennessee	45.8
Colorado	40.0	Kentucky	47.8	New Jersey	33.4	Texas	47.7
Connecticut	54.6	Louisiana	59.7	New Mexico	30.2	Utah	29.1
Delaware	43.2	Maine	51.4	New York	37.0	Vermont	46.2
District of		Maryland	44.6	North Carolina	30.4	Virginia	33.8
Columbia	34.4	Massachusetts	54.0	North Dakota	59.5	Washington	37.8
Florida	23.1	Michigan	36.1	Ohio	49.3	West Virginia	32.3
Georgia	45.5	Minnesota	67.6	Oklahoma	59.1	Wisconsin	62.1
Hawaii	13.6	Mississippi	40.6	Oregon	30.2	Wyoming	37.8

TABLE B.4

Data set `cigarett`; variables `tar` and `nicotine`

3	.3	8	.9	16	1.3
14	1.2	12	1.3	11	1.0
16	1.3	13	1.2	16	1.3
10	1.0	15	1.3	16	1.3
18	1.4	12.9	.89	19	1.4
13	1.2	16	1.2	10	1.1
1	.1	17	1.3	16	1.2
10	.9	4	.4	16	1.2
16	1.3	13	1.1	9	1.0
13	1.1	3	.4	4	.5
15	1.2	1	.2	.7	.09
12	1.0	13	1.1		

TABLE B.5

Data set `brate`; variables `year` and `birthrt`

1940	7.1	1951	15.1	1962	21.9	1973	24.3	1984	31.0
1941	7.8	1952	15.8	1963	22.5	1974	23.9	1985	32.8
1942	8.0	1953	16.9	1964	23.0	1975	24.5	1986	34.2
1943	8.3	1954	18.7	1965	23.4	1976	24.3	1987	36.0
1944	9.0	1955	19.3	1966	23.3	1977	25.6	1988	38.5
1945	10.1	1956	20.4	1967	23.7	1978	25.7	1989	41.6
1946	10.9	1957	21.0	1968	24.3	1979	27.2	1990	43.8
1947	12.1	1958	21.2	1969	24.8	1980	28.4	1991	45.2
1948	12.5	1959	21.9	1970	26.4	1981	29.5	1992	45.2
1949	13.3	1960	21.6	1971	25.5	1982	30.0		
1950	14.1	1961	22.7	1972	24.8	1983	30.3		

TABLE B.6
Data set `ischemic`; variables `sbp`, `duration`, and `time`

170	640	105	122	438	252	168	780	440
128	670	118	158	270	270	134	1065	480
150	560	130	132	781	277	160	1080	540
148	510	150	130	802	278	154	308	562
160	212	178	175	360	300	125	860	570
154	260	180	140	1775	300	140	60	578
175	228	192	150	524	322	154	765	720
140	335	200	180	441	330	120	396	729
140	460	200	160	1084	345	162	328	780
120	440	201	122	505	360	140	540	1200
172	210	240	170	75	375	178	130	1430
178	359	245	150	823	386			

TABLE B.7
Data set `lowbwt`; variables `sbp`, `sex`, `tox`, `grmhem`, `gestage`, and `apgar5`

sbp	sex	tox	grmhem	gestage	apgar5	sbp	sex	tox	grmhem	gestage	apgar5	sbp	sex	tox	grmhem	gestage	apgar5
43	Male	No	No	29	7	62	Female	No	No	27	7	67	Female	No	No	27	8
51	Male	No	No	31	8	59	Female	No	No	27	8	40	Female	No	Yes	31	8
42	Female	No	No	33	0	36	Male	No	No	27	9	48	Female	No	No	26	8
39	Female	No	No	31	8	47	Female	No	No	32	8	36	Male	No	No	27	5
48	Female	Yes	No	30	7	45	Male	No	Yes	31	2	44	Male	No	No	27	6
31	Male	No	Yes	25	0	62	Female	No	Yes	28	5	53	Female	Yes	No	35	9
31	Male	Yes	No	27	7	75	Male	Yes	No	30	7	45	Female	Yes	No	28	6
40	Female	No	No	29	9	44	Male	No	No	29	0	54	Male	No	No	30	8
57	Female	No	No	28	6	39	Male	No	No	28	8	44	Male	Yes	No	31	2
64	Female	Yes	No	29	9	48	Female	No	Yes	31	7	42	Male	No	No	30	5
46	Female	No	No	26	7	43	Female	Yes	No	27	6	50	Female	No	No	27	0
47	Female	No	Yes	30	6	19	Female	No	Yes	25	4	48	Female	No	No	25	5
63	Female	No	No	29	8	63	Male	No	No	30	7	29	Female	No	Yes	25	5
56	Female	No	No	29	1	42	Male	No	No	28	6	30	Female	No	Yes	26	2
49	Male	No	No	29	8	44	Female	No	No	28	9	36	Female	No	No	29	0
87	Male	No	No	29	7	25	Female	No	No	25	8	44	Female	No	No	29	0
46	Female	No	No	29	8	26	Female	No	No	23	8	46	Female	Yes	No	34	9
66	Female	No	No	33	8	27	Male	No	No	27	9	51	Male	Yes	No	30	4
42	Female	Yes	No	33	8	35	Male	No	No	28	8	51	Male	No	No	29	5
52	Female	No	No	29	7	40	Male	No	No	27	7	43	Male	Yes	No	33	7
51	Male	No	No	28	7	44	Female	No	No	27	6	48	Male	No	No	30	5
47	Female	No	No	30	9	66	Male	No	No	26	8	52	Male	No	No	29	8
54	Male	No	No	27	4	59	Female	No	No	25	3	43	Male	No	No	24	6
64	Male	No	No	33	9	24	Female	No	No	23	7	42	Male	Yes	No	33	8
37	Female	No	No	32	7	40	Female	No	Yes	26	3	48	Male	No	Yes	25	5
36	Male	Yes	No	28	3	49	Female	No	No	24	5	49	Female	Yes	No	32	8
45	Female	No	Yes	29	7	53	Male	Yes	No	29	9	62	Male	Yes	No	31	7
39	Male	No	No	28	7	45	Female	No	No	29	9	45	Male	No	No	31	9
29	Female	No	No	29	4	50	Male	No	Yes	27	8	51	Female	Yes	Yes	31	6
61	Female	No	No	30	3	64	Male	No	No	30	7	52	Male	No	No	29	8
53	Male	No	No	31	7	48	Female	No	No	30	6	47	Male	Yes	No	32	5
64	Female	No	No	30	7	48	Female	No	Yes	32	4	40	Female	Yes	No	33	8
35	Female	No	No	31	6	58	Female	Yes	No	33	7	50	Female	No	No	28	7
34	Male	No	No	29	9												

TABLE B.8

Data set dthrate; variables state, age, popn, and deaths

Maine	0-4	75037	1543
Maine	5-9	79727	148
Maine	10-14	74061	104
Maine	15-19	68683	153
Maine	20-24	60575	224
Maine	25-34	105723	413
Maine	35-44	101192	552
Maine	45-54	90346	980
Maine	55-64	72478	1476
Maine	65-74	46614	2433
Maine	75+	22396	3056
South Carolina	0-4	205076	4905
South Carolina	5-9	240750	446
South Carolina	10-14	222808	410
South Carolina	15-19	211345	901
South Carolina	20-24	166354	1073
South Carolina	25-34	219327	1910
South Carolina	35-44	191349	2377
South Carolina	45-54	143509	2862
South Carolina	55-64	80491	2667
South Carolina	65-74	40441	2486
South Carolina	75+	16723	2364

Data set us1940; variables age and popn

0-4	.0801
5-9	.0811
10-14	.0892
15-19	.0937
20-24	.0880
25-34	.1621
35-44	.1392
45-54	.1178
55-64	.0803
65-74	.0484
75+	.0201

TABLE B.9

Data set lifeexp; variables year, male, and female

year	male	female	year	male	female
1940	60.8	65.2	1980	70	77.4
1950	65.6	71.1	1981	70.4	77.8
1960	66.6	73.1	1982	70.8	78.1
1970	67.1	74.7	1983	71.0	78.1
1971	67.4	75.0	1984	71.1	78.2
1972	67.4	75.1	1985	71.1	78.2
1973	67.6	75.3	1986	71.2	78.2
1974	68.2	75.9	1987	71.4	78.3
1975	68.8	76.6	1988	71.4	78.3
1976	69.1	76.8	1989	71.7	78.5
1977	69.5	77.2	1990	71.8	78.8
1978	69.6	77.3	1991	72.0	78.9
1979	70.0	77.8	1992	72.3	79.1

TABLE B.10

Data set `liferace`; variables `year`, `wmale`, `wfemale`, `bmale`, and `bfemale`

1970	68.0	75.6	60.0	68.3
1971	68.3	75.8	60.5	68.9
1972	68.3	75.9	60.4	69.1
1973	68.5	76.1	60.9	69.3
1974	69.0	76.7	61.7	70.3
1975	69.5	77.3	62.4	71.3
1976	69.9	77.5	62.9	71.6
1977	70.2	77.9	63.4	72.0
1978	70.4	78.0	63.7	72.4
1979	70.8	78.4	64.0	72.9
1980	70.7	78.1	63.8	72.5
1981	71.1	78.4	64.5	73.2
1982	71.5	78.7	65.1	73.7
1983	71.7	78.7	65.4	73.6
1984	71.8	78.7	65.6	73.7
1985	71.9	78.7	65.3	73.5
1986	72.0	78.8	65.2	73.5
1987	72.2	78.9	65.2	73.6
1988	72.3	78.9	64.9	73.4
1989	72.7	79.2	64.8	73.5

TABLE B.11

Data set `diabetes`; variables `fcg`, `sens`, and `spec`

3.9	.971	.229
4.4	.954	.447
5.0	.920	.619
5.6	.872	.724
6.1	.744	.873
6.7	.647	.938
7.2	.536	.970
7.8	.493	.984
8.3	.430	.992
8.9	.379	.994

TABLE B.12

Data set `heart`; variables `trtment`, `pdi`, and `mdi`

0	80	74	1	98	96	1	98	109	1	92	119	0	98	106
1	118	124	1	92	126	1	98	91	0	105	115	0	98	98
1	122	109	0	80	91	1	115	120	1	104	100	1	93	107
0	98	78	1	98	106	1	98	96	0	75	58	0	77	109
0	98	91	0	86	91	1	80	70	0	87	102	1	105	103
0	111	130	1	92	117	1	92	86	0	110	115	0	105	103
1	111	119	0	86	91	0	86	81	0	93	131	1	87	97
0	82	115	1	105	86	0	111	100	0	63	86	0	105	140
1	99	112	1	115	118	1	110	102	0	104	110	0	99	94
0	86	119	0	105	98	0	98	83	1	104	112	0	117	106
1	92	100	1	118	117	0	82	97	0	98	98	0	92	109
0	122	115	1	92	115	1	93	102	0	111	103	0	134	131
1	98	117	0	92	130	1	99	97	0	111	117	0	86	100
0	78	108	0	70	91	0	86	115	1	124	122	1	75	89
0	92	115	1	117	115	1	92	103	1	114	118	0	87	82
0	86	112	0	80	96	0	86	103	0	98	70	1	115	107
1	90	82	0	80	96	0	60	78	1	92	109	1	80	91
0	92	142	1	98	98	0	130	114	0	105	112	1	80	112
0	104	102	1	98	117	1	122	109	1	86	122	0	105	103
1	92	130	0	117	103	1	98	107	1	86	119	1	52	78
0	105	103	0	.	109	0	80	89	1	92	117	0	92	108
1	98	93	1	92	122	1	92	130	0	70	56	1	105	109
1	80	96	0	63	96	0	80	112	1	120	122	0	66	93
1	50	50	1	115	105	1	122	112	0	105	115	0	92	100
1	87	112	0	71	100	1	98	83	0	99	89	1	110	97
0	67	106	1	98	122	0	98	122	1	99	107	0	87	100
1	80	93	0	111	117	1	70	103	1	111	109	0	87	112
0	76	110	0	111	122	0	60	115	1	98	98			
1	98	117	0	63	103	1	98	119	1	93	120			

TABLE B.13

Data set `bed`; variables `bed80`, `bed86`, and `state`

4.7	4.2	1	5.0	4.7	35	7.4	7.2	18
4.4	4.0	3	4.6	4.0	37	6.0	5.9	20
3.8	3.5	5	5.9	5.7	39	3.6	3.4	22
4.5	4.3	7	3.6	4.4	41	7.3	7.7	24
4.8	4.6	9	3.1	2.9	43	5.5	5.1	26
4.5	4.2	11	3.1	2.7	45	3.9	3.4	28
4.4	4.0	13	3.1	3.0	47	5.1	4.4	30
5.7	5.0	15	3.6	3.1	49	5.5	5.3	32
5.7	5.1	17	3.1	2.6	51	5.3	5.2	34
5.5	6.5	19	3.9	3.3	2	4.8	4.5	36
5.8	5.0	21	4.4	4.3	4	4.7	3.8	38
3.6	3.3	23	3.5	3.1	6	3.7	3.3	40
4.1	3.7	25	4.2	3.9	8	4.2	3.5	42
4.2	3.5	27	4.7	4.4	10	3.6	3.1	44
4.6	4.3	29	5.1	4.5	12	4.2	3.7	46
4.5	4.4	31	4.9	4.5	14	3.5	3.1	48
5.1	5.0	33	5.7	5.2	16	2.7	2.4	50

TABLE B.14
Data set cad; variables age and center

60	1	73	1	65	3
66	1	56	2	61	3
65	1	65	2	63	3
55	1	65	2	59	3
62	1	63	2	42	3
70	1	57	2	53	3
51	1	47	2	63	3
72	1	72	2	65	3
58	1	56	2	60	3
61	1	52	2	57	3
71	1	75	2	62	3
41	1	66	2	70	3
70	1	62	2	73	3
57	1	68	2	63	3
55	1	75	2	55	3
63	1	60	2	52	3
64	1	73	2	58	3
76	1	63	2	68	3
74	1	64	2	70	3
54	1	67	3	72	3
58	1	56	3	45	3

TABLE B.15
Data set program; variables group, baseline, and followup

2	16.50	18.02	1	28.61	24.34
2	29.60	29.68	1	27.56	27.71
2	24.80	19.27	1	32.21	22.15
2	31.11	27.35	1	25.22	21.33
2	26.65	23.70	1	26.44	23.76
2	16.66	17.73	1	28.93	28.93
2	28.06	25.74	1	22.26	16.39
2	9.85	12.44	1	29.55	26.15
2	20.37	15.64	1	22.67	19.70
2	26.66	28.76	1	25.78	19.54
2	28.13	27.35	1	15.41	13.49
2	26.85	26.39	1	28.03	28.47
2	25.71	25.15	1	23.90	21.52
2	24.09	24.16	1	15.82	13.99
2	23.25	25.13	1	19.09	16.84
2	21.87	18.64	1	24.51	23.02

TABLE B.16
Data set insure; variables stage and group (This data set consists of a large number of repeated observations; rather than display them all, we summarize the outcomes.)

288	Stage 1 and Uninsured
338	Stage 2 and Uninsured
226	Stage 3 and Uninsured
41	Stage 4 and Uninsured
76	Stage 5 and Uninsured
13	Stage 1 and Insured
18	Stage 2 and Insured
34	Stage 3 and Insured
6	Stage 4 and Insured
11	Stage 5 and Insured

TABLE B.17
Data set prison; variables hiv and ivdu (This data set consists of a large number of repeated observations; rather than display them all, we summarize the outcomes.)

61	Positive and Yes
27	Positive and No
75	Negative and Yes
312	Negative and No

TABLE B.18
Data set `angio`; variables
`site` and `appropo` (This data
set consists of a large number
of repeated observations;
rather than display them all,
we summarize the outcomes.)

452	Site 1 and Appropriate
63	Site 1 and Equivocal
113	Site 1 and Inappropriate
394	Site 2 and Appropriate
35	Site 2 and Equivocal
85	Site 2 and Inappropriate
434	Site 3 and Appropriate
21	Site 3 and Equivocal
80	Site 3 and Inappropriate

TABLE B.19
Data set `alcohol`; variables
`genques` and `alcques` (This
data set consists of a large
number of repeated observations;
rather than display them all, we
summarize the outcomes.)

82	Nondrinker and Nondrinker
39	Drinker and Nondrinker
4	Nondrinker and Drinker
112	Drinker and Drinker

TABLE B.20
Data set `stenosis`; variables
`smoke`, `disease`, and `sex`
(This data set consists of a
large number of repeated
observations; rather than
display them all, we summarize
the outcomes.)

37	Yes and Yes and Male
24	Yes and No and Male
25	No and Yes and Male
20	No and No and Male
14	Yes and Yes and Female
19	Yes and No and Female
29	No and Yes and Female
47	No and No and Female

TABLE B.21
Data set `water`;
variables
`fluoride` and
`caries`

0	810
0	673
0	722
.1	706
.1	823
.1	1037
.1	772
.2	733
.2	703
.3	652
.4	556
.5	444
.6	412
.9	343
1.2	258
1.2	303
1.2	281
1.3	323
1.8	252
1.9	236
2.6	246

Data set actions; variables state, rank91, rank92, rank93, rank94, and rank95

Alabama	31	30	29	43.5	30
Alaska	1	7	8	2	8
Arizona	22	22	16	17	10
Arkansas	29	18	26	28	23
California	37	42	32	34.5	20
Colorado	17	8	6	12	5
Connecticut	30	35	36	42	27
Delaware	16	51	43	48	48
District of Columbia	45	45	51	51	50
Florida	27	21	25	25	22
Georgia	4	9	10	8	4
Hawaii	41	50	46	50	51
Idaho	34.5	23	37.5	30	36
Illinois	36	36	31	40	45
Indiana	15	14	7	16	28
Iowa	3	2	5	7	2
Kansas	25	20	37.5	22	46
Kentucky	5	16	4	4	14
Louisiana	7	12	11	18	13
Maine	46	44	41	33	32
Maryland	42.5	27	19	21	29
Massachusetts	48	46	45	37	40
Michigan	40	40	35	34.5	21
Minnesota	28	31	33	45	39
Mississippi	6	6	9	9	1
Missouri	12	13	12	13	37
Montana	19	10	14	3	18
Nebraska	39	38	50	15	41.5
Nevada	26	25	20	31	11
New Hampshire	44	47	47	49	49
New Jersey	20	28	18	19	25
New Mexico	33	33	49	43.5	15
New York	49	39	34	29	17
North Carolina	42.5	34	40	36	35
North Dakota	13	5	3	10	34
Ohio	23	19	22.5	24	9
Oklahoma	2	1	2	5	12
Oregon	14	24	22.5	20	16
Pennsylvania	47	48	48	47	43
Rhode Island	50.5	41	42	26	26
South Carolina	11	11	15	14	44
South Dakota	50.5	32	13	11	33
Tennessee	38	49	44	38	31
Texas	21	29	28	23	19
Utah	18	43	39	46	38
Vermont	10	15	17	39	6
Virginia	32	37	30	32	41.5
Washington	24	17	24	27	24
West Virginia	8	3	1	6	7
Wisconsin	34.5	26	27	41	47
Wyoming	9	4	21	1	3

Data set twenty; variables length and weight

41	1360	32	880
40	1490	39	1130
38	1490	38	1140
38	1180	39	1350
38	1200	37	950
32	680	39	1220
33	620	38	980
38	1060	42	1480
30	1320	39	1250
34	830	38	1250

Data set africa; variables country, fertrate, and contra

Benin	7.1	9
Botswana	5.0	33
Burundi	6.8	9
Cameroon	6.4	2
Cote d'Ivoire	7.4	3
Ghana	6.4	13
Kenya	6.7	27
Lesotho	5.8	7
Liberia	6.6	7
Mali	6.7	5
Nigeria	6.2	6
Rwanda	8.5	10
Senegal	6.6	11
Sudan	5.9	6
Togo	6.2	34
Uganda	7.3	5
Zimbabwe	5.7	43

Data set miner; variables rate and year

.2419	1	.0904	7
.1732	2	.0792	8
.1361	3	.0701	9
.1108	4	.0891	10
.0996	5	.0799	11
.0952	6	.1084	12

TABLE B.26

Data set hospital; variables state, expadm, los, and salary

Connecticut	3328	7.8	16957	Arkansas	1859	6.6	12737
Maine	2810	7.7	14317	Florida	2780	7.5	14840
Massachusetts	4105	8.9	15696	Georgia	2156	6.7	12923
New Hampshire	2487	7.1	13542	Kentucky	1974	6.7	13504
Rhode Island	3380	8.5	14947	Louisiana	2412	6.3	14169
Vermont	2478	8.2	13432	Mississippi	1772	7.0	11928
Delaware	2957	8.2	15717	North Carolina	2196	7.5	13070
District of				South Carolina	2134	7.4	12973
Columbia	4612	8.7	18700	Tennessee	2176	7.2	12923
Maryland	3210	8.3	15213	Virginia	2575	7.9	14107
New York	2712	8.4	15573	West Virginia	2248	7.2	13731
New Jersey	3607	9.7	16657	Arizona	3101	6.5	15307
Pennsylvania	3194	8.5	15256	New Mexico	2626	6.7	14573
Illinois	3351	8.0	16872	Oklahoma	2425	6.5	13916
Indiana	2592	7.8	13946	Texas	2272	6.6	13847
Michigan	3351	8.0	16635	Colorado	2782	7.1	16618
Ohio	3007	8.1	15492	Idaho	2072	6.5	13211
Wisconsin	2724	8.3	14951	Montana	2152	8.3	14635
Iowa	2361	8.0	13579	Utah	2384	5.4	14192
Kansas	2600	7.8	14107	Wyoming	2009	5.5	14686
Minnesota	2730	9.4	14503	California	3886	6.5	18611
Missouri	2915	8.0	14950	Nevada	3500	6.3	17898
Nebraska	2448	8.4	13584	Oregon	2660	5.9	15578
North Dakota	2277	8.6	13559	Washington	2524	5.7	15617
South Dakota	2059	8.7	12508	Alaska	3633	5.9	23594
Alabama	2174	7.2	12834	Hawaii	2780	8.3	14758

TABLE B.27

Data set detroit; variables year, homicide, police, umemp, register, and weekly

1961	8.60	260.35	11.0	215.98	117.18
1962	8.90	269.80	7.0	180.48	134.02
1963	8.52	272.04	5.2	209.57	141.68
1964	8.89	272.96	4.3	231.67	147.98
1965	13.07	272.51	3.5	297.65	159.85
1966	14.57	261.34	3.2	367.62	157.19
1967	21.36	268.89	4.1	616.54	155.29
1968	28.03	295.99	3.9	1029.75	131.75
1969	31.49	319.87	3.6	786.23	178.74
1970	37.39	341.43	7.1	713.77	178.30
1971	46.26	356.59	8.4	750.43	209.54
1972	47.24	376.69	7.7	1027.38	240.05
1973	52.33	390.19	6.3	666.50	258.05

TABLE B.28
Data set `dialysis`; variables `perito`, `age`, `sex`, and `race`

perito	age	sex	race		perito	age	sex	race		perito	age	sex	race
Yes	67	Female	Non-white		Yes	57	Female	Non-white		Yes	31	Female	White
Yes	62	Female	Non-white		No	28	Female	Non-white		No	61	Male	White
Yes	59	Male	Non-white		No	65	Male	White		No	57	Female	White
Yes	65	Male	White		No	61	Male	White		Yes	62	Male	White
Yes	54	Female	White		Yes	60	Male	Non-white		Yes	48	Female	White
Yes	45	Male	White		No	38	Male	White		No	59	Male	White
No	56	Female	White		Yes	43	Male	Non-white		Yes	58	Male	White
Yes	49	Male	White		Yes	54	Male	White		Yes	49	Female	Non-white
Yes	42	Male	White		Yes	49	Female	White		No	68	Male	White
Yes	51	Female	White		Yes	71	Female	White		No	68	Male	White
Yes	52	Male	White		Yes	25	Female	Non-white		Yes	64	Male	White
Yes	70	Female	White		No	59	Male	White		No	49	Female	White
No	42	Male	White		Yes	61	Male	White		No	26	Male	White
No	53	Male	White		Yes	45	Female	White		Yes	63	Male	White
No	32	Female	White		Yes	29	Male	White		Yes	46	Male	Non-white
Yes	50	Female	Non-white										

TABLE B.29
Data set `cyto`; variables `time`, `censor`, and `group`

time	censor	group		time	censor	group		time	censor	group		time	censor	group		time	censor	group
11	1	1		126	1	1		229	1	1		1	1	2		61	1	2
26	1	1		142	1	1		261	1	1		1	1	2		82	1	2
35	1	1		149	1	1		362	1	1		1	1	2		90	1	2
60	1	1		191	1	1		368	0	1		1	1	2		121	1	2
89	1	1		204	1	1		387	0	1		16	1	2		162	1	2
101	1	1		213	1	1		400	0	1		47	1	2				

TABLE B.30
Data set `bladder`; variables `time`, `group`, `censor`, and `number`

time	group	censor	number		time	group	censor	number		time	group	censor	number		time	group	censor	number		time	group	censor	number
0	1	0	1		2	1	1	2		3	1	1	1		10	2	0	1		22	2	1	1
1	1	0	2		25	1	1	2		6	1	1	2		13	2	0	1		4	2	1	1
4	1	0	1		29	1	0	2		3	1	1	1		3	2	1	2		24	2	1	1
7	1	0	1		29	1	0	2		9	1	1	1		1	2	1	2		41	2	0	2
10	1	0	1		29	1	0	1		18	1	1	1		18	2	0	1		41	2	0	1
6	1	1	1		28	1	1	2		49	1	0	2		17	2	1	2		1	2	1	1
14	1	0	1		2	1	1	2		35	1	1	1		2	2	1	1		44	2	0	1
18	1	0	1		3	1	1	1		17	1	1	2		17	2	1	1		2	2	1	1
5	1	1	2		12	1	1	2		3	1	1	1		22	2	0	1		45	2	0	2
12	1	1	1		32	1	0	2		59	1	0	1		25	2	0	2		2	2	1	2
23	1	0	2		34	1	0	1		2	1	1	2		25	2	0	2		46	2	0	2
10	1	1	2		36	1	0	1		5	1	1	2		25	2	0	1		49	2	0	2
3	1	1	1		29	1	1	1		2	1	1	2		6	2	1	1		50	2	0	1
3	1	1	1		37	1	0	2		1	2	0	2		6	2	1	1		4	2	1	1
7	1	1	2		9	1	1	1		1	2	0	1		2	2	1	1		54	2	0	2
3	1	1	1		16	1	1	1		5	2	1	1		26	2	1	2		38	2	1	1
26	1	0	2		41	1	0	2		9	2	0	2		38	2	0	1		59	2	0	2
1	1	1	1																			

Index